COMPREHENSIVE SERIES IN PHOTOCHEMISTRY
& PHOTOBIOLOGY

Series Editors

Donat P. Häder
Professor of Botany

and

Giulio Jori
Professor of Chemistry

European Society for Photobiology

COMPREHENSIVE SERIES IN PHOTOCHEMISTRY
& PHOTOBIOLOGY

Series Editors: Donat P. Häder and Giulio Jori

Titles in this Series

COMPREHENSIVE SERIES IN PHOTOCHEMISTRY
& PHOTOBIOLOGY – VOLUME 3

Photoreceptors and Light Signalling

Editor

Alfred Batschauer

FB Biologie/Pflanzenphysiologie
Philipps-Universitaet
Marburg
Germany

advancing the chemical sciences

ISBN 0-85404-311-X

A catalogue record for this book is available from the British Library

Published by The Royal Society of Chemistry,
Thomas Graham House, Science Park, Milton Road,
Cambridge CB4 0WF, UK
Registered Charity Number 207890

For further information see our web site at www.rsc.org

Typeset by Charlesworth, Huddersfield, West Yorkshire, UK
Printed and bound by Sung Fung Offset Binding Co. Ltd, Hong Kong

Preface for the ESP series in Photochemical and Photobiological Sciences

"It's not the substance, it's the dose which makes something poisonous!" When Paracelsius, a German physician of the 14th century made this statement he probably did not think about light as one of the most obvious environmental factors. But his statement applies as well to light. While we need light for example for vitamin D production too much light might cause skin cancer. The dose makes the difference. These diverse findings of light effects have attracted the attention of scientists for centuries. The photosciences represent a dynamic multidisciplinary field which includes such diverse subjects as behavioral responses of single cells, cures for certain types of cancer and the protective potential of tanning lotions. It includes photobiology and photochemistry, photomedicine as well as the technology for light production, filtering and measurement. Light is a common theme in all these areas. In recent decades a more molecular centered approach changed both the depth and the quality of the theoretical as well as the experimental foundation of photosciences.

An example of the relationship between global environment and the biosphere is the recent discovery of ozone depletion and the resulting increase in high energy ultraviolet radiation. The hazardous effects of high energy ultraviolet radiation on all living systems is now well established. This discovery of the result of ozone depletion put photosciences at the center of public interest with the result that, in an unparalleled effort, scientists and politicians worked closely together to come to international agreements to stop the pollution of the atmosphere.

The changed recreational behavior and the correlation with several diseases in which sunlight or artificial light sources play a major role in the causation of clinical conditions (e.g. porphyrias, polymorphic photodermatoses, *Xeroderma pigmentosum* and skin cancers) have been well documented. As a result, in some countries (e.g. Australia) public services inform people about the potential risk of extended periods of sun exposure every day. The problems are often aggravated by the phototoxic or photoallergic reactions produced by a variety of environmental pollutants, food additives or therapeutic and cosmetic drugs. On the other hand, if properly used, light-stimulated processes can induce important beneficial effects in biological systems, such as the elucidation of several aspects of cell structure and function. Novel developments are centered around photodiagnostic and phototherapeutic modalities for the treatment of cancer, artherosclerosis, several autoimmune diseases, neonatal jaundice and others. In addition, classic research areas such as vision and photosynthesis are still very active. Some of these developments are unique to photobiology, since the peculiar physico-chemical properties of electronically excited biomolecules often lead to the promotion of reactions which are characterized by high levels of selectivity in space and time. Besides the biologically centered areas, technical developments have paved the way for the harnessing of solar energy to produce warm water and electricity or the development of environmentally friendly techniques for addressing problems of large social impact (e.g. the decontamination of polluted waters). While also in use in Western countries, these techniques are of great interest for developing countries.

The European Society for Photobiology (ESP) is an organization for developing and coordinating the very different fields of photosciences in terms of public knowledge and scientific interests. Due to the ever increasing demand for a comprehensive overview of the photosciences the ESP decided to initiate an encyclopedic series, the "Comprehensive Series in Photochemical and Photobiological Sciences". This series is intended to give an in-depth coverage over all the very different fields related to light effects. It will allow investigators, physicians, students, industry and laypersons to obtain an updated record of the state-of-the-art in specific fields, including a ready access to the recent literature. Most importantly, such reviews give a critical evaluation of the directions that the field is taking, outline hotly debated or innovative topics and even suggest a redirection if appropriate. It is our intention to produce the monographs at a sufficiently high rate to generate a timely coverage of both well established and emerging topics. As a rule, the individual volumes are commissioned; however, comments, suggestions or proposals for new subjects are welcome.

Donat-P. Häder and Giulio Jori
Spring 2002

Volume preface

Light is one of the most important environmental factors for living organisms, providing them in the case of photosynthetic organisms with energy, and information about their surroundings such as day and night cycles. This information is then used either to change behaviour or physiology. Therefore it is not surprising that, in all kingdoms, most species are able to sense light through so-called sensory photoreceptors. However, these photoreceptors are not only able to distinguish between light on and light off, but together can also use the total information that is present in the light. This information includes (i) the irradiance, (ii) the colour or spectral distribution, (iii) the direction of light, and (iv) the polarisation of light.

In principle, the irradiance can be measured by determining how often the photoreceptor is excited during a specified unit of time. This, of course, depends on the absorption cross section of the photoreceptor and how fast it reaches its ground state after excitation. The colour, or wavelength, of the photon can be sensed either by a complex photoreceptor such as phytochrome or by the combination of different photoreceptors. The absorption spectrum of the photoreceptor (and in particular the chemical nature of its chromophore) determines whether the photon can be detected. The ability to sense the direction of light can be governed by measuring a light gradient within the cell or – in multicellular organisms – within a tissue which depends on comparing light intensities in space. The movement of organisms through areas of different light intensity can also be used to sense the direction of light by measuring changes in light intensity over time. The ability to sense the polarisation of light probably depends on a fixed orientation of the photoreceptor (e.g. at membranes).

All photoreceptors known to date consist of the following: A protein moiety and one or several chromophore(s) which are covalently or non-covalently bound to the protein. If additional photoreceptors are identified in the future, it is very unlikely that they will disobey this rule since the protein by itself is not able to absorb light (at least in the visible region) and thus needs the chromophore. In principle, the chromophore can also originate from the protein as for the green fluorescent protein although this is not a sensory photoreceptor. The chromophore, with its conjugated π-electron system, can be excited with photons of longer wavelengths, or lower energy, such as those present in the visible region (400–760 nm). The protein moiety is required to transduce the primary light signal to downstream components. A possible exception to this rule could be UV-B photoreceptors, which have not been characterised at the molecular level so far.

It might be a bit surprising that only a small number of chromophore classes have been found in photoreceptors. However, one can argue from this small number that only a few chromophores are particularly well suited for photoreceptor function. These chromophore classes are: retinals, present in rhodopsins; linear tetrapyrroles, present in phytochromes and related photo-receptors from bacteria; thiol-ester linked 4-OH-cinnamic acid, present in xanthopsins (with the photoactive yellow protein as the archetype of this family); the flavins FAD and FMN, present in cryptochromes and photo-tropins, respectively; and the pterin 5,10-methenyltetrahydrofolate, present as a second chromophore in cryptochromes. Whereas some photoreceptor families have a wide distribution, such as the rhodopsins that are present in Bacteria, Archea, and Eukarya, others seem to have a very limited distribution, such as the phototropins that, so far, have only been found in plants. However, very recently phototropin-like proteins were identified in Bacteria [A. Losi et al. (2002). *Biophys. J.*, **82**, 2627–26349]. Further research might change this picture even more, an example being the phytochromes, which were originally thought to be typical plant photoreceptors. In recent years, genome projects have led to the identification of photoreceptors in cyanobacteria and even in non-photosynthetic eubacteria, which are related to phytochromes. It is also likely that additional photoreceptors will be found in the future. The progress in identifying novel photoreceptors is seen, for example, in the case of the plant blue-light photoreceptors. Before 1993, none were molecularly characterised or cloned, but with the use of molecular biology and genetic methods both the cryptochromes and the phototropins were then identified within a short time period. In the meantime, interacting partner proteins had already been found, well-characterised and, for phototropin, a photocycle had been demonstrated. Shortly after the discovery of cryptochromes in plants they were also identified in animals and humans through characterisation of mutants in circadian entrainment (*Drosophila*) and from the results of genome projects (human).

While writing this book, a novel blue-light receptor was described [M. Iseki et al. (2002). *Nature*, **415**, 1047–1051], which mediates the photoavoidance response in the unicellular flagellate *Euglena gracilis*. This blue-light receptor is a flavin-containing adenylyl cyclase and thus represents the third class of blue-light receptors identified within one decade.

Photobiology and research on photoreceptors and light-signalling is an interdisciplinary field using a broad range of methods such as action spectros-copy, various methods for protein purification, the whole range of molecular biological and genetic methods, and uncountable numbers of spectroscopic methods from absorption and fluorescence spectroscopy to X-ray diffraction for solving the structure of photoreceptors. Intimate knowledge of the struc-ture and function of photoreceptors can thus only be reached through the combined effort of scientists from physics, chemistry and biology.

As outlined above, some photoreceptors have been known for many decades whereas others have been identified very recently. It is thus not surprising that the depth of knowledge and understanding of photoreceptor function, structure and signalling is quite different for the various photoreceptors. For example, rhodopsins and xanthopsins are already very well understood at the atomic level, whereas structural data still seems far away for other photoreceptors. In contrast, the structure and the photocycle of photoactive yellow protein is very well known but, still, the physiological role of this photoreceptor is not well understood.

Such differences in our knowledge of the structure, photochemistry, signalling and physiological responses of the different photoreceptors is, of course, also reflected in the twelve chapters of this book. However, I believe that this is not a disadvantage but reflects the current status of photoreceptor and light-signalling analysis, and demonstrates the broad range of experimental approaches towards one goal, which is the full understanding of photoreceptor function all the way down to the atomic level.

The chapters of this book cover all known photoreceptors, with the exception of the above-mentioned *Euglena* blue-light receptor and those candidates for which photoreceptor function has not unambiguously been shown. Examples for such candidates exist in fungi.

I am aware that much more knowledge about photoreceptors and light signalling will be available after publication of this book, due to the very fast progress in this field. Consequently, the authors have updated their chapters even during editing so that most of the very recent results are included. I'm very happy and grateful for the involvement of the authors in making it possible for all of the chapters to be written by leading experts in their respective fields. I thank the authors for the time they have invested in writing their chapters and in answering the burning questions from the editor.

Finally, it is my hope that this book will not only be of worth to experts but that it can also attract biology, chemistry and physics students to this fascinating and interdisciplinary research field.

Alfred Batschauer

Contributors

Joachim Bentrop
Universität Karlsruhe
Zoologie I, Zell- und Neurobiologie
Haid-und-Neu-Str. 9
76131 Karlsruhe
Germany

Silvia E. Braslavsky
Max-Planck-Institut für
  Strahlenchemie
Postfach 101356
45470 Mülheim an der Ruhr
Germany

Wim Crielaard
Laboratory for Microbiology
Swammerdam Institute for Life
  Sciences
Bio Centrum
University of Amsterdam
Nieuwe Achtergracht 166
1018 TV Amsterdam
The Netherlands

Werner Deininger
Universität Regensburg
Institut für Biochemie
Universitätsstr. 31
93094 Regensburg
Germany

Paul Devlin
Dept. Life Sciences
Kings College London
Franklin-Wilkins Building
150 Stamford St.
London SE1 8WA
United Kingdom

Martin Engelhard
Max-Planck-Institut für Molekulare
  Physiologie
Otto-Hahn-Str. 11
44227 Dortmund
Germany

Oliver P. Ernst
Universitätsklinikum Charité
Humboldt Universität zu Berlin
Institut für Medizinische Physik und
  Biophysik
Schumann Str. 20–21
10098 Berlin
Germany

Markus Fuhrmann
Universität Regensburg
Institut für Biochemie
Universitätsstr. 31
93094 Regensburg
Germany

Wolfgang Gärtner
Max-Planck-Institut für
  Strahlenchemie
Stiftstr. 34–36
45470 Mülheim an der Ruhr
Germany

Thomas Gensch
Laboratory for Cellular Signal
  Processing
Forschungszentrum Jülich
52425 Jülich
Germany

Andrea Haker
Laboratory for Microbiology
Swammerdam Institute for Life
  Sciences
Bio Centrum
University of Amsterdam
Nieuwe Achtergracht 166
1018 TV Amsterdam
The Netherlands

Klaus Harter
Institut für Biologie II
Universität Freiburg
Schänzlestr. 1
79104 Freiburg
Germany

Peter Hegemann
Universität Regensburg
Institut für Biochemie
Universitätsstr. 31
93094 Regensburg
Germany

Klaas J. Hellingwerf
Laboratory for Microbiology
Swammerdam Institute for Life
  Sciences
Bio Centrum
University of Amsterdam
Nieuwe Achtergracht 166
1018 TV Amsterdam
The Netherlands

Johnny Hendriks
Laboratory for Microbiology
Swammerdam Institute for Life
  Sciences
Bio Centrum
University of Amsterdam
Nieuwe Achtergracht 166
1018 TV Amsterdam
The Netherlands

Klaus Peter Hofmann
Universitätsklinikum Charité
Humboldt Universität Berlin
Institut für Medizinische Physik und
  Biophysik
Schumann Str. 20-21
10098 Berlin
Germany

Jon Hughes
Pflanzenphysiologie
Justus-Liebig-Universität Giessen
Senkenbergstr. 3
35390 Giessen
Germany

Suneel Kateriya
Universität Regensburg
Institut für Biochemie
Universitätsstr. 31
93094 Regensburg
Germany

Eva Kevei
Hungarian Academy of Sciences
Plant Biology Institute
Biological Research Centre
Temesvari krt.62
H-6726 Szeged
Hungary

Tilman Lamparter
Freie Universität Berlin
Institut für Biologie/
  Pflanzenphysiologie
Königin-Luise-Str. 12–16
14195 Berlin
Germany

Chentao Lin
University of California, Los Angeles
Department of Molecular, Cell &
  Developmental Biology
P.O. Box 951606
Los Angeles, CA 90096-1606
USA

Ferenc Nagy
Hungarian Academy of Sciences
Plant Biology Institute
Biological Research Centre
Temesvari krt.62
H-6726 Szeged
Hungary

Krzysztof Palczewski
University of Washington, Seattle
Department of Ophthalmology,
  Pharmacology, and Chemistry

WA 98195
USA

Reinhard Paulsen
Universität Karlsruhe
Zoologie I, Zell- und Neurobiologie
Haid-und-Neu-Str. 9
76131 Karlsruhe
Germany

Michael Salomon
Vertis Biotechnologie AG
Lise-Meitner Strasse 30
D-85354 Freising Weihenstephan
Germany

May Santiago-Ong
University of California, Los Angeles
Department of Molecular, Cell &
  Developmental Biology
P.O. Box 951606
Los Angeles, CA 90096-1606
USA

Eberhard Schäfer
Institut für Biologie II
Universität Freiburg
Schänzlestr. 1
79104 Freiburg
Germany

Georg Schmies
Max-Planck-Institut für Molekulare
  Physiologie
Otto-Hahn-Str. 11
44227 Dortmund
Germany

Michael Van der Horst
Laboratory for Microbiology
Swammerdam Institute for Life
  Sciences
Bio Centrum
University of Amsterdam
Nieuwe Achtergracht 166
1018 TV Amsterdam
The Netherlands

Masamitsu Wada
Tokyo Metropolitan University
Biology
Minami Osawa 1-1
Hachioji-Shi
Tokyo, 192-0397
Japan

Ansgar A. Wegener
Max-Planck-Institut für Molekulare
  Physiologie
Otto-Hahn-Str. 11
44227 Dortmund
Germany

Contents

Abbreviations

B, bathorhodopsin
BphP, bacteriophytochrome photoreceptor
BR, bacteriorhodopsin
BSI, blue-shifted intermediate
BV, biliverdin IX-a
CCA, complementary chromatic adaption
CD, circular dichroism spectroscopy
cDNA, complementary DNA
cFR, constant far-red light
Chop1, channel opsin 1
Cop, chlamyopsin
CpH1, cyanobacterial phytochrome 1
Cry (or CRY), cryptochrome
E-PYP, PYP from *Ectothiorhodospira halophila*
FL, full length
FMN, flavin mononucleotide
FR, far-red light
FTIR, Fourier-transform infrared
FTR, Fourier-transform Raman spectroscopy
GFP, green fluorescent protein
GPCR, G-protein-coupled receptor
Gtbc, bc heterodimer subunit of Gt
Gtα, α subunit of Gt
Gt, transducin (retinal G-protein)
GUS, β-glucuronidase
HIR, high irradiance response
HOOP, hydrogen out-of-plane
HR, halorhodopsin
Htp, halobacterial transducer protein
L, lumirhodopsin
LADS, lifetime-associated difference spectra
LFR, low fluence response
MI, metarhodopsin I
MII, metarhodopsin II
NMR, nuclear magnetic resonance spectroscopy
PφB, phytochromobilin
PAS, photoacoustic spectroscopy
PBD, photothermal beam deflection
PC, phosphatidylcholine
PCB, phycocyanobilin
PE, phosphatidylethanolamine
PEB, phycoerythrobilin

PEC, phycoerythrocyanin
Pfr (or P_{fr}), far-red-adsorbing state of phytochrome
Phot (or PHOT), phototropin
Phy (or PHY), phytochrome
Pr (or P_r), red-adsorbing state of phytochrome
PS, phosphatidylserine
PSB, protonated Schiff base
PYP, photoactive yellow protein
R*, light-activated rhodopsin
R, rhodopsin
RK, rhodopsin kinase
ROS, rod outer segment
RPE, retinal pigment epithelial cells
SB, Schiff base
SDM, site-directed mutagenesis
SPR, surface plasmon resonance spectroscopy
SR, sensory rhodopsin
TG, thermal grating
VLIR, very low fluence response
Vop, volvoxopin

Chapter 1

Archeabacterial phototaxis

Martin Engelhard, Georg Schmies and Ansgar A. Wegener

Table of Contents

Abstract

Phototaxis in Archaea is regulated by the two receptors sensory rhodopsin I and sensory rhodopsin II which are closely related to the two ion pumps halorhodopsin (HR) and bacteriorhodopsin (BR). These seven helix membrane proteins are activated by light which induces an all-*trans* to 13-*cis* isomerisation of the retinal chromophore bound via a protonated Schiff base to helix G. The signal invoked by these reactions triggers structural changes in cognate halobacterial transducers of rhodopsin. The cytoplasmic domains of these membrane proteins are homologous to that of eubacterial chemoreceptors which activate proteins of the two-component signalling cascade. The similarities between the phototaxis machinery with the two-component signalling chain on the one hand and between the photoreceptors with the ion pumps BR and HR on the other direct the present review. The first part addresses the physiological response of the *H. salinarum* towards light and the underlying protein network. The next section focuses on the shared properties of receptors and ion pumps such as structural similarities and common principles of the light activated reactions. Finally, the molecular mechanism of signal transfer from the photoreceptor to the transducer is discussed.

1.1 Introduction

Bacteria and Archaea have survived the most dramatic environmental changes that have occurred since their first appearance, three billion years ago. They have occupied almost every ecological niche available, including extremes such as high temperatures at acidic or alkaline conditions. One reason for their endurance is their ability to respond adequately and precisely to environmental changes either genetically or by a locomotive answer. The information flow from the external input across the plasma membrane to the activation of the physiological signal is based on the so-called two-component signalling system that has been found in all three domains of life (for recent reviews on eukaryotic and prokaryotic two-component system see, e.g., [1–6]). This signalling pathway consists of sensors, which receive and transmit the external stimuli to cytoplasmic proteins, including both a histidine and an aspartate kinase (hence the name) which function as transmitter and receiver, the latter regulating the physiological response on the level of genes, proteins, or the cellular motor. The input signal can be quite diverse, ranging from magnetic fields, gravity, or osmolarity to chemicals, starvation, or photons, to name a few.

The two-component signalling cascade has been thoroughly investigated for the chemosensory system of *Escherichia coli*, *Salmonella typhimurium*, and related enteric bacteria. In recent years a similar signalling cascade from the archaeal *Halobacterium salinarum* has been discovered while analysing the mechanism of phototaxis. These archaea have been of particular interest since the discovery of bacteriorhodopsin (BR), the light-activated proton pump, in the early 1970s [7]. The wealth of available information on the function and structure of BR has been reviewed (e.g. [8]; see also a special issue of *Biochem. Biophys. Acta*, **1460** (2000) with a comprehensive discussion of BR and related pigments). Various three-dimensional structures of the BR ground state [9–11] and intermediates [11–13] (reviewed in [14]) are now accessible and provide a basis for the understanding of the molecular mechanism of the light-activated proton transfer. Furthermore, this data is important in elucidating signal transduction as exemplified in the sensory rhodopsins.

During these investigations on BR three other retinylidene proteins were discovered. Halorhodopsin (HR), an ion pump like BR (both reviewed e.g. in [15] and [16]), was first described and named by Mukohata and co-workers [17]. In subsequent work, HR has been recognised as an inward directed chloride pump [18] and the amino acid sequence has been determined [19]. Since 2000 the tertiary structure of HR has been available at 1.8 Å resolution [20]. The other two pigments, sensory rhodopsin I (SRI) and sensory rhodopsin II (HsSRII), are responsible for the phototaxis of the bacteria and enable them to seek optimal light conditions for the functioning of the ion pumps HR and BR (SRI) and to avoid photo-oxidative stress (HsSRII) [21] (Figure 1). The earliest

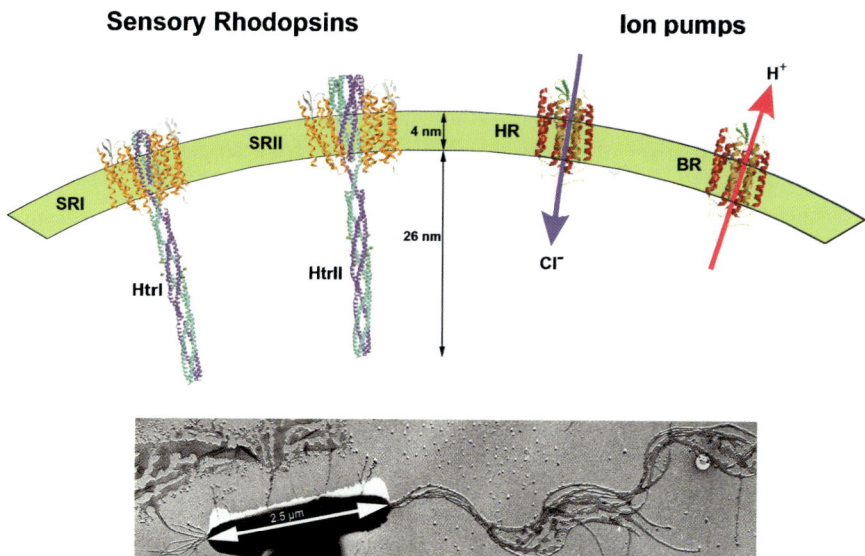

Figure 1. The four archaeal rhodopsins as molecular models. The structures depicted were taken from Sass et al. (BR) [13], Luecke et al. (for SRI and HsSRII the structure of NpSRII was taken [124]) and Kolbe et al. (HR) [20]. The receptors SRI and HsSRII are bound to their cognate transducers, forming a 2:2 complex. For the dimeric structure of the transducer the model of the serine chemotaxis receptor was taken [87]. The models are not drawn to scale. Approximate distances are indicated. In the lower panel an electron microscopic picture of *H. salinarum* is depicted, showing the bacterium with its polarly inserted flagella. [Electron micrograph adapted from [42]].

report on the phototactic behaviour of *H. salinarum* was published in 1975 [22] although the involvement of retinal proteins was only recognised in subsequent work [23–26]. At about the same time it was demonstrated that methylation of membrane proteins is involved in the photosensory and chemosensory behaviour of *H. salinarum* [27–29] which suggested that a sensory pathway similar to that in *E. coli* exists.

Research into halobacterial photosensing made a decisive step forward when Spudich and Spudich isolated HR-deficient mutants [30]. These so-called flux mutants were obtained by exciting HR in cells in which a small proton leak had been introduced with a protonophore. The method selects for mutants which escape cytoplasmic acidification. In such a way isolated mutants lacking BR as well as HR were used for phototaxis studies. The photo-sensory behaviour of these bacteria was unimpaired, demonstrating that neither BR nor HR are involved in phototaxis [31] (however, see below for more recent experiments on BR as photosensor) [32]. The authors identified a retinal-containing protein absorbing between 580 and 590 nm. It was named 'slow rhodopsin-like pigment' (later renamed as sensory rhodopsin I; SRI) because of its photocycle turnover of 800 ms, in contrast to that of about 10 ms for BR or HR.

On light excitation SR forms, in analogy to the BR-photocycle, a long-lived intermediate with a fine-structured absorption band with a maximum at 373 nm. This species is also photoactive and has been correlated with the photophobic response of *H. salinarum*. The notion that the same photoreceptor is responsible for both the repellent as well as the attractant responses has been further elaborated by the same authors [33]. The observations were summarised in a mechanism of colour sensing mediated by a single receptor (SRI). The essence of the model is the discrimination between visible and UV light by one- and two-photon processes, respectively. The absorption of a photon ($\lambda > 500$ nm) by SRI triggers the photocycle, which results in the activation of the attractant signal transduction chain. However, in the presence of both visible and UV light the long-lived intermediate (S_{373}) is excited and the repellent signalling cascade is turned on. This proposal of Spudich and Bogomolni was confirmed in later work and is now the accepted explanation for the colour discrimination of *H. salinarum*.

During further work on the halobacterial phototaxis, another repellent pigment was identified, named phoborhodopsin (pR) [34] or sensory rhodopsin II (HsSRII) [35]. HsSRII covers the blue–green region of the spectrum. Its photocycle, like that of SRI, is quite slow and also contains, similar to BR, an M-like intermediate. Contrary to SRI, this pigment induces in *H. salinarum* only a single answer to light, i.e. a photophobic response. The four archaeal rhodopsins detected in *H. salinarum* are depicted in Figure 1. The corresponding amino acid sequences are shown in Figure 2.

The amino acid sequences of the two sensory rhodopsins have been determined [36,37]. Additionally, the primary structures of HsSRII from the archaeal species *Natronobacterium pharaonis* (NpSRII) and *Haloarcula vallismortis* are available [38]. The amino acid sequences of the SRs reveal

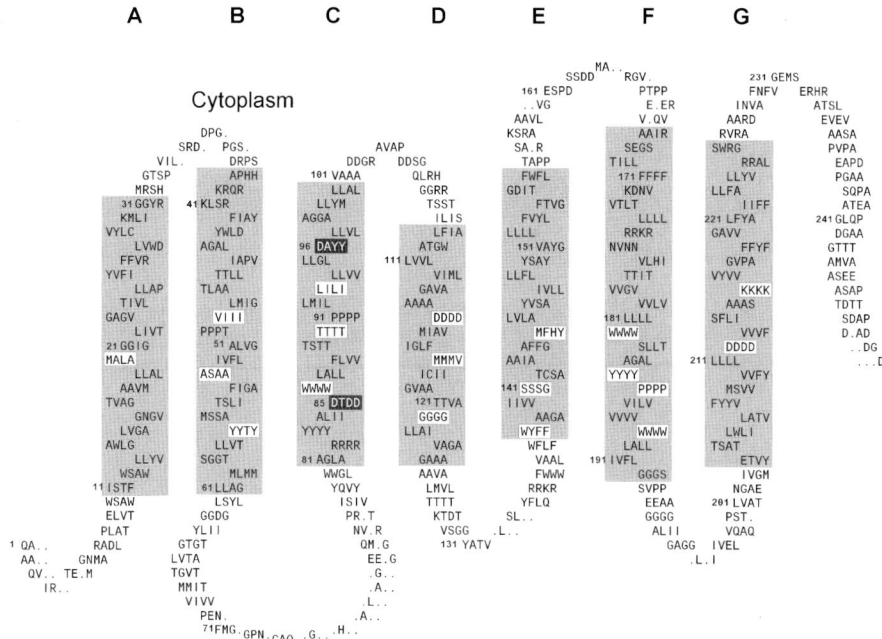

Figure 2. Alignment of the amino acid sequences of the four archaeal rhodopsins in the order BR, HR, SRI, and HsSRII. Numbers present the sequence position in BR. Amino acids forming the retinal binding site are marked by a white background. The two crucial positions in the cytoplasmic channel (96) and extracellular channel (85) are highlighted by a black background. The cytoplasmic side of the membrane is at the top. The amino acid sequences were taken from the SwissProt data bank. Accession numbers: BR, M11720; HR, P16102; SRI, X51682; HsSRII, U62676.

considerable homologies to those of BR and HR. Positions defining the retinal binding site are usually identical; however, the site which in BR is intimately involved in the proton uptake from the cytoplasm (Asp96) is changed to an aromatic amino acid. It has been proposed that this change interferes with an efficient reprotonation of the Schiff base. Indeed, the NpSRII mutant F86D displays an unperturbed M-decay, although the overall turnover is unchanged [39,40].

The sequence determinations revealed upstream of the *sopI* and *sopII* loci open reading frames correspond to the halobacterial transducer of rhodopsin (Htr) [37,38,41]. Both *sop* and their corresponding *Htr* genes are under the control of the same promoter. Sequence alignments with the chemotactic receptors from enteric bacteria revealed considerable homologies, especially in regions important for signalling and adaptation processes. This observation connects the well-known two-component system with the signalling chain in phototaxis, combining our knowledge of the two separate fields of archaeal phototaxis and eubacterial signal transduction.

1.2   Two-component systems in Archaea

1.2.1   Chemotaxis in H. salinarum

The rod-like *H. salinarum* are up to 6 μm long and 0.5 μm in diameter (see Figure 1) [42]. The bacterium is propelled by polarly inserted motor-driven right-handed helical flagella. The forward swimming direction is reversed by switching the motor from a clockwise to a counter clockwise rotation, which is under the control of cytoplasmic factors. The swimming pattern of halobacteria without stimulus is like a random walk. Forward swimming periods are interrupted by a short stop and a reversal of direction. Angular changes are thereby caused by Brownian motion or by mechanical obstacles in the path of the cells.

The cells respond to various chemicals, e.g. arginine, leucine, or dipeptides such as Met-Val, as attractants and also to phenol, indole or benzoate as repellents [43,44]. From more than eighty compounds tested six amino acids and seven peptides turned out to be attractants whereas three substances were shown to induce phobic responses [44]. Recently, the genome of *Halobacterium* species NRC-1 has been sequenced [45]. In this project, at least 17 homologous methyl-accepting taxis transducers (halobacterial transducer proteins, Htps [46]) have been recognised whereas, for comparison, *E. coli* contains only 5 taxis receptors. Originally, in a screening with oligo nucleotides comprising consensus sequences of the signal domain of eubacterial methyl-accepting proteins 13 genes encoding Htps were identified [47]. The primary amino acid sequences clearly showed that the group of Htps includes not only transmembrane receptors but also soluble cytoplasmic proteins. In a couple of instances the proteins could be functionally assigned [37,41,44,48–50]. For example, the membrane-bound transducer HtrVIII is an oxygen sensor and involved in the aerotaxis of the cells [51]. An interesting cytoplasmic arginine sensor has been shown to be physiologically coupled to an arginine:ornithine antiporter [44]. Further evidence was provided that branched chain amino acids like leucine or valine are sensed by BasT [50]. It should be noted that the phototactic transducer HtrII displays a dual function as a photophobic as well as a serine receptor. The latter property is conferred by an extracellular domain inserted between the two transmembrane helices [52].

The signal transduction chain consists of proteins of the two-component signalling chain. Genome analysis of *Halobacterium* species NRC-1 revealed the complete set of *Bacillus subtilis* che gene homologues with the exception of CheZ [45]. The adaptor protein CheW, the histidine kinase CheA, the response regulators CheY, and CheB, as well as CheJ had been described earlier by Oesterhelt and Rudolph [53,54]. The picture emerging from these data indicate a similar signal transduction chain to that described for enteric bacteria (see Figure 3). As in *E. coli* an adapter molecule (CheW) is attached to the cytoplasmic domain of the transducer. An attractant or repellent stimulus deactivates or activates the histidine kinase CheA [54] which is bound in an unknown fashion to the signalling domain of the transducer. CheA phosphorylates

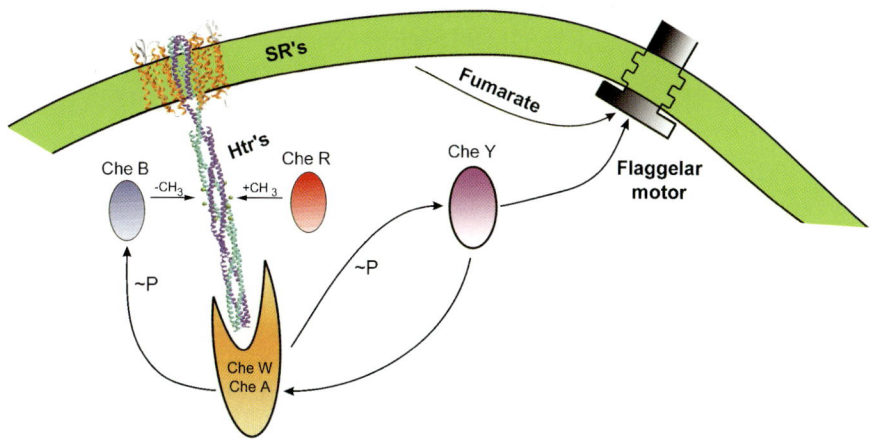

Figure 3. A model of the two-component system in archaea with NpSRII as an example (the structures were taken from [87,124], not drawn to scale). Excitation of NpSRII by light activates the histidine kinase CheA which becomes phosphorylated. Subsequently, CheA transfers ~P onto the response regulators CheY or CheB. CheY~P is the switch factor for the flaggellar motor. Thereby the periods between reversals are shortened, resulting in a photophobic response of the bacteria. Fumarate was recognised as a second switch factor. If CheB is phosphorylated, CheB~P catalyses the demethylation of glutamic acid residues (small balls), thus enabling the cells to adapt to constant input of stimuli. An intrinsic phosphatase activity of CheY and CheB deactivates the two proteins.

CheY the switch factor of the flagellar motor, thereby reversing the rotational direction of the motor. Low concentrations of CheY~P prolong the swimming period between reversals. CheA can not only activate CheY but also CheB, which functions as a methylesterase of methylated Asp or Glu residues that flank the signalling domain. CheR, a constitutively active methyltransferase, re-methylates the carboxylates. Thereby, depending on the input of attractant and repellent impulses the methylation level is altered. These methylation/demethylation reactions are involved in the adaptation of the bacteria to constant stimuli and have been studied in great detail for enteric bacteria (see Figure 3, for a model of the signal transduction chain). It should be noted that an overall chemo- and phototactic signal integration occurs in *H. salinarum* [25,55]. For more information on the two-component signalling cascade in enteric bacteria and the adaptation processes the reader is referred to recent reviews (e.g. [56,57]).

A second sensing system, triggered by the activated transducer, has been discovered that relies on fumarate [58,59]. This metabolite operates as a second messenger and acts together with CheY at the flagellar motor. This principle is not unique to archaea, it has also been demonstrated in eubacteria [60,61].

Several *H. salinarum* mutants, defective in taxis, displayed distinct phenotypes corresponding to a mutant missing photoreceptors, a mutant defective in CheR, and mutants on the level of the methylesterase and intracellular signalling [62,63]. Especially, the phototaxis mutant Pho81 proved to be useful for the homologous and heterologous expression of sensory rhodopsins (e.g. [64–67]).

1.2.2   Phototaxis of Halobacteria

A gradient of an external stimulus such as light alters the regular intervals of forward swimming, stopping, and backward swimming by prolonging or reducing the swimming period [22,25,68]. For example, cells swimming up a gradient of light (>500 nm) will increase the period between reversals. If they move against this gradient the interval between reversals will be shortened. Both behavioural responses result in a net movement towards the source of light. Gradients of light at wavelengths below 500 nm produce the opposite effect, consequently the bacteria move away from this kind of irradiation. The responsible pigments are SRI, which guides the halobacteria towards favourable light conditions and, in a two-photon process, away from UV-light [33], as well as HsSRII which enables the bacteria to seek the dark when the oxygen supply is plentiful, thus avoiding photooxidative stress [21].

The phototactic behaviour of halobacteria has been studied by visual tracking of single cells through a microscope (e.g. [22,69]) or subsequently by computer tracking and motion analysis techniques (e.g. [70]). In the 1980s, Stoeckenius and co-workers developed a rapid population method to determine action spectra [71] which the authors successfully used for identifying sensory rhodopsin II [35]. The first paper to describe negative phototaxis other than the SRI-mediated blue light repellent response was published by Takahashi et al. [72], who described a mutant that displayed only negative phototaxis with a maximum of the action spectrum at about 475 nm. In subsequent work the same group isolated the responsible pigment, which they named phoborhodopsin [34]. Other groups also described a fourth rhodopsin in *H. salinarum* [73,74].

Once halobacteria reach areas of constant stimulus influx, the cells adapt to these conditions and resume spontaneous switching. The apparatus behind this adaptation process includes those proteins involved in the methylation and demethylation of Glu or Asp residues located in the cytoplasmic domain of the transducer. These reactions were first described by Schimz [28,29] who discovered that a stepwise increase of orange light results in the liberation of methanol, the product of the demethylation reaction. Conversely, a decrease in light intensity or an increase of UV-light reduces the methanol release. The author also provided evidence that membrane proteins are the target of the methylation reaction. In further work the reactions were analysed in more detail [43,75–77]. Perazzona and Spudich identified the methylation sites in HtrI and HtrII by mutagenic substitutions of Glu residues selected from consensus sequences [77]. Indeed, replacing the Glu265-Glu266 pair in HtrI and the homologous Glu513-Glu514 couple in HtrII by alanine eliminated the methylation of these transducers. The physiology of methylation/demethylation was further investigated by Marwan et al. who proposed a kinetic model of photosensory adaptation that relies on receptor deactivation [78]. It was suggested, in accordance with other publications, that a reversible methylation of HtrI – itself a membrane protein – is the chemical basis for the sensory adaptation.

Halobacterial colour sensing, such as the visual perception of higher eukaryots, can adjust to intensity changes in incoming light. The sensitivity of this process, however, is not yet known. Conversely, measurements of stimulus response curves revealed that *H. salinarum* can detect a single photon [79].

The cell synthesises sensory rhodopsin II constitutively. In contrast, the biosynthesis of SRI was shown – like BR and HR – to be induced by decreasing oxygen tension in a cell culture [73]. This repertoire of light-sensing pigments, which includes SRI with its dual function (photo-attractant response with maximum at 587 nm and photo-repellent answer at 373 nm) and the photophobic receptor HsSRII (λ_{max} = 490 nm), enables the bacteria to seek, at low oxygen concentrations, optimal light conditions for the functioning of the two ion pumps BR and HR. With ample oxygen supply *H. salinarum* solely relies on oxidative phosphorylation. Possible photo-oxidative damage can be avoided because HsSRII with an absorption maximum matching that of sun light directs the cells towards the dark [21,80].

1.2.3 Receptor/transducer complexes

The incoming extracellular signal, which can be of either chemical or physical nature, has to reach the cytoplasm to activate the two-component system. The interface between the transmembrane signalling complex and the cellular chemotactic proteins is provided by the cytoplasmic domain of the receptors, and in the case of phototaxis by their cognate transducers. A comparison of the primary sequences of 29 proteins from 16 different species, which also included archaeal Htr's [38,41] (excluding HtrII, whose sequence had not yet been published [37]) revealed a consensus secondary structure consisting mostly of α-helices. A seven-residue repeat (a-b-c-d-e-f-g) with hydrophobic residues in positions a and d indicated a coiled/coil arrangement of the helices [81]. This domain structure was also recognised for HtrI using a sequence alignment and crosslinking studies of single Cys substitutions into selected sites of the membrane domain of HtrI [82]. In this latter work it has also been shown that HtrI forms a dimer whose interface is sensitive to receptor photoactivation. A dimer structure of HtrII from the alkalophilic archaea *N. pharaonis* has been deduced from electron paramagnetic resonance (EPR) investigations [83,84]. The interaction of archaeal transducers with their cognate sensory rhodopsins has been analysed [64,85] and it could be proven that specificity is determined by their transmembrane helices [86].

Since the sequence homology between the archaeal transducers and the bacterial receptors leads to similar secondary structure predictions, the tertiary structure of the cytoplasmic domain of the serine chemotaxis receptor provided by Kim and co-workers [87] can also be taken as a model for the phototaxis transducers. By analogy one can deduce that the cytoplasmic part of a transducer dimer is a distinct four-helix bundle formed by the association of two helical hairpins. This rod extends about 200 Å into the cytoplasm with three different functional sections recognisable (Figure 3). At the membrane-distal end, a kinase-interaction region is responsible for the interaction with CheW and CheA. Approaching the cytoplasmic membrane a methylation region follows which is involved with the adaptation processes and probably binds CheB and CheR. A linker element connecting the methylation region with the transmembrane domain is, so far, structurally not very well charac-terised. An alignment of various linker sequences suggests two amphipathic helices [88]. A structural characterisation of this part of the transducer will certainly be the key to understanding the transmembrane signal transduction and activation of the cytoplasmic two-component system. The Htrs region might also harbour the recognition site for their cognate photoreceptors SRI and HsSRII [89–91].

The functionality of a complex between NpSRII with a truncated transducer devoid of most of its cytoplasmic domain is unimpaired, as shown in studies involving the binding of the transducer to the receptor using blue native gel electrophoresis and isothermal calorimetry experiments [83]. Additional infor-mation about the functionality of the complex comes from electrophysiological measurements [92]. Previous work demonstrated that the innate capability of NpSRII to pump protons on light excitation is blocked by the binding of its cognate transducer [93,94]. In his thesis Schmies has demonstrated that a trun-cated transducer consisting of the N-terminal amino acid sequence from 1 to 113 does indeed block the proton transfer in NpSRII, indicating a functional complex [92].

Chemoreceptors form heterogeneous clusters primarily at the poles of the bacterial cell [95]. For the archaeal phototaxis transducers and chemoreceptors no such information is available. It would be important to know whether such complex structures are also established in Archaea and, if they are, whether the components are recruited from the chemotaxis as well as from the phototaxis branch.

1.3   Properties of sensory rhodopsins and the receptor/transducer complex

An important prerequisite for the analysis of the structural and biophysical properties of photoreceptors and their transducers is their availability. Originally, sensors were prepared from their natural host [96–99]. However, due to their low cellular concentration – in wild-type *H. salinarum* there are only 2000–3000 copies of SRI [100] – molecular genetic tools had to be applied. A homologous expression system was introduced that allowed the overexpression of SRI [101–103] as well as the heterologous expression of NpSRII [104]. A decisive step forward was taken when it became possible to functionally express NpSRII in *E. coli* [105]. This method proved to be successful not only for the facile preparation of NpSRII and HR [106] but also for SRI [107] using a His-tag as an affinity label. In a similar way a truncated form of NpHtrII could also be prepared [108,109].

1.3.1 Primary sequences

The functional and structural properties of sensory rhodopsins are determined by their primary structures. A two-dimensional structural map of NpSRII at 6.9 Å [110] and more recently high-resolution structures are available. Already the two-dimensional map clearly demonstrates the structural similarity between NpSRII and BR, which had already been deduced from an alignment of the corresponding primary sequences [36–38,111]. The highest percentage of homology is found at those sites which constitute the retinal binding pocket (sequences of ~30 homologous rhodopsins have been published so far [112]). This is also true for an archaeal rhodopsin-like pigment (NOP-1) detected from *Neurospora crassa* [113,114], the first example of an archaeal rhodopsin discovered outside its own kingdom. Interestingly, according to its amino acid sequence it is more similar to BR than to SRI or HsSRII (e.g. the position of Asp96 (BR) is conserved); however, its photochemistry resembles that of sensory rhodopsins [114] (see Figure 2 for a sequence alignment of the four archaetypical rhodopsins BR, HR, SRI, and HsSRII). Most recently archaeal rhodopsins have also been discovered in marine microbial populations. Apparently, these pigments (named proteorhodopsins) function as light-driven proton pumps involved in phototrophy [115,116].

Comparing primary sequences of sensors, proton pumps, and halide pumps, obvious differences are connected to proton release and to the proton uptake channels (Figure 4). In BR the key residues are Asp85 and Asp96, which are crucially involved in the proton pump mechanism. After light excitation the Schiff base proton is transferred to Asp85, with concomitant release of a proton into the extracellular buffer. Once the salt bridge between the protonated Schiff base and the negatively charged Asp85 has been broken in the so-called M-state a protein switch can occur, which alters the accessibility of the Schiff base from the extracellular channel towards the cytoplasmic channel. Thus, the Schiff base can be reprotonated from the cytoplasmic side via Asp96, thereby completing the vectorial proton transfer across the membrane (see Lanyi for a detailed discussion of proton transfer reactions in BR [8]). In SRs the Asp 96 is replaced by an aromatic residue, thus interfering with an optimal proton transfer from the cytoplasm to the Schiff base (see below). A scheme of the proton transfer steps, comparing BR and NpSRII, is depicted in Figure 4.

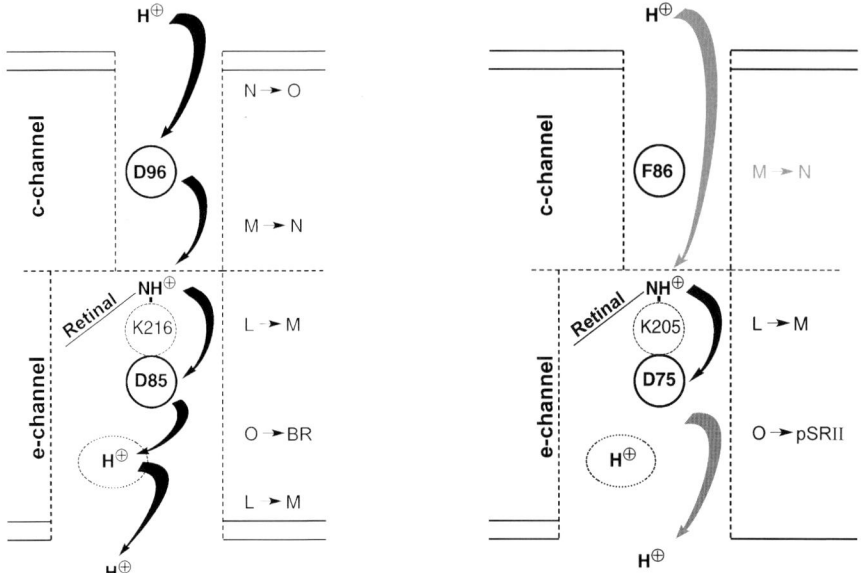

Figure 4. Comparison of proton transfer steps in BR (left) and NpSRII (right). After light excitation the proton from the Schiff base is, during the L→M transition, transferred to an Asp residue (Asp85; Asp75). The time course and the mechanism of the susequent steps are different for BR and NpSRII. Whereas in BR Asp96 donates its proton to the Schiff base, this reaction is not possible for NpSRII (as well as for SRI and HsSRII) because an aromatic residue (F86) has replaced Asp96. Instead the reprotonation has to occur directly from the cytoplasm. The proton release to the extracellular medium is also different for the two pigments. Whereas BR releases the proton during the L→M transition, in NpSRII this only happens in the last step of the photocycle (O→NpSRII). The proton in the circle depicts sites connected to a hydrogen-bonded network with an excess proton. Abbreviations: c-channel, cytoplasmic channel; e-channel, extracellular channel.

1.3.2 Absorption spectra

All rhodopsins, if excited with light corresponding to their absorption maxima, display a characteristic photocycle. These maxima are for BR, HR and SRI above 560 nm. An exception is observed for HsSRII, which absorbs maximally at 490 nm (the homologous protein from *N. pharaonis* has a maximum at 500 nm; see Figure 5).

The reason for this reduced opsin shift is not yet clear. The opsin shift is a measure (in cm^{-1}) of the protein´s influence on the chromophore absorption maximum. The colour regulation in BR has been explained by a synergistic effect of a 6-s-*trans* bond at the ß-ionone ring (as is found in SRI [117]) together with a complex counterion at the protonated Schiff base [118]. Experiments with retinal analogues indicated that in HsSRII and/or NpSRII the retinal binding site is more restricted than in BR [119]. A planarisation

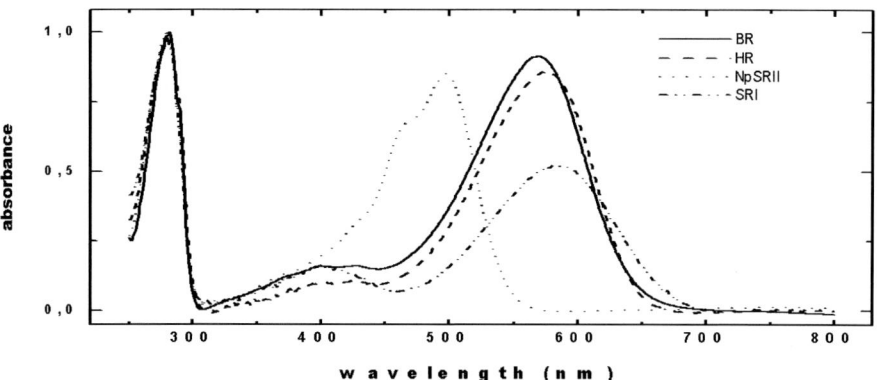

Figure 5. Absorption spectra of BR (solid), HR (dash), NpSRII (dot), and SRI (dash, dot, dot). The spectrum of SRI was measured at pH 6, displaying only maxima at 400 and 587 nm. The spectra are corrected for the contribution of light scattering [216].

of the retinal ring with respect to the polyene chain could account for the opsin shift as well as for the fine structure of the absorption maximum only observed in the known HsSRIIs [120]. The structural basis for these observations has been sought in those amino acid residues of the retinal binding site that differ from those of the other archaeal rhodopsins. However, mutational studies were unable to narrow the cause to particular amino acids [121–123]. Even the simultaneous replacement of 10 amino acids from the retinal binding pocket by corresponding residues from BR did not result in a substantial batho-chromic shift (N. Kamo, personal communication). These failures to explain the absorption properties of HsSRII or NpSRII indicate that our knowledge of colour regulation in retinal proteins is not adequate, thus further experi-ments are mandatory. However, a recent crystal structure of NpSRII at 2.4 Å [124] provides new insight into the colour regulation in NpSRII. It appears that the main contribution to the blue-shift is the longer distance of the guanidinium group of Arg72 from the Schiff base as compared to that of Arg82 in BR.

The absorption maximum of NpSRII is slightly dependant on external conditions. Lowering the pH to 3.5 shifts the maximum to 325 nm ($pK = 5.6$; detergent solubilised NpSRII has a pK of < 4) [125,126]. These observations can be explained by the protonation of the Schiff base counter-ion Asp75 which, under physiological conditions, is deprotonated [127]. Removal of this anion from the protonated Schiff base – which can be accomplished either by its protonation or by its mutation into a neutral amino acid – exerts a bathochromic shift of the same order of ~900 cm^{-1} for BR, pHR, NpSRII, and HsSRII [125,128–130]. The addition of highly concentrated solutions of chloride to the acid form of NpSRII reverses the bathochromic shift [125,126].

A similar dependency of the absorption maximum on the external condi-tions is found for SRI. At neutral pH three maxima, at 587 nm (SRI$_{587}$), 550 nm (SRI$_{550}$), and 400 nm (SRI$_{400}$), are detected [32,33,107,131–133] (Figure 5).

The action spectrum of the photo-attractant response of *H. salinarum* corresponds to the maximum at 587 nm [22]. Fourier-transform infrared (FTIR) data have revealed that the counterion (Asp76) of the Schiff base is protonated [134], as it was also proposed for Asp 85 in the acidified (or deionised) purple membrane [135,136]. Congruent with this observation is the fact that the SRI-mutants D76N and D76A are fully functional as phototaxis receptors [134]. The pK of Asp76 is about 7.2, leading to the species absorbing at 550 nm [32,137–139], which turned out to be a light-driven proton pump (see below). There is so far no obvious explanation for the maximum at about 400 nm, although a deprotonated Schiff base might be responsible. It is interesting to note that the binding of the transducer increases the pK of Asp76 to 8.7 [131]. The characteristic pH of the natural habitat of *H. salinarum* is at about 7.5, implying complete occupancy of the 587 nm state.

1.3.3   Photocycle of sensory rhodopsins

The chromophore in SRI and NpSRII is all-*trans* retinal [140,141] bound via a Schiff base linkage to a Lys residue on helix G. In NpSRII a *trans*-13-*cis* isomerisation of the retinal chromophore (this so-called light/dark adaptation was first observed in BR) cannot occur because the retinal binding site only accepts all-*trans* retinal but not 13-*cis* retinal [142]. In HsSRII from *H. salinarum* the fraction of all-*trans* retinal has been determined to be 80% [98]. For SRI the light-induced all-*trans* to 13-*cis* isomerisation is a prerequisite for its functioning [143,144]. In the initial state SRI contains almost only all-*trans* retinal (95%) which is shifted to 93% 13-*cis* retinal in the M-state [145]. The resonance Raman spectra are neither altered by the mutation D76N nor by the complexation with the transducer HtrI [145].

Steric constraints in the retinal binding pocket have been deduced from experiments using retinal analogues [146–148]. The photoactive site of SRI and BR was probed by a set of 24 retinal analogues [149]. This investigation revealed differences in the protein environment near the retinal 13-methyl group and near the β-ionone ring. It was proposed that the 13-methyl group–protein interaction functions as a trigger for SRI activation. A similar proposal has been made for the retinal 9-methyl group in mammalian rhodopsin [150].

After light excitation the sensory rhodopsins thermally relax back to the original state through several intermediates. Generally, these photocycles are quite similar to that of BR, consisting of the canonical intermediates K,L,M,N and O states ([151] and literature therein); schemes of the photocycles are depicted in Figure 6. The K-intermediate of NpSRII is formed within 5 ps [133]. For SRI (at pH 6; protonated Asp76) a slow biexponential absorbance change indicates a long-lived excited state. Photoacoustic measurements for both pigments revealed volume changes at the stage of the K-intermediate that are considerably larger than those observed for BR [132,152,153]. The NpSRII mutant D75N does not influence the production of

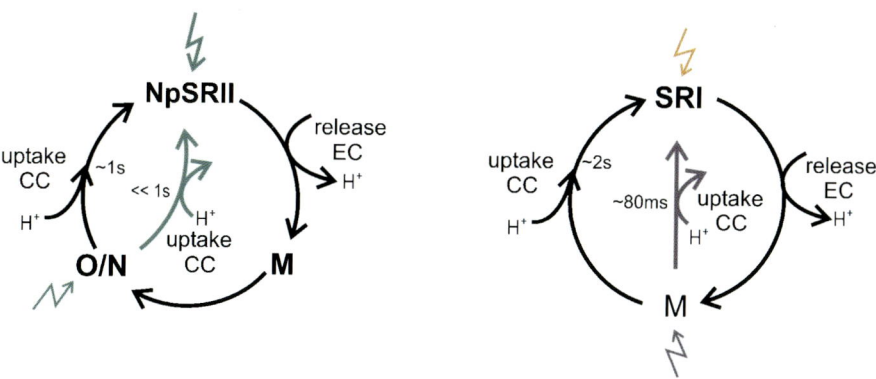

Figure 6. Scheme of the photocycles of SRI and NpSRII, depicting the major intermediates. Included are the proton uptake and release steps. Abbreviations: CC, cytoplasmic channel; EC, extracellular channel.

the K-like intermediate but strongly affects its relaxation pathway [154]. Interestingly, the same volume change of 10 mL mol^{-1} was determined for the NpSRII-transducer complex [109].

Early on, it was obvious that an M-like intermediate was present in the photocycles of both pigments, SRI [155,156] and HsSRII [34]. In the case of SRI the investigation of the photocycle is hampered by the fact that two species SRI$_{587}$ and SRI$_{550}$ are present at physiological pH. Bogomolni and Spudich – assuming a unidirectional, unbranched reaction scheme–identified two precursor intermediates (K: S$_{610}$; L: S$_{560}$) which are followed by the long-lived M-like state (S$_{373}$) [157]. The latter intermediate, possessing a lifetime of about 750 ms, can absorb a second photon which short-cuts the cycle and is responsible for the UV-sensitive negative phototaxis. Apparently, the photocycle of both SRI$_{587}$ and SRI$_{550}$ contain an intermediate with a deprotonated Schiff base, i.e. S$_{373}$. Since in SRI$_{587}$ the proton acceptor of the Schiff base proton (Asp76) is already protonated another group has to take over this role, which might be His166 (see below, [158]).

After excitation by light S$_{373}$ relaxes back to the ground state via a red-shifted intermediate (S$_{510}$) with a half-life of 80 ms. These reactions have been studied in more detail by Bogomolni and co-workers [159,160]. In an interesting observation Manor et al. described the influence of the membrane potential on the photocycling rates of SRI, BR and HR [161], which might be important for the function of SRI under physiological conditions, which displays a proton motive force of about −250 mV [162–164].

The photochemical cycle of HsSRII is not very well characterised due to its sensitivity towards external conditions [98]. Nevertheless, the few data available show a photocycle with a sequence of intermediates corresponding to blue-shifted L, M and O intermediates [35,96,165,166]. FTIR experiments proved that Asp73, the counter-ion of the Schiff base, is the Schiff base proton acceptor [167]. Interestingly, the O-intermediate (SII$_{540}$) is characterised by an all-*trans* retinal conformation, indicating that the reisomerisation from 13-*cis* has already occurred. More information about the photocycle is available for the homologous protein NpSRII [125,168–171]. Summarising these data, a photocycle similar to that of bacteriorhodopsin is obtained (Figure 6). However, the turnover rate of 1.2 s is much slower than that of BR. FTIR data provided evidence that, in NpSRII, Asp75 becomes protonated during the formation of the M-intermediate [127] which can be photoconverted into the initial state [172].

Interestingly, in the BR-mutant D96N the slow photocycle can be enhanced by the addition of azide which accelerates the reprotonation of the Schiff base [173]. Similarly, in NpSRII the M-decay rate is substantially increased by the addition of azide although the overall turnover did not change [39,174]. A similar acceleration of the M-decay can be obtained if residues from the cytoplasmic or extracelluar proton conducting channel are mutated to residues found in BR [39,40].

1.3.4   Proton transfer reactions of sensory rhodopsins

Investigating the light-induced proton transfer reactions in sensory rhodopsins provided important insight into the signalling mechanism of the photo-receptors. However, only recent progress in the biochemical accessibility of the SRs and the application of different electrophysiological techniques has led to a common picture of how the photoreceptors transport protons across the plasma membrane and how their transducers influence this process. For SRI and NpSRII it is evident that in the transducer-free state the sensors act as outwardly directed proton pumps and that their corresponding transducers HtrI and NpHtrII inhibit this pumping specifically [39,66,93,94,137,138, 175,176]. Although the capability of HsSRII to pump protons has been questioned [177], the latest results indicate that NpSRII, like SRI, acts as an outward directed proton pump [93,94].

1.3.4.1   Receptors as proton pumps

Due to their slow photocycle turnover ($t \sim 1$ s), SRI and HsSRII are less effective pumps than BR and HR ($t \sim 10$ ms) [178]. The main differences between the ion pumps and the receptors are found in the second part of the photocycle. The molecular events in this part of the photocycle are the reprotonation of the Schiff base (M-decay), the subsequent 13-*cis*/all-*trans* isomerisation of the retinal and the deprotonation of the counter-ion of the Schiff base. From this comparison of the photocycles of the SRs and BR the proton release during M-formation should be as fast as in BR but the proton uptake (M-decay) should be decelerated considerably. A sequence alignment (Figure 2) supports this hypothesis since the Schiff base proton donor Asp96 in BR is replaced by aromatic residues in all SRs [36–38]. It should be noted that in Nop-1, the recently discovered eukaryotic archaeal rhodopsin, this Asp residue is conserved. Apparently, this site does not solely govern the photocycle turn-over, as was also demonstrated by mutational studies on NpSRII and SRI (see below).

SRI was the first of the two receptors for which proton pumping was demonstrated. Bogomolni as well as Haupts and their co-workers analysed pH changes in suspensions of SRI-containing vesicles or *H. salinarum* cells and observed an acidification upon illumination [137,138]. More detailed mechanistic information was provided later on by time-resolved measurements of photocurrents using the BLM-technique with reconstituted SRI samples [175] and voltage-clamp recordings from *Xenopus* oocytes after the injection of mRNA encoding the *sopI* gene [93]. The results can be summarised as follows. At neutral pH there are two species (SRI_{587} and SRI_{550}, see above) which can be excited by orange light. Apparently, SRI_{550}, the species with a deprotonated Asp76, contributes exclusively to the electrogenic photocycle, as can be concluded from the action spectrum and the pH-dependency of the voltage-clamp photocurrents. The action spectrum for the light-induced membrane potential shows a maximum at 550 nm [137] and the pH dependent amplitudes of the voltage-clamp signals coincide with the titration of Asp76 with a pK_a of 7.2 [93].

The thermal M-decay can be accelerated from 1 s to 80 ms by the absorption of a second "blue" photon [33,159], resulting in an enhanced proton pumping activity [93,138,175]. Consequently, under white light ("natural") conditions SRI is a two-photon driven pump. This amplifying effect of blue light is unique for SRI. For BR and NpSRII (see below) blue light has exactly the opposite effect as it quenches the photocurrent [39,179]. Apparently, the molecular switch which changes the Schiff base orientation from the extracellular side in M_1 to the cytoplasmic side in M_2 is faster in SRI than in BR and NpSRII, thereby accumulating M_2. Hence, in SRI the additional blue light would mainly excite M_2, which is followed by a proton uptake from the cytoplasm but not, as was demonstrated for BR, from the extracellular side after excitation of M_1. Therefore one can conclude that SRI, contrary to BR, is a two-photon driven proton pump.

For HsSRII as a proton pump the results are ambiguous. Sasaki and Spudich investigated light-induced pH changes in vesicles or open membrane sheets solutions containing HsSRII [177]. Under continuous illumination the authors observed an increase in pH to a constant level which returned back to the initial value after switching off the light. From their results they concluded that the proton uptake and release occurs from the same (extracellular) side, resulting in a non-electrogenic photocycle. The light induced increase of pH was explained by a steady-state mixture of photo-intermediates which have picked up a proton from the bulk. Conversely, Schmies and co-workers detected from *Xenopus* oocytes, with more sensitive voltage-clamp recordings, a photocurrent that proves outwardly directed proton pumping by HsSRII [93]. Although there is no explanation for this contradiction the photocycle data indicate a similar reaction pathway of HsSRII and BR, strongly supporting an electrogenic photocycle.

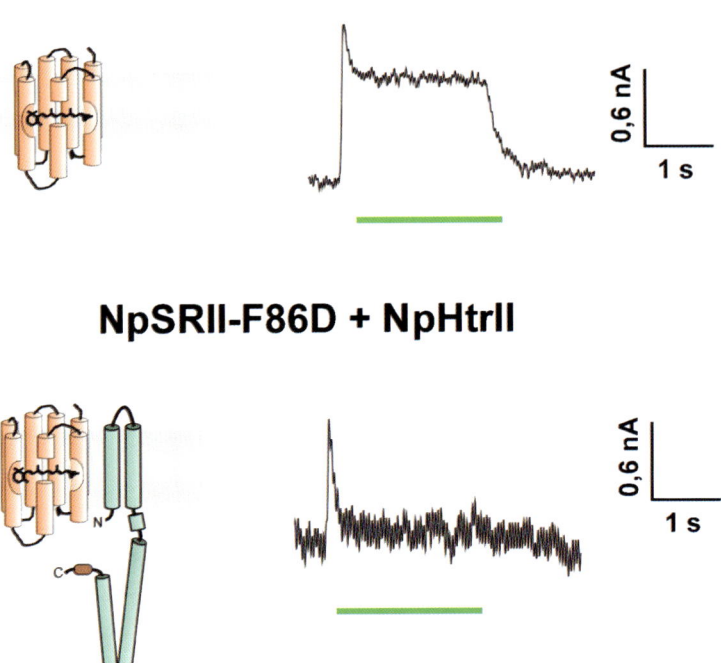

Figure 7. Photocurrent traces of the NpSRII mutant F86D without and in the presence of NpHtrII (data taken from [92]). The currents were measured using the oocyte system which was clamped at −20 mV. The transient photocurrent is seen in the first part of the traces directly after the onset of illumination.

Recently proton transport in NpSRII has been analysed in greater detail. Schmies et al. [39,93] demonstrated an outwardly directed proton transport by NpSRII (see Figure 7). Kamo and co-workers confirmed the proton pumping by using a proton-sensitive electrochemical SnO_2 cell and pH-measurements in suspensions of NpSRII-containing vesicles [94,176]. The efficiency of the light-induced photocurrent is about 100 times weaker than that of BR. It can be enhanced by replacing F86 by an Asp residue (in BR, the corresponding Asp96 serves as proton donor to the Schiff base) or by the addition of an external proton donor like azide [39]. Interestingly, these modifications do not increase the photocycle turnover which still lasts about 1s [39,174] although the addition of azide accelerates one step in the photocycle, namely the repro-tonation of the Schiff base [40]. Since in these examples the turnover has not changed as compared to that of the wild-type the increased amplitudes of the photocurrents must originate from other sources. A proposal was put forward by Schmies et al. [39,93] who assume a two-photon process which short-cuts the photocycle thereby enhancing the pump activity. The second photon has to excite an intermediate which follows M because an excitation of M quenches pumping in a similar manner as observed for BR (see above).

The proton transfer steps resulting in a vectorial transport of protons across the membrane can be formally separated into two parts. First, during the L–M transition the proton from the Schiff base is transferred to the counter-ion Asp75 as in BR and SRI (BR-Asp85, SRI-Asp76) [127,134,180–182]. This fast charge movement towards the extracellular side is represented in the transient photocurrent (Figure 7). From this experiment it cannot directly be concluded that the proton has already been released into the bulk phase. However, Iwamoto et al. [40] and Sasaki et al. (for HsSRII [177]) have provided evidence that the proton release from the membrane occurs later during the O-decay. The second step, the reprotonation of the Schiff base from the cytoplasm is not resolved under steady state conditions, but the occurrence of a continuous photocurrent proves the net transport of protons across the membrane. These data show that both HsSRII and NpSRII are proton pumps, albeit not very efficient.

The oocyte experiments allow the measurement of the photocurrent in the presence of an applied membrane potential. At about -120 mV both the transient and the photostationary current disappear in NpSRII [92]. For comparison, proton pumping in BR vanishes at about -250 mV [162,183]. Since the membrane potential ($\Delta\Psi$) in *N. pharaonis* is about -250 mV [164] and the proton motive force in *H. salinarum* is of the same order [162,163], a physiological relevance of the proton translocation can be ruled out. Nevertheless, the steep voltage-dependence of the SR's photocurrents indicates charge movements accompanied by large conformational changes. This is in line with models describing the signal transfer between photoreceptors and transducers which assume that the tilting of helix F triggers the activation of the transducers signalling domain (see 1.4.2, [21,108]).

1.3.4.2   Proton transfer reactions in the receptor/transducer complexes
The observation that sensory rhodopsins can function as proton pumps led to the question whether the binding of their cognate transducers influences this property. Indeed, Bogomolni et al. reported this effect for the SRI/HtrI complex [184], using the same experimental approach as they applied to demonstrate proton pumping by uncomplexed SRI [137]. In the latter experiment the illumination of sealed vesicles containing SRI resulted in an acidification of the bath medium. However, no pH-change was observed for vesicles with an incorporated SRI/HtrI-complex. This was confirmed by voltage-clamp recordings using *Xenopus* oocytes [93]. Similar results were also obtained for the mutant NpSRII-F86D (which displays an enhanced photocurrent). In an experiment similar to the pH-measurements of Bogomolni et al. [137], Sudo et al. [94] verified the voltage-clamp data for NpSRII/NpHtrII. For HsSRII, Sasaki et al. [177] did not detect an overall proton transport in HsSRII. Consequently, the HsSRII/HtrII complex should not show vectorial proton transfer.

It is important to note that the binding of the transducer to its receptor only affects the photostationary but not the transient photo-current [93]. Therefore, the fast proton reactions are not inhibited and neutralisation of the Schiff base–counter-ion pair can still occur. Therefore, the proton transfer reactions, which lead to a disruption of the salt bridge between the protonated Schiff base and its counter-ion, might be important in the formation of the signalling state, as Spudich and co-workers pointed out (e.g. [130,185]).

In two different studies, the specificity of receptor/transducer interactions was demonstrated. With the oocyte system, cross co-expression of SRI with NpHtrII and vice versa, NpSRII with HtrI, does not alter the receptors' photocurrent, indicating that no complex between SRI/NpHtrII and NpSRII/ HtrI is formed, or at least no specific interaction takes place [93]. In another approach Spudich and co-workers prepared transducer chimeras between HtrI and HtrII in which different transducer domains were combined [86]. After expression of the chimeric signalling complexes in *H. salinarium* the authors analysed the phototaxis of these cells and concluded that the receptors interact specifically with their corresponding transducers. Whereas the cytoplasmic domains can be exchanged, SRI and HsSRII need the two transmembrane helices of their cognate transducers HtrI and HtrII, respectively to mediate a correct physiological response.

The question still arises, which properties of the functional complex are responsible for the inhibitory effect of the transducers and is it an important feature of the signalling mechanism? Data obtained so far indicate that these specific SR/Htr interactions are likely to be located in the cytoplasmic part of the membrane. Spudich and co-workers concluded that HtrI closes the cyto- plasmic channel of SRI because the reprotonation rate of the SB (M-decay) in SRI/HtrI is insensitive to the pH, but becomes pH-sensitive after removal of the transducer [65,184,186]. A closure of the cytoplasmic channel of SRI by HtrI would reduce the accessibility of protons from the cytosol which explains the inhibition of the pumping. However, an alternative explanation might also be possible. If, as discussed above, SRI_{550} exclusively contributes to the pumping, the shift of the SR_{550}/SRI_{587} equilibrium towards almost 100% SRI_{587} upon HtrI binding would automatically abolish the photocurrent. Certainly, the two explanations are not exclusive and both mechanisms might occur simultanously. For NpSRII, the pK_a of the SB-counter-ion Asp75 is, at 5.6, too low for a $-COOH \Leftrightarrow -COO^-$ equilibrium under physiological conditions [125]. It follows that the inhibition of the proton pump on binding of NpHtrII is not due to a non-pumping species.

1.3.5   *Properties of the SR/Htr complex*

The binding of HtrI to SRI has large kinetic effects on the SRI photocycle, a result which provided initial evidence for the formation of a functional complex [65,85]. A smaller but distinct influence on the lifetime of the O-intermediate in HsSRII on binding its HtrII has been described [187].

Contrary to these results no effects on the photocycle kinetics of NpSRII were observed if a shortened transducer was bound [108]. In respect of the inhibition of the proton pump by the transducer [92] and a binding constant in the 100 nmol range [83] it was concluded that a functional complex is formed.

The observation that proton pumping is inhibited and photocycle turnover is altered in the SRI/HtrI complex has been utilised to elucidate the interaction of SRI with its transducer HtrI, including also mutational studies. A minimal structural unit comprised the receptor and the N-terminal transducer sequence (1–159) [89,91]. Function-perturbing mutations in SRI and HtrI altered the rate of S_{373} (M) formation which was interpreted as a modulation of the electrostatic interactions of the protonated Schiff base and an optimisation of the photocycle by the transducer [184].

Interesting mutations in SRI presented by Spudich and co-workers involved Asp201 and His166 [82,158,188]. In [188] Olson et al. replaced Asp residues at five positions of SRI by site-specific mutagenesis. It turned out that Asp201 is most vital for the attractant signalling function, which is changed to a repellant response when this residue is substituted by the isosteric asparagine. The authors point out that this result genetically separates the attractant and repellant response of the bacteria. One proposed explanation assumes that in the dark the signalling complex is locked in an inappropriately attractant adapted state, similar to the situation of repetitive stimulation by orange light [189,190].

From mutational studies on His166 Zhang and Spudich concluded that this residue plays a role in the proton pathway after deprotonation of the Schiff base, the modulation of SRI photoreaction kinetics by HtrI and is important in phototaxis signalling [158]. Since under physiological conditions the protonated Asp76 is not available to accept the proton His166 might be an alternate site. Only this reaction sequence would lead to the formation of the signalling state capable of activating HtrI. Indeed, His166 replacements have conformational effects on the structure of HtrI at position 64 [82].

These results were explained by a two-conformation equilibrium model introduced by Spudich and Lanyi as a unified mechanism for ion pumping and signal transduction [191]. According to this proposal SRI consists of an equilibrium mixture of two conformers which have similar properties to the closed and open channel conformers of BR. Orange light shifts the equilibrium towards the attractant (A) state whereas in the two-photon cycle the repellent (R) conformer is populated. In support of this model are observations concerning the "orange-light inverted" phenotype of some SRI-mutants like D201N [188] or H166A [158] and the HtrI-mutant E56Q [192]. Apparently, in all these mutants the equilibrium is shifted in the dark towards the A conformer, exhibiting a repellent response to both one-photon and two-photon activation. Further support comes from a suppressor mutational analysis on mutants at these two sites (D201N, H166S, H166A) as well as on the Htr mutant E56Q [193]. Assuming that the effects of these single site mutants can be reversed to wild-type properties the authors screened for second site mutations that restored the attractant response. Fifteen such mutants were identified with three suppressor mutations at the cytoplasmic side of helix F and G of SRI and the other 12 mutations in HtrI clustering at the cytoplasmic end of TM2. These sites certainly are intimately involved in the binding surface of the receptor to the transducer.

1.4   Molecular mechanism of the signal transfer

1.4.1   The receptor-transducer interaction

The interaction of receptors with their cognate transducer as outlined above has been assessed by phototaxis, photocycle, and proton pump measurements. A modulating effect on the photocyle kinetics of sensory rhodopsin was frequently considered as characteristic of this interaction extensively exemplified by SRI and HtrI (for a review see [80]). Also for the green light receptor of *H. salinarum* HsSRII Sasaki & Spudich [187] observed significant acceleration of the "M" to "O" to ground state transition upon binding of the transducer. The authors suggest that the nearby transducer produces changes in the hydrogen network around Asp73, thereby accelerating proton transfer reactions. In contrast the photochemistry of NpSRII is not altered in the presence of the transducer [108].

Most of the cytoplasmic domain of HtrI can be deleted without altering the photochemical properties of the SRI/HtrI complex [89,91]. These experiments indicated that the specificity of the SRI/HtrI interaction is confined to the two transmembrane helices and a hydrophilic stretch of 90 residues subsequent to the cytoplasmic end of TM-2. As already pointed out, the specificity to recognise the cognate receptor is restricted to the two transmembrane helices of the transducers [86]. Despite the high sequence homologies among the archaebacterial phototransducers [46,82] on the one hand and the eubacterial chemotaxis receptors [81] on the other hand, it is not yet possible to further narrow the interaction domain (see note on pg 39). For NpSRII the binding to NpHtrII is, with a $K_D < 160$ nM, quite strong. This data was obtained by titrating a solubilised truncated transducer (1–157) to the receptor and analysing the complexation by blue native gel chromatography and isothermal titration calorimetry [83]. The stoichiometry was determined to be 1:1. The calorimetric experiments allowed the calculation of ΔC_p according to Kirchhoff's law. The large negative value of -1.7 kJ mol^{-1} K^{-1} might be the result of the removal of protein surface area from exposure to the solvent as one would expect during a complex assembly [194–197]. The binding of a truncated transducer to NpSRII induces the transition from random coil to α-helix as revealed by CD spectroscopy [83] which is comparable to the α-helical content of the structurally related aspartate receptor [198].

Obtaining structural information at a molecular level is crucial for an understanding of the transmembrane signal transfer. To extract this data and to acquire knowledge on the dynamics of the process EPR spectroscopy has been applied. Previous studies on rhodopsin and BR have demonstrated the general applicability of the method to clarify e.g., domain fold and light-activated kinetics of mobility and/or distance changes between two spin labels (reviewed in [199]). The method relies on site-directed spin labeling (SDSL) of the protein under investigation. The spin-label is introduced via single cysteine mutants at positions of interest. The shape of the EPR spectrum reflects the re-orientational motion of the nitroxide side chain, which depends on the interaction with neighbouring protein structures and – if a second spin label is present – provides distances between the two paramagnetic centres [199–201]. Further information can be gained from measurements of samples in the presence of freely diffusing paramagnetic probes (e.g. oxygen or Cr^{3+}). These experiments can differentiate between water, lipid bilayer, or protein interior accessibility of a particular protein side chain [202]. Additionally, conformational changes can be monitored with a time resolution of about 1 ms [203].

This methodology has been applied extensively to BR [204,205] and more recently to NpSRII [84,108]. Sequential spin labeling of helical turns on helices F and G of NpSRII allowed to deduce the topology of the cytoplasmic extensions of these helices [108]. Comparing these data with those obtained for BR [202,206], it becomes evident that helices F and G are not only similarly oriented to each other and to the other parts of the protein but also have the same boundary separating cytoplasmic residues as those immersed in the membrane. Co-expression of NpSRII with a truncated fragment of NpHtrII affects the accessibility and the mobility of outward oriented spin-labeled residues on helices F and G as a result of direct physical contact with the transducer molecule. Therefore these helices are located within the binding surface of the photoreceptor with its transducer [108].

The crystal structure of Luecke et al. [124] confirmed most of the assignments made by EPR [108], demonstrating the potential for studying membrane proteins with this particular technique. Concerning the interaction between NpSRII and its transducer, NpHtrII, the crystal structure revealed a tyrosine residue (Tyr199) which sticks out from the lipid-facing surface of helix G. The authors note that this Tyr is an 'excellent candidate for transducer binding in the SRII-HtrII complex in *N. pharaonis* membranes'.

To build up a detailed model of the membrane embedded transduction complex the SDSL approach was extended to the cytoplasmic parts of both transmembrane helices (TM1 and TM2) of the transducer [84]. The results reveal a quaternary complex between two copies of truncated-Htr and NpSRII each with an apparent two-fold symmetry (Figure 8). The core is composed of two transmembrane helices of the transducer. This structure is in agreement with cross-linking experiments on the HtrI/SRI complex which demonstrate the dimeric nature of HtrI [82] and has now been confirmed by the crystal structure of the complex (see note on pg 39). Moreover, the formation of a pseudo four helix bundle in the transmembrane region proves the hypothesis that archaeal phototransducers resemble structural features of the eubacterial chemoreceptors (MCP) for which a four helical bundle was also proposed for the dimer (see e.g. [87]).

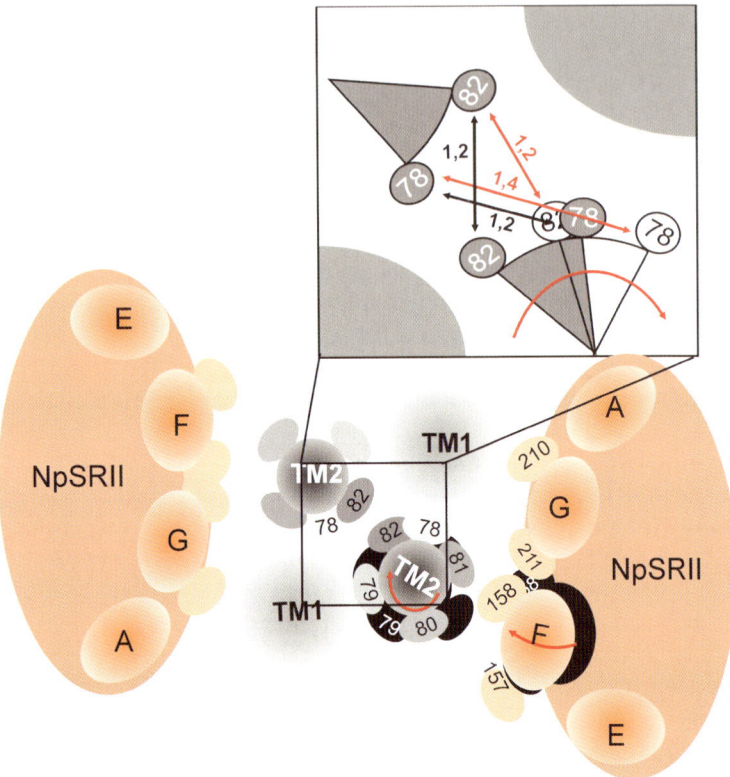

Figure 8. Schematic illustration of light-induced conformational changes within the transmembrane region of the 2:2-complex of NpSRII with NpHtrII viewed from the cytoplasm [84]. According to distance changes in the signalling M-state conformation helix F moves outwardly in the direction of the neighbouring TM2, which in turn is rotated clockwise as indicated by the red arrows. Black areas represent the original positions. The inset shows a close up of the dimer interface, suggesting the relative orientations of V78R1 and L82R1 in the dark (grey) and the light-activated states (white). The numbers at the arrows depict the distances (in nm) between corresponding residues in the dark (black) and light (red) states. It should be noted that a small piston-like movement of TM2 cannot be excluded.

1.4.2   Molecular mechanism of the signal transfer

Conformational alterations within protein interfaces play a key role in activating signal transduction chains mediated by membrane receptors such as receptor tyrosine kinases [207] or two-component systems in chemo- and phototaxis [1]. Especially, the signal transfer mediated by seven helix receptors such as rhodopsin is of great medical interest and the elucidation of the mechanism of archaeal photoreception might shed light on these questions.

Conformational changes of pharaonis sensory rhodopsin are already mani-
fest in the early K state of the photocycle [127,152]. Taking the EPR data into
account, the structural alterations might occur in the vicinity of helix F.
In BR an outward movement of the cytoplasmic end of this helix was originally proposed based on electron diffraction studies [208]. Recent crystallographic analysis of BR [209,210] suggested that structural changes in the
cytoplasmic parts of helix G occur during the transition of M and N. Supporting results were obtained by Steinhoff and co-workers [205,206] who observed
transient mobility changes of residues in the interface of helices F and G. For
NpSRII analogous experiments also indicate an outward tilting of helix F;
however, it is correlated with the early steps of the photocycle and sustained
in time over at least three orders of magnitude until the O-intermediate decays
to the initial state. A similar movement of helix F (in the rhodopsin nomenclature VII) was described for rhodopsin [211]. Apparently, this region of the
seven helix receptors seems to be generally involved in transmembrane signal
transduction.

The question of the assignment of the signalling state to photocycle intermediates has been addressed by phototaxis experiments using retinal analogues.
For SRI the attractant signalling is governed by the lifetime of the S373 intermediate rather than by the frequency of photocycling [190]. In similar experiments Yan et al. provided evidence that the signalling site is activated during
formation of M, but is only reset by the decay of O [212]. This is in line with
EPR measurements using the NpSRII/NpHtrII complex, which indicated that
the rearrangement of TM2 to its original conformation occurs during the last
step of the photocycle [84].

Conformational changes were analysed in more detail by comparing the
inter-residual distances in the original ground state and in the M-state within
the complex between NpSRII and its truncated transducer [84]. On light excitation of the NpSRII/HtrII complex, significant distance changes are observed
between the two TM2 and between TM2 and helix F of NpSRII, confirming
the reported movement of the cytoplasmic part of helix F towards the transducer [108]. The pattern of the inter-helical distance changes allowed the establishment of a model of protein–protein signal transfer. The flap-like motion of
helix F induces a clockwise rotation of TM2 (Figure 8). This rotary mechanism
(a combination with small piston-like movement cannot be excluded) might
be the trigger for the activation of the cytoplasmic two-component system.
Supporting evidence comes from experiments on the aspartate receptor from
S. typhimurium, which shows that upon binding of a substrate the periplasmic
extension of one TM2 moves towards the cell interior by about 1.6 Å and tilts
with an angle of 5° [213]. Moreover, Ottemann et al. [214] provided evidence of
a piston-like movement (<2.5 Å) of TM2. An interesting observation was
made on the time course of the rotation in comparison to the helix F movement [84]. Clearly, the two structural changes were out of phase in the second
half of the photocycle: TM2 reaches its original location only after helix F
(as well as the chromophore) has returned back to the resting state. This
decoupling permits the system to modulate the activation/deactivation of the

transducer by changing the state of methylation or by allosteric regulation within a receptor/transducer network [215].

1.5   Outlook

Future research on the mechanism of the signal transfer from the photorecep-tor to the cytoplasmic two-component system will be guided by structural information. Primarily, the structure of the sensory rhodopsins as well as that of the receptor/transducer complexes will be of great importance. Previous work on the crystallisation of BR and HR provides the technique required to accomplish this goal. These structures will allow a first detailed insight into the receptor–transducer binding interface (see note on pg 39). The exploitation of structural information of intermediates might give direct evidence for the mechanism of signal transfer and will supplement the results from EPR, FTIR and other kinetic experiments. These investigations showed that the disruption of the salt bridge between the protonated Schiff base and its counter-ion leads to conformational changes involving helix F, which in turn triggers a rotation of TM2. It will be of fundamental interest to understand the mechanism of signal transfer from the membrane to the cytoplasmic tip of the transducer, which are about 20 nm apart. This knowledge will not only be important for the phototaxis and chemotaxis receptors, but will have general implications for transmembrane signalling such as in the case of, e.g., tyrosine kinases.

Acknowledgements

Our own work described here was supported by the Deutsche Forschungs-gemeinschaft and the Max-Plank Society. We thank T. Savopol and R. Seidel for critically reading the manuscript.

References

1. J.J. Falke, R.B. Bass, S.L. Butler, S.A. Chervitz, M.A. Danielson (1997). The two-component signaling pathway of bacterial chemotaxis – a molecular view of signal transduction by receptors, kinases, and adaptation enzymes. *Annu. Rev. of Cell & Develop. Biol.*, **13**, 457–512.
2. M.C. Pirrung (1999). Histidine kinases and two-component signal transduction systems. *Chem. & Biol.*, **6**, R167–R175.
3. W.F. Loomis, A. Kuspa, G. Shaulsky (1998). Two-component signal transduction systems in eukaryotic microorganisms. *Curr. Opin. Microbiol.*, **1**, 643–648.
4. J.B. Stock, M.G. Surette, M. Levit, P. Park (1995). Two-component signal trans-duction systems: Structure-function relationships and mechanisms of catalysis. In: J.A. Hoch, T.J. Silhavy (Eds), *Two-Component Signal Transduction* (pp. 25–51). American Society for Microbiology, Washington, DC.

5. V.L. Robinson, D.R. Buckler, A.M. Stock (2000). A tale of two components: a novel kinase and a regulatory switch. *Nat. Struct. Biol.*, **7**, 626–633.

6. S. Aizawa, C.S. Harwood, R.J. Kadner (2000). Signaling components in bacterial locomotion and sensory reception. *J. Bacteriol.*, **182**, 1459–1471.

7. D. Oesterhelt, W. Stoeckenius (1971). Rhodopsin-like protein from the purple membrane of *Halobacterium halobium*. *Nature – New Biol.*, **233**, 149–152.

8. J.K. Lanyi (2000). Molecular mechanism of ion transport in bacteriorhodopsin: Insights from crystallographic, spectroscopic, kinetic, and mutational studies. *J. Phys. Chem. B*, **104**, 11441–11448.

9. H. Luecke, B. Schobert, H.T. Richter, J.P. Cartailler, J.K. Lanyi (1999). Structure of bacteriorhodopsin at 1.55 Ångstrom resolution. *J. Mol. Biol.*, **291**, 899–911.

10. E. Pebay-Peyroula, G. Rummel, J.P. Rosenbusch, E.M. Landau (1997). X-ray structure of bacteriorhodopsin at 2.5 Ångstroms from microcrystals grown in lipidic cubic phases. *Science*, **277**, 1676–1681.

11. K. Edman, P. Nollert, A. Royant, H. Belrhali, E. Pebay-Peyroula, J. Hajdu, R. Neutze, E.M. Landau (1999). High-resolution X-ray structure of an early intermediate in the bacteriorhodopsin photocycle. *Nature*, **401**, 822–826.

12. H. Luecke (2000). Atomic resolution structures of bacteriorhodopsin photocycle intermediates: the role of discrete water molecules in the function of this light-driven ion pump. *Biochem. Biophys. Acta*, **1460**, 133–156.

13. H.J. Sass, G. Büldt, R. Gessenich, D. Hehn, D. Neff, R. Schlesinger, J. Berendzen, P. Ormos (2000). Structural alterations for proton translocation in the M state of wild-type bacteriorhodopsin. *Nature*, **406**, 649–653.

14. J.K. Lanyi (2000). Crystallographic studies of the conformational changes that drive directional transmembrane ion movement in bacteriorhodopsin. *Biochem. Biophys. Acta*, **1459**, 339–345.

15. D. Oesterhelt (1995). Structure and function of halorhodopsin. *Isr. J. Chem.*, **35**, 475–494.

16. J.K. Lanyi, G. Váró (1995). The photocycles of bacteriorhodopsin. *Isr. J. Chem.*, **35**, 365–385.

17. A. Matsuno-Yagi, Y. Mukohata (1980). ATP synthesis linked to light-dependent proton uptake in a rad mutant strain of Halobacterium lacking bacteriorhodopsin. *Arch. Biochem. Biophys.*, **199**, 297–303.

18. B. Schobert, J.K. Lanyi (1982). Halorhodopsin is a light-driven chloride pump. *J. Biol. Chem.*, **257**, 10306–10313.

19. A. Blanck, D. Oesterhelt (1987). The halo-opsin gene. II. Sequence, primary structure of halorhodopsin and comparison with bacteriorhodopsin. *EMBO J.*, **6**, 265–273.

20. M. Kolbe, H. Besir, L.O. Essen, D. Oesterhelt (2000). Structure of the light-driven chloride pump halorhodopsin at 1.8 Å resolution. *Science*, **288**, 1390–1396.

21. J.L. Spudich (1998). Variations on a molecular switch – transport and sensory signalling by archaeal rhodopsins. *Mol. Microbiol.*, **28**, 1051–1058.

22. E. Hildebrand, N. Dencher (1975). Two photosystems controlling behavioural responses of *Halobacterium halobium*. *Nature*, **257**, 46–48.

23. N. Dencher (1978). Light-induced behavioral reactions of *Halobacterium halobium*: Evidence for two rhodopsins acting as photopigments. In: *Energetics and Structure of Halophilic Microorganisms* (pp. 67–88) Elsevier/North Holland Biomedical Press.

24. N.A. Dencher, E. Hildebrand (1979). Sensory transduction in *Halobacterium halobium*: retinal protein pigment controls UV-induced behavioral response. *Z. Naturforsch., Teil C*, **34**, 841–847.

25. J.L. Spudich, W. Stoeckenius (1979). Photosensory and chemosensory behaviour of *Halobacterium halobium*. *Photobiochem. Photobiophys.*, **1**, 43–53.

26. W. Sperling, A. Schimz (1980). Photosensory retinal pigments in *Halobacterium halobium*. *Biophys. Struct. Mechanism*, **6**, 165–169.

27. J.L. Spudich, W. Stoeckenius (1980). Protein modification reactions in *Halobacterium* photosensing. *Fed. Proc. FASEB*, **39**, 1972.

28. A. Schimz (1981). Methylation of membrane proteins is involved in chemosensory and photosensory behavior of *Halobacterium halobium*. *FEBS Lett.*, **125**, 205–207.

29. A. Schimz (1982). Localization of the methylation system involved in sensory behaviour of *Halobacterium halobium* and its dependence on calcium. *FEBS Lett.*, **139**, 283–286.

30. E.N. Spudich, J.L. Spudich (1982). Control of transmembrane ion fluxes to select halorhodopsin-deficient and other energy-transduction mutants of *Halobacterium halobium*. *Proc. Natl. Acad. Sci. U.S.A.*, **79**, 4308–4312.

31. R.A. Bogomolni, J.L. Spudich (1982). Identification of a third rhodopsin-like pigment in phototactic *Halobacterium halobium*. *Proc. Natl. Acad. Sci. U.S.A.*, **79**, 6250–6254.

32. J.L. Spudich, R.A. Bogomolni (1983). Spectroscopic discrimination of the three rhodopsin-like pigments in *Halobacterium halobium* membranes. *Biophys. J.*, **43**, 243–246.

33. J.L. Spudich, R.A. Bogomolni (1984). Mechanism of colour discrimination by a bacterial sensory rhodopsin. *Nature*, **312**, 509–513.

34. H. Tomioka, T. Takahashi, N. Kamo, Y. Kobatake (1986). Flash spectroscopic indentification of a fourth rhodopsin-like pigment in *Halobacterium halobium*. *Biochem. Biophys. Res. Commun.*, **139**, 389–395.

35. E.K. Wolff, R.A. Bogomolni, P. Scherrer, B. Hess, W. Stoeckenius (1986). Color discrimination in halobacteria: Spectroscopic characterization of a second sensory receptor covering the blue-green region of the spectrum. *Proc. Natl. Acad. Sci. U.S.A.*, **83**, 7272–7276.

36. A. Blanck, D. Oesterhelt, E. Ferrando, E.S. Schegk, F. Lottspeich (1989). Primary structure of sensory rhodopsin I, a prokaryotic photoreceptor. *EMBO J.*, **8**, 3963–3971.

37. W.S. Zhang, A. Brooun, M.M. Mueller, M. Alam (1996). The primary structures of the archaeon *Halobacterium salinarium* blue light receptor sensory rhodopsin II and its transducer, a methyl-accepting protein. *Proc. Natl. Acad. Sci. U.S.A.*, **93**, 8230–8235.

38. R. Seidel, B. Scharf, M. Gautel, K. Kleine, D. Oesterhelt, M. Engelhard (1995). The primary structure of sensory rhodopsin II: A member of an additional retinal protein subgroup is coexpressed with its transducer, the halobacterial transducer of rhodopsin II. *Proc. Natl. Acad. Sci. U.S.A.*, **92**, 3036–3040.

39. G. Schmies, B. Lüttenberg, I. Chizhov, M. Engelhard, A. Becker, E. Bamberg (2000). Sensory rhodopsin II from the haloalkaliphilic *Natronobacterium pharaonis*: Light-activated proton transfer reactions. *Biophys. J.*, **78**, 967–976.

40. M. Iwamoto, K. Shimono, M. Sumi, N. Kamo (1999). Positioning proton-donating residues to the Schiff-base accelerates the M-decay of *pharaonis* phoborhodopsin expressed in *Escherichia coli*. *Biophys. Chem.*, **79**, 187–192.

41. V.J. Yao, J.L. Spudich (1992). Primary structure of an archaebacterial transducer, a methyl-accepting protein associated with sensory rhodopsin I. *Proc. Natl. Acad. Sci. U.S.A.*, **89**, 11915–11919.

42. M. Alam, D. Oesterhelt (1984). Morphology, function and isolation of halobacterial flagella. *J. Mol. Biol.*, **176**, 459–475.
43. M. Alam, M. Lebert, D. Oesterhelt, G.L. Hazelbauer (1989). Methyl-accepting taxis proteins in *Halobacterium halobium*. *EMBO J.*, **8**, 631–639.
44. K.F. Storch, J. Rudolph, D. Oesterhelt (1999). Car: a cytoplasmic sensor responsible for arginine chemotaxis in the archaeon *Halobacterium salinarum*. *EMBO J.*, **18**, 1146–1158.
45. W.V. Ng, S.P. Kennedy, G.G. Mahairas, B. Berquist, M. Pan, H.D. Shukla, S.R. Lasky, N.S. Baliga, V. Thorsson, J. Sbrogna, S. Swartzell, D. Weir, J. Hall, T.A. Dahl, R. Welti, Y.A. Goo, B. Leithauser, K. Keller, R. Cruz, M.J. Danson, D.W. Hough, D.G. Maddocks, P.E. Jablonski, M.P. Krebs, C.M. Angevine, S. DasSarma (2000). Genome sequence of *Halobacterium species* NRC-1. *Proc. Natl. Acad. Sci. U.S.A.*, **97**, 12176–12181.
46. J. Rudolph, B. Nordmann, K.F. Storch, H. Gruenberg, K. Rodewald, D. Oesterhelt (1996). A family of halobacterial transducer proteins. *FEMS Microbiol. Lett.*, **139**, 161–168.
47. W.S. Zhang, A. Brooun, J. McCandless, P. Banda, M. Alam (1996). Signal transduction in the Archaeon *Halobacterium salinarium* is processed through three subfamilies of 13 soluble and membrane-bound transducer proteins. *Proc. Natl. Acad. Sci. U.S.A.*, **93**, 4649–4654.
48. A. Brooun, J. Bell, T. Freitas, R.W. Larsen, M. Alam (1998). An archaeal aerotaxis transducer combines subunit I core structures of eukaryotic cytochrome c oxidase and eubacterial methyl-accepting chemotaxis proteins. *J. Bacteriol.*, **180**, 1642–1646.
49. A. Brooun, W.S. Zhang, M. Alam (1997). Primary structure and functional analysis of the soluble transducer protein htrxi in the archaeon *Halobacterium salinarium*. *J. Bacteriol.*, **179**, 2963–2968.
50. M.V. Kokoeva, D. Oesterhelt (2000). BasT, a membrane-bound transducer protein for amino acid detection in *Halobacterium salinarum*. *Mol. Microbiol.*, **35**, 647–656.
51. S.B. Hou, R.W. Larsen, D. Boudko, C.W. Riley, E. Karatan, M. Zimmer, G.W. Ordal, M. Alam (2000). Myoglobin-like aerotaxis transducers in archaea and bacteria. *Nature*, **403**, 540–544.
52. S.B. Hou, A. Brooun, H.S. Yu, T. Freitas, M. Alam (1998). Sensory rhodopsin II transducer HtrII is also responsible for serine chemotaxis in the archaeon *Halobacterium salinarium*. *J. Bacteriol.*, **180**, 1600–1602.
53. J. Rudolph, D. Oesterhelt (1996). Deletion analysis of the che operon in the archaeon *Halobacterium salinarium*. *J. Mol. Biol.*, **258**, 548–554.
54. J. Rudolph, D. Oesterhelt (1995). Chemotaxis and phototaxis require a CheA histidine kinase in the archaeon *Halobacterium salinarium*. *EMBO J.*, **14**, 667–673.
55. E. Hildebrand, A. Schimz (1986). Integration of photosensory signals in *Halobacterium halobium*. *J. Bacterioleriol.*, **167**, 305–311.
56. J. Stock, M. Levit (2000). Signal transduction: hair brains in bacterial chemotaxis. *Curr. Biol.*, **10**, R11–R14.
57. A.M. Stock, V.L. Robinson, P.N. Goudreau (2000). Two-component signal transduction. *Annu. Rev. Biochem.*, **69**, 183–215.
58. M. Montrone, W. Marwan, H. Grünberg, S. Musseleck, C. Starostzik, D. Oesterhelt (1993). Sensory rhodopsin-controlled release of the switch factor fumarate in *Halobacterium salinarium*. *Mol. Microbiol.*, **10**, 1077–1085.
59. W. Marwan, W. Schäfer, D. Oesterhelt (1990). Signal transduction in *Halobacterium* depends on fumarate. *EMBO J.*, **9**, 355–362.

60. R. Barak, I. Giebel, M. Eisenbach (1996). The specificity of fumarate as a switching factor of the bacterial flagellar motor. *Mol. Microbiol.*, **19**, 139–144.

61. M. Montrone, M. Eisenbach, D. Oesterhelt, W. Marwan (1998). Regulation of switching frequency and bias of the bacterial flagellar motor by CheY and fumarate. *J. Bacteriol.*, **180**, 3375–3380.

62. S.A. Sundberg, M. Alam, M. Lebert, J.L. Spudich, D. Oesterhelt, G.L. Hazelbauer (1990). Characterization of *Halobacterium halobium* mutants defective in taxis. *J. Bacteriol.*, **172**, 2328–2335.

63. S.A. Sundberg, R.A. Bogomolni, J.L. Spudich (1985). Selection and properties of phototaxis-deficient mutants of *Halobacterium halobium*. *J. Bacteriol.*, **164**, 282–287.

64. E. Ferrando-May, M. Krah, W. Marwan, D. Oesterhelt (1993). The methyl-accepting transducer protein HtrI is functionally associated with the photorecep-tor sensory rhodopsin I in the archaeon *Halobacterium salinarium*. *EMBO J.*, **12**, 2999–3005.

65. E.N. Spudich, J.L. Spudich (1993). The photochemical reactions of sensory rhodopsin I are altered by its transducer. *J. Biol. Chem.*, **268**, 16095–16097.

66. K.D. Olson, J.L. Spudich (1993). Removal of the transducer protein from sensory rhodopsin I exposes sites of proton release and uptake during the receptor photocycle. *Biophys. J.*, **65**, 2578–2585.

67. B. Lüttenberg, E.K. Wolff, M. Engelhard (1998). Heterologous coexpression of the blue light receptor NpSRII and its transducer NpHtrII from *Natro-nobacterium pharaonis* in the *Halobacterium salinarium* strain pho81/w restores negative phototaxis. *FEBS Lett.*, **426**, 117–120.

68. W. Marwan, D. Oesterhelt (1990). Quantitation of photochromism of sensory rhodopsin-I by computerized tracking of *Halobacterium halobium* cells. *J. Mol. Biol.*, **215**, 277–285.

69. H. Tomioka, T. Takahashi, N. Kamo, Y. Kobatake (1986). Action spectrum of the photoattractant response of *Halobacterium halobium* in early logarithmic growth phase and the role of sensory rhodopsin. *Biochim. Biophys. Acta*, **884**, 578–584.

70. T. Takahashi, Y. Kobatake (1982). Computer-linked automated method for measurement of the reversal frequency in phototaxis of *Halobacterium halobium*. *Cell Struct. Funct.*, **7**, 183–192.

71. W. Stoeckenius, E.K. Wolff, B. Hess (1988). A rapid population method for action spectra applied to *Halobacterium halobium*. *J. Bacteriol.*, **170**, 2790–2795.

72. T. Takahashi, H. Tomioka, N. Kamo, Y. Kobatake (1985). A photosystem other than PS370 also mediates the negative phototaxis of *Halobacterium halobium*. *FEMS Microbiol. Lett.*, **28**, 161–164.

73. W. Marwan, D. Oesterhelt (1987). Signal formation in the halobacterial photo-phobic response mediated by a fourth retinal protein /(P480). *J. Mol. Biol.*, **195**, 333–342.

74. E.N. Spudich, S.A. Sundberg, D. Manor, J.L. Spudich (1986). Properties of a second sensory receptor protein in *Halobacterium halobium* phototaxis. *Proteins*, **1**, 239–246.

75. B. Nordmann, M.R. Lebert, M. Alam, S. Nitz, H. Kollmannsberger, D. Oesterhelt, G.L. Hazelbauer (1994). Identification of volatile forms of methyl groups released by *Halobacterium salinarium*. *J. Biol. Chem.*, **269**, 16449–16454.

76. E.N. Spudich, T. Takahashi, J.L. Spudich (1989). Sensory rhodopsins I and II modulate a methylation/demethylation system in *Halobacterium halobium* phototaxis. *Proc. Natl. Acad. Sci. U.S.A.*, **86**, 7746–7750.

77. B. Perazzona, J.L. Spudich (1999). Identification of methylation sites and effects of phototaxis stimuli on transducer methylation in *Halobacterium salinarum*. *J. Bacteriol.*, **181**, 5676–5683.
78. W. Marwan, S.I. Bibikov, M. Montrone, D. Oesterhelt (1995). Mechanism of photosensory adaptation in *Halobacterium salinarium*. *J. Mol. Biol.*, **246**, 493–499.
79. W. Marwan, P. Hegemann, D. Oesterhelt (1988). Single photon detection by an archaebacterium. *J. Mol. Biol.*, **199**, 663–664.
80. W.D. Hoff, K.H. Jung, J.L. Spudich (1997). Molecular mechanism of photo-signaling by archaeal sensory rhodopsins. *Annu. Rev. Biophys. Biomol. Struct.*, **26**, 223–258.
81. H. Le Moual, D.E. Koshland (1996). Molecular evolution of the C-terminal cytoplasmic domain of a superfamily of bacterial receptors involved in taxis. *J. Mol. Biol.*, **261**, 568–585.
82. X.-N. Zhang, J.L. Spudich (1998). HtrI is a dimer whose interface is sensitive to receptor photoactivation and His-166 replacements in sensory rhodopsin I. *J. Biol. Chem.*, **273**, 19722–19728.
83. A.A. Wegener (2000). Untersuchungen zur Wechselwirkung des archaebakteriellen Lichtrezeptors NpSRII mit seinem Transducerprotein NpHtrII. *Thesis*, University of Dortmund, Germany.
84. A.A. Wegener, J.P. Klare, M. Engelhard (2001). Structural insights into the early steps of receptor-transducer signal transfer in archaeal phototaxis. *EMBO J.*, **20**, 5312–5319.
85. M. Krah, W. Marwan, A. Verméglio, D. Oesterhelt (1994). Phototaxis of *Halobacterium salinarium* requires a signalling complex of sensory rhodopsin I and its methyl-accepting transducer HtrI. *EMBO J.*, **13**, 2150–2155.
86. X.N. Zhang, J. Zhu, J.L. Spudich (1999). The specificity of interaction of archaeal transducers with their cognate sensory rhodopsins is determined by their transmembrane helices. *Proc. Natl. Acad. Sci. U.S.A.*, **96**, 857–862.
87. K.K. Kim, H. Yokota, S.H. Kim (1999). Four-helical-bundle structure of the cytoplasmic domain of a serine chemotaxis receptor. *Nature*, **400**, 787–792.
88. S.B. Williams, V. Stewart (1999). Functional similarities among two-component sensors and methyl-accepting chemotaxis proteins suggest a role for linker region amphipathic helices in transmembrane signal transduction. *Mol. Microbiol.*, **33**, 1093–1102.
89. M. Krah, W. Marwan, D. Oesterhelt (1994). A cytoplasmic domain is required for the functional interaction of SRI and HtrI in archaeal signal transduction. *FEBS Lett.*, **353**, 301–304.
90. V.J. Yao, E.N. Spudich, J.L. Spudich (1994). Identification of distinct domains for signaling and receptor interaction of the sensory rhodopsin I transducer, HtrI. *J. Bacteriol.*, **176**, 6931–6935.
91. B. Perazzona, E.N. Spudich, J.L. Spudich (1996). Deletion mapping of the sites on the htrI transducer for sensory rhodopsin I interaction. *J. Bacteriol.*, **178**, 6475–6478.
92. G. Schmies (2001). Spektroskopische und elektrophysiologische Untersuchung der beiden archaebakteriellen Photorezeptor/Transducer-Komplexe. *Thesis*, Universität Dortmund, Germany.
93. G. Schmies, M. Engelhard, P.G. Wood, G. Nagel, E. Bamberg (2001). Electro-physiological characterization of specific interactions between bacterial sensory rhodopsins and their transducers. *Proc. Natl. Acad. Sci. U.S.A.*, **98**, 1555–1559.

94. Y. Sudo, M. Iwamoto, K. Shimono, M. Sumi, N. Kamo (2001). Photo-induced proton transport of pharaonis phoborhodopsin (sensory rhodopsin II) is ceased by association with the transducer. *Biophys. J.*, **80**, 916–922.

95. J.R. Maddock, L. Shapiro (1993). Polar location of the chemoreceptor complex in the *Escherichia coli* cell. *Science*, **259**, 1717–1723.

96. H. Tomioka, T. Takahashi, N. Kamo, Y. Kobatake (1986). Flash spectrometric identification of a fourth rhodopsin-like pigment in *Halobacterium halobium*. *Biochem. Biophys. Res. Commun.*, **139**, 389–395.

97. E.S. Schegk, D. Oesterhelt (1988). Isolation of a prokaryotic photoreceptor: sensory rhodopsin from halobacteria. *EMBO J.*, **7**, 2925–2933.

98. B. Scharf, B. Hess, M. Engelhard (1992). Chromophore of sensory rhodopsin II from *Halobacterium halobium*. *Biochemistry*, **31**, 12486–12492.

99. B. Scharf, B. Pevec, B. Hess, M. Engelhard (1992). Biochemical and photochemical properties of the photophobic receptors from *Halobacterium halobium* and *Natronobacterium pharaonis*. *Eur. J. Biochem.*, **206**, 359–366.

100. J. Otomo, W. Marwan, D. Oesterhelt, H. Desel, H. Uhl (1989). Biosynthesis of the two halobacterial light sensors P480 and sensory rhodopsin and variation in gain of their signal transduction chains. *J. Bacteriol.*, **171**, 2155–2159.

101. M.P. Krebs, E.N. Spudich, H.G. Khorana, J.L. Spudich (1993). Synthesis of a gene for sensory rhodopsin I and its functional expression in *Halobacterium halobium*. *Proc. Natl. Acad. Sci. U.S.A.*, **90**, 3486–3490.

102. E. Ferrando-May, B. Brustmann, D. Oesterhelt (1993). A *C*-terminal truncation results in high-level expression of the functional photoreceptor sensory rhodopsin I in the archaeon *Halobacterium salinarium*. *Mol. Microbiol.*, **9**, 943–953.

103. M.P. Krebs, E.N. Spudich, J.L. Spudich (1995). Rapid high-yield purification and liposome reconstitution of polyhistidine-tagged sensory rhodopsin I. *Protein Express. Purific.*, **6**, 780–788.

104. B. Lüttenberg (1998). Heterologe Expression von sensorischem Rhodopsin II aus *Natronobacterium pharaonis* in *Halobacterium salinarium*. *Thesis*, Universität Münster, Germany.

105. K. Shimono, M. Iwamoto, M. Sumi, N. Kamo (1997). Functional expression of pharaonis phoborhodopsin in *Escherichia coli*. *FEBS Lett.*, **420**, 54–56.

106. I.P. Hohenfeld, A.A. Wegener, M. Engelhard (1999). Purification of histidine tagged bacteriorhodopsin, *pharaonis* halorhodopsin and *pharaonis* sensory rhodopsin II functionally expressed in *Escherichia coli*. *FEBS Lett.*, **442**, 198–202.

107. G. Schmies, I. Chizhov, M. Engelhard (2000). Functional expression of His-tagged sensory rhodopsin I in *Escherichia coli*. *FEBS Lett.*, **466**, 67–69.

108. A.A. Wegener, I. Chizhov, M. Engelhard, H.J. Steinhoff (2000). Time-resolved detection of transient movement of helix F in spin-labelled pharaonis sensory rhodopsin II. *J. Mol. Biol.*, **301**, 881–891.

109. A. Losi, A.A. Wegener, M. Engelhard, S.E. Braslavsky (2001). Enthalpy-entropy compensation in a photocycle: The K-to-L transition in sensory rhodopsin II from *Natronobacterium pharaonis*. *J. Am. Chem. Soc.*, **123**, 1766–1767.

110. E.R. Kunji, E.N. Spudich, R. Grisshammer, R. Henderson, J.L. Spudich (2001). Electron crystallographic analysis of two-dimensional crystals of sensory rhodopsin II: a 6.9 Å projection structure. *J. Mol. Biol.*, **308**, 279–293.

111. J. Soppa, J. Duschl, D. Oesterhelt (1993). Bacterioopsin, haloopsin, and sensory opsin I of the halobacterial isolate *Halobacterium* sp. strain SG1: Three new members of a growing family. *J. Bacteriol.*, **175**, 2720–2726.

112. Y. Mukohata, K. Ihara, T. Tamura, Y. Sugiyama (1999). Halobacterial rhodopsins. *J. Biochem. (Tokyo)*, **125**, 649–657.

113. J.A. Bieszke, E.L. Braun, L.E. Bean, S.C. Kang, D.O. Natvig, K.A. Borkovich (1999). The nop-1 gene of *Neurospora crassa* encodes a seven transmembrane helix retinal-binding protein homologous to archaeal rhodopsins. *Proc. Natl. Acad. Sci. U.S.A.*, **96**, 8034–8039.

114. J.A. Bieszke, E.N. Spudich, K.L. Scott, K.A. Borkovich, J.L. Spudich (1999). A eukaryotic protein, NOP-1, binds retinal to form an archaeal rhodopsin-like photochemically reactive pigment. *Biochemistry*, **38**, 14138–14145.

115. O. Béjà, L. Aravind, E.V. Koonin, M.T. Suzuki, A. Hadd, L.P. Nguyen, S. Jovanovich, C.M. Gates, R.A. Feldman, J.L. Spudich, E.N. Spudich, E.F. DeLong (2000). Bacterial rhodopsin: Evidence for a new type of phototrophy in the sea. *Science*, **289**, 1902–1906.

116. O. Béjà, E.N. Spudich, J.L. Spudich, M. Leclerc, E.F. DeLong (2001). Proteorhodopsin phototrophy in the ocean. *Nature*, **411**, 786–789.

117. D.R. Baselt, S.P.A. Fodor, R. Van der Steen, J. Lugtenburg, R.A. Bogomolni, R.A. Mathies (1989). Halorhodopsin and sensory rhodopsin contain a C6–C7 s-trans retinal chromophore. *Biophys. J.*, **55**, 193–196.

118. J. Hu, R.G. Griffin, J. Herzfeld (1994). Synergy in the spectral tuning of retinal pigments: Complete accounting of the opsin shift in bacteriorhodopsin. *Proc. Natl. Acad. Sci. U.S.A.*, **91**, 8880–8884.

119. J. Hirayama, Y. Imamoto, Y. Shichida, T. Yoshizawa, A.E. Asato, R.S.H. Liu, N. Kamo (1994). Shape of the chromophore binding site in *pharaonis* phoborhodopsin from a study using retinal analogs. *Photochem. Photobiol.*, **60**, 388–393.

120. T. Takahashi, B. Yan, P. Mazur, F. Derguini, K. Nakanishi, J.L. Spudich (1990). Color regulation in the archaebacterial phototaxis receptor phoborhodopsin (sensory rhodopsin II). *Biochemistry*, **29**, 8467–8474.

121. K. Shimono, M. Iwamoto, M. Sumi, N. Kamo (2000). Effects of three characteristic amino acid residues of pharaonis phoborhodopsin on the absorption maximum. *Photochem. Photobiol.*, **72**, 141–145.

122. K. Shimono, M. Iwamoto, M. Sumi, N. Kamo (1998). V108M mutant of pharaonis phoborhodopsin – substitution caused no absorption change but affected its M-state. *J. Biochem. (Tokyo)*, **124**, 404–409.

123. J.Y. Zhu, E.N. Spudich, M. Alam, J.L. Spudich (1997). Effects of substitutions D73E, D73N, D103N and V106M on signaling and pH titration of sensory rhodopsin II. *Photochem. Photobiol*, **66**, 788–791.

124. H. Luecke, B. Schobert, J.K. Lanyi, E.N. Spudich, J.L. Spudich (2001). Crystal structure of sensory rhodopsin II at 2.4 Å: Insights into color tuning and transducer interaction. *Science*, **293**, 1499–1503.

125. I. Chizhov, G. Schmies, R. Seidel, J.R. Sydor, B. Lüttenberg, M. Engelhard (1998). The photophobic receptor from *Natronobacterium pharaonis*–temperature and pH dependencies of the photocycle of sensory rhodopsin II. *Biophys. J.*, **75**, 999–1009.

126. K. Shimono, M. Kitami, M. Iwamoto, N. Kamo (2000). Involvement of two groups in reversal of the bathochromic shift of pharaonis phoborhodopsin by chloride at low pH. *Biophys. Chem.*, **87**, 225–230.

127. M. Engelhard, B. Scharf, F. Siebert (1996). Protonation changes during the photocycle of sensory rhodopsin II from *Natronobacterium pharaonis*. *FEBS Lett.*, **395**, 195–198.

128. Y. Kimura, A. Ikegami, W. Stoeckenius (1984). Salt and pH-dependent changes of the purple membrane absorption spectrum. *Photochem. Photobiol.*, **40**, 641–646.

129. B. Scharf, M. Engelhard (1994). Blue halorhodopsin from *Natronobacterium pharaonis*: Wavelength regulation by anions. *Biochemistry*, **33**, 6387–6393.

130. E.N. Spudich, W.S. Zhang, M. Alam, J.L. Spudich (1997). Constitutive signaling by the phototaxis receptor sensory rhodopsin II from disruption of its proto-nated Schiff base Asp-73 interhelical salt bridge. *Proc. Natl. Acad. Sci. U.S.A.*, **94**, 4960–4965.

131. K.D. Olson, P. Deval, J.L. Spudich (1992). Absorption and photochemistry of sensory rhodopsin I: pH effects. *Photochem. Photobiol.*, **56**, 1181–1187.

132. A. Losi, S.E. Braslavsky, W. Gärtner, J.L. Spudich (1999). Time-resolved absorption and photothermal measurements with sensory rhodopsin I from *Halobacterium salinarium*. *Biophys. J.*, **76**, 2183–2191.

133. I. Lutz, A. Sieg, A.A. Wegener, M. Engelhard, I. Boche, M. Otsuka, D. Oesterhelt, J. Wachtveitl, W. Zinth (2001). Primary reactions of sensory rhodopsins. *Proc. Natl. Acad. Sci. U.S.A.*, **98**, 962–967.

134. P. Rath, K.D. Olson, J.L. Spudich, K.J. Rothschild (1994). The Schiff base counterion of bacteriorhodopsin is protonated in sensory rhodopsin I: Spectro-scopic and functional characterization of the mutated proteins D76N and D76A. *Biochemistry*, **33**, 5600–5606.

135. O. Bousché, E.N. Spudich, J.L. Spudich, K.J. Rothschild (1991). Conforma-tional changes in sensory rhodopsin I: Similarities and differences with bacteriorhodopsin, halorhodopsin, and rhodopsin. *Biochemistry*, **30**, 5395–5400.

136. G. Metz, F. Siebert, M. Engelhard (1992). Asp[85] is the only internal aspartic acid that gets protonated in the M intermediate and the purple-to-blue transition of bacteriorhodopsin: A solid-state ^{13}C CP-MAS NMR investigation. *FEBS Lett.*, **303**, 237–241.

137. R.A. Bogomolni, W. Stoeckenius, I. Szundi, E. Perozo, K.D. Olson, J.L. Spudich (1994). Removal of transducer HtrI allows electrogenic proton translocation by sensory rhodopsin I. *Proc. Natl. Acad. Sci. U.S.A.*, **91**, 10188–10192.

138. U. Haupts, C. Haupts, D. Oesterhelt (1995). The photoreceptor sensory rhod-opsin I as a two-photon-driven proton pump. *Proc. Natl. Acad. Sci. U.S.A.*, **92**, 3834–3838.

139. P. Rath, E.N. Spudich, D.D. Neal, J.L. Spudich, K.J. Rothschild (1996). Asp76 is the Schiff base counterion and proton acceptor in the proton-translocating form of sensory rhodopsin I. *Biochemistry*, **35**, 6690–6696.

140. S.P.A. Fodor, R. Gebhard, J. Lugtenburg, R.A. Bogomolni, R.A. Mathies (1989). Structure of the retinal chromophore in sensory rhodopsin I from resonance Raman spectroscopy. *J. Biol. Chem.*, **264**, 18280–18283.

141. Y. Imamoto, Y. Shichida, J. Hirayama, H. Tomioka, N. Kamo, T. Yoshizawa (1992). Chromophore configuration of *pharaonis* phoborhodopsin and its isomerization on photon absorption. *Biochemistry*, **31**, 2523–2528.

142. J. Hirayma, N. Kamo, Y. Imamoto, Y. Shichida, T. Yoshizawa (1995). Reason for the lack of light-dark adaptation in *pharaonis* phoborhodopsin: Reconstitution with 13-*cis*-retinal. *FEBS Lett.*, **364**, 168–170.

143. B. Yan, T. Takahashi, R. Johnson, F. Derguini, K. Nakanishi, J.L. Spudich (1990). All-*trans*/13-*cis* isomerization of retinal is required for phototaxis signaling by sensory rhodopsins in *Halobacterium halobium*. *Biophys. J.*, **57**, 807–814.

144. M. Tsuda, B. Nelson, C.H. Chang, R. Govindjee, T.G. Ebrey (1985). Characterisation of the chromophore of the third rhodopsin-like pigment of *Halobacterium halobium* and its photoproduct. *Biophys. J.*, **47**, 721–724.

145. U. Haupts, W. Eisfeld, M. Stockburger, D. Oesterhelt (1994). Sensory rhodopsin I photocycle intermediate SRI_{380} contains 13-*cis* retinal bound via an unprotonated Schiff base. *FEBS Lett.*, **356**, 25–29.

146. J.L. Spudich, D.A. McCain, K. Nakanishi, M. Okabe, N. Shimizu, H. Rodman, B. Honig, R.A. Bogomolni (1986). Chromophore/protein interaction in bacterial sensory rhodopsin and bacteriorhodopsin. *Biophys. J.*, **49**, 479–483.

147. B. Yan, T. Takahashi, D.A. McCain, V.J. Rao, K. Nakanishi, J.L. Spudich (1990). Effects of modifications of the retinal β-ionone ring on archaebacterial sensory rhodopsin I. *Biophys. J.*, **57**, 477–483.

148. B. Yan, A. Xie, G.U. Nienhaus, Y. Katsuta, J.L. Spudich (1993). Steric constraints in the retinal binding pocket of sensory rhodopsin I. *Biochemistry*, **32**, 10224–10232.

149. B. Yan, K. Nakanishi, J.L. Spudich (1991). Mechanism of activation of sensory rhodopsin I: Evidence for a steric trigger. *Proc. Natl. Acad. Sci. U.S.A.*, **88**, 9412–9416.

150. U.M. Ganter, E.D. Schmid, D. Perez-Sala, R.R. Rando, F. Siebert (1989). Removal of the 9-methyl group of retinal inhibits signal transduction in the visual process. A Fourier transform infrared and biochemical investigation. *Biochemistry*, **28**, 5954–5962.

151. I. Chizhov, D.S. Chernavskii, M. Engelhard, K.H. Müller, B.V. Zubov, B. Hess (1996). Spectrally silent transitions in the bacteriorhodopsin photocycle. *Biophys. J.*, **71**, 2329–2345.

152. A. Losi, A.A. Wegener, M. Engelhard, W. Gärtner, S.E. Braslavsky (1999). Time-resolved absorption and photothermal measurements with recombinant sensory rhodopsin II from *Natronobacterium pharaonis*. *Biophys. J.*, **77**, 3277–3286.

153. D. Zhang, D. Mauzerall (1996). Volume and enthalpy changes in the early steps of bacteriorhodopsin photocycle studied by time-resolved photoacoustics. *Biophys. J.*, **71**, 381–388.

154. A. Losi, A.A. Wegener, M. Engelhard, W. Gärtner, S.E. Braslavsky (2000). Aspartate 75 mutation in sensory rhodopsin II from *Natronobacterium pharaonis* does not influence the production of the K-like intermediate, but strongly affects its relaxation pathway. *Biophys. J.*, **78**, 2581–2589.

155. A.R. Bogomolni, J.L. Spudich (1982). Identification of a third rhodopsin-like pigment in phototactic *Halobacterium halobium*. *Proc. Natl. Acad. Sci. U.S.A.*, **79**, 6250–6254.

156. T. Takahashi, M. Watanabe, N. Kamo, Y. Kobatake (1985). Negative phototaxis from blue light and role of third rhodopsin-like pigment in *Halobacterium cutirubrum*. *Biophys. J.*, **48**, 235–240.

157. R.A. Bogomolni, J.L. Spudich (1987). The photochemical reactions of bacterial sensory rhodopsin-I. Flash photolysis study in the one microsecond to eight second time window. *Biophys. J.*, **52**, 1071–1075.

158. X.N. Zhang, J.L. Spudich (1997). His(166) is critical for active-site proton transfer and phototaxis signaling by sensory rhodopsin I. *Biophys. J.*, **73**, 1516–1523.

159. T.E. Swartz, I. Szundi, J.L. Spudich, R.A. Bogomolni (2000). New photointermediates in the two photon signaling pathway of sensory rhodopsin-I. *Biochemistry*, **39**, 15101–15109.

160. I. Szundi, T.E. Swartz, R.A. Bogomolni (2001). Multicolored protein conforma-
     tion states in the photocycle of transducer-free sensory rhodopsin-I. *Biophys. J.*,
     80, 469–479.
161. D. Manor, C.A. Hasselbacher, J.L. Spudich (1988). Membrane potential
     modulates photocycling rates of bacterial rhodopsins. *Biochemistry*, **27**, 5843–
     5848.
162. H. Michel, D. Oesterhelt (1980). Electrochemical proton gradient across the cell
     membrane of *Halobacterium halobium*: effect of N,N′-dicyclohexylcarbodiimide,
     relation to intracellular adenosine triphosphate, adenosine diphosphate, and
     phosphate concentration, and influence of the potassium gradient. *Biochemistry*,
     19, 4607–4614.
163. H. Michel, D. Oesterhelt (1980). Electrochemical proton gradient across the cell
     membrane of *Halobacterium halobium*: Comparison of the light-induced increase
     with the increase of intracellular adenine triphosphate under steady-state
     illumination. *Biochemistry*, **19**, 4615–4619.
164. R. Wittenberg (1995). Charakterisierung der Elektronentransportkette und
     Untersuchungen zur Bioenergetik in *Natronobacterium pharaonis. Thesis*, Ruhr-
     Universität Bochum, Germany.
165. H. Tomioka, N. Kamo, T. Takahashi, Y. Kobatake (1984). Photochemical
     intermediate of third rhodopsin-like pigment in *Halobacterium halobium* by
     simultaneous illumination with red and blue light. *Biochem. Biophy. Res.
     Commun.*, **123**, 989–994.
166. Y. Shichida, Y. Imamoto, T. Yoshizawa, T. Takahashi, H. Tomioka, N. Kamo,
     Y. Kobatake (1988). Low-temperature spectrophotometry of phoborhodopsin.
     FEBS Lett., **236**, 333–336.
167. V. Bergo, E.N. Spudich, K.L. Scott, J.L. Spudich, K.J. Rothschild (2000). FTIR
     analysis of the SII_{540} intermediate of sensory rhodopsin II: Asp73 is the Schiff
     base proton acceptor. *Biochemistry*, **39**, 2823–2830.
168. Y. Imamoto, Y. Shichida, T. Yoshizawa, H. Tomioka, T. Takahashi, K.
     Fujikawa, N. Kamo, Y. Kobatake (1991). Photoreaction cycle of phobor-
     hodopsin studied by low-temperature spectrophotometry. *Biochemistry*, **30**,
     7416–7424.
169. Y. Imamoto, Y. Shichida, J. Hirayama, H. Tomioka, N. Kamo, T. Yoshizawa
     (1992). Nanosecond laser photolysis of phoborhodopsin from *Natronobacterium
     pharaonis*: Appearance of KL and L intermediates in the photocycle at room
     temperature. *Photochem. Photobiol.*, **56**, 1129–1134.
170. J. Hirayama, Y. Imamoto, Y. Shichida, N. Kamo, H. Tomioka, T. Yoshizawa
     (1992). Photocycle of phoborhodopsin from haloalkaliphilic bacterium
     (*Natronobacterium pharaonis*) studied by low-temperature spectrophotometry.
     Biochemistry, **31**, 2093–2098.
171. M. Miyazaki, J. Hirayama, M. Hayakawa, N. Kamo (1992). Flash photolysis
     study on pharaonis phoborhodopsin from a haloalkaliphilic bacterium
     (*Natronobacterium pharaonis*). *Biochim. Biophys. Acta*, **1140**, 22–29.
172. S.P. Balashov, M. Sumi, N. Kamo (2000). The M intermediate of pharaonis
     phoborhodopsin is photoactive. *Biophys. J.*, **78**, 3150–3159.
173. J. Tittor, C. Soell, D. Oesterhelt, H.-J. Butt, E. Bamberg (1989). A defective pro-
     ton pump, point-mutated bacteriorhodopsin Asp96→Asn is fully reactivated by
     azide. *EMBO J.*, **8**, 3477–3482.
174. K. Takao, T. Kikukawa, T. Araiso, N. Kamo (1998). Azide accelerates the decay
     of M-intermediate of pharaonis phoborhodopsin. *Biophys. Chem.*, **73**, 145–153.

175. U. Haupts, E. Bamberg, D. Oesterhelt (1996). Different modes of proton translocation by sensory rhodopsin I. *EMBO J.*, **15**, 1834–1841.

176. M. Iwamoto, K. Shimono, M. Sumi, K. Koyama, N. Kamo (1999). Light-induced proton uptake and release of *pharaonis* phoborhodopsin detected by a photoelectrochemical cell. *J. Phys. Chem. B*, **103**, 10311–10315.

177. J. Sasaki, J.L. Spudich (1999). Proton circulation during the photocycle of sensory rhodopsin II. *Biophys. J.*, **77**, 2145–2152.

178. J.L. Spudich, R.A. Bogomolni (1984). Mechanism of colour discrimination by a bacterial sensory rhodopsin. *Nature*, **312**, 509–513.

179. Z. Dancshazy, L.A. Drachev, P. Ormos, K. Nagy, V.P. Skulachev (1978). Kinetics of the blue light-induced inhibition of photoelectric activity of bacteriorhodopsin. *FEBS Lett.*, **96**, 59–63.

180. H.J. Butt, K. Fendler, E. Bamberg, J. Tittor, D. Oesterhelt (1989). Aspartic acids 96 and 85 play a central role in the function of bacteriorhodopsin as a proton pump. *EMBO J.*, **8**, 1657–1663.

181. M.S. Braiman, T. Mogi, T. Marti, L.J. Stern, H.G. Khorana, K.J. Rothschild (1988). Vibrational spectroscopy of bacteriorhodopsin mutants: Light-driven proton transport involves protonation changes of aspartic acid residues 85, 96, and 212. *Biochemistry*, **27**, 8516–8520.

182. K. Gerwert, B. Hess, J. Soppa, D. Oesterhelt (1989). Role of aspartate-96 in protein translocation by bacteriorhodopsin. *Proc. Natl. Acad. Sci. U.S.A.*, **86**, 4943–4947.

183. G. Nagel, B. Kelety, B. Mockel, G. Büldt, E. Bamberg (1998). Voltage dependence of proton pumping by bacteriorhodopsin is regulated by the voltage-sensitive ratio of M-1 to M-2. *Biophys. J.*, **74**, 403–412.

184. K.H. Jung, E.N. Spudich, P. Dag, J.L. Spudich (1999). Transducer-binding and transducer-mutations modulate photoactive-site-deprotonation in sensory rhodopsin I. *Biochemistry*, **38**, 13270–13274.

185. J. Sasaki, J.L. Spudich (2000). Proton transport by sensory rhodopsins and its modulation by transducer-binding. *Biochem. Biophys. Acta*, **1460**, 230–239.

186. S.-W. Chiu, L.K. Nicholson, M.T. Brenneman, S. Subramaniam, Q. Teng, J.A. McCammon, T.A. Cross, E. Jakobsson (1991). Molecular dynamics computations and solid state nuclear magnetic resonance of the gramicidin cation channel. *Biophys. J.*, **60**, 974–978.

187. J. Sasaki, J.L. Spudich (1998). The transducer protein HtrII modulates the lifetimes of sensory rhodopsin II photointermediates. *Biophys. J.*, **75**, 2435–2440.

188. K.D. Olson, X.-N. Zhang, J.L. Spudich (1995). Residue replacements of buried aspartyl and related residues in sensory rhodopsin I: D201N produces inverted phototaxis signals. *Proc. Natl. Acad. Sci. U.S.A.*, **92**, 3185–3189.

189. E. Hildebrand, A. Schimz (1987). Role of the response oscillator in inverse responses of *Halobacterium halobium* to weak light stimuli. *J. Bacteriol.*, **169**, 254–259.

190. D.A. McCain, L.A. Amici, J.L. Spudich (1987). Kinetically resolved states of the *Halobacterium halobium* flagellar motor switch and modulation of the switch by sensory rhodopsin I. *J. Bacteriol.*, **169**, 4750–4758.

191. J.L. Spudich, J.K. Lanyi (1996). Shuttling between two protein conformations – the common mechanism for sensory transduction and ion transport. *Curr. Opin. Cell Biol.*, **8**, 452–457.

192. K.H. Jung, J.L. Spudich (1996). Protonatable residues at the cytoplasmic end of transmembrane helix-2 in the signal transducer HtrI control photochemistry and function of sensory rhodopsin I. *Proc. Natl. Acad. Sci. U.S.A.*, **93**, 6557–6561.

193. K.H. Jung, J.L. Spudich (1998). Suppressor mutation analysis of the sensory rhodopsin I-transducer complex–insights into the color-sensing mechanism. *J. Bacteriol.*, **180**, 2033–2042.

194. J.M. Sturtevant (1977). Heat capacity and entropy changes in processes involving proteins. *Proc. Natl. Acad. Sci. U.S.A.*, **74**, 2236–2240.

195. J.R. Livingstone, R.S. Spolar, M.T. Record (1991). Contribution to the thermo-dynamics of protein folding from the reduction in water-accessible nonpolar surface area. *Biochemistry*, **30**, 4237–4244.

196. R. Varadarajan, P.R. Connelly, J.M. Sturtevant, F.M. Richards (1992). Heat capacity changes for protein–peptide interactions in the ribonuclease S system. *Biochemistry*, **31**, 1421–1426.

197. R.S. Spolar, M.T. Record (1994). Coupling of local folding to site-specific binding of proteins to DNA. *Science*, **263**, 777–784.

198. D.L. Foster, S.L. Mowbray, B.K. Jap, D.E. Koshland (1985). Purification and characterization of the aspartate chemoreceptor. *J. Biol. Chem.*, **260**, 11706–11710.

199. W.L. Hubbell, D.S. Cafiso, C. Altenbach (2000). Identifying conformational changes with site-directed spin labeling. *Nat. Struct. Biol.*, **7**, 735–739.

200. W.L. Hubbell, A. Gross, R. Langen, M.A. Lietzow (1998). Recent advances in site-directed spin labeling of proteins. *Curr. Opin. Struc. Biol.*, **8**, 649–656.

201. H.J. Steinhoff, N. Radzwill, W. Thevis, V. Lenz, D. Brandenburg, A. Antson, G. Dodson, A. Wollmer (1997). Determination of interspin distances between spin labels attached to insulin: comparison of electron paramagnetic resonance data with the X-ray structure. *Biophys. J.*, **73**, 3287–3298.

202. M. Pfeiffer, T. Rink, K. Gerwert, D. Oesterhelt, H.J. Steinhoff (1999). Site-directed spin-labeling reveals the orientation of the amino acid side-chains in the E-F loop of bacteriorhodopsin. *J. Mol. Biol.*, **287**, 163–171.

203. C. Altenbach, T. Marti, H.G. Khorana, W.L. Hubbell (1990). Transmembrane protein structure: Spin labeling of bacteriorhodopsin mutants. *Science*, **248**, 1088–1092.

204. T. Rink, J. Riesle, D. Oesterhelt, K. Gerwert, H.J. Steinhoff (1997). Spin-labeling studies of the conformational changes in the vicinity of D36, D38, T46, and E161 of bacteriorhodopsin during the photocycle. *Biophys. J.*, **73**, 983–993.

205. N. Radzwill, K. Gerwert, H.-J. Steinhoff (2001). Time-resolved detection of transient movement of helices F and G in doubly spin-labeled bacteriorhodopsin. *Biophys. J.*, **80**, 2856–2866.

206. T. Rink, M. Pfeiffer, D. Oesterhelt, K. Gerwert, H.J. Steinhoff (2000). Unravel-ing photoexcited conformational changes of bacteriorhodopsin by time resolved electron paramagnetic resonance spectroscopy. *Biophys. J.*, **78**, 1519–1530.

207. W.J. Fantl, D.E. Johnson, L.T. Williams (1993). Signalling by receptor tyrosine kinases. *Annu. Rev. Biochem.*, **62**, 453–481.

208. S. Subramaniam, M. Gerstein, D. Oesterhelt, R. Henderson (1993). Electron dif-fraction analysis of structural changes in the photocycle of bacteriorhodopsin. *EMBO J.*, **12**, 1–8.

209. H. Luecke, B. Schobert, H.T. Richter, J.P. Cartailler, J.K. Lanyi (1999). Structural changes in bacteriorhodopsin during ion transport at 2 Ångstrom resolution. *Science*, **286**, 255–261.

210. J. Vonck (2000). Structure of the bacteriorhodopsin mutant F219L N intermediate revealed by electron crystallography. *EMBO J.*, **19**, 2152–2160.

211. D.L. Farrens, C. Altenbach, K. Yang, W.L. Hubbell, H.G. Khorana (1996). Requirement of rigid-body motion of transmembrane helices for light activation of rhodopsin. *Science*, **274**, 768–770.

212. B. Yan, T. Takahashi, R. Johnson, J.L. Spudich (1991). Identification of signaling states of a sensory receptor by modulation of lifetimes of stimulus-induced conformations: The case of sensory rhodopsin II. *Biochemistry*, **30**, 10686–10692.

213. S.A. Chervitz, J.J. Falke (1996). Molecular mechanism of transmembrane signaling by the aspartate receptor – A model. *Proc. Natl. Acad. Sci. U.S.A.*, **93**, 2545–2550.

214. K.M. Ottemann, W. Xiao, Y.K. Shin, D.E.J. Koshland (1999). A piston model for transmembrane signaling of the aspartate receptor. *Science*, **285**, 1751–1754.

215. J. Stock, S. Da Re (1999). A receptor scaffold mediates stimulus-response coupling in bacterial chemotaxis. *Cell Calcium*, **26**, 157–164.

216. I. Chizhov, I., M. Engelhard (2001). Temperature and halide dependence of the photocycle of halorhodopsin from *Natronobacterium pharaonis*. *Biophys. J.*, **81**, 1600–1612.

Note: After the writing of this chapter another structure of NpSRII was published (A. Royant, P. Nollert, K. Edman, R. Neutze, E.M. Landau, E. Pebay-Peyroula, J. Navarro (2001). X-ray structures of sensory rhodopsin II at 2.1 Å resolution. *Proc. Natl. Acad. Sci. USA*, **98**, 10131–10136). In addition the crystal structure of the receptor/transducer couple is now available (V.I. Gordeliy, J. Labahn, R. Moukhametzianov, R. Efremov, J. Granzin, R. Schlesinger, G. Büldt, T. Savapol, A.J. Scheidig, J.P. Klare, M. Engelhard (2002). Molecular basis of transmembrane signalling by sensory rhodopsin II-transducer complex. *Nature*, **419**, 484–487.

Chapter 2

Invertebrate rhodopsins

Joachim Bentrop and Reinhard Paulsen

Table of contents

Abstract

The primary events in phototransduction in invertebrate photoreceptor cells
are initiated at rhodopsins which share structural features common to the subgroup of G-protein coupled receptors specialized for the absorption of light.
Evaluation of cephalopod rhodopsin indicates that the seven transmembrane
helices of invertebrate rhodopsins are spaced around the retinal chromophore
in a 3D structure very similar to that of vertebrate rhodopsins. Nevertheless,
from a multitude of information obtained from studies of phototransduction
pathways in model systems from the phyla of molluscs (squid, octopus, scallop) and arthropods (*Drosophila*, *Limulus*, crayfish) it is obvious that evolution
has led to some divergence (i) within invertebrates and (ii) between vertebrates
and invertebrates. These modifications are manifested at the level of rhodopsin
function as well as in the individual steps of the signalling cascade initiated
through photon absorption by rhodopsin. This divergence specifically concerns the formation of stable metarhodopsin states, the coupling to different
types of G-proteins, the mechanisms of amplification and termination of
signalling, the supramolecular organization of the signalling cascade and
non-visual, intracellular signalling pathways for morphogenesis, membrane
renewal and apoptosis.

2.1   Introduction

The general acceptance of "rhodopsin" as a generic designation for a visual
pigment is a consequence of the rapidly increasing progress in the cloning and
sequencing of rhodopsins and the application of genetical, gene technological
and molecular physiological methods. These studies indicate that a rhodopsin
serves as the primary light receptor protein in the visual systems of all animals
investigated so far, independent of the structural and functional complexity,
of the optical apparatus, or of the neuronal networks which animals have
developed to analyse and process the information encoded in a light signal.
The reason why visual perception is generally mediated by a rhodopsin is seen
in a common origin of all visual systems, which depends on the the action
of two key genes. These genes, which are proposed to have already interacted
in a prototypic photoreceptor are, first, a gene coding for a rhodopsin (opsin),
which has been aquired to allow photons to be absorbed, and, second, a gene
that directs rhodopsin expression. From this gene equipment of a photoreceptor prototype the existing diversity of visual systems may have evolved by
intercalary evolution. *Pax 6* genes have been shown to operate as master control genes for eye morphogenesis and divergent rhodopsin genes are expressed
in a terminal step of photoreceptor differentiation. The latter specify which
wavelengths of light will be absorbed [1–4].

The structure of rhodopsins is remarkably conserved throughout evolution:
animal rhodopsins consist of an apoprotein, designated as opsin, and a

chomophore (11-*cis*-retinaldehyde or a closely related form of retinal). A key feature of rhodopsin is the folding of its single amino acid chain into a secondary structure with seven transmembrane α helices (Figure 1). Rhodopsins are therefore classified with 7 TM receptors. The chromophore is, without any known exception, covalently linked to the side chain of a lysine located in transmembrane helix seven.

In view of the structural conservation of rhodopsins from simple invertebrate organisms to man, one may ask whether it makes sense to differentiate between an invertebrate and a vertebrate subgroup of visual pigments. Since a taxon called "invertebrates" does not exist, whereas the vertebrates constitute a well-defined group within the phylum of chordata, there might be primarily practical reasons to individually deal with the rhodopsins expressed in invertebrate photoreceptors. At present, information is available on the amino acid sequences of more than 60 vertebrate rhodopsins and of a similar number of rhodopsins from invertebrates. This situation is likely to shift rapidly in favour of the invertebrate rhodopsins. From the number of species described to date, one may estimate that about 10^7 different invertebrate rhodopsins exist, with a single species harbouring as many as 16 rhodopsins [5]. In relation to about 10^5 different rhodopsins of vertebrates, more than 99% of the existing rhodopsin genes are expected to be expressed in the photosensitive cells of invertebrates. This abundance is likely to have led to a considerably higher divergence in the structure and function of rhodopsin than is indicated by the current level of information, which mainly stems from vertebrate rhodopsins. Thus, only a close look at the invertebrates will provide the information on what is general and what is special in the function of rhodopsins.

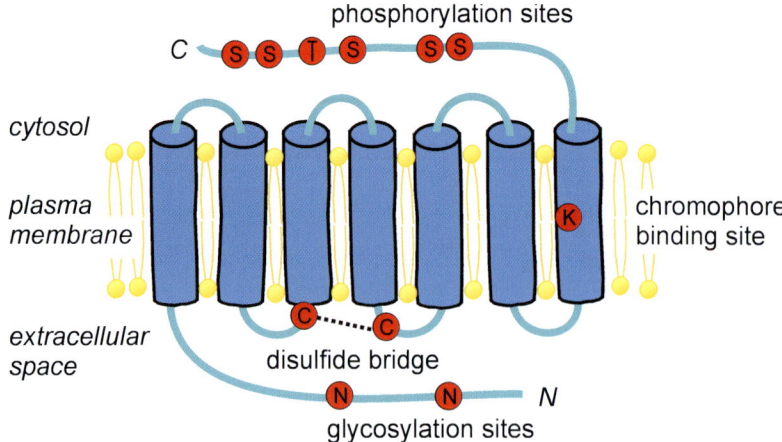

Figure 1. Secondary structure model of invertebrate rhodopsins. The amino acid chain is folded into seven transmembrane α helices. The C-terminus is located intracellularly, the N-terminus extracellularly. Functionally important domains, as discussed in the text, are highlighted.

The current state of knowledge about invertebrate rhodopsins is determined primarily by work directed to study rhodopsin-related events in phototransduction in a limited number of model systems. One such system is the compound eye of the fruitfly *Drosophila*, which provides a model system that can be dissected by a combination of genetical, molecular biological and physiological methods in a way unmatched by other visual systems [6–13]. Visual systems like the eyes of cephalopods (octopus, squid) provide the opportunity to study rhodopsin functions with sophisticated biophysical and biochemical methods, due to the availability of large amounts of rhodopsin-containing photoreceptor membranes.

Comparative studies of different invertebrate visual systems show that evolutionary modifications have occurred to optimize eyes to particular types of visual input. These modifications concern the photochemistry of rhodopsin, the mobility of rhodopsin, its spectral tuning by sensitizing pigments and the type of G-proteins activated upon a light stimulus. Furthermore, non-phototransducing functions of the visual pigment are concerned, e.g. membrane targeting and endocytosis of rhodopsin or its role in triggering apoptosis.

The main focus of this review are aspects of invertebrate rhodopsin structure and function as well as rhodopsin-related events in the activation and control of phototransduction. Some of these topics have also been reviewed recently elsewhere [13–18].

2.2   Molecular evolution of invertebrate rhodopsins

The analysis of over 60 genes coding for the opsin moiety of invertebrate rhodopsins yields a phylogenetic tree as depicted in Figure 2. Invertebrate rhodopsins are members of the rhodopsin superfamily within the phylogenetically related hyperfamily of G-protein coupled receptors (GPCRs) [19]. DNA sequence data for the protein coding regions of rhodopsin genes have, apart from the information on rhodopsin functions, provided useful data for the molecular taxonomy of hymenopteran insects [20]. Despite the rapidly increasing sequence information yielded by several invertebrate genome projects as well as by the visual system-oriented research focussing on rhodopsin itself, there are still gaps in the knowledge of rhodopsin evolution. A more detailed understanding is likely to be gained if sequence data become available from lower invertebrates. This would allow to link evolution of metazoan rhodopsins more directly to unicellular eucaryotes, for example the evolution of rhodopsin in the green algae *Chlamydomonas* and *Volvox* (see Chapter 4).

One gap has recently been narrowed by the cloning of a rhodopsin expressed in the larval eye spot of an "invertebrate" chordate, the tunicate *Cionia* [21,22]. In phylogenetic trees depicting the evolutionary relationships of rhodopsins on the basis of homologies in their primary structure, *Ci-opsin1* segregates into a subgroup with the rhodopsins from vertebrates and man to form a new

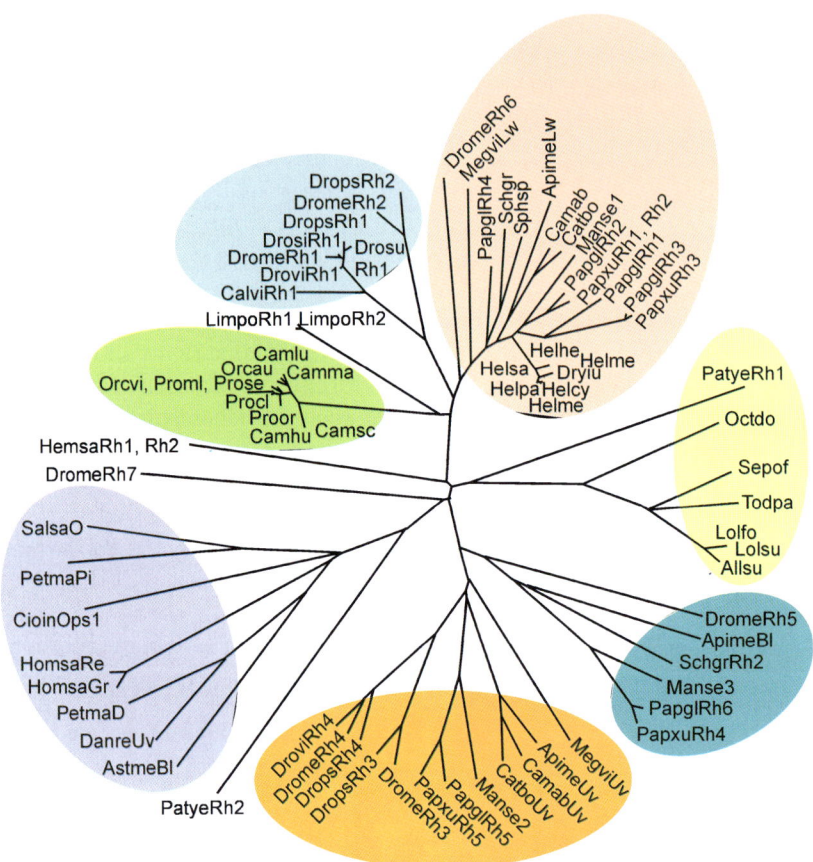

Figure 2. Evolutionary relationship of invertebrate rhodopsins. Phylogenetic tree (amino acid sequence) of invertebrate rhodopsins and selected chordate rhodopsins. Included are all invertebrate rhodopsins whose complete amino acid sequence was determined by May 2002. In alphabetical order: abbreviation, species (and name of rhodopsin), accession number: Allsu, *Allotheutis subulata*, S71931; ApimeLw, *Apis mellifera* long-wavelength rhodopsin, Q17063; ApimeBl, *Apis mellifera* blue-sensitive rhodopsin, AAC13415, AAC 47455; ApimeUV, *Apis mellifera* UV-sensitive rhodopsin, AAC13418; Astme, *Astyanax mexicanus* blue-sensitive rhodopsin, P51471; CalviRh1, *Calliphora vicina* opsin Rh1, P22269; Camsc, *Cambarellus shufeldtii*, O16018; Camhu, *Cambarus hubrichti*, O18312; Camlu, *Cambarus ludovicianus*, O16017; Camma, *Cambarus maculatus*, O1915; Camab, *Camponotus abdominalis* opsin, Q17292; CamabUV, *Camponotus abdominalis* UV-sensitive rhodopsin, AAC050920; Catbo, *Cataglyphis bombycinus*, Q17296; CatboUV, *Cataglyphis bombycinus* UV-sensitive rhodopsin, AAC05091; CioinOps1, *Cionia intestinalis* opsin1, BAB68391; DanreUV, *Danio rerio* UV-sensitive rhodopsin; DromeRh1-Rh1, *Drosophila melanogaster* rhodopsins Rh1–Rh7, P06002, P08099, P04950, P08255, P91657, O01668, AAF49949, resp.; DropsRh1-Rh4, *Drosophila pseudoobscura* rhodopsins Rh1–Rh4, P28678, P28679, P28680, P29404, resp.; DrosiRh1, *Drosophila simulans* rhodopsin Rh1, AAB31030; DrosuRh1, *Drosophila subobscura* rhodopsin Rh1, AAB87898; DroviRh1, Rh4, *Drosophila virilis* rhodopsin Rh1, Rh4, AAB31031, O17646, resp.; Dryiu, *Dryas iulia*, AAK58111; Helcy,

Figure 2. *continued*

Heliconius cydno, AAK58109; Helhe, *Heliconius hewitsoni*, AAK58106; Helme, *Heliconius melpomene*, AAK13246; Helpa, *Heliconius pachinus*, AAK58110; Helsa, *Heliconius sapho*, AAK58108; Helsa, *Heliconius sara*, AAK58107; HemsaRh1, Rh2, *Hemigrapsus sanguineus* compound eye opsins Rh1, Rh2, Q25157, Q25158, resp.; HomsaGr, *Homo sapiens* green-sensitive rhodopsin, P04001; HomsaRe, *Homo sapiens* red-sensitive rhodopsin, P04000; LimpoRh1, *Limulus polypemus* lateral eye opsin, P35360; LimpoRh2, *Limulus polyphemus* ocellar opsin, P35361; Lolsu, Loligo subulata, Q17094; Lolfo, *Loligo forbesi*, P24603; Manse1–3, *Maduca sexta* opsin 1–3, AAD11964, AAD11965, AAD11966, resp.; MegviLV, *Megoura viciae* long-wavelength-like opsin, AAG17119; MegviUV, *Megoura viciae* UV-wavelength-like opsin, AAG17120; Octdo, *Octopus dofleini*, P09241; Orcau, *Oronectes australis*, O18418; Orcvi, *Oronectes virilis*, O16019; PapglRh1–RH6, *Papilio glaucus* rhodopsin Rh1–Rh6, AAD34220, AAD34221, AAD29445, AAD34224, AAD34222, AAD34223, resp.; PapxuRh1–Rh5, *Papilio xuthus* rhodopsin Rh1–RH5, BAA31721, BAA31722, BAA32723, BAA93469, resp.; PatyeRh1, *Patinopecten yessoensis* rhodopsin G$_q$-coupled, O15973; PatyeRh2, *Patinopecten yessoensis* rhodopsin G$_o$-coupled, O15974; PetmaPi, *Petromyzon marinus* pineal opsin, AAV41240; PetmaD, *Petromyzon marinus* rhodopsin, Q98980; Procl, *Procambarus clarkii*, P35356, Proml, *Procambarus milleri*, O16020; Prose, *Procambarus seminolae*, O18486; Proor, *Procambarus orcinus*, O18485; Prose, *Procambarus seminolae*, O18486; SalsaO, *Salmo Salar*, ancient opsin, O13018; SchgrRh1, Rh2, *Schistocerca gregaria* rhodopsin Rh1,Rh2, Q94741, Q26495; Sepof, Sepia officinalis, O16005; Sphsp, *Sphodromantis* sp., P35362; Todpa, *Todarodes pacificus*, P31356;

clade for chordates [21] (Figure 2). This clade is distinct from that of mollusc rhodopsins, in particular of cephalopod rhodopsins. The rhodopsins of arthropods, e.g. those of xiphosura (*Limulus*), crustacea (crabs and crayfish) and a number of insects are clustered into two larger rhodopsin subfamilies which differ in wavelength absorption properties. One clade consists of long wavelength (i.e. yellow, green and blue-green light) absorbing rhodopsins, the other assembles rhodopsins absorbing at shorter wavelengths, i.e. UV and blue light. These two subgroups can be further refined with respect to the spectral tuning of rhodopsin [23–26]. One lesson from such a comparison is that complex visual achievements which rely on rhodopsin divergence, like UV- and colour vision, have evolved independently several times in invertebrates and in vertebrates [26–30].

Figure 2 also depicts two exceptions to the general observation that a rhodopsin segregates with a clade representative for the taxon from which it originated. The outgrouped rhodopsins are PatyeRh2 (SCOP2) [31], a rhodopsin localized to ciliary photoreceptors of the scallop *Patinopecten yessoensis*, and an orphan rhodopsin, Rh7 of *Drosophila melanogaster*, with unknown expression pattern and function. The latter was identified by the *Drosophila* genome project [32]. Both rhodopsin homologs cluster in the vicinity of the chordate rhodopsins. What are the determinants in the primary structure of rhodopsins that underlie such a divergence? It has been shown that cytoplasmic and extracellularly located ends of rhodopsin are more variable than the 7 TM regions [33]. The tuning of a rhodopsin´s absorption maximum to a given

wavelength is apparently based on a number of conserved interactions, particularly between the chromophore and the amino acid side chains within the
7 TM-helices of rhodopsin. Indeed, if one takes into account the tertiary structure of rhodopsin and establishes a phylogenetic tree solely on the basis
of sequence homology within the 7 TM helices, a slightly different picture
appears: The divergence of rhodopsins is more restricted than observed with the
holorhodopsin, and, as expected, PatyeRh2 and DromeRh7 now cluster with
subgroups of rhodopsins from molluscs and insects, respectively (Figure 3).

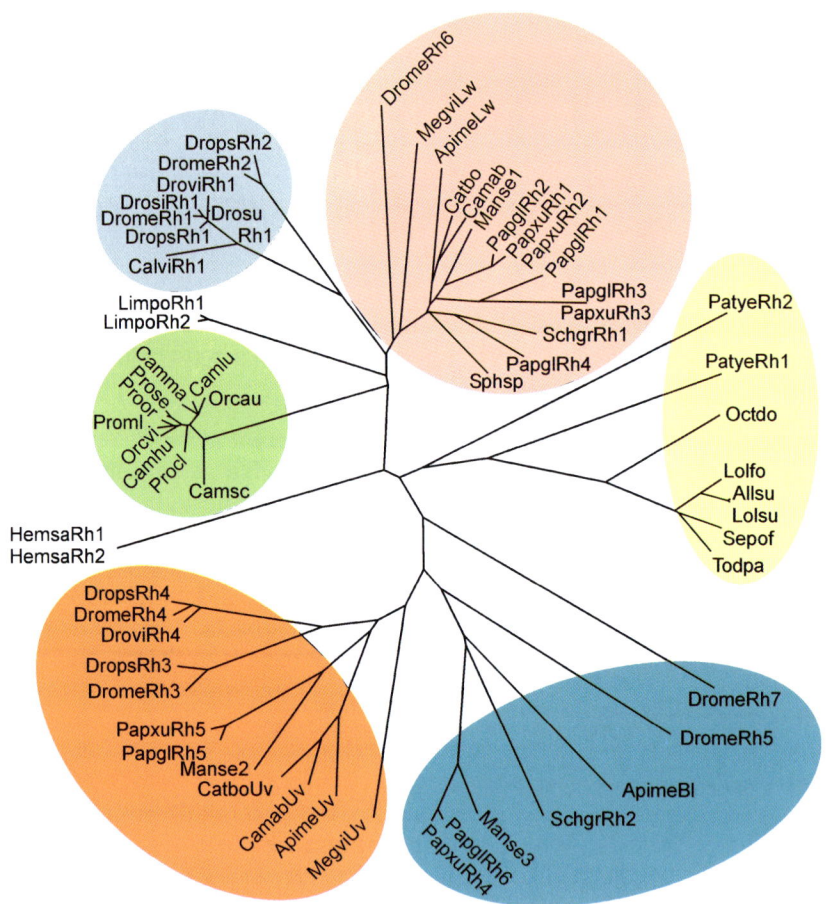

Figure 3. Evolutionary relationship of invertebrate rhodopsin transmembrane
domains. The amino acid sequences of invertebrate rhodopsins were modified, such
that peptides were constructed consisting only of the linked seven transmembrane
domains, the C- and the N-terminus as well as intra- and extracellular loops of
rhodopsins were removed for that purpose. In the phylogenetic tree of these peptides,
Patye2 and DromeRh7 cluster with subgroups of rhodopsins from molluscs and insects,
respectively. Abbreviations are as in Figure 2.

The determinants leading to an outgroup position in the phylogenetic tree in Figure 2 have been eliminated and must therefore reside in the extra-7TM sections of these rhodopsin sequences. An obvious divergence concerns the C-termini of both rhodopsins. The C-terminus of DromeRh7 is considerably longer than that of any other insect rhodopsin. PateyRh2, the scallop rhodopsin which is expressed in a ciliary photoreceptor cell type [32,34]–contrary to the majority of invertebrate rhodopsins which are expressed in rhabdomeral photoreceptors–also posseses an extended C-terminus. This C-terminus, how-ever, does not exhibit a particular homology to the extended C-termini of other molluscs, for example of the cephalopod rhodopsins. It may be that in PatyeRh2 extra-7 TM sites are conserved for initiation of a transduction mecha-nism operating in ancestral photoreceptors of both molluscs and chordates. Following that line of arguments, one would expect that DromeRh7 is coupled to a transducin-like G-protein, like the one utilized in chordate phototransduc-tion, or that it does not serve a signal transducing function at all.

2.3   Key motifs of invertebrate rhodopsins

A detailed structure comparison of invertebrate rhodopsins has recently been performed by Gärtner [14]. To highlight some of the key motifs for rhodopsin function in the following chapters of this review, a coding sequence compari-son is carried out for *Drosophila melanogaster*. This is the only insect species in which six out of seven expressed opsins genes (*rh1* to *rh6*) have not only been sequenced, but the corresponding rhodopsins have been functionally charac-terized as well [35–41]. The *Drosophila* rhodopsins compared in the alignment of Figure 4 are expressed in the compound eye (Rh1, Rh3 to Rh6) and in extra-compound eye structures such as the adult ocellar photoreceptors and the testis (Rh2) as well as in larval photoreceptors (Rh1, Rh3, Rh4) [42]. As mentioned above, Rh7 identified by the *Drosophila* genome project [32] is still in an orphan state due to the lack of data for the expression pattern of the gene and the function of the protein. In the alignment depicted in Figure 4, the 7TM-structure of the *Drosophila* rhodopsins is founded on hydrophobicity calculations [14]. Experimental studies exploring the transmembrane location of the 7 transmembrane helices of distinct *Drosophila* rhodopsins in more detail have not yet been carried out.

2.3.1   Positionally conserved domains

The positions of domains conserved in Rh1 to Rh6 of *Drosophila* are summa-rized in the structure model shown in Figure 4. These domains are of particular interest since the light-activated state of each rhodopsin activates the same visual G-protein (Gq) with the same efficiency, regardless of the structural differences determining the particular absorption properties of a rhodopsin. This has been demonstrated most convincingly by the functional expression of rhodopsins Rh2 to Rh6 in photoreceptor cells R1-6 in place of the intrinsic

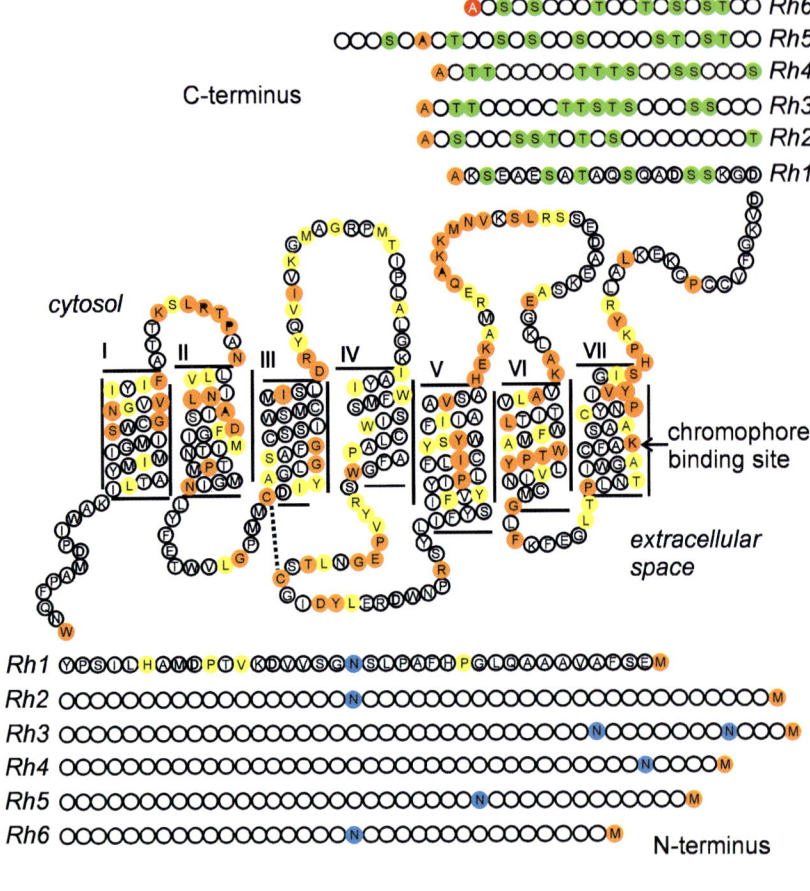

Figure 4. Conserved amino acids and functional domains in *Drosophila* rhodopsins. Structure model of *Drosophila* Rh1 rhodopsin, amino acids are shown in one letter code. The C-termini and N-termini of Rh2 to Rh6 are aligned with those of Rh1. Conserved amino acids and functional domains are highlighted in the figure and explained in the text.

Rh1 rhodopsin [43]. Thus, these rhodopsins share a common interface for the interaction with the G-protein (Gq) coupled to Rh1. As all of the ectopically expressed rhodopsins are properly targeted to the rhabdomeral phototransduction compartment in R1-6 photoreceptor cells, they have to provide a common interface for cellular targeting as well as a common interface for the activation and control of phototransduction. This conservation in the structure of rhodopsin extends beyond the rhodopsins expressed in *Drosophila* photoreceptors, as the heterologous expression of opsin genes encoding the UV- and blue-light absorbing rhodopsins of the honey bee (*Apis mellifera*) and of a locust rhodopsin (*Schistocerca gregarina*) fully rescues signal transduction in null-mutants of *Drosophila* Rh1 [44,45].

As indicated in Figure 4, highly conserved amino acids in *Drosophila* rhodopsins are located in cytoplasmic loops i1 to i3 and in extracellular loop e2. A similar scheme of overlapping conserved sites is observed if rhodopsins from other insect species are included in the structure comparison, e.g. the rhodopsins cloned from the honey bee [45,46] or from the sphingid moth *Manduca sexta* [25]. In contrast to vertebrate rhodopsins (see Chapter 3), no unequivocal evidence is available on the role of conserved amino acids with regard to forming an interface with the visual G-protein and arrestin isoforms, which control the active rhodopsin state (metarhodopsin). The finding that mutations of highly conserved amino acids (L81Q, N86I) in the first cytoplasmic loop and in the extracellular loop e2 (E194K, G195) of *Drosophila* Rh1 arrest rhodopsin synthesis in a nascent state suggests that this loop harbours information required in initial steps of the targeting process to the photoreceptor membrane [47]; for summaries see [48,49]. In addition to extended domains, amino acids are positionally conserved at single sites in all rhodopsins. This holds particularly for two cysteines (C123 and C200 of *Drosophila* Rh1, Figure 4) which form a structure-stabilizing disulfide bridge on the extracellular surface of each rhodopsin molecule [14].

2.3.2   The chromophore binding site

Positional conservation in particular includes the chromophore binding site, a lysine residue in transmembrane helix VII (corresponding to K319 of *Drosophila* Rh1 in Figure 4). In invertebrate rhodopsins, this site is occupied by a variety of retinaldehyde-related chromophores in the 11-*cis* conformation (retinal, 3-hydroxyretinal, 4-hydroxyretinal) [50–53].

Raman resonance measurements in cepahalopod rhodopsin [54,55] and in an insect UV-rhodopsin [56], as well as infrared and cryogenic spectroscopy [57,58], indicate that the chromophore is covalently linked to the chromophore binding site via the protonated state of the Schiff base, as is the case in vertebrates. In vertebrates, the protonated Schiff base of rhodopsin is stabilized by a glutamic acid residue (E113 of bovine rhodopsin) which serves as a negatively charged counter-ion to the positively charged chromophore [59–61]. In most invertebrate rhodopsins, a conserved tyrosine residue (Y126 of *Drosophila* Rh1, see Figure 4) has been regarded as the prime candidate for the Schiff base counter-ion [14]. A denaturation–reprotonation study with cephalopod (*Octopus*) rhodopsin, however, does not support this view [62]. In the absence of other candidates for a corresponding counter-ion, it is thus proposed that Schiff base protonation of invertebrate rhodopsin does not require a stabilizing salt bridge with a negative charge [62].

Accordingly, a profound difference exists between vertebrate and inverte-brate rhodopsins. This has to be seen in the context that cephalopod rhodopsin and most of the other insect rhodopsins investigated so far form, upon light absorption, a relatively stable metarhodopsin state [50,63–66]. This state repre-sents the active state of invertebrate rhodopsin, or at least a state spectrally closely related to the active state. From this state, in which the chromophore is still attached to opsin via a protonated Schiff base, the initial rhodopsin state is regained by photoregeneration via an intermediate with the chromophore in the 11-*cis* configuration [63,64].

For *Drosophila* and related insects, the formation of a stable protonated Schiff base bond between the chromophore and opsin is a primary structural requirement for proper processing of newly synthesized rhodopsin and its trans-port to the rhabdomeral photoreceptor membrane. In 11-*cis*-3-hydroxyretinal-deficient flies, opsin synthesis is blocked at the post-translational level, in a nascent opsin state [67]. Opsin molecules which, due to the absence of an attached chromophore, cannot adopt the properly folded structure of a mature rhodopsin, are targeted into a degradation pathway. As a consequence, the opsin/rhodopsin density in the phototransducing rhabdomeral membrane compartment becomes drastically reduced [68]. The requirement for a stabilized form of chromophore attachment extends beyond synthesis, targeting, activa-tion and regeneration of rhodopsin. The renewal of rhodopsin involves selective internalization of metarhodopsin [69,70], and thus also relies on a stable metarhodopsin with a protonated Schiff base-bound chromophore.

In the visual cycle of vertebrates, Schiff base deprotonation and proton translocation to the counter-ion is part of the molecular mechanism of rhodop-sin activation [71,72] (see Chapter 3). Relaxation of the opsin structure, which initiates Schiff base hydrolysis, the subsequent release of all-*trans* retinal from opsin and, finally, the formation of a new Schiff base bond between opsin and 11-*cis* retinal are the hallmarks of the chemical regeneration of vertebrate rhodopsin (see Chapter 3). In invertebrates, it is questionable whether the Schiff base ever becomes deprotonated during a regular rhodopsin cycle. Basic steps of that cycle involve transitions which terminate at rhodopsin states with chromophore configurations possessing a stabilized, protonated Schiff base bond. How this stability is achieved is not yet known. Apart from unidentified intramolecular interactions, rhodopsin stability might be affected by intermo-lecular interactions. To this end it has been shown *in situ* that the interaction with arrestin significantly enhances the life-time of metarhodopsin. *In vivo*, the life-time of metarhodopsin is also extended in the absence of arrestin [73], which suggests the existence of additional stabilization mechanisms.

2.3.3   *Three-dimensional structure*

Despite the unique properties *Drosophila* offers for mutation analyses to solve structure–function relationships of invertebrate rhodopsin, advanced biophysical techniques are still limited to invertebrate species that provide sufficient rhodopsin for purification and crystallization. A recent study which sheds light on the three-dimensional structure of an invertebrate rhodopsin employed cryo-electronmicroscopy and image processing of 2D-crystals of squid rhodopsin [74]. Crystallization was achieved with a C-terminally truncated form of squid rhodopsin. The proteolytic truncation removed a proline-rich extension from the C-terminus which is unique for rhodopsins of cephalopods [75]. The projection structure of the crystallized squid rhodopsin at 8 Å resolution revealed a high similarity in the rhodopsin topology to that of vertebrate rhodopsin at a comparable resolution [74,76,77]. Apart from some differences in the packing of the transmembrane helices and the presence of a well-ordered structure in loop i3, the fit of the 3D map of cephalopod rhodopsin with the 3D structure of bovine rhodopsin agrees with the evolutionary and structural conservation of rhodopsins from invertebrates to man. An interesting result derived from crystallization studies of squid rhodopsin concerns the existence of a linear lattice contact between rhodopsin molecules. This may be the structural basis for an ordered alignment of rhodopsins in the microvillar membranes of invertebrate rhabdomeral photoreceptors. The fixed orientation of invertebrate rhodopsin in such a lattice may account for the ability to analyse the plane of polarized light [74]. The finding of a lateral lattice organization of squid rhodopsin may imply that the remaining members of the phototransduction cascade are also organized into superstructures [78]. In *Drosophila*, such a multimeric signalling complex is organized via the scaffolding-protein INAD (see Section 2.5.1).

2.3.4 Post-translational modification: palmitoylation, glycosylation and phosphorylation

The excellent fit between the 3D structures of squid and bovine rhodopsin extends to the location of a short, transverse α-helix at the cytoplasmic surface of cephalopod rhodopsin, which appears to be part of the intermolecular docking domain [74]. Helix VIII is located close to a site at which the C-terminus of rhodopsin may interact with the phospholipid bilayer via S-palmitoylation of two adjacent cysteine residues. These cysteine residues are conserved in many, but not all, invertebrate rhodopsins, as well as in vertebrate rhodopsins [14,79,80]. Moreover, mutation of the cysteines present at the C-terminus of *Drosophila* Rh1 (see Figure 4) is without functional consequences (Bentrop unpublished).

Two other sites for post-translational modifications are observed in all invertebrate rhodopsins. All rhodopsins harbour at their extracellularly located N-terminal peptide at least one consensus sequence for N-glycosylation. On the cytoplasmic side, serine and threonine residues are located close to the C-terminus, which may serve as sites for multiple phosporylation by a receptor kinase. As indicated in Figure 4, there is no clear cut indication for a positional conservation, either of the sites for N-glycosylation or of the putative phosphorylation sites.

Post-translational N-glycosylation of invertebrate rhodopsin has been demonstrated for cephalopod rhodopsin [81] as well as for insect rhodopsin [82,83]. Site-directed mutagenesis of the putative glycosylation sites in *Drosophila* Rh1 indicated that only Asn 20 (see Figure 4), but not a second putative site (Asn 169), becomes glycosylated. The mutation (N20L) leads to the accumulation of a nascent state of rhodopsin and causes photoreceptor degeneration [83,84], possibly by disturbing the interaction of nascent rhodopsin with the chaperone NinaA [85,86]. The functional importance of rhodopsin glycosylation for folding, sorting, and transport is highlighted by the finding that Rh1 adopts the gylcosylated state only transiently. Thus, while cephalopod (*Octopus*) rhodopsin remains equipped with a rather unique oligosaccharide side chain after being incorporated into the rhabdomeral photoreceptor membrane, mature Rh1 rhodopsin in the rhabdomeric membrane of fly photoreceptors is deglycosylated [68,83]. For *Octopus* rhodopsin it has been recently shown by mass spectrometry that in addition to the N-glycan, which is conserved in 7TM receptors, two *N*-acetylgalactosamine residues are O-linked near the N-terminus [87]. The functional relevance of these glycosylations remains to be elucidated.

Phosphorylation of vertebrate rhodopsin was recognized as the first light-triggered enzymatic modification of any rhodopsin [88–90]. It has been shown subsequently that multiple phosphorylation of vertebrate rhodopsin at C-terminally located serine residues is an essential step in the deactivation of the light-activated rhodopsin state. Phosphorylation of the active state (MII) leads to a rapid high-affinity interaction of the regulator protein, arrestin, with MII, which hinders MII from further activating the G-protein, transducin, (see Chapter 3). Thus, in the vertebrate visual cycle, rhodopsin phosphorylation is directly linked to the function of arrestin in regulating the active state of rhodopsin.

Cephalopod rhodopsin [91,92] and fly rhodopsin [93–96] also undergo a light-dependent phosphorylation-dephosphorylation cycle (Figure 5). In these rhodopsins, a single site (*Octopus* [97]) or multiple sites (flies [94]) become light-dependently phosphorylated and dephosphorylated. The light-dependent steps in this phosphorylation cycle are induced by the conversion of rhodopsin (P-state) into metarhodopsin (M-state). Conversion of P into M also induces the binding of Arrestin2 (Arr2), one of the arrestin isoforms expressed in photoreceptors (Figure 5). Dephosphorylation of M is induced by photoregeneration of M into P, as a result of the release of arrestin, which allows the phosphorylated M to interact with protein phosphatase [94]. Thus, a link between Arr2 function and light-activated phosphorylation of rhodopsin in fly photoreceptors exists at the level of the regulation of rhodopsin dephosphorylation. The distinct and profound difference regarding the function of rhodopsin phosphorylation in fly photoreceptors is that here phosphorylation *per se* is not a prerequisite for binding of arrestin to the active M-state. This is supported by the finding that Arr2 still binds to the M-state of *Drosophila* Rh1 and suppresses the activity of metarhodopsin even if the C-terminal peptide containing the phosphorylation sites is removed [98].

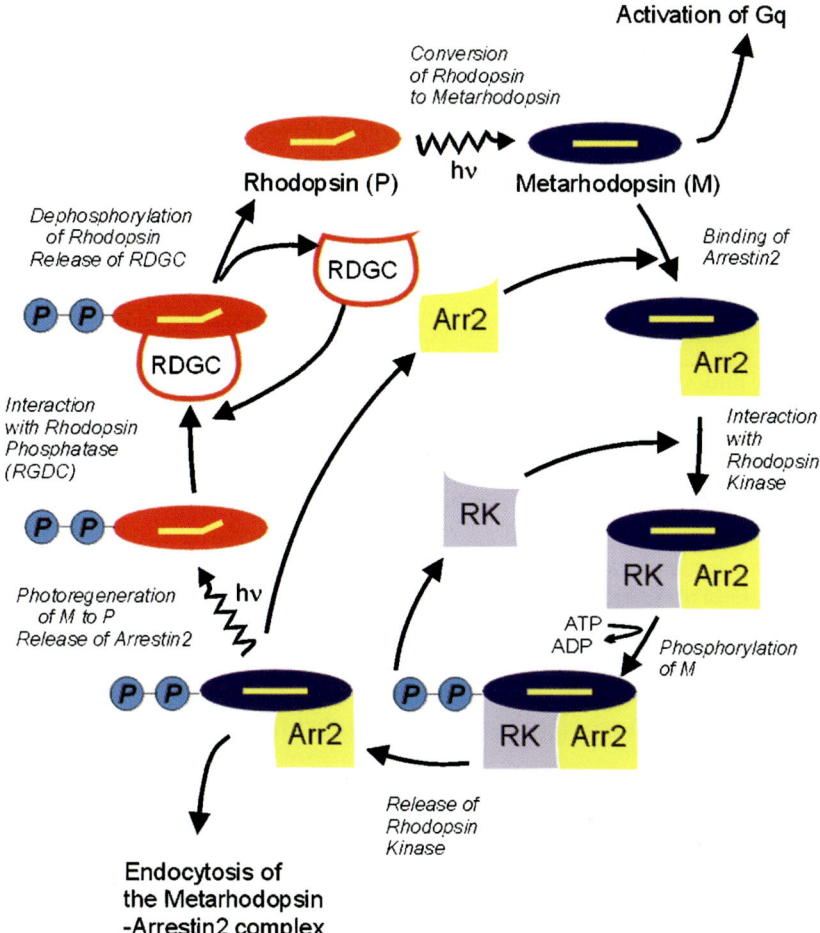

Figure 5. Activation, inactivation and endocytosis of *Drosophila* rhodopsin. Absorption of a photon causes a conformational change in the rhodopsin molecule (P) and triggers the formation of the active metarhodopsin state (M). Red ellipses show rhodopsin (chromophore in the 11-*cis* configuration), blue ellipses show metarhodopsin (chromophore in the all-*trans* configuration). Active M triggers the phototransduction cascade by activation of the visual G-protein Gq. Binding of Arrestin2 (Arr2) inactivates M and is followed by phosphorylation of M through a yet unidentified rhodopsin kinase (RK). Arrestin-bound M is subject to Clathrin-mediated endocytosis. Upon photon capture, phosphorylated M is re-converted into phosphorylated P and releases bound Arr2. This enables phosphorylated P to be dephosphorylated through the action of the Retinal Degeneration C protein (= rhodopsin phosphatase), yielding the initial rhodopsin state.

Instead, phosphorylation of the M-state is part of a signalling mechanism that induces the endocytosis of M, an important step in the renewal pathway of rhodopsin. This pathway comprises of the integration of newly synthesized rhodopsin into the photoreceptor membrane and the endocytosis of meta-rhodopsin molecules. The renewal cycle guarantees a high level of rhodopsin in the photoreceptor membrane in a situation in which photoregeneration of rhodopsin cannot take place. The signalling capacity of the phosphorylated metarhodopsin-Arr2 complex (Figure 5) is not limited to metarhodopsin endocytosis. A defect in rhodopsin dephosphorylation, resulting from a null mutation in the *rdgC* gene which encodes the rhodopsin phosphatase, induces photoreceptor apoptosis [98]. Apoptosis is also induced if the release of Arr2 from this complex is prevented [70]. The apparent switch in the function from an endocytosis control mechanism to a function in the control of the active rhodopsin state may have occurred in parallel with the development of a mechanism for the chemical regeneration of rhodopsin by vertebrates. Rhodopsin is here regenerated by the exchange of all-*trans* retinal for 11-*cis* retinal. This step takes place at opsin molecules still located in the photo-receptor disc membrane. In vertebrates, a need for selective labelling of meta-rhodopsin for endocytosis no longer exists as photoreceptor disc membranes are shedded from the photoreceptor tips as a whole, and are then phagocytized by the pigment epithelium (reviewed in [99]).

2.4   Spectral characteristics of invertebrate rhodopsins

2.4.1   *Spectral absorption of rhodopsins and formation of photointermediates*

Invertebrates exhibit a considerable variation in the types of different rhodop-sins expressed by a species. Analysis of the genome of the soil-living nematode worm *Caenorhabditis elegans* suggests that it may not express a rhodopsin at all, while the visual system of a crustacean, the mantis shrimp *Haptosquilla*, might harbour of up to 16 distinct rhodopsins [5]. Rhodopsins with different absorption characteristics are generally found in invertebrates with eyes capable of colour vision. The compound eye of *Drosophila*, which expresses in total five rhodopsins, provides such a multi-input system, as do the eyes of bees, ants, butterflies, moths, and many other insects. Another rhodopsin, Rh2, is expressed in ocelli located at the vortex of the head. In *Drosophila*, the expression pattern of rhodopsins is correlated with a highly regular positioning of the photoreceptor cells within each ommatidium of the eye. The outer photoreceptor cells R1-6 express the same rhodopsin (Rh1). The two central cells, R7 and R8, of the ommatidium join their individual rhabdomeres to form a rhabdomere located centrally in the open rhabdom (Figure 6). R7 and R8 cells appear to be specialized for colour vision. The central photoreceptors are functionally subdivided into two pairs which either express the combina-tion Rh3/Rh5 or the combination Rh4/Rh6 (Figure 6). The developmental mechanisms underlying the type of rhodopsin patterning realized here is a topic of current research [100,101].

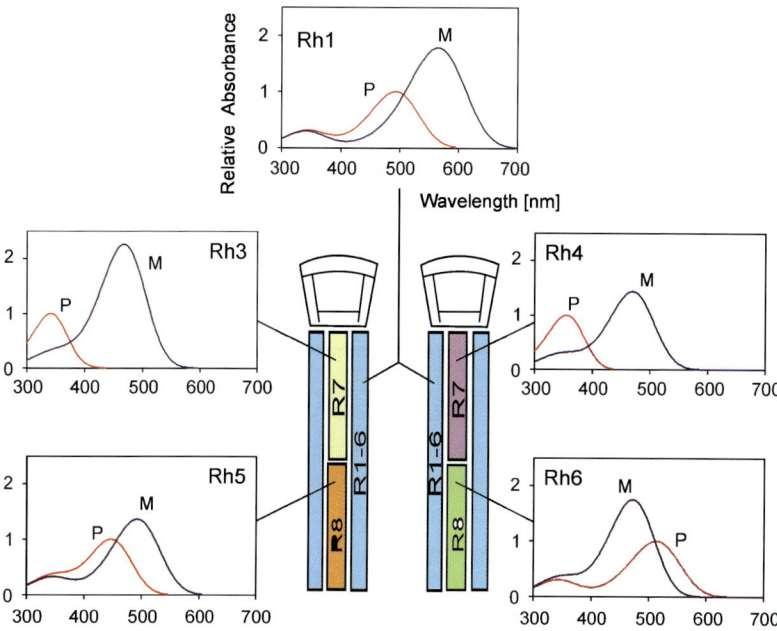

Figure 6. Expression pattern of rhodopsins in the *Drosophila* compound eye. The central panel indicates the expression pattern of the visual pigments in photoreceptor cells of the ommatidium. The absorption spectra of rhodopsins are shown in red, those of metarhodopsins in blue.

It has been already pointed out that the photointermediate sequence initiated by absorption of a photon by an invertebrate rhodopsin molecule terminates in a relatively long-lived metarhodopsin state. For the stable M-state of fly Rh1 [102], as well as for other stable invertebrate metarhodopsins, this state exists in a pH-dependent equilibrium with an alkaline form of metarhodopsin [63,64]. The shift in maximum absorption of acid M (565 nm) to alkaline M (about 380 nm) suggests that the retinylidene Schiff base becomes deprotonated. However, none of the numerous microspectrophotometric measurements performed with different insect species provided any evidence for the *in vivo* formation of the alkaline metarhodopsin form. The intracellular pH, ion composition, protein–protein interactions, etc. apparently favour the formation of stable acid metarhodopsin. Investigations of the transition states, which are assumed by an invertebrate rhodopsin after photon absorption, suggest that primary events in the formation of the photointermediate bathorhodopsin are similar but not identical to those occurring in vertebrate rhodopsins [103]. At later steps the homology in transition states is even less obvious. In cephalopods, a spectrally distinct late intermediate, named mesorhodopsin, is formed prior to the formation of stable acid metarhodopsin; however, the complete sequence of transition states that finally leads to the formation of long-lived metarhodopsin has not been determined. Analysis of time-resolved transient grating signals in *Octopus*, which were shown to represent chromophore-independent protein dynamics, has revealed the transition of the late intermediate mesorhodopsin into a transient form of acid metarhodopsin. Transient acid M itself is transformed into stable acid metarhodopsin in a spectrally silent transition with a time constant of 180 μs [104]. It is assumed that in cephalopods this transient acid M represents the G-protein activating state of rhodopsin [62].

As a result of the formation of stable metarhodopsin, irradiation of invertebrate photoreceptors always establishes an equilibrium between rhodopsin and metarhodopsin. Accordingly, invertebrate rhodopsins are characterized sufficiently only by indicating the absorption of rhodopsin as well as the absorption of its stable M-state. Both rhodopsin states constitute a photoconvertible system in which the amount of rhodopsin/metarhodopsin present is determined by the wavelength of incident light and the individual spectral characteristics of both states [17,63,64]. The absorption properties of the rhodopsin/metarhodopsin systems of *Drosophila* (Figure 7) reveal some principles that hold also for other invertebrate rhodopsins. The absorbance coefficient of metarhodopsin is higher than that of the corresponding rhodopsin by a factor of up to 1.8 [82]. The absorption maxima of metarhodopsins derived from rhodopsins absorbing below 500 nm are bathochromically shifted in relation to the absorption of the corresponing rhodopsins (Rh1, Rh3, Rh4, Rh5). Rhodopsins absorbing maximally at wavelengths above 500 nm, like Rh6, are converted into metarhodopsins with hypsochromically shifted absorption maxima [17]. In some cases, for example in cephalopods, the absorption spectra of rhodopsin and metarhodopsin more or less overlap [66]. Accordingly, two classes of rhodopsin/metarhodopsin systems exist, one in which metarhodopsin can be photoconverted into rhodopsin by 100% and one in which this is not the case. Whether this has consequences for the rhodopsin renewal mechanisms, in which the rhodopsin content of a photoreceptor membrane is maximized light-dependently, is not yet known. There is, however, evidence that the hypsochromically shifted M-state of *Drosophila* Rh6 does not accumulate in an amount comparable to that of the other types of rhodopsin expressed in the *Drosophila* eye [24]. This might indicate that, in the case of Rh6, less efficient photoregeneration is compensated by a more efficient mechanism of rhodopsin turnover or by chemical regeneration of rhodopsin.

2.4.2   Spectral tuning of rhodopsin and metarhodopsin

The rhodopsin/metarhodopsin systems of the compound eye of *Drosophila* and of the ocelli, cover a wavelength range of 350 nm from the red to the UV spectral range (Figure 7). The mechanism underlying spectral tuning of rhodopsins is a highly active research field of functional genomics. So far, this topic has been investigated and discussed primarily on the basis of information on the spectral absorption of vertebrate rhodopsins [29,105]. A primary mode to modulate spectral properties of rhodopsins is seen in the interaction of a

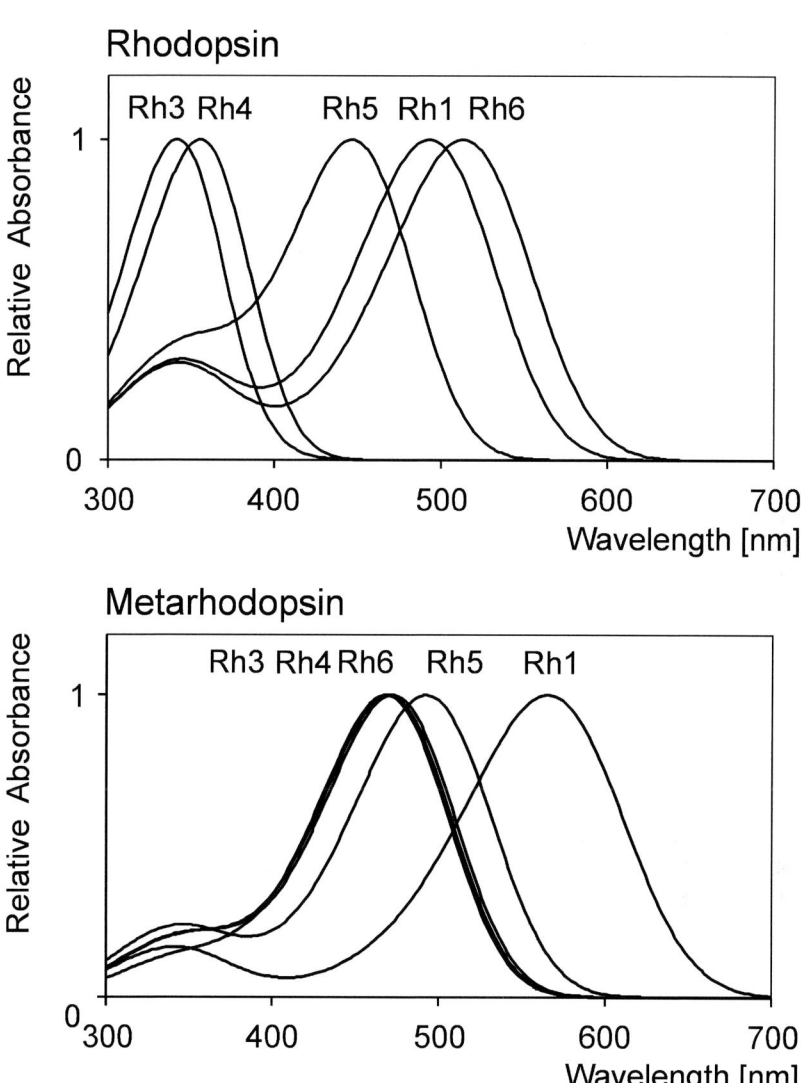

Figure 7. Spectral absorptions of the rhodopsin and metarhodopsin states of visual pigments expressed in the *Drosophila* compound eye. Calculated nomograms of visual pigments. Top: rhodopsin (P-) states, maximum absorption (nm) Rh1 – 493, Rh3 – 341, Rh4 – 358, Rh5 – 492, Rh6 – 513. Bottom: metarhodopsin (M-) states: Rh1 – 565, Rh3 – 468, Rh4 – 470, Rh5 – 492, Rh6 – 472.

negatively charged amino acid side chain, which is properly positioned in the 3D structure of the respective opsin to serve as a counter-ion for the posively charged retinylidene Schiff base. In the absence of a distinct charged counter-ion, the required electron density may be provided by several properly positioned amino acid side chains [14]. Spectral tuning of rhodopsins in invertebrates is a complex challenge as one has not only to explain the bathochromic and hypsochromic shifts in the absorption spectra of rhodopsins but also the absorption shifts that occur upon the formation of stable metarhodopsin states. The rhodopsin systems present in the compound eye of *Drosophila* cover a wavelength range of about 350 nm (Figure 7), while the absorption of the corresponding stable metarhodopsins spreads over about 400 nm from the red to the UV spectral range. Rh2, the rhodopsin expressed in the photoreceptors of ocelli, absorbs light within the same range and is character-ized by a maximum absorption of rhodopsin at 418 nm and of metarhodopsin at 506 nm [43,106,107]. So far, there is no evidence that 7TM proteins of the rhodopsin type are responsible for photon absorption outside the wavelength limits given by the rhodopsins/metarhodopsins expressed in *Drosophila*. Thus, the visual system of *Drosophila* provides the full set of wavelenth regulations realized in other invertebrate visual systems. To obtain information on the mechanism of spectral tuning of rhodopsins in *Drosophila*, Britt et al. [108] used germline transformation to generate transgenic flies that express chimeric rhodopsin molecules. By systematically replacing transmembrane domains of Rh1 with the corresponding regions of Rh2 and vice versa, they were able to shift the spectral properties between the absorption limits given by Rh1 and Rh2, as determined by ERG recordings and microspectrophotometry. Tuning of the native rhodopsins to other wavelengths was only observed with chimeric rhodopsins in which multiple novel transmembrane segments were introduced. The study revealed that the absorption of the rhodopsin and metarhodopsin state is tuned independently, and that spectral tuning of rhodopsin occurs as a coordinated process involving more than one region of opsin.

Octopus rhodopsin is the only invertebrate visual pigment for which highly resolved 3D structures are available, which could help to select amino acids possibly involved in wavelength regulation [74]. However, since *Octopus* is not accessible to mutagenesis studies, one is restricted to the analysis of pri-mary, secondary and tertiary structure data. In an intraspecific comparison of 2D structures, as shown in Figure 2, one expects to find candidates involved in spectral tuning among the non-conserved amino acids located near the chromophore. A more refined selection may be achieved by comparing the structure of phylogenetically closely related insect rhodopsins with similar absorption characteristics. Gärtner [14] has provided such an analysis in which the distinct variations in the primary/secondary structure revealed by sequence alignments of invertebrate rhodopsins were evaluated together with spectral tuning data obtained for vertebrate rhodopsins. This evaluation included results of computer simulations, of studies using polyene model com-pounds, as well as biochemical and biophysical studies of recombinantly expressed rhodopsins. Such comparative studies help to define amino acid residues which possibly account for spectral shifts, e.g. those allowing the detection of UV-light [14]. They do not, however, eliminate the need to test the deduced function of particular amino acids in spectral tuning by heterologous expression of correspondingly mutated rhodopsins.

Variations in the chromophore structure of fly rhodopsins, i.e. exchange of 3-hydroxyretinal versus retinal, do not lead to a significant change in the absorption [109]. However, 4-hydroxyretinal, the chromophore of rhodopsin from the cephalopod *Watsenia scintillans*, may well contribute to a blue-shift in the absorption of this rhodopsin [53]. Instead, flies have invented remarkable alternative mechanisms to optimize the wavelength range for light absorption in a photoreceptor cell. The spectral absorption of the rhodopsin expressed in the photoreceptors R1-6 is enhanced in the UV spectral range by the interaction of rhodopsin with a sensitizing pigment identified as 3-hydroxyretinol [110–112], reviewed in [113]. The interaction is not restricted to Rh1, as other rhodopsins, if ectopically expressed in photoreceptors R1-6 of *Drosophila*, also show this interaction [43]. Analysis of the fine structure of the sensitivity peak emerging in the UV leads to the proposal that the rhodopsin undergoes a rigid interaction with the 13-*cis* isomer of 3-hydroxyretinol [114], the exact binding sites are, however, not yet known. Finally, there remains the possibility that the spectral absorption of photoreceptor cells is broadened by the expression of more than one rhodopsin, as has been reported for the butterfly *Papilio xuthus* [115]. Whether dual expression of rhodopsins is realized in other invertebrates or occurs as the result of a defect in the control of rhodopsin patterning remains to be investigated.

2.5 Signalling pathways coupling to activated invertebrate rhodopsins

2.5.1 *Activation of phototransduction*

In view of the prototypical structure of a GPCR, there is no doubt that in phototransduction the transition of rhodopsin into an active metarhodopsin state is transmitted to downstream components of the phototransduction cascade via interaction with a heterotrimeric G-protein. Retinal-binding proteins which are distantly related to rhodopsin, for example the retinochromes of cephalopods, show an overall topology that is similar to that of rhodopsins. Retinochromes, however, act as photoisomerases. They are integral members of a shuttle system which provides retinal in its 11-*cis* configuration for rhodopsin synthesis and regeneration [116,117]. The non-transducing function of retinochromes is clearly mirrored in the absence of domains for G-protein interaction in the cytoplasmic loops i2 and i3 of this protein [14].

The current state of research suggests that intercalary evolution of rhodopsin–G-protein coupling in invertebrates has assembled at least two distinct types of phototransduction cascades. The first pathway operates in depolarizing, microvillar (rhabdomeral) photoreceptors, the second in hyperpolarizing, ciliary photoreceptors. In microvillar photoreceptors, rhodopsin couples upon light activation to the Gq subtype known to activate phospholipase C (PLCβ) as effector enzyme. The sequential interaction of rhodopsin, Gq and PLC has been firmly established for microvillar photoreceptors of cephalopods,

crustacea and insects [118–134]. The genes encoding the three subunits (*dgq,*
dbe, dge) of the visual Gq-protein of fly photoreceptors [127,135–137] and
cephalopod photoreceptors [138–141] have been cloned and sequenced. A
sequence comparison between the visual G-protein subunits of *Drosophila* and
the subunits of other G-proteins identifies the visual G-protein as Gq subtype
[127,137]. The primary structure of Gqβ bears no function related sequence
conservation [136]. The amino acid sequence of Gqγ, however, reveals a distant relationship to gamma-subunits of vertebrate transducins. Thus, rhodopsin and Gqγ are members of phototransduction pathways that are conserved
irrespective of the photoreceptor cell type and the effector enzyme activated.
Drosophila mutants in the genes coding for the Gqα and Gqβ subunits demonstrate that Gqα is required for the activation of phototransduction while Gqβ
is essential for the interaction of Gqβ with metarhodopsin [142]. The crucial
role of phospholipase C in phototransduction was demonstrated most clearly
in *Drosophila* by isolating the *norpA* gene, which encodes for a phospholipid-
specific phosphodiesterase abundantly expressed in the retina [129,130]. Strong
alleles of *norpA* completely abolish the light response [143,144].

Thus, up to the point of PLC activation, the initial stages of the visual
cascade in microvillar photoreceptors appear to be rather similar. There is,
however, no final answer to the essential question whether the transduction
mechanisms diverge among invertebrates at later stages of the phototransduction cascade. PLC activated by Gq is known to catalyse the hydrolysis of
the membrane lipid phosphatidyl inositol 4,5-bisphosphate (PIP_2) into inositol
1,4,5-trisphosphate (IP_3) and diacylglycerol (DAG) [13,18]. Both IP_3 and DAG
have been implicated in the light-initiated opening of cation channels. There
is evidence indicating that the opening of cation channels in the ventral nerve
photoreceptor of *Limulus* involves an IP_3-mediated calcium release from
IP_3-sensitve calcium stores [145]. The expression of a cyclic GMP gated channel subunit in these cells raises the possibility that cGMP is involved in phototransduction [146]. In *Drosophila*, the opening of the light-activated channel
has been attributed to the messenger functions of DAG and DAG metabolites,
in particular of polyunsaturated fatty acids (PUFAs) [147,148], while PUFAs
appear not to activate phototransduction in *Limulus* [145]. The ion channels
activated in response to a light stimulus have been unequivocally identified in
fly photoreceptors – they are members of the TRP protein family [149–154].

Discussion of the general design and the functional role of individual stages
downstream of Gq activation recently took a new turn after it had been shown
that in fly photoreceptors the proteins involved in phototransduction are
assembled into a supramolecular signalling complex (Figure 8). This complex
is organized by the scaffold protein INAD (inactivation no afterpotential) via
the binding of protein ligands to its PDZ domains [155–158]. The functional
role of the proteins assembled in this complex is indicated by the phototransduction defects caused by mutations of the respective genes. The ligands make
up the *norpA* (*no receptor potential A*) encoded PLC, the *inaC* (*inactivation*
no afterpotential) encoded eye-specific protein kinase C (ePKC) and the *trp*
(*transient receptor potential*) encoded major light-activated cation channel

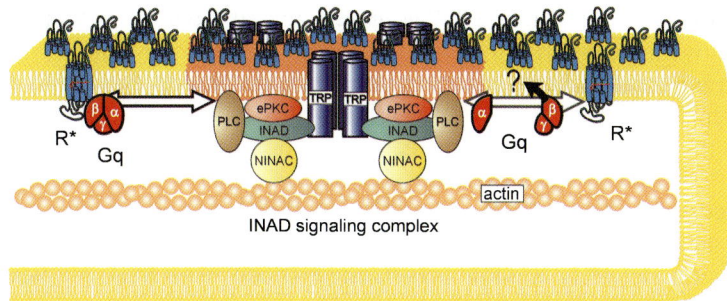

Figure 8. Model for the molecular design of the phototransduction machinery in *Drosophila* photoreceptor cells. A light-activated rhodopsin (R*, left) interacts with the heterotrimeric G-protein, G_q, which results in the dissociation of G_q into $G_{q\alpha}$-GTP and $G_{q\beta\gamma}$ (right). The activated $G_{q\alpha}$ serves as a molecular shuttle which transmits activation of the visual pigment to the target enzyme PLCβ. PLCβ is anchored via the PDZ-domain protein INAD to the so-called INAD signalling complex, which further contains an eye-specific protein kinase C (ePKC) and the ion channel TRP. Through the unconventional myosin NINAC, the signalling complex is anchored to the actin cytoskeleton. One core complex, forming an intact TRP ion channel, is composed of four TRP molecules and the corresponding numbers of INAD, ePKC and PLC. The second messengers, IP_3 and diacylglyerol, generated by the action of PLCβ, are thought to activate and modulate the influx of Ca^{2+} ions through the TRP channels. Model modified from [12] and [163]; see these references for an in-depth discussion.

TRP. INAD has been shown to interact via homomeric interactions with other INAD molecules. In this way the core proteins associated by INAD may become organized in a more extended web [159]. Further transduction-relevant proteins reported to bind to INAD are the unconventional myosin (NINAC, neither inactivation nor afterpotential), the second light-activated channel protein (TRPL, trp-like) and even rhodopsin [159].

Figure 8 summarizes some important features of the structure and function of the INAD signalling complex. INAD, PLC, ePKC and TRP can be reliably isolated together by co-immunoprecipitation experiments in a rather invariable 1:1:1:1 stoichiometry [155]. The pattern of their binding to the five PDZ domains of INAD as summarized by Huber [154] has been evaluated in detail. That these proteins constitute a functional unit is clearly demonstrated by the finding that ePKC present in the isolated complex catalyses the DAG- and calcium-dependent phosphorylation of TRP as well as of INAD [155,160]. They are, therefore, regarded as the core complex of a larger transduction unit (transducisome) [9] which would include about 15–25 TRP channels in tetrameric form. In addition, the complex might be linked to actin filaments located in the microvillar lumen via the binding of the unconventional myosin NINAC to INAD. Direct proof for this interaction has, however, not yet been presented. Calmodulin (CaM), which has been shown to interact with the various CaM binding sites present on the members of the complex [161], is omitted from the scheme. In Figure 8 it is also emphasized that rhodopsin is present in the membrane in a large excess to INAD signalling complexes. Estimates on the basis of a homomeric TRP channel composition suggest that a single microvillus may contain only about 25 INAD signalling complexes [162]. The shuttle function for the information transfer from photon capture by an individual rhodopsin molecule to INAD-linked PLC is attributed to Gqα. Light-dependent as well as light-independent activation of Gq reveals that Gqα but not Gqβγ interacts with INAD-linked PLC [163]. If the hydrolysis of GTP is prevented by replacing GTP with GTP-γ-S, the INAD-linked PLC molecules form a stable high-affinity complex with Gqα [163]. There is, however, no evidence that under these conditions (activated) rhodopsin is also complexed with INAD. The effect of the activation of INAD-linked PLC on the inositol phospholipid composition in the vicinity of the complex has not yet been explored. The activated membrane patch in Figure 8 indicates that activation of PLC and the localized ion influx through INAD-linked TRP channels are likely to create transient inhomogeneities in the inositol lipid derived messengers as well as in Ca^{2+}, which would have consequences for excitation as well as termination of the visual response [164].

Some of the implications of assembling key members of the phototransduction cascade into a supramolecular complex are: the members of the core complex become correctly targeted in a defined stoichiometry to the microvillar photoreceptor membrane and remain retained in this compartment [165–167]. Furthermore, the close proximity of proteins involved in the generation and control of the visual response appears to ensure signal amplification, high specificity and sensitivity of signalling as well as high speed of signalling, i.e. short latencies for activation and termination of light responses [9,18,157]. It has been concluded that a single complex (transducisome) represents the functional unit for the generation of a quantum bump, the response of which is elicited by absorption of a single photon at a rhodopsin molecule [9]. Whether the organization principle of phototransduction in fly photoreceptors also holds for other microvillar photoreceptor systems is not yet known. It is to be expected that signalling complexes constitute the basis for high speed signalling primarily in the photoreceptors of insects. Thus, in addition to modifications in the type of second messenger there may be another point of divergence in invertebrate phototransduction mechanisms, which concerns the organization of signalling cascade components in heteromultimeric protein complexes.

Invertebrates have developed at least one rhodopsin-activated phototransduction pathway which does not involve the activation of PLC. Such a pathway is realized in hyperpolarizing ciliary photoreceptors of molluscs, such as the scallop. Cumulative evidence suggests that light-activated rhodopsin here couples to a Go subtype [34] which in turn activates guanylate cyclase as effector enzyme [168]. In fact hyperpolarization of these ciliary photoreceptors has been shown not to result from the blocking of cGMP gated ion channels as in vertebrate photoreceptors but from the opening K^+ selective ion channel by cyclic GMP [169]. Thus, as in vertebrate photoreceptors, the light-activated conductance is controlled by cyclic GMP (see Chapter 3), but the activation parameters for the key enzymes involved in the control of cyclic GMP levels, cGMP phosphodiesterase and guanylate cyclase, have been reverted.

2.5.2   Termination of phototransduction

Equally important as the activation of phototransduction is the efficient inactivation of each step of the transduction cascade. Only highly effective inactivation mechanisms enable a transduction cascade to repeatedly transmit information with high temporal fidelity. The necessity for an effective deactivation of the active rhodopsin state is particularly evident in invertebrate photoreceptors in which light-absorption triggers the formation of a long-lived active metarhodopsin state [63]. Most of the information on deactivation of active metarhodopsin comes from studies on dipteran flies (*Drosophila*, *Calliphora*), from which two genes encoding visual system-specifically expressed arrestins, arrestin 1 (Arr1) and arrestin 2 (Arr2), have been isolated [170–173]. Analysis of *arr1*- and *arr2*-mutants in *Drosophila* showed that both arrestin isoforms contribute to the termination of the phototransduction cascade [174]. Arr2, the major arrestin form present in the photoreceptor cell, binds light-dependently with high affinity to metarhodopsin (Figure 9) [73,95,96,175]. This interaction constitutes the rate-limiting step in the overall termination of the light response [174]. Wild-type *Drosophila* contain more rhodopsin molecules than Arr2 molecules. Therefore, irradiation of Rh1-containing photoreceptors with blue light, which shifts about 70% of Rh1 rhodopsin into metarhodopsin (Figure 9), creates a large excess of activated metarhodopsin over Arr2. Under these circumstances, i.e. in a situation unlikely to occur under normal light conditions, the photoreceptor cell generates a sustained electrical response. This prolonged depolarizing afterpotential (PDA) terminates very slowly after the cessation of light, unless metarhodopsin is photoconverted back into rhodopsin [64,176–179] (Figure 9). The conversion of about 20% of Rh1 rhodopsin present in the fly photoreceptor into metarhodopsin is sufficient to elicit a PDA. It is thus concluded that the molar ratio of Rh1 to Arr2 within a photoreceptor cell is around 5 : 1 [178]. In flies, the PDA may last from minutes to hours, which indicates that the stable M-state itself and not a transient intermediate has the ability to excite the photoreceptor for a rather long time.

Both arrestins are phosphorylated light-dependently by a Ca^{2+}/calmodulin-dependent kinase (CaM Kinase) [173,180–182]. For Arr2, this phosphorylation was shown to be a prerequisite for its release from the visual pigment after photoreconversion of M into P [175]. But, as indicated in Figure 5, in a distinct difference to the arrestin-mediated deactivation of vertebrate rhodopsin, phosphorylation of metarhodopsin in flies is not a prerequisite for binding of Arr2 [8,95,98]. It seems rather that binding of Arr2, similar to β-arrestin binding to β-adrenergic receptors [183–185], is part of a recruitment mechanism for rhodopsin internalization allow metarhodopsin to become phosphorylated by a rhodopsin kinase [95] and preventing metarhodopsin dephosphorylation by the receptor phosphatase RDGC [96,98,186] (see also Section 2.3.2). In a second distinct difference to the situation in vertebrate photoreceptors, phosphorylation of metarhodopsin, at least in flies, is not linked to receptor inactivation [187]. Since visual arrestins have also been isolated from

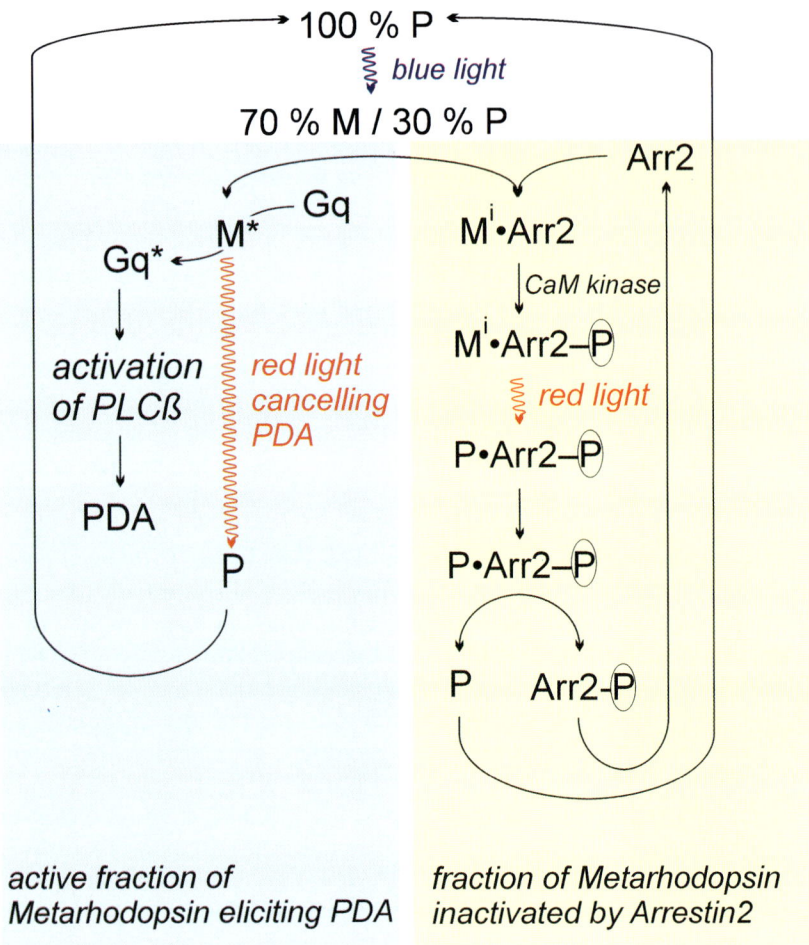

Figure 9. Inactivation of rhodopsin, and generation of the prolonged depolarizing afterpotential (PDA). Irradiation of fly (*Drosophila*, *Calliphora*) Rh1 containing photoreceptor cells with blue light establishes a photoequilibrium containing about 70% metarhodopsin (M) and 30% rhodopsin (P). Due to the limited amount of Arrestin 2 (Arr2), only 30% of the total M can be inactivated (M^i) by binding of Arr2 (yellow field); 70% M remains in the active conformation (M*) and gives rise to a sustained electrical response of the receptor cell (PDA) through activation of the visual G-protein (Gq), followed by activation of phospholipase Cβ (PLCβ) (blue field). Both, M* and M^i, are reconverted into (inactive) P by irradiation with red light. Following photoreconversion of M into P, Arr2 can only be released from the visual pigment if it is in the phosphorylated state (Arr2-P). Phosphorylation of Arr2 is a light-dependent reaction catalysed by Ca^{2+}/calmodulin-dependent kinase (CaM Kinase). The sequence of Arr2 binding and Arr2 phosphorylation, as indicated here, is hypothetical. The Arr2 phosphatase has not been identified.

invertebrate species other than flies (*Loligo pelagi* [Mayeenuddin, unpublished], *Limulus polyphemus* [188], *Ascalaphus macaronius* [189]), arrestin-mediated metarhodopsin inactivation seems to be a general mechanism for turning off activated metarhodopsin in invertebrate photoreceptors.

A deactivation mechanism of phototransduction is also required for the active Gqα-subunit, at the next stage of the cascade. Generally the inactive, GDP bound state of G-proteins is restored by the intrinsic GTPase activity of Gqα. Experiments directed to investigate Gqα interaction with INAD linked PLC (Figure 8) indicated that Gqα does not reach this state unless it has interacted with PLC. This suggested that INAD-linked PLC has the function of a GTPase activating protein (GAP) [163], as has been previously shown for other phospholipase C types [190,191]. Analyses of the response termination in *inaD* and *norpA* mutants of *Drosophila* not only revealed that INAD-linked PLC has the dual role of an effector enzyme of the phototransduction cascade and of a negative regulator of G-protein activity, they also demonstrated that the occupance of a sufficient number of binding sites on the INAD signalling complex is a prerequisite for the high temporal and intensity resolution of visual responses [192].

Acknowledgements

We thank Dr Armin Huber for critically reading the manuscript.

References

1. G. Halder, P. Callaerts, W.J. Gehring (1995). Induction of ectopic eyes by targeted expression of the eyeless gene in Drosophila. *Science*, **267**, 1788–1792.
2. P. Callaerts, G. Halder, W.J. Gehring (1997). PAX-6 in development and evolution. *Annu. Rev. Neurosci.*, **20**, 483–532.
3. W.J. Gehring, K. Ikeo (1997). Pax 6: mastering eye morphogenesis and eye evolution. *Trends. Genet.*, **15**, 371–377.
4. W.J. Gehring (2002). The genetic control of eye development and its implications for the evolution of the various eye-types. *Int. J. Dev. Biol.*, **46**, 65–73.
5. T.W. Cronin, M. Jarvilehto, M. Weckstrom, A.B. Lall (2000). Tuning of photoreceptor spectral sensitivity in fireflies (Coleoptera: Lampyridae). *J. Comp. Physiol. [A]*, **186**, 1–12.
6. W.L. Pak (1995). Drosophila in vision research. The *Friedenwald* Lecture. *Invest. Ophthalmol. Vis. Sci.*, **36**, 2340–2357.
7. R. Ranganathan, D.M. Malicki, C.S. Zuker (1995). Signal transduction in Drosophila photoreceptors. *Annu. Rev. Neurosci.*, **18**, 283–317.
8. K. Scott, C. Zuker (1997). Lights out: deactivation of the phototransduction cascade. *Trends Biochem. Sci.*, **22**, 350–354.
9. S. Tsunoda, C.S. Zuker (1999). The organization of INAD-signaling complexes by a multivalent PDZ domain protein in *Drosophila* photoreceptor cells ensures sensitivity and speed of signaling. *Cell Calcium*, **26**, 165–171.

10. C. Montell (2000). Regulation of *Drosophila* visual transduction through a supramolecular signaling complex. In: P.M. Conn, A.R. Means (Eds), *Principles of Molecular Regulation* (pp. 85–97). Humana Press, Totowa, N.J.

11. R. Paulsen, M. Bähner, A. Huber, M. Schillo, S. Schulz, R. Wottrich, J. Bentrop (2001). The molecular design of a visual cascade: Molecular stages of phototransduction in Drosophila. In: C. Musio (Ed.), *Vision: The Approach of Biophysics and Neurosciences* (pp. 41–59). World Scientific, Singapore.

12. R. Paulsen, M. Bähner, J. Bentrop, M. Schillo, S. Schulz, A. Huber (2001). The molecular design of a visual cascade: Assembly of the Drosophila phototransduction pathway into a supramolecular signaling complex. In: C. Musio (Ed.), *Vision: The Approach of Biophysics and Neurosciences* (pp. 60–73). World Scientific, Singapore.

13. R.C. Hardie, P. Raghu (2001).Visual transduction in *Drosophila*. *Nature*, 186–193.

14. W. Gärtner (2000). Invertebrate visual pigments. In: D.G. Stavenga, W.J. de Grip, E.N.J. Pugh (Eds), *Handbook of Biological Physics* (pp. 297–388). Elsevier Science B.V., Amsterdam.

15. I.M. Pepe (2001). Recent advances in our understanding of rhodopsin and phototransduction. *Prog. Retin. Eye Res.*, **20**, 733–759.

16. B. Minke, R.C. Hardie (2000). Genetic Dissection of *Drosophila* Phototransduction. In: D.G. Stavenga, W.J. DeGrip, E.N.J. Pugh (Eds), *Handbook of Biological Physics* (pp. 449–525). Elsevier Science B.V., Amsterdam.

17. D.G. Stavenga, J. Oberwinkler, M. Postma (2000). Modeling Primary Visual Processes in Insect Photoreceptors. In: D.G. Stavenga, W.J. DeGrip, E.N.J. Pugh (Eds), *Handbook of Biological Physics* (pp. 527–574). Elsevier Science B.V., Amsterdam, London, New York, Oxford, Paris, Shannon, Tokyo.

18. R.C. Hardie (2001). Phototransduction in *Drosophila melanogaster*. *J. Exp. Biol.*, **204**, 3403–3409.

19. T.P. Sakmar (1998). Rhodopsin: a prototypical G protein-coupled receptor. *Prog. Nucleic. Acid. Res. Mol. Biol.*, **59**, 1–34.

20. J.S. Ascher, B.N. Danforth, S. Ji (2001). Phylogenetic utility of the major opsin in bees (Hymenoptera: Apoidea): a reassessment. *Mol. Phylogenet. Evol.*, **19**, 76–93.

21. T. Kusakabe, R. Kusakabe, I. Kawakami, Y. Satou, N. Satoh, M. Tsuda (2001). Ci-opsin1, a vertebrate-type opsin gene, expressed in the larval ocellus of the ascidian Ciona intestinalis. *FEBS Lett.*, **506**, 69–72.

22. T. Kusakabe, R. Yoshida, I. Kawakami, R. Kusakabe, Y. Mochizuki, L. Yamada, Shin, Y. Kohara, N. Satoh, et al., (2002). Gene expression profiles in tadpole larvae of Ciona intestinalis. *Dev. Biol.*, **242**, 188–203.

23. W.H. Chou, A. Huber, J. Bentrop, S. Schulz, K. Schwab, L.V. Chadwell, R. Paulsen, S.G. Britt (1999). Patterning of the R7 and R8 photoreceptor cells of *Drosophila*: evidence for induced and default cell-fate specification. *Development*, **126**, 607–616.

24. E. Salcedo, A. Huber, S. Henrich, L.V. Chadwell, W.H. Chou, R. Paulsen, S.G. Britt (1999). Blue-and green-absorbing visual pigments of Drosophila: ectopic expression and physiological characterization of the R8 photoreceptor cell-specific Rh5 and Rh6 rhodopsins. *J. Neurosci.*, **19**, 10716–10726.

25. M.R. Chase, R.R. Bennett, R.H. White (1997). Three opsin-encoding cDNAS from the compound eye of Manduca sexta. *J. Exp. Biol.*, **200**, 2469–2478.

26. A.D. Briscoe, L. Chittka (2001). The evolution of color vision in insects. *Annu. Rev. Entomol.*, **46**, 471–510.

27. F. Pichaud, A. Briscoe, C. Desplan (1999). Evolution of color vision. *Curr. Opin. Neurobiol.*, **9**, 622–627.

28. S. Yokoyama, Y. Shi (2000). Genetics and evolution of ultraviolet vision in vertebrates. *FEBS Lett.*, **486**, 167–172.

29. S. Yokoyama (2000). Molecular evolution of vertebrate visual pigments. *Prog. Retin. Eye Res.*, **19**, 385–419.

30. Y. Shi, F.B. Radlwimmer, S. Yokoyama (2001). Molecular genetics and the evolution of ultraviolet vision in vertebrates. *Proc. Natl. Acad .Sci. U.S.A.*, **98**, 11731–11736.

31. D. Kojima, A. Terakita, T. Ishikawa, Y. Tsukahara, A. Maeda, Y. Shichida (1997). A novel Go-mediated phototransduction cascade in scallop visual cells. *J. Biol. Chem.*, **272**, 22979–22982.

32. M.D. Adams, S.E. Celniker, R.A. Holt, C.A. Evans, J.D. Gocayne, P.G. Amanatides, S.E. Scherer, P.W. Li, R.A. Hoskins, et al., (2000). The genome sequence of *Drosophila melanogaster*. *Science*, **287**, 2185–2195.

33. J.P. Carulli, D.M. Chen, W.S. Stark, D.L. Hartl (1994). Phylogeny and physiology of *Drosophila* opsins. *J. Mol. Evol.*, **38**, 250–262.

34. D. Kojima, A. Terakita, T. Ishikawa, Y. Tsukahara, A. Maeda, Y. Shichida (1997). A novel Go-mediated phototransduction cascade in scallop visual cells. *J. Biol. Chem.*, **272**, 22979–22982.

35. J.E. O'Tousa, W. Baehr, R.L. Martin, J. Hirsh, W.L. Pak, M.L. Applebury (1985). The *Drosophila* ninaE gene encodes an opsin. *Cell*, **40**, 839–850.

36. C.S. Zuker, A.F. Cowman, G.M. Rubin (1985). Isolation and structure of a rhodopsin gene from *D. melanogaster*. *Cell*, **40**, 851–858.

37. C.S. Zuker, C. Montell, K. Jones, T. Laverty, G.M. Rubin (1987). A rhodopsin gene expressed in photoreceptor cell R7 of the *Drosophila* eye: homologies with other signal-transducing molecules. *J. Neurosci.*, **7**, 1550–1557.

38. C. Montell, K. Jones, C. Zuker, G. Rubin (1987). A second opsin gene expressed in the ultraviolet-sensitive R7 photoreceptor cells of *Drosophila melanogaster*. *J. Neurosci.*, **7**, 1558–1566.

39. W.H. Chou, K.J. Hall, D.B. Wilson, C.L. Wideman, S.M. Townson, L.V. Chadwell, S.G. Britt (1996). Identification of a novel Drosophila opsin reveals specific patterning of the R7 and R8 photoreceptor cells. *Neuron*, **17**, 1101–1115.

40. D. Papatsenko, G. Sheng, C. Desplan (1997). A new rhodopsin in R8 photo-receptors of *Drosophila*: evidence for coordinate expression with Rh3 in R7 cells. *Development*, **124**, 1665–1673.

41. A. Huber, S. Schulz, J. Bentrop, C. Groell, U. Wolfrum, R. Paulsen (1997). Molecular cloning of *Drosophila* Rh6 rhodopsin: the visual pigment of a subset of R8 photoreceptor cells. *FEBS Lett.*, **406**, 6–10.

42. J.A. Pollock, S. Benzer (1988). Transcript localization of four opsin genes in the three visual organs of *Drosophila*; RH2 is ocellus specific. *Nature*, **333**, 779–782.

43. E. Salcedo, A. Huber, S. Henrich, L.V. Chadwell, W.H. Chou, R. Paulsen, S.G. Britt (1999). Ectopic expression and physiological characterization of the R8 photoreceptor cell-specific Rh5 and Rh6 rhodopsins of *Drosophila*. *J. Neurosci.*, **24**, 10716–10726.

44. A. Engels, H. Reichert, W.J. Gehring, W. Gartner (2000). Functional expression of a locust visual pigment in transgenic *Drosophila* melanogaster. *Eur. J. Biochem.*, **267**, 1917–1922.

45. S.M. Townson, B.S. Chang, E. Salcedo, L.V. Chadwell, N.E. Pierce, S.G. Britt (1998). Honeybee blue- and ultraviolet-sensitive opsins: cloning, heterologous expression in *Drosophila*, and physiological characterization. *J. Neurosci.*, **18**, 2412–2422.

46. H.Y. Chang, D.F. Ready (2000). Rescue of photoreceptor degeneration in rhodopsin-null *Drosophila* mutants by activated rac1. *Science*, **290**, 1978–1980.

47. J. Bentrop, K. Schwab, W.L. Pak, R. Paulsen (1997). Site-directed mutagenesis of highly conserved amino acids in the first cytoplasmic loop of *Drosophila* Rh1 opsin blocks rhodopsin synthesis in the nascent state. *EMBO J.*, **16**, 1600–1609.

48. T. Washburn, J.E. O'Tousa (1989). Molecular defects in *Drosophila* rhodopsin mutants. *J. Biol. Chem.*, **264**, 15464–15466.

49. J. Bentrop (1998). Rhodopsin mutations as the cause of retinal degeneration. Classification of degeneration phenotypes in the model system *Drosophila melanogaster*. *Acta Anat.*, **162**, 85–94.

50. R. Hubbard, R.C.C. StGeorge (1958). The rhodopsin system of the squid. *J. Gen. Physiol.*, **41**, 501–528.

51. R. Paulsen, J. Schwemer (1983). Biogenesis of blowfly photoreceptor membranes is regulated by 11-cis-retinal. *Eur. J. Biochem.*, **137**, 609–614.

52. K. Vogt (1983). Is the fly visual pigment a rhodopsin? *Z. Naturforsch. [C]*, **38**, 329–333.

53. S. Matsui, M. Seidou, I. Uchiyama, N. Sekiya, K. Hiraki, K. Yoshihara, Y. Kito (1988). 4-Hydroxyretinal, a new visual pigment chromophore found in the bioluminescent squid, *Watasenia scintillans*. *Biochim. Biophys. Acta*, **966**, 370–374.

54. T. Kitagawa, M. Tsuda (1980). Resonance Raman spectra of octopus acid and alkaline metarhodopsins. *Biochim. Biophys. Acta*, **624**, 211–217.

55. C. Pande, A. Pande, K.T. Yue, R. Callender, T.G. Ebrey, M. Tsuda (1987). Resonance Raman spectroscopy of octopus rhodopsin and its photoproducts. *Biochemistry*, **26**, 4941–4947.

56. C. Pande, H. Deng, P. Rath, R.H. Callender, J. Schwemer (1987). Resonance raman spectroscopy of an ultraviolet-sensitive insect rhodopsin. *Biochemistry*, **26**, 7426–7430.

57. S. Nishimura, H. Kandori, M. Nakagawa, M. Tsuda, A. Maeda (1997). Structural dynamics of water and the peptide backbone around the Schiff base associated with the light-activated process of octopus rhodopsin. *Biochemistry*, **36**, 864–870.

58. B.W. Vought, E. Salcedo, L.V. Chadwell, S.G. Britt, R.R. Birge, B.E. Knox (2000). Characterization of the primary photointermediates of *Drosophila* rhodopsin. *Biochemistry*, **39**, 14128–14137.

59. E.A. Zhukovsky, D.D. Oprian (1989). Effect of carboxylic acid side chains on the absorption maximum of visual pigments. *Science*, **246**, 928–930.

60. T.P. Sakmar, R.R. Franke, H.G. Khorana (1989). Glutamic acid-113 serves as the retinylidene Schiff base counterion in bovine rhodopsin. *Proc. Natl. Acad. Sci. U.S.A.*, **86**, 8309–8313.

61. J. Nathans (1990). Determinants of visual pigment absorbance: identification of the retinylidene Schiff's base counterion in bovine rhodopsin. *Biochemistry*, **29**, 9746–9752.

62. M. Nakagawa, T. Iwasa, S. Kikkawa, M. Tsuda, T.G. Ebrey (1999). How vertebrate and invertebrate visual pigments differ in their mechanism of photoactivation. *Proc. Natl. Acad. Sci. U.S.A.*, **96**, 6189–6192.

63. K. Hamdorf, R. Paulsen, J. Schwemer (1973). Photoregeneration and sensitivity control of photoreceptors in invertebrates. In: H. Langer (Ed.), *Biochemistry and Physiology of Visual Pigments* (pp. 155–166). Springer Verlag, Berlin.

64. K. Hamdorf (1979). The Physiology of Invertebrate Visual Pigments. In: H. Autrum (Ed.), *Handbook of Sensory Physiology*, (pp. 145–224). Springer-Verlag, Berlin.

65. P. Hillman, S. Hochstein, B. Minke (1983). Transduction in invertebrate photoreceptors: role of pigment bistability. *Physiol. Rev.*, **63**, 668–772.

66. D.G. Stavenga, J. Schwemer (1984). Visual Pigments of Invertebrates. In: M.A. Ali (Ed.), *Photoreception and Vision in Invertebrates* (pp. 11–61). Plenum Publishing, New York.

67. K. Ozaki, H. Nagatani, M. Ozaki, F. Tokunaga (1993). Maturation of major *Drosophila* rhodopsin, ninaE, requires chromophore 3-hydroxyretinal. *Neuron*, **10**, 1113–1119.

68. A. Huber, U. Wolfrum, R. Paulsen (1994). Opsin maturation and targeting to rhabdomeral photoreceptor membranes requires the retinal chromophore. *Eur. J. Cell Biol.*, **63**, 219–229.

69. J. Schwemer (1984). Renewal of visual pigment in photoreceptors of blowfly. *J. Comp. Physiol. [A]*, **154**, 535–547.

70. P.G. Alloway, L. Howard, P.J. Dolph (2000). The formation of stable rhodopsin-arrestin complexes induces apoptosis and photoreceptor cell degeneration. *Neuron*, **28**, 129–138.

71. T.G. Ebrey (2000). pKa of the protonated Schiff base of visual pigments. *Methods Enzymol.*, **315**, 196–207.

72. T. Okada, O.P. Ernst, K. Palczewski, K.P. Hofmann (2001). Activation of rhodopsin: new insights from structural and biochemical studies. *Trends. Biochem. Sci.*, **26**, 318–324.

73. A. Kiselev, S. Subramaniam (1994). Activation and regeneration of rhodopsin in the insect visual cycle. *Science*, **266**, 1369–1373.

74. A. Davies, B.E. Gowen, A.M. Krebs, G.F. Schertler, H.R. Saibil (2001). Three-dimensional structure of an invertebrate rhodopsin and basis for ordered alignment in the photoreceptor membrane. *J. Mol. Biol.*, **314**, 455–463.

75. C. Venien-Bryan, A. Davies, K. Langmack, J. Baverstock, A. Watts, D. Marsh, H. Saibil (1995). Effect of the C-terminal proline repeats on ordered packing of squid rhodopsin and its mobility in membranes. *FEBS Lett.*, **359**, 45–49.

76. J.M. Baldwin, G.F. Schertler, V.M. Unger (1997). An alpha-carbon template for the transmembrane helices in the rhodopsin family of G-protein-coupled receptors. *J. Mol. Biol.*, **272**, 144–164.

77. K. Palczewski, T. Kumasaka, T. Hori, C.A. Behnke, H. Motoshima, B.A. Fox, T. Le, I, D.C. Teller, T. Okada, et al. (2000). Crystal structure of rhodopsin: A G protein-coupled receptor. *Science*, **289**, 739–745.

78. J.S. Lott, J.I. Wilde, A. Carne, N. Evans, J.B. Findlay (1999). The ordered visual transduction complex of the squid photoreceptor membrane. *Mol. Neurobiol.*, **20** 61–80.

79. Y. Ovchinnikov, N.G. Abdulaev, A.S. Zolotarev, I.D. Artamonov, I.A. Bespalov, A.E. Dergachev, M. Tsuda (1988). Octopus rhodopsin. Amino acid sequence deduced from cDNA. *FEBS Lett.*, **232**, 69–72.

80. M. Nakagawa, T. Iwasa, S. Kikkawa, T. Takao, Y. Shimonishi, M. Tsuda (1997). Identification of two palmitoyl groups in octopus rhodopsin. *Photochem. Photobiol.*, **65**, 187–191.

81. Y. Zhang, T. Iwasa, M. Tsuda, A. Kobata, S. Takasaki (1997). A novel mono-antennary complex-type sugar chain found in octopus rhodopsin: occurrence of the Galβ1–4Fuc group linked to the proximal N-acetylamine residue of the trimannosyl core. *Glycobiology*, **7**, 1153–1158.

82. A. Huber, D.P. Smith, C.S. Zuker, R. Paulsen (1990). Opsin of *Calliphora* peripheral photoreceptors R1–6. Homology with *Drosophila* Rh1 and posttranslational processing. *J. Biol. Chem.*, **265**, 17906–17910.

83. K. Katanosaka, F. Tokunaga, S. Kawamura, K. Ozaki (1998). N-linked glycosylation of *Drosophila* rhodopsin occurs exclusively in the amino-terminal domain and functions in rhodopsin maturation. *FEBS Lett.*, **424**, 149–154.

84. J.E. O'Tousa (1992). Requirement of N-linked glycosylation site in *Drosophila* rhodopsin. *Vis. Neurosci.*, **8**, 385–390.

85. E.K. Baker, N.J. Colley, C.S. Zuker (1994). The cyclophilin homolog NinaA functions as a chaperone, forming a stable complex in vivo with its protein target rhodopsin. *EMBO J.*, **13**, 4886–4895.

86. S. Schneuwly, R.D. Shortridge, D.C. Larrivee, T. Ono, M. Ozaki, W.L. Pak (1989). *Drosophila* ninaA gene encodes an eye-specific cyclophilin (cyclosporine A binding protein). *Proc. Natl. Acad. Sci. U.S.A.*, **86**, 5390–5394.

87. M. Nakagawa, T. Miyamoto, R. Kusakabe, S. Takasaki, T. Takao, Y. Shichida, M. Tsuda (2001). O-Glycosylation of G-protein-coupled receptor, octopus rhodopsin. Direct analysis by FAB mass spectrometry. *FEBS Lett.*, **496**, 19–24.

88. H. Kühn, W.J. Dreyer (1972). Light dependent phosphorylation of rhodopsin by ATP. *FEBS Lett.*, **20**, 1–6.

89. D. Bownds, J. Dawes, J. Miller, M. Stahlman (1972). Phosphorylation of frog photoreceptor membranes induced by light. *Nat. New Biol.*, **237**, 125–127.

90. R.N. Frank, H.D. Cavanagh, K.R. Kenyon (1973). Light-stimulated phosphorylation of bovine visual pigments by adenosine triphosphate. *J. Biol. Chem.*, **248**, 596–609.

91. R. Paulsen, I. Hoppe (1978). Light-activated phosphorylation of cephalopod rhodopsin. *FEBS Lett.*, **96**, 55–58.

92. M. Tsuda, T. Tsuda, H. Hirata (1989). Cyclic nucleotides and GTP analogues stimulate light-induced phosphorylation of octopus rhodopsin. *FEBS Lett.*, **257**, 38–40.

93. R. Paulsen, J. Bentrop (1984). Reversible phosphorylation of opsin induced by irradiation of blowfly retinae. *J. Comp. Physiol. [A]*, **155**, 39–45.

94. J. Bentrop, R. Paulsen (1986). Light-modulated ADP-ribosylation, protein phosphorylation and protein binding in isolated fly photoreceptor membranes. *Eur. J. Biochem.*, **161**, 61–67.

95. J. Bentrop, A. Plangger, R. Paulsen (1993). An arrestin homolog of blowfly photoreceptors stimulates visual-pigment phosphorylation by activating a membrane-associated protein kinase. *Eur. J. Biochem.*, **216**, 67–73.

96. A. Plangger, D. Malicki, M. Whitney, R. Paulsen (1994). Mechanism of arrestin 2 function in rhabdomeric photoreceptors. *J. Biol. Chem.*, **269**, 26969–26975.

97. H. Ohguro, N. Yoshida, H. Shindou, J.W. Crabb, K. Palczewski, M. Tsuda (1998). Identification of a single phosphorylation site within octopus rhodopsin. *Photochem. Photobiol.*, **68**, 824–828.

98. A. Kiselev, M. Socolich, J. Vinos, R.W. Hardy, C.S. Zuker, R. Ranganathan (2000). A molecular pathway for light-dependent photoreceptor apoptosis in *Drosophila*. *Neuron*, **28**, 139–152.

99. J. Nguyen-Legros, D. Hicks (2000). Renewal of photoreceptor outer segments and their phagocytosis by the retinal pigment epithelium. *Int. Rev. Cytol.*, **196**, 245–313.

100. F. Pichaud, C. Desplan (2001). A new visualization approach for identifying mutations that affect differentiation and organization of the Drosophila ommatidia. *Development*, **128**, 815–826.

101. F. Pichaud, J. Treisman, C. Desplan (2001). Reinventing a common strategy for patterning the eye. *Cell*, **105**, 9–12.

102. R. Paulsen (1984). Spectral characteristics of isolated blowfly rhabdoms. *J. Comp. Physiol. [A]*, **155**, 47–55.
103. L. Huang, H. Deng, Y. Koutalos, T. Ebrey, M. Groesbeek, J. Lugtenburg, M. Tsuda, R.H. Callender (1997). A resonance Raman study of the C=C stretch modes in bovine and octopus visual pigments with isotopically labeled retinal chromophores. *Photochem. Photobiol.*, **66**, 747–754.
104. Y. Nishioku, M. Nakagawa, M. Tsuda, M. Terazima (2001). A spectrally silent transformation in the photolysis of octopus rhodopsin: a protein conformational change without any accompanying change of the chromophore's absorption. *Biophys. J.*, **80**, 2922–2927.
105. G.G. Kochendoerfer, S.W. Lin, T.P. Sakmar, R.A. Mathies (1999). How color visual pigments are tuned. *Trends. Biochem. Sci.*, **24**, 300–305.
106. R. Feiler, W.A. Harris, K. Kirschfeld, C. Wehrhahn, C.S. Zuker (1988). Targeted misexpression of a *Drosophila* opsin gene leads to altered visual function. *Nature*, **333**, 737–741.
107. R. Feiler, R. Bjornson, K. Kirschfeld, D. Mismer, G.M. Rubin, D.P. Smith, M. Socolich, C.S. Zuker (1992). Ectopic expression of ultraviolet-rhodopsins in the blue photoreceptor cells of *Drosophila*: visual physiology and photochemistry of transgenic animals. *J. Neurosci.*, **12**, 3862–3868.
108. S.G. Britt, R. Feiler, K. Kirschfeld, C.S. Zuker (1993). Spectral tuning of rhodopsin and metarhodopsin in vivo. *Neuron*, **11**, 29–39.
109. W. Gärtner, D. Ullrich, K. Vogt (1991). Quantum yield of CHAPSO-solubilized rhodopsin and 3-hydroxy retinal containing bovine opsin. *Photochem. Photobiol.*, **54**, 1047–1055.
110. K. Kirschfeld, N. Franceschini (1977). Evidence for a sensitising pigment in fly photoreceptors. *Nature*, **269**, 386–390.
111. B. Minke, K. Kirschfeld (1979). The contribution of a sensitizing pigment to the photosensitivity spectra of fly rhodopsin and metarhodopsin. *J. Gen. Physiol.*, **73**, 517–540.
112. K. Vogt, K. Kirschfeld (1983). Sensitizing pigment in the fly. *Biophys. Struct. Mech.*, **9**, 319–328.
113. K. Kirschfeld (1986). Activation of Visual Pigment: Chromophore Structure and Function, In: H. Stieve (Ed), *The Molecular Mechanism of Photoreception* (pp. 31–49). Springer-Verlag, Berlin.
114. K. Hamdorf, P. Hochstrate, G. Höglund, M. Moser, S. Sperber, P. Schlecht (1992). Ultra-violet sensitizing pigment in blowfly photoreceptors R1–6: probable nature and binding sites. *J. Comp. Physiol. [A]*, **171**, 601–615.
115. J. Kitamoto, K. Sakamoto, K. Ozaki, Y. Mishina, K. Arikawa (1998). Two visual pigments in a single photoreceptor cell: identification and histological localization of three mRNAs encoding visual pigment opsins in the retina of the butterfly *Papilio xuthus*. *J. Exp. Biol.*, **201**, 1255–1261.
116. S.L. Fong, P.G. Lee, K. Ozaki, R. Hara, T. Hara, C.D. Bridges (1988). IRBP-like proteins in the eyes of six cephalopod species–immunochemical relationship to vertebrate interstitial retinol-binding protein (IRBP) and cephalopod retinal-binding protein. *Vision Res.*, **28**, 563–573.
117. A. Terakita, R. Hara, T. Hara (1989). Retinal-binding protein as a shuttle for retinal in the rhodopsin-retinochrome system of the squid visual cells. *Vision Res.*, **29**, 639–652.
118. M. Tsuda, T. Tsuda (1990). Two distinct light regulated G-proteins in octopus photoreceptors. *Biochim. Biophys. Acta*, **1052**, 204–210.

119. J.D. Pottinger, N.J. Ryba, J.N. Keen, J.B. Findlay (1991). The identification and purification of the heterotrimeric GTP-binding protein from squid (*Loligo forbesi*) photoreceptors. *Biochem. J.*, **279**, 323–326.

120. T. Suzuki, A. Terakita, K. Narita, K. Nagai, Y. Tsukahara, Y. Kito (1995). Squid photoreceptor phospholipase C is stimulated by membrane Gq alpha but not by soluble Gq alpha. *FEBS Lett.*, **377**, 333–337.

121. T. Suzuki, K. Narita, A. Terakita, E. Takai, K. Nagai, Y. Kito, Y. Tsukahara (1999). Regulation of squid visual phospholipase C by activated G-protein alpha. *Comp. Biochem. Physiol. A. Mol. Integr. Physiol.*, **122**, 369–374.

122. S. Kikkawa, K. Tominaga, M. Nakagawa, T. Iwasa, M. Tsuda (1996). Simple purification and functional reconstitution of octopus photoreceptor Gq, which couples rhodopsin to phospholipase C. *Biochemistry*, **35**, 15857–15864.

123. L.H. Mayeenuddin, C. Bamsey, J. Mitchell (2001). Retinal phospholipase C from squid is a regulator of Gq alpha GTPase activity. *J. Neurochem.* **78**, 1350–1358.

124. A. Terakita, T. Hariyama, Y. Tsukahara, Y. Katsukura, H. Tashiro (1993). Interaction of GTP-binding protein Gq with photoactivated rhodopsin in the photoreceptor membranes of crayfish. *FEBS Lett.*, **330**, 197–200.

125. A. Terakita, H. Takahama, T. Hariyama, T. Suzuki, Y. Tsukahara (1998). Light-regulated localization of the beta-subunit of Gq-type G-protein in the crayfish photoreceptors. *J. Comp. Physiol. [A]*, **183**, 411–417.

126. O. Devary, O. Heichal, A. Blumenfeld, D. Cassel, E. Suss, S. Barash, C.T. Rubinstein, B. Minke, Z. Selinger (1987). Coupling of photoexcited rhodopsin to inositol phospholipid hydrolysis in fly photoreceptors. *Proc. Natl. Acad. Sci. U.S.A.*, **84**, 6939–6943.

127. Y.J. Lee, M.B. Dobbs, M.L. Verardi, D.R. Hyde (1990). dgq: a *Drosophila* gene encoding a visual system-specific G alpha molecule. *Neuron*, **5**, 889–898.

128. K. Scott, A. Becker, Y. Sun, R. Hardy, C. Zuker (1995). Gq alpha protein function *in vivo*: genetic dissection of its role in photoreceptor cell physiology. *Neuron*, **15**, 919–927.

129. B.T. Bloomquist, R.D. Shortridge, S. Schneuwly, M. Perdew, C. Montell, H. Steller, G. Rubin, W.L. Pak (1988). Isolation of a putative phospholipase C gene of *Drosophila*, norpA, and its role in phototransduction. *Cell*, **54**, 723–733.

130. R.D. Shortridge, J. Yoon, C.R. Lending, B.T. Bloomquist, M.H. Perdew, W.L. Pak (1991). A *Drosophila* phospholipase C gene that is expressed in the central nervous system. *J. Biol. Chem.*, **266**, 12474–12480.

131. S. Schneuwly, M.G. Burg, C. Lending, M.H. Perdew, W.L. Pak (1991). Properties of photoreceptor-specific phospholipase C encoded by the norpA gene of *Drosophila melanogaster*. *J. Biol. Chem.*, **266**, 24314–24319.

132. R.R. McKay, D.M. Chen, K. Miller, S. Kim, W.S. Stark, R.D. Shortridge (1995). Phospholipase C rescues visual defect in norpA mutant of *Drosophila melanogaster*. *J. Biol. Chem.*, **270**, 13271–13276.

133. M.T. Pearn, L.L. Randall, R.D. Shortridge, M.G. Burg, W.L. Pak (1996). Molecular, biochemical, and electrophysiological characterization of *Drosophila* norpA mutants. *J. Biol. Chem.*, **271**, 4937–4945.

134. Z. Selinger, B. Minke (1988). Inositol lipid cascade of vision studied in mutant flies. *Cold Spring Harb. Symp. Quant. Biol.*, **53** (Pt 1), 333–341.

135. Y.J. Lee, S. Shah, E. Suzuki, T. Zars, P.M. O'Day, D.R. Hyde (1994). The *Drosophila* dgq gene encodes a G alpha protein that mediates phototransduction. *Neuron*, **13**, 1143–1157.

136. S. Yarfitz, G.A. Niemi, J.L. McConnell, C.L. Fitch, J.B. Hurley (1991). A G beta protein in the *Drosophila* compound eye is different from that in the brain. *Neuron*, **7**, 429–438.

137. S. Schulz, A. Huber, K. Schwab, R. Paulsen (1999). A novel Ggamma isolated from *Drosophila* constitutes a visual G protein gamma subunit of the fly compound eye. *J. Biol. Chem.*, **274**, 37605–37610.

138. N.J. Ryba, J.B. Findlay, J.D. Reid (1993). The molecular cloning of the squid (*Loligo forbesi*) visual Gq-alpha subunit and its expression in *Saccharomyces cerevisiae*. *Biochem. J.*, **292**, 333–341.

139. T. Iwasa, T. Yanai, M. Nakagawa, S. Kikkawa, S. Obata, J. Usukura, M. Tsuda (2000). G protein α subunit genes in *Octopus* photoreceptor cells. *Zool. Sci.*, **17**, 711–716.

140. N.J. Ryba, J.D. Pottinger, J.N. Keen, J.B. Findlay (1991). Sequence of the beta-subunit of the phosphatidylinositol-specific phospholipase C-directed GTP-binding protein from squid (*Loligo forbesi*) photoreceptors. *Biochem. J.*, **273**, 225–228.

141. J.S. Lott, N.J. Ryba, J.D. Pottinger, J.N. Keen, A. Carne, J.B. Findlay (1992). The gamma-subunit of the principal G-protein from squid (*Loligo forbesi*) photoreceptors contains a novel N-terminal sequence. *FEBS Lett.*, **312**, 241–244.

142. P.J. Dolph, S.H. Man, S. Yarfitz, N.J. Colley, J.R. Deer, M. Spencer, J.B. Hurley, C.S. Zuker (1994). An eye-specific G beta subunit essential for termination of the phototransduction cascade. *Nature*, **370**, 59-61.

143. Y. Hotta, S. Benzer (1970). Genetic dissection of the *Drosophila* nervous system by means of mosaics. *Proc. Natl. Acad. Sci. U.S.A.*, **67**, 1156–1163.

144. W.L. Pak, J. Grossfield, K.S. Arnold (1970). Mutants of the visual pathway of *Drosophila melanogaster*. *Nature*, **227**, 518–520.

145. A. Fein, S. Cavar (2000). Divergent mechanisms for phototransduction of invertebrate microvillar photoreceptors. *Vis. Neurosci.*, **17**, 911–917.

146. F.H. Chen, A. Baumann, R. Payne, J.E. Lisman (2001). A cGMP-gated channel subunit in *Limulus* photoreceptors. *Vis. Neurosci.*, **18**, 517–526.

147. S. Chyb, P. Raghu, R.C. Hardie (1999). Polyunsaturated fatty acids activate the *Drosophila* light-sensitive channels TRP and TRPL. *Nature*, **397**, 255–259.

148. P. Raghu, N.J. Colley, R. Webel, T. James, G. Hasan, M. Danin, Z. Selinger, R.C. Hardie (2000). Normal phototransduction in *Drosophila* photoreceptors lacking an InsP(3) receptor gene. *Mol. Cell Neurosci.*, **15**, 429–445.

149. C. Montell, G.M. Rubin (1989). Molecular characterization of the *Drosophila* trp locus: a putative integral membrane protein required for phototransduction. *Neuron*, **2**, 1313–1323.

150. R.C. Hardie, B. Minke (1992). The *trp* gene is essential for a light-activated Ca^{2+} channel in *Drosophila* photoreceptors. *Neuron*, **8**, 643–651.

151. A.M. Phillips, A. Bull, L.E. Kelly (1992). Identification of a *Drosophila* gene encoding a calmodulin-binding protein with homology to the trp phototransduction gene. *Neuron*, **8**, 631–642.

152. R.C. Hardie, B. Minke (1993). Novel Ca^{2+} channels underlying transduction in *Drosophila* photoreceptors: implications for phosphoinositide-mediated Ca^{2+} mobilization. *Trends. Neurosci.*, **16**, 371–376.

153. B.A. Niemeyer, E. Suzuki, K. Scott, K. Jalink, C.S. Zuker (1996). The *Drosophila* light-activated conductance is composed of the two channels TRP and TRPL. *Cell*, **85**, 651–659.

154. A. Huber (2001). Scaffolding proteins organize multimolecular protein complexes for sensory signal transduction. *Eur. J. Neurosci.*, **14**, 769–776.

155. A. Huber, P. Sander, R. Paulsen (1996). Phosphorylation of the InaD gene product, a photoreceptor membrane protein required for recovery of visual excitation. *J. Biol. Chem.*, **271**, 11710–11717.

156. B.H. Shieh, M.Y. Zhu (1996). Regulation of the TRP Ca^{2+} channel by INAD in *Drosophila* photoreceptors. *Neuron*, **16**, 991–998.

157. S. Tsunoda, J. Sierralta, Y. Sun, R. Bodner, E. Suzuki, A. Becker, M. Socolich, C.S. Zuker (1997). A multivalent PDZ-domain protein assembles signalling complexes in a G-protein-coupled cascade. *Nature*, **388**, 243–249.

158. J. Chevesich, A.J. Kreuz, C. Montell (1997). Requirement for the PDZ domain protein, INAD, for localization of the TRP store-operated channel to a signaling complex. *Neuron*, **18**, 95–105.

159. X.Z. Xu, A. Choudhury, X. Li, C. Montell (1998). Coordination of an array of signaling proteins through homo- and heteromeric interactions between PDZ domains and target proteins. *J. Cell Biol.*, **142**, 545–555.

160. A. Huber, P. Sander, A. Gobert, M. Bahner, R. Hermann, R. Paulsen (1996). The transient receptor potential protein (Trp), a putative store-operated Ca^{2+} channel essential for phosphoinositide-mediated photoreception, forms a signaling complex with NorpA, InaC and InaD. *EMBO J.*, **15**, 7036–7045.

161. X.Z. Xu, P.D. Wes, H. Chen, H.S. Li, M. Yu, S. Morgan, Y. Liu, C. Montell (1998). Retinal targets for calmodulin include proteins implicated in synaptic transmission. *J. Biol. Chem.*, **273**, 31297–31307.

162. M. Bähner, S. Frechter, N. Da Silva, B. Minke, R. Paulsen, A. Huber (2002). Light-regulated subcellular translocation of *Drosophila* TRPL channels induces long-term adaptation and modifies the light-induced current. *Neuron*, **34**, 83–93.

163. M. Bähner, P. Sander, R. Paulsen, A. Huber (2000). The visual G protein of fly photoreceptors interacts with the PDZ domain assembled INAD signaling complex via direct binding of activated Galpha(q) to phospholipase cbeta. *J. Biol. Chem.*, **275**, 2901–2904.

164. R.C. Hardie, P. Raghu, S. Moore, M. Juusola, R.A. Baines, S.T. Sweeney (2001). Calcium influx via TRP channels is required to maintain PIP2 levels in *Drosophila* photoreceptors. *Neuron*, **30**, 149–159.

165. S. Tsunoda, Y. Sun, E. Suzuki, C. Zuker (2000). Independent anchoring and assembly mechanisms of INAD signaling complexes in Drosophila photoreceptors. *J. Neurosci.* **21**, 150–158.

166. H.S. Li, C. Montell (2000). TRP and the PDZ protein, INAD, form the core complex required for retention of the signalplex in *Drosophila* photoreceptor cells. *J. Cell Biol.*, **150**, 1411–1422.

167. A. Huber, G. Belusic, N. Da Silva, M. Bähner, G. Gerdon, K. Draslar, R. Paulsen (2000). The *Calliphora* rpa mutant lacks the PDZ domain-assembled INAD signalling complex. *Eur. J. Neurosci.*, **12**, 3909–3918.

168. M.P. Gomez, E. Nasi (2000). Light transduction in invertebrate hyperpolarizing photoreceptors: possible involvement of a Go-regulated guanylate cyclase. *J. Neurosci.*, **20**, 5254–5263.

169. M.P. Gomez, E. Nasi (1995). Activation of light-dependent K^+ channels in ciliary invertebrate photoreceptors involves cGMP but not the $IP3/Ca^{2+}$ cascade. *Neuron*, **15**, 607–618.

170. D.R. Hyde, K.L. Mecklenburg, J.A. Pollock, T.S. Vihtelic, S. Benzer (1990). Twenty Drosophila visual system cDNA clones: one is a homolog of human arrestin. *Proc. Natl. Acad. Sci. U.S.A.*, **87**, 1008–1012.

171. D.P. Smith, B.H. Shieh, C.S (1990). Zuker Isolation and structure of an arrestin gene from Drosophila. *Proc. Natl. Acad. Sci. U.S.A.*, **87**, 1003–1007.

172. H. LeVine, D.P. Smith, M. Whitney, D.M. Malicki, P.J. Dolph, G.F. Smith, W. Burkhart, C.S. Zuker (1990). Isolation of a novel visual-system-specific arrestin: an in vivo substrate for light-dependent phosphorylation. *Mech. Dev.*, **33**, 19–25.

173. T. Yamada, Y. Takeuchi, N. Komori, H. Kobayashi, Y. Sakai, Y. Hotta, H. Matsumoto (1990). A 49-kilodalton phosphoprotein in the Drosophila photoreceptor is an arrestin homolog. *Science*, **248**, 483–486.

174. P.J. Dolph, R. Ranganathan, N.J. Colley, R.W. Hardy, M. Socolich, C.S. Zuker (1993). Arrestin function in inactivation of G protein-coupled receptor rhodopsin in vivo. *Science*, **260**, 1910–1916.

175. P.G. Alloway, P.J. Dolph (1999). A role for the light-dependent phosphorylation of visual arrestin. *Proc. Natl. Acad. Sci. U.S.A.*, **96**, 6072–6077.

176. K. Hamdorf, S. Razmjoo (1977). The prolonged depolarizing afterpotential and its contribution to the understanding of photoreceptor function, *Biophys. Struct. Mech.*, **3**, 163–170.

177. W.L. Pak (1979). Study of photoreceptor function using *Drosophila* mutants. In: X.O. Breakfield (Ed.), *Neurogenetics: Genetic Approaches to the Nervous System* (pp. 67–99). Elsevier, North-Holland.

178. B. Minke (1986). Photopigment-dependent Adaptation in Invertebrates–Implication for Vertebrates. In: H. Stieve (ed.), *The Molecular Mechanism of Photoreception* (pp. 241–265). Springer-Verlag, Berlin.

179. K. Hamdorf, R. Paulsen, J. Schwemer (1989). Insect Photoreception: I. Primary Mechanisms of Visual Excitation. In: H.C. Lüttgau, R. Necker (Eds), *Biological Signal Processing* (pp. 64–82). VCH Verlagsgemeinschaft, Weinheim.

180. H. Matsumoto, T. Yamada (1991). Phosrestins I and II: arrestin homologs which undergo differential light-induced phosphorylation in the Drosophila photo-receptor in vivo. *Biochem. Biophys. Res. Commun.*, **177**, 1306–1312.

181. H. Matsumoto, B.T. Kurien, Y. Takagi, E.S. Kahn, T. Kinumi, N. Komori, T. Yamada, F. Hayashi, K. Isono, et al. (1994), Phosrestin I undergoes the earliest light-induced phosphorylation by a calcium/calmodulin-dependent protein kinase in Drosophila photoreceptors. *Neuron*, **12**, 997–1010.

182. E.S. Kahn, H. Matsumoto (1997). Calcium/calmodulin-dependent kinase II phosphorylates *Drosophila* visual arrestin. *J. Neurochem.*, **68**, 169–175.

183. S.S. Ferguson, W.E. Downey, A.M. Colapietro, L.S. Barak, L. Menard, M.G. Caron (1996). Role of beta-arrestin in mediating agonist-promoted G protein-coupled receptor internalization. *Science*, **271**, 363–366.

184. O.B.J. Goodman, J.G. Krupnick, F. Santini, V.V. Gurevich, R.B. Penn, A.W. Gagnon, J.H. Keen, J.L. Benovic (1996). Beta-arrestin acts as a clathrin adaptor in endocytosis of the beta2- adrenergic receptor. *Nature*, **383**, 447–450.

185. J. Zhang, S.S.G. Ferguson, L.S. Barak, L. Menard, M.G. Caron (1996). Dynamin and beta-arrestin reveal distinct mechanisms for G protein-coupled receptor internalization. *J. Biol. Chem.*, **271**, 18302–18305.

186. T. Byk, M. Bar-Yaacov, Y.N. Doza, B. Minke, Z. Selinger (1993). Regulatory arrestin cycle secures the fidelity and maintenance of the fly photoreceptor cell. *Proc. Natl. Acad. Sci. U.S.A.*, **90**, 1907–1911.

187. J. Vinos, K. Jalink, R.W. Hardy, S.G. Britt, C.S. Zuker (1997). A G protein-coupled receptor phosphatase required for rhodopsin function. *Science*, **277**, 687–690.

188. W.C. Smith, R.M. Greenberg, B.G. Calman, M.M. Hendrix, L. Hutchinson, L.A. Donoso, B.A. Battelle (1995). Isolation and expression of an arrestin cDNA from the horseshoe crab lateral eye. *J. Neurochem.*, **64**, 1–13.

189. J. Bentrop, M. Schillo, G. Gerdon, K. Draslar, R. Paulsen (2001). UV-light-dependent binding of a visual arrestin 1 isoform to photoreceptor membranes in a neuropteran (*Ascalaphus*) compound eye. *FEBS Lett.*, **493**, 112–116.

190. G. Berstein, J.L. Blank, D.Y. Jhon, J.H. Exton, S.G. Rhee, E.M. Ross (1992). Phospholipase C-beta 1 is a GTPase-activating protein for Gq/11, its physiologic regulator. *Cell*, **70**, 411–418.

191. G.H. Biddlecome, G. Berstein, E.M. Ross (1996). Regulation of phospholipase C-beta1 by Gq and m1 muscarinic cholinergic receptor. Steady-state balance of receptor-mediated activation and GTPase-activating protein-promoted deactivation. *J. Biol. Chem.*, **271**, 7999–8007.

192. B. Cook, M. Bar-Yaacov, B. Cohen, R.E. Goldstein, Z. Paroush, Z. Selinger, B. Minke (2000). Phospholipase C and termination of G-protein-mediated signalling *in vivo*. *Nat. Cell Biol.*, **2**, 296–301.

Chapter 3

Vertebrate rhodopsin

Oliver P. Ernst, Klaus Peter Hofmann and Krzysztof Palczewski

Table of contents

Abstract

Rhodopsin, described more than 120 years ago as the visual pigment of the retina, is a transmembrane protein composed of the apoprotein opsin and the covalently linked chromophore 11-*cis*-retinal. It is highly expressed in rod cells, where it localizes to plasma and internal membranes of the rod outer segment, a specific cellular compartment dedicated for transformation of light energy into biochemical reactions. Absorption of light by the chromophore triggers transient conformational changes of the apoprotein, which in turn initiates the G-protein mediated enzymatic cascade of reactions, termed photo-transduction, that result in neuronal signaling. Rhodopsin is also the best-studied member of a large group of cell-surface receptors that signal through G-proteins and therefore are called G-protein-coupled receptors (GPCRs). Unique members of the GPCR superfamily are involved in a vast variety of specific cellular signal transduction processes including visual, taste and odor perceptions and sensing a variety of hormones. These receptors share a common seven-transmembrane α-helical structure and use the binding energy of extracellular chemical ligands for stabilization of an active receptor confor-mation. Thus, conformational changes of GPCRs allow transmission of the extracellular signal, across the plasma membrane, into the cell. Elucidation of the crystal structure of rhodopsin and characterization of fundamental aspects of the photoactivation mechanism paved the way for better understanding of other GPCRs. In this review, we describe the first steps in seeing, comprising light-induced activation of rhodopsin, and its interaction with proteins of the phototransduction cascade.

3.1   Introduction

In 1878, Kühne and co-workers recognized that vision originates from the absorption of light by visual pigments [1,2]. These pigments are membrane-bound photoreceptor proteins composed of the apoprotein opsin and a retinal chromophore. In the retina of vertebrates, two main types of photoreceptor cells, rod and cone cells, are present. The rods are responsible for scotopic vision and several sub-types of cones for photopic vision. Rod visual pigment rhodopsin is a 40 kD integral membrane protein, which consists of the apoprotein opsin containing seven helices spanning the membrane and the prosthetic group 11-*cis*-retinal. The color of rhodopsin and its response to light arises from the covalent linkage of the 11-*cis*-retinal chromophore. The chro-mophore is linked via a protonated Schiff base to Lys[296] in helix VII, yielding a broad absorption with a maximum at 500 nm (ε_{500} = 40,000 cm^{-1} M^{-1}) that matches the solar spectrum. The human retina contains three sub-types of cone pigments that have distinct sensitivity to different wavelengths of visible light: blue, green and red pigments, with absorption maxima of 424, 530 and 560 nm, respectively [3]. In principle, all visual pigments convert light energy into changes in the protein conformation, and in turn trigger intracellular reactions that ultimately lead to a neuronal impulse [4]. The prosthetic group that absorbs light (i.e. the chromophore) undergoes isomerization after photon absorption [5], and transmits the light energy to the chromophore–receptor complex, where it is initially stored as an energetically unfavorable conformation of the chromophore and an unstable tertiary conformation of the polypeptide chain. The signaling state of the receptor is then reached by a subsequent thermal relaxation process. The present chapter focusses on the most extensively studied visual pigment, rhodopsin.

Rod cells are capable of detecting single quanta [6]. This ultimate sensitivity is achieved as a consequence of high probability of absorption of the incoming light, efficient photochemical reaction, a rapid, reproducible and greatly amplified intracellular signal transduction and a high signal-to-noise ratio of the overall transduction process. The visual system evolved just to perform such a task. A prerequisite of such signaling properties is for rhodopsin to have an extremely low dark activity. In the time domain of the electrical response, no spontaneous activation is tolerated from any of the 10^8 inactive rhodopsin molecules present in a photoreceptor cell. The estimated lifetime of the inactive state of rhodopsin is >10 years (see e.g. [4]). However, when rhodopsin is photoactivated, it initiates the transduction cascade with maximal quantum efficiency.

The phototransduction system is composed of the G-protein transducin (Gt), named according to its rod cell-specific expression of the α-subunit, and the effector, a cGMP-specific phoshodiesterase (reviewed in [7]). Therefore, rhodopsin is considered to be a member of a large group of transmembrane proteins of similar topology, termed G-protein-coupled receptors (GPCRs). Upon activation of GPCRs by ligand binding, or in the case of visual pigments by photon-induced alteration in the conformation of the bound ligand, the cytoplasmic surface of GPCRs becomes competent for G-protein binding, leading to subsequent catalytic GDP/GTP exchange on the α-subunit and G-protein activation. In general, GPCRs serve to respond to chemical signals and transmit them across biological membranes by coupling to heterotrimeric guanine nucleotide-binding proteins (G-proteins), which in turn, modulate effector protein activity and thereby affect second messenger levels (reviewed by [8]).

3.2   Phototransduction

3.2.1   The disk membrane

Photoactivation of rhodopsin, which is the only light-sensitive reaction in rods, and interaction with Gt are the first two steps in phototransduction. This term denotes a sequence of protein–protein interactions and biochemical reactions, which are initiated by photoactivation of rhodopsin. All reactions are localized on the disk membranes of the rod outer segment (ROS) (Figure 1). The human

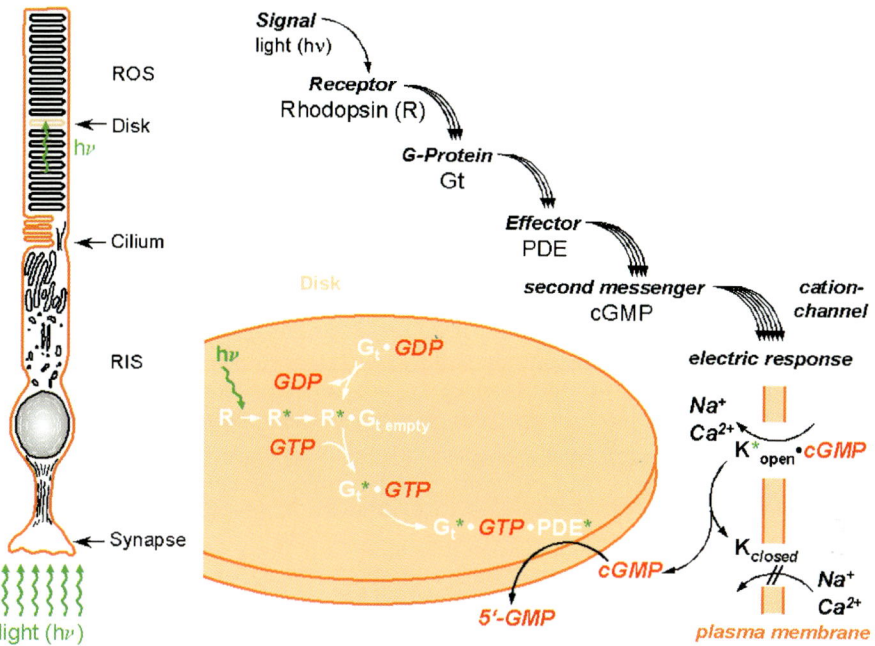

Figure 1. Phototransduction in the rod cell of the retina. Left: diagram of retinal rod cell, showing the direction of incoming light and cellular compartments of this highly differentiated neuron. The ROS, connected with the inner segment (RIS) through a narrow cilium, is densely packed with a stack of disk membranes. These internal vesicles contain an integral membrane protein, rhodopsin. Right: absorption of photon by rhodopsin's chromophore causes isomerization of 11-*cis*-retinylidene to all-*trans*-retinylidene. A sequence of protein–protein interactions between the G-protein-coupled receptor (rhodopsin, R), a G-protein (transducin, Gt), and an effector [a cGMP specific phosphodiesterase (PDE)] is the mechanism of visual transduction that allows conversion of the photon signal into the biochemical cascade of events. The signal is initially amplified on the level of the receptor, because hundreds of molecules of Gt interact with a single activated rhodopsin molecule, and on the level of the effector, as a consequence of its catalytic property. Light-activated rhodopsin (R*) is mobile through diffusion along the two-dimensional disk membrane plane and, when it encounters Gt, induces GDP release from the G-protein, and forms a transient nucleotide-free R*•Gt complex. GTP, present in the ROS, dissociates this complex immediately by binding into the nucleotide-binding pocket of the Gtα-subunit. Activated GTP-bound Gtα is then capable of activating the effector by binding to its inhibitory subunits. The activated effector reduces the level of cGMP and, in turn, leads to the closure of cGMP-gated cation channels. A decreased influx of cations through the channel causes hyperpolarization of the plasma membrane, spreading the electrical signal to the synaptic terminal, and a subsequent decrease in the transmitter release (see text for details).

retina contains over 100 million photoreceptor cells [4] – highly differentiated post-mitotic neurons that consist of an inner and outer segment and an axonal part with its synaptic ending. In the retina of warm-blooded animals, ROS is connected via a thin, ~1.5 μm in diameter, cilium to the cell body.

Rhodopsin is embedded within the membranes of ROS, which are arranged in a long, closely spaced stack of approximately one thousand isolated disk-like saccules, termed disks. Gt is associated with the cytoplasmic surface of the disk membranes. To provide an effective target for light, rhodopsin accounts for half of the dry weight of the disk membranes. Crucial for the function of R* is the high fluidity of the disk membrane. The high fluidity is a result of the large amounts of highly unsaturated (22:6n–3) acyl chains in its major phospholipids, phosphatidylcholine (PC), phosphatidylserine (PS) and phosphatidylethanolamine (PE) [9]. A lack of polyunsaturated fatty acids causes abnormalities in visual function [10,11]. Also, the high fluidity and other properties of the membrane's special composition may be crucial for the anchoring of G-protein and effector to the membrane surface, and thus for proper signal transduction in the disk membrane. Understanding the role of membrane properties in signal transduction is, at the physicochemical level, still not well understood. However, with regard to photoreceptor function, there are two identified properties of membranes: the formation of the active intermediate and the transport of the hydrophobic retinal ligand.

The fluidity of the membrane enables fast lateral and rotatory diffusion of rhodopsin, with trajectories over the membrane surface in the order of one second. All protein–protein interactions, on which the G-protein coupled signal transmission relies, are localized on the surface of the disk membrane. Molecular recognition between these proteins is bound to active phases, which result from the intramolecular processes and the uptake of cofactors such as GTP (see [12]).

3.2.2   G-protein and the effector activation

In its inactive, GDP bound state (Gt•GDP), the heterotrimeric Gt holoprotein (Gtαβγ) is peripherally bound to the disk membrane by weak hydrophobic and ionic interactions [13–15]. The first step of nucleotide exchange catalysis is the collisional interaction between light-activated rhodopsin (R*) and Gt•GDP (step 1, Figure 1). This interaction triggers the release of GDP and subsequent formation of a stable R*•Gt complex with an empty nucleotide binding site on the Gtα-subunit (step 2). Binding of GTP to the Gtα-subunit within the R*•Gt complex enables a conformational change (step 3) that induces the release of active Gt•GTP (Gt*) from the receptor (step 4) and the (simultaneous or immeasurably delayed) separation of the α- and βγ-subunits (Gtα•GTP and Gtβγ). In vitro, activation is accompanied by an immediate (delay < 1 ms, [16]) dissociation of both Gtα•GTP and Gtβγ from the disk membrane. The high rate of R*-catalyzed nucleotide exchange leads to the rapid accumulation of Gt*. The visual system utilizes the Gα-subunit to relay the signal to the effector. Active Gtα•GTP, in turn, binds to the effector cGMP-specific PDE; this stoichiometric, non-catalytic interaction occurs within less than 5 ms [16]. The interaction keeps the PDE active, and hydrolysis of cGMP leads to the closure of several hundred cGMP-gated ion channels which control the flow of Na^+ and Ca^{2+} ions into the photoreceptor ROS (reviewed in [17]). The resulting hyperpolarization of the rod plasma membrane inhibits the release of glutamate neurotransmitter at the synapse, which establishes the light signal that is transmitted to the brain via the *nervus opticus* [4].

The rising phase of the electrical response is dictated by the diffusional encounter between R* and Gt on the disk membrane. A 50% reduction of receptor density in the rod disks causes acceleration of both response onset and recovery of flash responses [18]. Nucleotide exchange catalysis in Gt by R* establishes a first step amplification, because one R* can activate several hundred G-proteins. This has been demonstrated for isolated disk membranes in vitro [12], but the exact number is not yet known in vivo.

3.2.3   Deactivation

To terminate signal transduction and to allow repeated and/or graded excitation of the cell, each single step of the transduction cascade must be properly deactivated. This happens via interactions with regulatory proteins. Rapid shut-off of active rhodopsin does not happen by thermal decay of the active conformation, but rather by concurrent interaction with rhodopsin kinase (RK) and eventually phosphorylation of the receptor. Phosphorylated rhodopsin enables interaction with arrestin and by this shut-off of the signal for the G-protein. Inactivation of Gt•GTP results from the hydrolysis of bound GTP in the nucleotide binding site of the Gtα-subunit, while Gt is bound to the γ-subunit of PDE. Other proteins regulate and accelerate this reaction [19]. To terminate the hyperpolarization of the rod cell membrane, the cGMP level in the cytoplasm must be restored by resynthesis of cGMP, which is regulated by a feedback mechanism. Since the Na^+/Ca^{2+}-K^+ exchanger continues to extrude Ca^{2+} from the ROS, the concentration of intracellular Ca^{2+} decreases. This leads to activation of the Ca^{2+}-binding proteins, the guanylate cyclase-activating proteins (GCAP1/GCAP2 and GCAP3), which in turn activate the enzyme guanylate cyclase. The resulting rise in cGMP causes the cGMP-gated channels to reopen and, consequently, causes the Na^+/Ca^{2+} influx to terminate the hyperpolarization and decrease the guanylate cyclase activity by negative feedback inhibition (for details see [17,20,21]).

3.3   Structure of bovine rhodopsin

Rhodopsin is the only GPCR that utilizes photochemistry to generate an activating ligand in situ, by converting the chromophore from its inverse agonistic form into its agonistic form. Tethering the ligand to the protein via a Lys side chain serves two purposes: (1) the inverse agonist 11-*cis*-retinal stabilizes the inactive receptor conformation and (2) permanent occupancy of the binding pocket ensures maximal sensitivity to light and a fast response necessary for phototransduction.

3.3.1 Overall topology

The 2.8 Å crystal structure of rhodopsin, the first high-resolution structure of a GPCR with a bound ligand, has been solved and refined [22,23]. The predicted seven transmembrane helices and the covalent linkage between 11-cis-retinal and opsin, via Lys[296] in helix VII, was confirmed. Rhodopsin shows an ellipsoidal shape (view parallel to the disk membrane plane, Figures 3 and 4) with a seven transmembrane helix region, and an extracellular and intracellular region each consisting of three interhelical loops and a terminal tail (NH$_2$ or COOH, respectively). Approximately equal amounts of the protein mass are distributed to the two solvent-exposed regions (Figures 2 and 3); however, the degree of association of the polypeptide segments is different in both regions. Only a few interactions are seen in the cytoplasmic region, whereas the extracellular parts associate significantly with each other, making the second extracellular loop (E-II) fit tightly within a limited space inside the bundle of helices (Figure 4). A twisted β-hairpin is formed by part of this loop, creating a plug upon which the retinal lies. The β-strand of loop E-II is connected to helix III by a highly conserved disulfide bridge among GPCRs between Cys[187] (loop E-II) and Cys[110] (helix III). Another feature of the extracellular region is its post-translational modification: N-glycosylation at Asn[2] and Asn[15].

The seven transmembrane helices of rhodopsin contain a mixture of α- and 3$_{10}$-helices, and vary in length from 19 to 34 residues, in the degree of bending, kinking and twisting, and also in the tilt angles with respect to the expected membrane surface (for details see [22,23]). A significant feature found in the crystal structure is a strong distortion by one of the most conserved residues among GPCRs, Pro[267] in helix VI, causing the largest helix bend (36°) seen in the structure [23]. This is of special importance, as movement of the cytoplasm-facing part of helix VI is involved in formation of the active receptor conformation [24,25], and mutations at this position exert long-range effects on the structure of the third cytoplasmic loop [26]. H-V and H-III are not significantly bent, although Pro and Gly residues are present. In contrast, a deviation of helix II from an ideal helix by 30° is due to the flexibility in the Gly-Gly sequence in the middle of this helix. Helix VII shows a considerable distortion and elongation in the region around the retinal attachment site Lys[296] and contains two prolines, Pro[291] and Pro[303]. The latter is part of the highly conserved NPxxY motif (Figure 2) found in many GPCRs. This motif, whose function is still unknown, might be involved in the formation of a structural domain that keeps the receptor in the inactive conformation. The structure revealed an additional, non-transmembrane helix. The short cytoplasmic helix VIII, adjacent to helix VII, runs parallel to the cytoplasmic surface and is terminated by palmitoylated cysteines (Cys[322] and Cys[323]), fixing the helix to the membrane (Figures 2 and 4). This cytoplasmic helix is part of the binding site for the

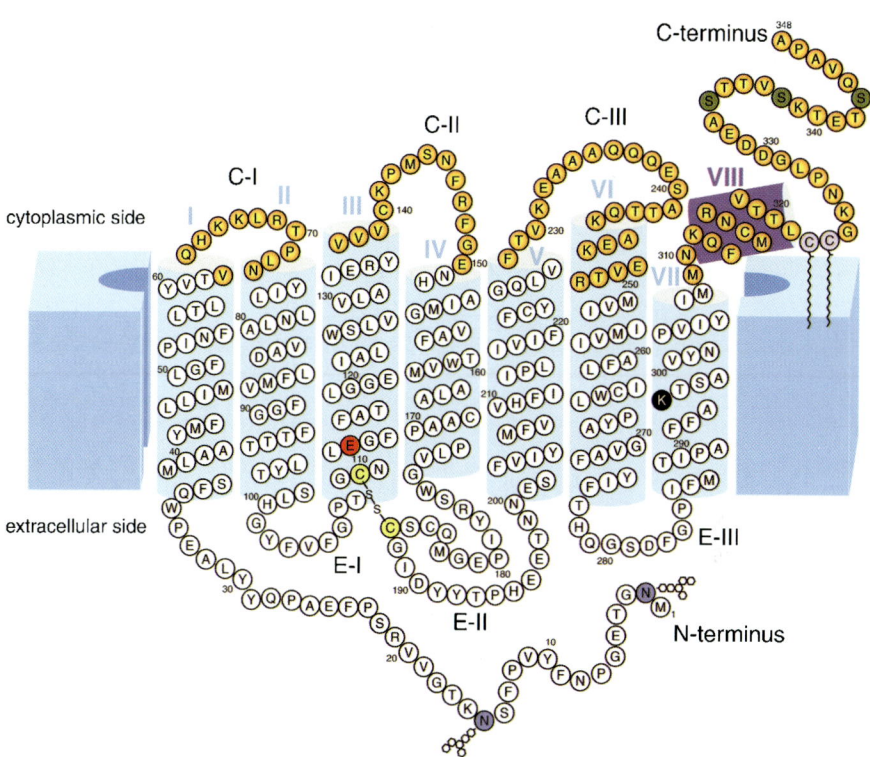

Figure 2. Two-dimensional model of bovine rhodopsin. Half of the polypeptide chain of the apoprotein is embedded in the disk membrane and forms seven transmembrane helices (represented by blue cylinders). A cytoplasmic amphiphatic helix is terminated by palmitoylated Cys^{322} and Cys^{323} (violet filled circles) and is represented as a purple cylinder. Visual pigments consist of the opsin apoprotein and the chromophore 11-*cis*-retinal, vitamin A aldehyde, which is attached via a protonated Schiff base linkage to the ε-amino group of a Lys side chain in helix VII (Lys^{296} is represented by a black circle). The counter-ion of the protonated Schiff base is Glu^{113} and is shown as a red circle. Photoisomerization of the chromophore to all-*trans*-retinal leads to activating conformational changes concomitant with exposure of binding sites for Gt at the cytoplasmic side of the receptor. The cytoplasmic surface consists of loops connecting successive helices and the C-terminus (shown as yellow filled circles with a red periphery). Loops C-II, C-III, and the cytoplasmic helix, are involved in interaction with Gt. The regulatory proteins RK and arrestin also bind to the cytoplasmic side although with different loop preferences (see text). Preferred phosphorylation sites for RK at rhodopsin's C-terminus are represented by brown filled circles. A highly conserved disulfide bridge stabilizes the inactive receptor conformation bridge (Cys residues involved are represented by yellow filled circles). Asn^2 and Asn^{15} at the N-terminus carry carbohydrate chains.

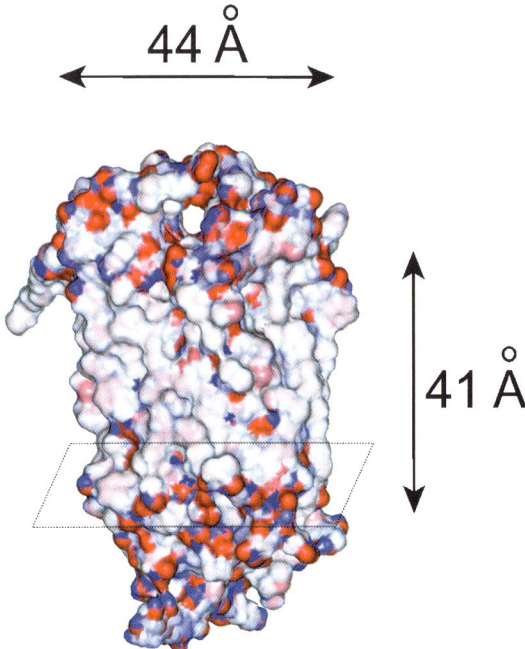

Figure 3. Charge distribution and dimensions of rhodopsin. The space-filling model shows an ellipsoidal shape of rhodopsin, 75 Å long and perpendicular to the membrane, while the transmembrane domain is 41 Å high. The length and width of the elliptic footprint on the plane at the middle of the membrane are roughly 45 and 37 Å, respectively [23]. The position of the chromophore is indicated by a plane. Negative charges are in red, positive charges are in blue.

C-terminus of the Gtα-subunit and plays a role in the regulation of Gtγ binding [27].

Rhodopsin's C-terminal penta-peptide region (residues 344–348), GVAPA, is essential for the translocation of newly synthesized rhodopsin molecules from the inner to the outer segment of the rod photoreceptor cell [28]. In GPCRs, inactivation of the activated receptor is frequently achieved by phosphorylation of the C-terminal Ser and Thr residues and subsequent binding of arrestin. The major phosphorylation sites on rhodopsin are Ser[338], Ser[343] and Ser[334] [29–31]. Although the structure of the C-terminus is poorly determined, H-bond interactions between this region and parts of the third cytoplasmic loop are conceivable. Photoactivation of rhodopsin would break these interactions, allowing binding of RK and phosphorylation of the hydroxyl groups [32].

The members of the rhodopsin subfamily, making up ~ 90% of all known GPCRs, share many key structural features (Figures 2 and 4), such as a disulfide bond between helix III and the extracellular region and a tripeptide sequence D(E)RY(W) located at the intracellular end of helix III. These

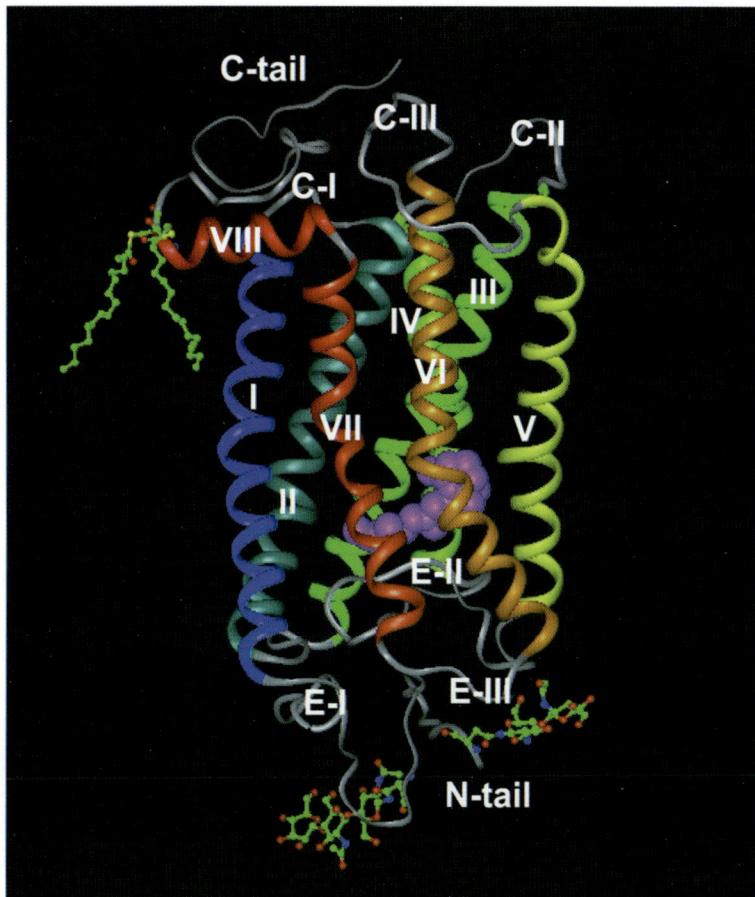

Figure 4. Crystal structure of bovine rhodopsin. The seven transmembrane helices are labeled I to VII, cytoplasmic and the extracellular loops as C-I, -II, -III and C-tail, and E-I, -II, -III and N-tail. A short cytoplasmic helix (VIII), corresponding to Lys[311]–Cys[322], is found between H-VII and the C-tail and runs parallel to the membrane surface. Two palmitoyl groups are attached to Cys[322] and Cys[323] and anchor the C-tail to the membrane. Carbohydrate chains are oriented toward the intradiskal (extracellular) face of rhodopsin, and are attached to Asn[2] and Asn[15].

residues are critical for proper protein folding and G-protein activation, respectively. There are also several other highly conserved residues with a frequency of occurrence >90% that define the rhodopsin family, such as an Asn-Asp pair located in helix I and helix II, respectively, Pro residues in helices V and VI, aromatic residues in helices IV and VI, and the NPxxY motif in helix VII. The crystal structure of bovine rhodopsin offers the most reliable model for these conserved features (Figures 2 and 4), for all members of this sub-family of GPCRs.

The area of rhodopsin projected into the membrane plane is <1500 Å². Inclusion of the cytoplasmic helix VIII elongates the cytoplasmic region, resulting in an area that is sufficiently large to dock a single trimeric Gt holoprotein on the surface. Receptor dimerization appears to be important for the function of GPCRs [33]. However, for rhodopsin, a lack of dimerization upon activation was concluded from the absence of light-induced changes of the diffusional speed within functional rods [34]. From spectroscopic measurements, monitoring the binding of Gt to R*, a 1:1 stoichiometry was determined [35] (see below); however, the dimer between two rhodopsin molecules was observed in the crystals of rhodopsin (Figure 5). These dimers are stabilized by hydrophobic interactions (Figure 5A) and by the dipole–dipole interaction (Figure 5B). Such dimerization in detergent implies that, in vivo, there would be repulsion rather than association of two or more rhodopsin molecules, unless phospholipids play an important role in compensating for this repulsion.

3.3.2   The inactive ground state

Vertebrate rhodopsin, like other visual pigments, contains 11-*cis*-retinal as a chromophore covalently bound to the ε-amino group of a lysine side chain via a Schiff base linkage. By its spectral properties, the Schiff base is likely to be protonated, but a protonated Schiff base of 11-*cis*-retinal formed from

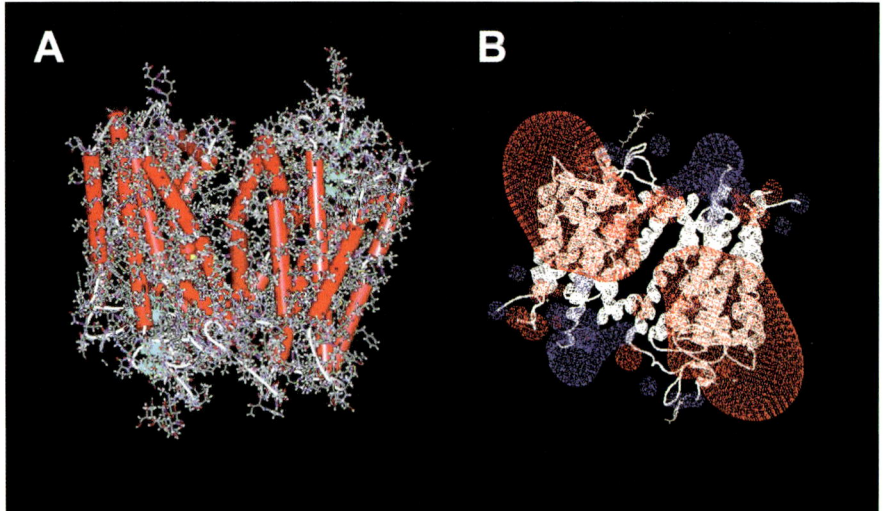

Figure 5.   Rhodopsin dimer in the asymmetric crystal unit. (A) The asymmetric unit of the rhodopsin crystals contains two rhodopsin molecules [22,23]. Helices are shown as red rods, β-strands are shown as blue arrows. (B) The dimer of rhodopsins within the asymmetric unit is held together by hydrophobic interactions and by dipole interactions between monomers. Red and blue represent negative and positive electric fields of the protein dimer.

n-butylamine in free methanol solution absorbs at 440 nm. Thus, it was recognized early that charged groups on opsin are required to tune the position of the spectral maximum of bound 11-*cis*-retinal (see [36]). The shift from 440 nm to that found in rhodopsin (500 nm) or color pigments is due to the interaction with the protein environment ("opsin shift") (see [36]). The environment of the chromophore is depicted in Figure 6. Important for spectral tuning is the position Glu[181], a residue in the β-hairpin plug structure brought into proximity of carbon C_{12} of the polyene chain of retinal due to the Cys[110]–Cys[187] disulfide bridge. In red and green color pigments Glu[181] is replaced by a

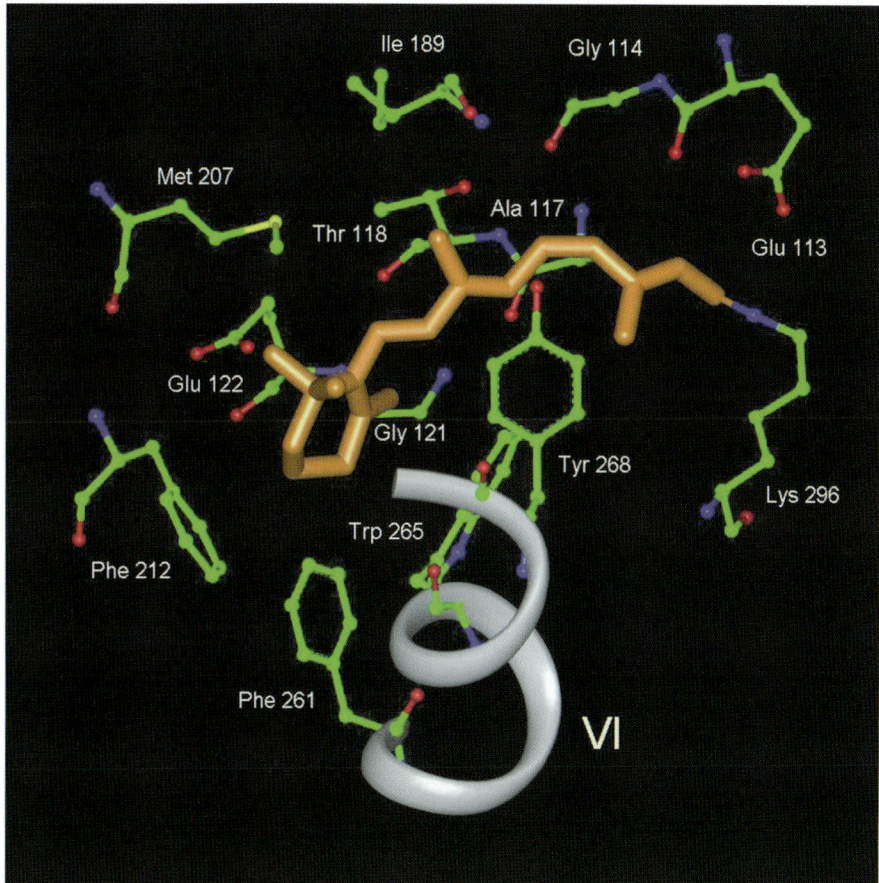

Figure 6. Chromophore-binding site of bovine rhodopsin. Schematic of the side chains surrounding the 11-*cis*-retinylidene group (drawn in orange), viewed upside down from within the disk membrane. The side chains are drawn as ball and sticks and colored by elements. Retinal forms a protonated Schiff base with Lys[296]. Glu[113] serves as counter-ion and forms a salt bridge with the protonated Schiff base. The β-ionone ring is mainly kept in place by residues Trp[265], Phe[261] (both from helix VI) and Glu[122] (from helix III).

histidine residue which, in combination with a chloride anion, is likely to be involved in the spectral tuning of these pigments (see [74]). Position Glu^{122} was implicated in regulating the rate of decay of the active species and the rate of regeneration [37].

The crystal structure of ground state bovine rhodopsin with bound inactivating 11-*cis*-retinal represents a template for GPCRs in their least active conformation. Compared to light-activated rhodopsin, the basal activity of free opsin is less than $1/10^6$ [38]. An important element in keeping the receptor in its inactive conformation is a salt bridge between Lys^{296}, the retinal attachment site in helix VII and its counter-ion, Glu^{113} in helix III, which is neutralized in active forms of rhodopsin [39]. Removal of the charge at either position, Lys^{296} or Glu^{113}, leads to increased basal activity of opsin, referred to as constitutive activity. Corresponding salt bridges between helix III and helix VII were shown to be critical for other GPCRs (see, for example, [40]). In rhodopsin, the salt bridge is formed between the protonated Schiff base nitrogen of Lys^{296} and Glu^{113}. The counter-ion increases the K_a of the Schiff base by several orders of magnitude (reviewed in [41]) and prevents its spontaneous hydrolysis. Binding of 11-*cis*-retinal to opsin, however, reduces the basal activity of free opsin further by a factor of 10^4 [38], classifying this retinal isomer as an inverse agonist. This reaction is exothermic ($\Delta H = -11$ kcal mol^{-1}, [42]) and is used to build up inactivating structural constraints seen in the ground state structure. This is reflected in the lowest crystallographic temperature factors of the structure in the region around the retinal Schiff base. The structure also revealed several H-bonded networks and hydrophobic interactions which connect neighboring helices, stabilizing the ground state (for details see [22,23,43]). Helix VII, which carries the retinal, shows most connections and interacts with all helices except helix IV and V. In contrast, helix VI is only connected to helix VII via hydrophobic interactions, which is what may allow light-induced helix movement, as mentioned earlier. Which of these interactions are caused by the presence of the 11-*cis*-retinal and which structural constraints are broken or change upon light-activation of rhodopsin are largely unknown.

3.4   Photoisomerization of rhodopsin

3.4.1   *Classical photoisomerization pathway*

The UV/Vis photointermediates of rhodopsin may be arranged in a (simplified) reaction scheme, which includes the approximate lifetime at room temperature and absorption maxima (in nm):

$$\begin{array}{ccccccc} ps & ns & ns & \mu s & ms & min \\ R\,(498) \rightarrow B\,(540) \rightleftharpoons BSI\,(477) \rightarrow L\,(497) \rightarrow MI\,(478) \rightleftharpoons MII\,(380) \rightleftharpoons MIII\,(465) \end{array}$$

$$\downarrow min$$

$$\text{opsin + all-}trans\text{-retinal (387)}$$

After illumination of rhodopsin (R), bathorhodopsin (B) is trapped below −140°C. Lumirhodopsin (L) is obtained by warming above this temperature, and metarhodopsin I (MI) begins to form above −40°C (see ref. [5]). Above −15°C, MI is in thermal equilibrium with metarhodopsin II (MII) [44,45], which decays slowly to metarhodopsin III (MIII) and/or to opsin and all-*trans*-retinal [46,47]. The blue-shifted intermediate (BSI), which does not accumulate at low temperatures, is only obtained in time-resolved measurements at room temperature [48].

As with other photoreceptors, photoproducts are denoted by their UV/Vis spectrophotometric properties. According to this convention, any 380 nm absorbing species (indicative of a deprotonated retinal Schiff base bond, see below) will be termed MII, and isochromic forms of MII will be denoted as subforms, as for example MIIa, MIIb, etc. An exception to this is the early isochromic MII-like species, which is termed MI_{380} [49]. It can be observed in detergent solution in considerable amounts, and is discussed in branched reaction schemes (for a review see [7]). A MII-like species with an absorption maxima at 470 nm due to a reprotonated Schiff base can be obtained at low pH and high anion concentrations [50].

3.4.2   Early events – storage of photon energy in bathorhodopsin

Photon absorption provides rhodopsin with approximately 55 kcal mol^{-1}. Two-thirds of this energy is stored in the photo-activated chromophore (all-*trans*-retinylidene) opsin complex [51], lifting the receptor from the dormant 11-*cis*-retinal/opsin conformation, via photorhodopsin, to bathorhodopsin. The dynamics of isomerization of the 11-*cis*-retinal protonated Schiff base (PSB) have been elucidated by femtosecond pump-probe experiments. In rhodopsin, the 11-*cis* → all-*trans* photoisomerization occurs on an ultrafast (femtosecond) timescale yielding the first photoproduct after only 200 fs [52–54]. In solution, the photoisomerization reaction of 11-*cis*-retinal-PSB is completed on the picosecond time scale [55]. Steric interaction of the chromophore with the protein and chromophore distortion is believed to be responsible for the extremely fast kinetics and the high photoreaction quantum yield of rhodopsin of ~0.67. Upon excitation of the chromophore, it is assumed that the molecule undergoes a nonadiabatic barrierless motion along a coordinate, which leads from the 11-*cis*-retinal-PSB excited state to the all-*trans*-retinal-PSB ground state. This appears to happen during a single torsional vibration [53]. The observed vibrational coherence in the photoproduct [53] argues that the isomerization coordinate on the excited state continues directly onto the ground state potential energy surface thus avoiding excited state equilibration. *Ab initio* methods have been used to study excited state dynamics of 3,5-pentadienal PSB as a model chromophore for 11-*cis*-retinal-PSB [56,57, and references therein]. Based on experimental data and computations it appears that during the first 25 fs the excited state 11-*cis*-retinal-PSB chromophore moves out of the Franck-Condon region along a mode which involves primarily stretching of the polyene chain [54,56]. It is assumed that the initial motion consists of an elongation of the double bond in the middle of the polyene chain associated with the change in bond order. Then, the 11-*cis*-retinal-PSB relaxes along a different coordinate towards a S_1 region where the excited (S_1) and ground (S_0) states conically intersect. The subsequent S_1 to S_0 decay occurs within 60 fs [54]. The intersection point has a twisted central double bond that provides a route for efficient nonadiabatic *cis*→*trans* isomerization. Thus, torsion around the central *cis*-configured double bond would set in only after the bond stretching has been completed. The computations on the 3,5-pentadienal PSB model chromophore provide information on the changes of charge distribution along the photoisomerization path and suggest that the S_1/S_0 intersection point is influenced by the charge distribution around the retinal chromophore. The x-ray structure reveals that Glu[181], in the extra-cellular plug domain, points towards the 11,12-ene of the chromophore and may be responsible for the exclusive isomerization around the $C_{11}=C_{12}$ double bond. A corresponding external point charge, which can perturb the electron distribution of the polyene chain, was postulated earlier [58].

In contrast, ring-constrained 11-*cis*-locked analogs, i.e. 11-*cis*-retinal analogs, with a bridge between C_{10} and C_{13} of 1–4 carbons which prevents isomerization around the $C_{11}=C_{12}$ bond, stabilize opsin in its inactive conformation (or minimally active), even under light conditions [59–61]. Rhodopsin, regenerated with a locked 11-ene via a six-membered ring (two-carbon bridge between C_{10} and C_{13}), readily undergoes photoisomerization, albeit not around the $C_{11}=C_{12}$ double bond. Several isomers (9-*cis*- and/or 13-*cis*-forms of locked 11-*cis*-retinal) can be extracted from the retinal binding pocket [62]. However, this photoisomerization does not lead to significant activation of Gt [61–63] or chromophore-induced conformational changes of the opsin moiety, as investigated by FTIR spectroscopy [62]. These results suggest that the main effect of the native chromophore *cis*→*trans* isomerization around $C_{11}-C_{12}$ is to impose strain on the chromophore, confining energy in the interaction of the chromophore with nearby residues. This allows transformation and storage of photon energy into chemical free energy, which is used to change the protein conformation into the active state. In a possible scenario compatible with the crystal structure data, isomerization by rotation around the Schiff base side of the $C_{11}-C_{12}$ double bond would lead to relocation of the polyene chain to a closer position to the side chain of Ser[186]. Steric restriction should limit the degree of rotation. This may explain part of the distortion in the all-*trans* configuration in bathorhodopsin and the unchanged Schiff base environment, in agreement with the spectroscopic data [64,65].

3.4.3   *Relaxation and steric trigger – lumirhodopsin and metarhodopsin I*

A large positive reaction enthalpy and reaction volume relative to the ground state accompanies the formation of the lumi (L) intermediate [66]. The enthalpy of reaction depends strongly on the hydrophobic environment (90 vs.

11 kJ mol^{-1} for washed membranes and dodecyl maltoside solution, respectively [67]). In the B→L transition (via BSI), stored energy is released rapidly through changes in the chromophore and its local environment and thermal dissipation. In L, most of the photon energy absorbed by rhodopsin has already transferred to the apoprotein [68].

Several spectroscopic investigations using cyclohexenyl ring (β-ionone)-modified retinal analogues suggested that, in this transition, the ring portion of the chromophore changes its interaction with nearby amino acid residues (for details see [69]). Photolabelling studies have indicated that the β-ionone ring, whose position is largely constrained in the ground state, is relocated in the B→L transition [70], thereby releasing chromophore distortion as seen in the largely reduced hydrogen out-of-plane modes in this intermediate [65]. It is thought that in this new configuration the β-ionone ring triggers the formation of the later, protonation- and G-protein-dependent MI and MII states, which form on the timescale of micro- and milliseconds. MII formation coincides with the eventual activation of the receptor. Accordingly, ring-modified rhodopsins showed a decrease in Gt activation [71]. The L→MI transition occurs with a decrease in enthalpy and entropy, while the MI→MII transition occurs with an increase in enthalpy and entropy [72], suggesting that the molecular events occurring in the L→MI transition are opposite in nature to those in the MI→MII transition. This may be explained by specific interactions built up in the chromophore region leading to a thermodynamically stable MI conformation. Conversely, loss of these interactions leads to an increased coupling between rhodopsin's hydrophobic core and its cytoplasmic domain and allows the MI→MII transition, which results in relaxation of the whole protein into the thermodynamically stable MII state [43,73,74]. Rhodopsin, regenerated with 11-*cis*-9-demethyl-retinal, forms MI but shows a significant shift in the MI⇌MII equilibrium towards the MI side with severe consequences for the ability of 9-demethyl rhodopsin to activate Gt. The activity of this pigment is reduced four- to five-fold, because fewer molecules enter the active MII state due to less constrainting chromophore–protein interactions and increased entropy in 9-demethyl MI [50,75]. Correct chromophore-protein interactions in MI are decisive for transition to MII, especially for accompanying proton transfer reactions which depend on the scaffolding function of all-*trans*-retinal in the MI state [75]. Transitions L→MI and MI→MII are influenced by rhodopsin's lipid environment. Detergents such as alkylglycosides and alkylmaltosides increase the entropy of the ground state and intermediates, driving rhodopsin's light-induced conformational changes to MII [76].

3.5   Metarhodopsisn II: the active photoproduct of rhodopsin

3.5.1   *Metarhodopsin II*

With the formation of MII, the intermediate capable of catalyzing nucleotide exchange in Gt is reached [77]. In this photoproduct, the Schiff base bond between chromophore and apoprotein is still intact but deprotonated [78]. Blocking Schiff base deprotonation by monomethylating the active-site lysine (Lys[296]) abolishes light-induced formation of a MII-like intermediate [79]. According to FTIR analysis, Schiff base deprotonation is mechanistically coupled to the protonation of Glu[113] [80]. Based on the definition of MII as any photoproduct that binds the chromophore via a deprotonated Schiff base, the intermediate is characterized by several special properties (for details see [7,73]) MII is formed with a large activation energy of >150 kJ mol^{-1}, both for the light-induced formation from the ground state (for references see [7]), and from MI by pressure jump [81]. The reaction enthalpy relative to the ground state (ΔH = 110 kJ mol^{-1}) was determined either directly [42] or was derived from the van t'Hoff equation (ΔH = 40 kJ mol^{-1} relative to its predecessor MI) [82]. The entropy increases largely in MII (37.9 cal K^{-1} mol^{-1} was determined for the MI$\rightleftharpoons$MII equilibrium at pH 7) [83]. The large entropy change and the MI$\rightarrow$MII reaction volume, of the order of 100 ml mol^{-1} [81] (dependent on the preparation), are apparently due to an unfolding of the protein, leading to exposure of binding sites for Gt and other signal proteins. In accordance, MII shows an enhanced susceptibility to partial digestion [84,85]. MII formation depends on the presence of an aqueous milieu (see [86–88]). Hydroxylamine and sodium borohydride can attack the Schiff base only after formation of MII [89–91], supporting a less constrained chromophore–protein interaction in this state. The formation of MII from MI is dependent on the osmotic pressure [92]. Osmotically sensitive regions of rhodopsin, which change their hydration during the MI/MII conversion, are narrow crevices or pockets [92]. MII formation causes perturbations in the UV-spectrum, indicating a more hydrophilic environment of aromatic residues [93,94], changes in linear and circular dichroism, interpreted as a rotation of the chromophore relative to the plane of the disk membrane [47,95], birefringence changes [96,97] and changes in near-infrared light scattering of the disk membranes [98,99].

UV-absorption changes of Trp[126] in helix III and Trp[265] in helix VI can be seen in detergent solution between the ground state and MII [94]. Whether these changes occur in earlier photoproducts is not known. Lack of a steric interaction between Trp[265] (Figure 6) and the β-ionone ring could cause the movement of helix VI [100]. This is in accordance with stationary linear dichroism measurements that show a Trp-residue changes its orientation ~30° during the MI$\rightarrow$MII transition [101].

MII formation is accompanied by changes of the membrane interfacial and transmembrane potential, the latter being positive relative to the aqueous exterior of the disk. Electrostatic potentials are measured in situ as a component of the "early receptor potential" [102,103], on lipid impregnated filter materials [104,105], and on lipid bilayers with photoreceptor membranes attached at one side [106]. The ERP can also be recorded directly in giant cells that heterologously express rhodopsin [107,108] and may, in combination with site-directed mutagenesis, offer new approaches to the relationship between electrostatics and structure of the receptor protein.

Vibrational spectroscopy indicates considerable alterations of the apopro-tein structure in MII [109–111]. Some bands in the infrared spectrum can be specifically assigned to carboxyl groups [86,112,113]. Time-resolved EPR spectra indicate small movements near the second cytoplasmic loop, kinetically correlated with MII formation [114,115]. Many mutations in rhodopsin have been reported to affect its functional activity, but only a few have been assigned to the formation of MII as a spectrally defined species, including the Schiff base counter-ion [116–118] and certain histidine residues [119,120]. All these data are consistent with considerable conformational changes that accompany MII formation – necessary for interaction with Gt or arrestin which are only able to bind after MII has formed.

3.5.2 Role of the hydrophobic environment and light-induced reorganization of disk membrane phospholipids

The native disk membrane contains approximately equimolar amounts of PC and PE phospholipids and lower proportions of PS ($\approx$14%) and phospha-tidylinositol (2.5%). The content of cholesterol in the disk membranes varies from the bottom to the top of the ROS [between 0.30 (bottom) and 0.05 (top), relative to the total lipid content] [121]. The phospholipids consist mainly of C_{22}-fatty acids which are polyunsaturated to a high percentage (22:6, but no 22:3 in the dominant species) [9,122,123]. In the native disk membrane envi-ronment, the MII photoproduct is formed in milliseconds, and the MI$\rightleftharpoons$MII equilibrium is well on the side of MII. This equilibrium and formation of MII is influenced by changes in the lipid composition of the disk membrane. MII formation depends on the presence of unsaturated lipids and on the fluid-ity of the phospholipid hydrocarbon chains. Thus, increasing the membrane rigidity by the addition of cholesterol or removal of lipid unsaturation, e.g. by reconstitution of the receptor in egg-lecithin vesicles, shifts the MI$\rightleftharpoons$MII equilibrium towards MI [124–126]. Similarly, in membranes with a reduced lipid content obtained by phospholipase C treatment or membranes containing short-chain, saturated lecithin, the decay of MI is retarded and yields almost no MII but rather predominantly free retinal plus opsin [127,128]. The ratio between the free energies of MI and MII is modulated by an interfacial tension-like interaction between rhodopsin and the bilayer. However, MII formation does not require a specific phospholipid head group [129], but it is enhanced in PE or PS bilayers [130], probably as a consequence of a surface charge effect [131].

In highly fluid detergent micelles (octyl glucoside or dodecyl maltoside), both an enhanced rate of MII formation and a shift of the equilibrium to MII are observed. In general terms, this means that the energy barrier for the MI$\rightarrow$MII conversion is lowered, and that photoproducts preceding the conver-sion are affected [76] (for the early MI$_{380}$ product see [132]). In octyl glucoside solubilized disk membranes, the activation free energy of MII formation is linearly dependent on the level of associated disk phospholipid [133], until the lipid/rhodopsin ratio of the native membrane is reached. This argues against a specific mode of interaction between rhodopsin (and/or MII) and the lipid. However, FTIR spectroscopy has indicated that a small amount of lipid may bind tightly to the receptor ground state, which alters its interaction in the transition to the active state MII [134].

In the disk membrane, lipids undergo a rapid flip-flop between outer and inner leaflet (half mean time <5 min) [135]. In the resulting equilibrium, PS has a distinct preference for the outer leaflet, whereas PC and PE show a small, if any, asymmetry (see [135,136] and references therein). Studies on osmotically intact disk vesicles of bovine ROS have shown that light-induced formation of the active MII state has an effect on the transbilayer redistribution of disk membrane phospholipids [137]. Redistribution was measured by bovine serum albumin extraction of spin-labelled PC-, PE- and PS-phospholipid analogs from the outer leaflet of the membrane. Upon photolysis of rhodopsin, a change in the redistribution of PS was found as seen by a fast transient (<10 min) enhancement of spin-labelled PS extraction. This effect was augmented by a peptide stabilizing MII, suggesting a direct release of one molecule PS per rhodopsin into the outer leaflet upon MII formation and subsequent redistribution between the leaflets. In the case of PC and PE, more complex kinetics were observed. In both cases there was a consistent prolonged period of reduced extraction (two lipids per rhodopsin in each case). The different phases of phospholipid reorganization after illumination are likely to be related to the formation and decay of the active rhodopsin species and to the subsequent regeneration process.

3.5.3 Formation of the signaling state

To discuss the signaling state, the MI$\rightleftharpoons$MII equilibrium in the reaction scheme (Section 3.4.1) has to be extended to incorporate its pH-dependency and to do justice to the known substates of MII:

$$\text{MI (478)} \ \rightleftharpoons \ \text{MIIa (380)} \ \overset{\text{H}^+}{\rightleftharpoons} \ \text{MIIb (380)}$$

The negative and positive enthalpies (ΔH) in forming MI and MII, respectively, indicate that molecular interactions built up in MI are lost upon transition to MII. To drive the conversion, the entropy must increase, and thus the overall disorder in the protein. In both MII states, MIIa and MIIb [76], the Schiff base bond of all-*trans*-retinal is still intact but deprotonated. The first protonation switch occurs in the MI to MIIa transition, which is accompanied by translocation of the Schiff base proton to the counter-ion Glu[113] (see [138]). At that stage the prosthetic group all-*trans*-retinal has the characteristics of a ligand agonist that facilitates MIIa formation by elevation of the free energy (ΔG) of MI. Although the disruption of the salt bridge shifts the conformational equilibrium towards the signaling state, there are other determinants of the active state as shown by mutagenesis of the counter-ion region [138]. Formation of MIIa may also release inactivating constraints among H-II, H-III, H-VI and H-VII due to changes in steric interaction between opsin and all-*trans*-retinal.

The second step, formation of MIIb, involves the protonation of Glu^{134} by proton uptake from the aqueous phase [76,138,139]. This residue, a part of the highly conserved E(D)RY motif in GPCRs, forms a salt bridge with the adjacent Arg^{135}, suggesting protonation of Glu^{134} as a mechanism to directly destabilize the constraints imposed by this salt bridge and to induce formation of the active cytoplasmic receptor surface. By FTIR spectroscopy, it could be shown that, in the complex with Gt, Glu^{134} of rhodopsin is protonated [139]. Mutation E134Q, eliminating the negative charge, is known to evoke constitutive activity of opsin [140] and abolishes light-induced proton uptake [141]. However, with bound 11-*cis*-retinal, the mutant rhodopsin is inactive but shows light-induced activity. One explanation could be that an activation mechanism, which is merely based on successive release of constraints, leads to formation of the catalytic receptor–Gt interface. This was approached by construction of mimics of the receptor surface, in which combinations of fragments corresponding to the cytoplasmic loops and/or carboxy-terminal tail of opsin were inserted onto a surface loop of thioredoxin [142]. These mimics showed binding to Gt of varying degrees, but low catalysis of nucleotide exchange in Gt. Full Gt activation requires both the whole opsin apoprotein and the retinal ligand, which controls even the last steps of activation in the native receptor [75]. However, conformational changes caused by release of inactivating constraints due to light-induced changes in steric interaction between opsin and all-*trans*-retinal and deprotonation of the Schiff base were measured by EPR and are seen in a dominant movement of the helix VI out of the helix bundle [24]. Similar helix movements upon receptor activation were shown for other GPCRs [143]. A corresponding movement of helix VI cannot be seen in the ground state of the E134Q mutant; however, EPR analysis has shown part of the conformational change around helix III [144]. Mechanistically, it would be interesting to learn whether the dominant movement of helix VI coincides with formation of MIIa or MIIb, and whether it can only occur as a consequence of the protonation changes at Glu^{113} and Glu^{134}.

To explain the sequential flow of events, we propose that, first, the N-terminal part of H-III moves into a position to induce proton transfer from the Schiff base to Glu^{113} (MIIa). The tandem of glycines would then allow amplified movement of the C-terminal part of this helix, thus inducing a larger structural change at the cytoplasmic surface, which is linked to the reprotonations and rearrangement events around the ERY tripeptide. This mechanism would require a separate conformational change, following neutralization of Glu^{113}, in agreement with a separate MIIb state. The MIIa/MIIb two-step proton translocation scheme is independently supported by the established two-step pumping or signaling processes in the archaeal rhodopsins.

3.5.4   Mechanistic insights from archaeal rhodopsins and photoreversal of metarhodopsin II

Archaea contain four retinal-binding proteins, two transport proteins for protons and chloride [called bacteriorhodopsin (BR) and halorhodopsin (HR), respectively], and the sensory rhodopsins (SRs) that mediate phototaxis responses (see also Chapter 1). As in rhodopsins, the chromophore, in this case all-*trans*-retinal, is bound to a lysine residue in H-VII. High-resolution structures of the ground states of BR, HR and SRII are known [145–147], while structural information on the active M_1 and M_2 states of BR (the possible analogs of MIIa and MIIb in rhodopsin) is available [148,149]. In all retinal proteins, a transient movement of H-VI is found that correlates with the opening of a cytoplasmic half-channel and the relocation of water in the hydrophobic environment of the half-channel. In BR and HR, protonation of a residue (Asp^{96} in BR) at the cytoplasmic border of H-III occurs in this context. The SRs interact constitutively with dimers of Htr transducer proteins by lateral helix–helix contact, thus forming a receptor that can bind and activate intracellular phosphoregulatory proteins [150]. Both SRs are proton pumps in their free state, but bound transducer proteins block the half-channel for proton transport while helix movement is still allowed to occur [151]. Thus, in the SR–transducer photoreceptor complex, part of the free energy used for the transport mode in free SR is channeled into a long-lived signaling state to account for its sensory function [150]. In these rhodopsins, both H^+ transfer near the Schiff base and H^+ uptake with movement of helix VI which might provide the trigger for transducer activation are also seen [151]. Therefore, it might be useful to consider this part of signal transmission in rhodopsin as a SR–transducer-like partial proton pump [76]. For archaeal rhodopsins, an extended H-bonded network arises [149,152] which might also be relevant for the mechanism of signal transmission, i.e. formation of MIIa and MIIb.

Like bacterial rhodopsin ground states, the MII intermediate carries all-*trans*-retinal in its retinal-binding pocket. Therefore the effect of blue light on this photoproduct was soon investigated. Among other photoproducts a measurable one absorbing at 500 nm is generated. The spectral characteristics and the accompanying proton release argued for a reversal of the activation process and photoregeneration of rhodopsin [153,154]. However, a new product with novel properties is formed rather than rhodopsin in ground state [155]. FTIR studies and retinal extraction showed that this product has the protein conformational characteristics of MII, which arises from MII by thermal decay, and still carries the all-*trans*-retinal isomer. The data indicates the presence of a "second switch" between active and inactive conformations that operates by photolysis but without stable isomerization around the $C_{11}=C_{12}$ double bond. It is not known whether transient or metastable isomerizations are involved in this pathway. This emphasizes the characteristic of the rhodopsin ground state, which in contrast to invertebrate rhodopsins (see also Chapter 2) is only accessible by metabolic regeneration with 11-*cis*-retinal and ensures by its exceedingly stable chromophore–protein interaction the low noise of the signal transduction process in rods. Under conditions of substantial bleaching, however, accumulation of the photoreverted product discussed above may influence bleaching adaptation phenomena and even may be involved in blue-light-induced retinal degeneration [156].

3.6   Interaction between photoactivated rhodopsin and G-protein

3.6.1   Stabilization of metarhodopsin II by Gt

In the absence of exogenous GDP and GTP, Gt can shift the MI/MII equilibrium resulting from photolysis of rhodopsin due to binding to MII. This "extra-MII" effect was first observed for native disk membranes [157] and follows the reaction scheme:

$$
R \xrightarrow{\text{light}} MI \rightleftharpoons MII \underset{GDP}{\overset{G\alpha\beta\gamma \bullet GDP}{\rightleftharpoons}} MII \bullet G\alpha\beta\gamma \xrightarrow{GTP} MII + G\alpha \bullet GTP + G\beta\gamma
$$

In the presence of GTP, the catalytic GDP/GTP exchange function of rhodopsin leads to dissociation of the nucleotide-free MII·Gt complex and dissociation of the Gt holoprotein into the activated Gtα subunit and the Gβγ dimer as a quasi-irreversible step because of the subsequent GTP hydrolysis step. The fast rate of MII-Gt dissociation requires millimolar GTP concentrations [12]. MII stabilization is abolished when GDP is bound [158,159]. Thus, either GDP or GTP can dissociate the nucleotide free MII-Gt complex. The scheme implies that Gt does not bind with significant affinity to MI or MIII, in agreement with other studies [77,79,160]. Gt binds non-cooperatively to MII with 1:1 stoichiometry, and shows a bell-shaped pH-dependence with a maximum at pH 7.6 [35]. For an in-depth analysis of the dissociation constant of the MII-Gt complex see [12]. Transient formation of extra MII [161] provides a direct measure of the Gt activation kinetics and activation energy (see [73]). The stabilization of MII is also observed with C-terminal peptides of Gt [Gtα(340–350) and Gtγ(60–71)farnesyl] and arrestin but not with RK (see below).

3.6.2   The rhodopsin–Gt interface

Early approaches employing biochemical techniques, including proteolysis and chemical modifications of rhodopsin and Gt, allowed the first insight into the main structural elements (for a review see [7]). Major progress was made with the development of heterologous expression systems for these proteins, allowing the investigation of mutant proteins by biochemical and biophysical means.

3.6.2.1   Binding sites at Gt

Another approach is to probe the interface by synthetic peptides from rhodopsin and Gt. Synthetic peptides derived from Gtα, including the C-terminal stretch [Gtα(340–350) and the one spanning residues 311–323, compete with Gt to form extra MII [162,163]. Interaction of these stretches with rhodopsin was confirmed by mutational analysis [164,165]. Besides sites in the Gα subunit, the C-terminal domain of Gtγ(60–71)farnesyl could be identified as directly stabilizing MII [166,167]. Both the C-terminal peptide sequence and the farnesyl moiety are specific determinants for the R*–Gt interaction [166]. The C-terminal tail of the Gtγ-subunit is masked in the Gtβγ dimer and becomes exposed on collisional coupling of the holoprotein to the receptor (see [168]). The crystal structure of Gt [169] revealed that both C-terminal regions of Gtα and Gtγ, although only partly resolved in the structure, are localized to a common surface of the Gt holoprotein (see [167,170]). By transferred nuclear Overhauser effect spectroscopy, a helical turn conformation followed by an open reverse turn was determined for the Gtα(340–350) peptide in the rhodopsin-bound state [171]. MII stabilization, with analogs of this peptide, provided information about individual residues contributing to the interaction with R* [172]. Binding of the peptides from the C-termini of Gtα and Gtγ was also confirmed by a photoregeneration assay [27]. FTIR difference spectroscopy was extended to the R*Gt interaction. Specific protonation changes are induced when Gt or C-terminal peptides of Gtα and Gtγ bind to rhodopsin and shift the MI/MII equilibrium [88,139,173,174].

3.6.2.2   Binding sites at rhodopsin

Loop structures of rhodopsin were shown to interact with Gt by competition with synthetic peptides and mutagenesis approaches. Only peptides from the second and third loop, and the cytoplasmic helix VIII (Figure 2), competed for Gt-dependent stabilization of MII [175]. Interaction of these loops with Gt was confirmed by nucleotide exchange catalysis of mutant rhodopsins [176–179], flash photolysis [27,176], light scattering [180] and fluorescence techniques [179]. The lack of peptide inhibition and the overall minor effect of point mutations in the first cytoplasmic loop suggested that the residues connecting helices I and II are not involved directly in recognition by Gt [181]. Sites at either the second or the third cytoplasmic loop, in conjunction with the fourth loop, appear to be sufficient to maintain the empty-site R*–Gt interaction. A more complex interaction pattern may be required to allow for the fast release of GDP from the Gt-rhodopsin collisional complex. GTP binding to the empty Gtα site, and the subsequent release of Gtα•GTP from rhodopsin, requires intact structures of both the second and the third loop [176,180]. It is not known, however, whether only loop structures contribute to the interaction surface. Light-induced exposure of the cytoplasmic end of transmembrane helix VII, as probed by specific binding of a monoclonal antibody recognizing this epitope, may argue for additional binding sites closer to the hydrophobic core of rhodopsin [182]. The fourth loop follows helix VII and is restricted by two palmitoylated cysteines anchoring it to the disk membrane. According to the X-ray structure, this loop adopts a cationic amphipathic helical conformation (helix VIII) with Lys[311] and Arg[314] facing the cytoplasm. The cytoplasmic helix VIII is attached to helix VII by a short linker (VIII in Figure 2). Interaction of this loop with Gt has been discussed previously (see [179]). Based on fluorescence studies with a corresponding loop peptide the interaction of this loop was specified for the βγ-subunit [183,184]. Further studies have confirmed interaction of the fourth loop with Gt [27, 178,179] and suggested that the N-terminal part of this loop is involved in binding of the C-terminus of Gtα and that this loop plays a structural role in binding of Gβγ [27,179]. Only minor effects on Gt activation by R* were seen when the palmitoyl modifications at Cys[322] and Cys[323] were removed, either using high concentratons of NH_2OH in the dark [185] or mutational substitution of the palmitoyl-anchoring cysteines by serines [179,186].

3.6.3   Conclusions

Although most binding sites on both rhodopsin and Gt are identified, a mutual assignment of the binding sites is not possible on the data available. Controversial assignments could be explained by the formation of receptor dimers (see [187]). Also, the hydrophobicity and charge pattern derived from the available ground state structures of both proteins can only give vague hints about how docking of the two proteins could occur. Critical residues, e.g. the ERY sequence, are buried in the structure. From site-directed spin labelling studies (see [188]) and other work it is known that the receptor undergoes large structural changes upon receptor activation, thereby exposing buried binding sites (see [7,187]). Similar changes are thought to occur in the G-protein upon receptor binding. In the absence of a clarifying crystal structure of the R*•Gt-complex, new approaches were devised to obtain insight into the R–Gt interface. The nuclear Overhauser effect between site-directed ^{19}F-labels on the cytoplasmic receptor surface can be used to measure light-induced conformational changes [189]. In another approach, site-directed Cys mutagenesis was used to introduce crosslinkers at specific sites of the rhodopsin surface. Covalent crosslinking of the R*•Gt-complex followed by trypsin degradation and mass spectrometric analysis can reveal the Gt-sequences binding to the labelled sites on rhodopsin [190,191].

Different models were proposed for temporal aspects and the mode of interaction. Based on kinetic studies a "sequential fit" mechanism involving a sequence of microscopic recognition via interacting C-termini of Gt and conformational interlocking was proposed [167]. By mutagenesis studies, the α5 helix was identified as a functional microdomain, affecting nucleotide release [192], and the unit consisting of Gtα's C-terminus/α5 helix/β6/α5 loop was suggested to constitute a dominant channel for transmission of the GPCR-induced conformational change leading to G-protein activation [193]. It appears that the receptor uses the G-protein's βγ heterodimer as a lever, tilting it to pull open the guanine nucleotide binding pocket of Gα [194].

3.7   Interactions between photoactivated rhodopsin and arrestin and rhodopsin kinase

3.7.1   Arrestin

Deactivation of light-activated rhodopsin involves phosphorylation of its C-terminus by RK, and subsequent binding of arrestin to block interaction with Gt. Photoactivated rhodopsin eventually decays into (phosphorylated) opsin and all-*trans*-retinal. The role of arrestin in this process was confirmed by knock-out studies [195]. Binding of arrestin is specific for the phosphorylated MII form with a K_D of 50 nM and a bimolecular on-rate of about 0.2 μM^{-1} s^{-1} [196,208]. The major phosphorylation sites of rhodopsin are Ser[338], Ser[343] and Ser[334] [29]. Under extrapolated cellular concentrations, the binding to rhodopsin is fast (of the order of 50 ms) and thus not rate limiting for the overall shut-off reaction sequence [197]. Moreover, self-association of arrestin was proposed as a regulation mechanism [198]. The affinity to opsin is low, as seen by the release of arrestin due to hydroxylamine-induced MII decay [208]. The rhodopsin–arrestin interaction may be short-lived in vivo; additional experiments in mice indicate that MII-arrestin complexes dissociate, to allow the reduction of all-*trans*-retinal to all-*trans*-retinol [199] and dephosphorylation of the receptor by a membrane-associated form of protein phosphatase 2A (PP2A) [32]. It remains to be studied how the necessary release of arrestin is induced. On contact with the phosphorylated C-terminus of rhodopsin, arrestin switches from the inactive to the active state which is capable of binding to rhodopsin [200]. Two crystal structures of arrestin are available [201,202] and show an interesting "double-cap" structure, formed by a N- and a C-terminal domain of arrestin, respectively. However, the two structures disagree in important details such as the location of the N- and C-termini. Both structures probably represent the inactive conformation of the molecule.

3.7.1.1   Arrestin–receptor interaction sites
By using synthetic peptides and phage display, arrestin residues 109–130 were found to be involved in the interaction with rhodopsin, and indications of additional sites were found [203]. By employing the spectroscopic "extra MII" assay and overlapping synthetic peptides, three regions of arrestin could be identified which competed with arrestin. These regions display higher affinity than the region previously identified [204]. The respective parts of the sequence are located in both the N- and C-terminal domains of the protein structure, i.e. both caps. Interaction from the side of the receptor appears to involve cytoplasmic loop domains of rhodopsin and a phosphorylation site at the C-terminus [205]. Synthetic peptides corresponding to rhodopsin's cytoplasmic loops competed against R*–arrestin interaction in a co-elution assay; the effect was strongest for a peptide from loop C-III and much weaker for a peptide from loop C-I [206]. Analysis of arrestin binding to rhodopsin mutants suggested an interaction of loops C-I and C-II with arrestin [207].

3.7.1.2 Conformational switch

Visual arrestin, and presumably all arrestins, share a conformational switch with G-proteins, operated by contact with the active receptor [170,200, 208–210]. Early indications came from the Arrhenius plots of the arrestin binding reaction monitored via MII stabilization which yielded a large activation energy, which can be interpreted as a conformational transition in domains of arrestin and/or rhodopsin, linked to the interaction [208]. Limited proteolysis of free and bound arrestin has indeed shown that binding to rhodopsin protects arrestin from the attack of the proteolytic enzyme. This was interpreted as an indication that arrestin bound to phosphorylated rhodopsin and assumes a conformation which is different from that of free arrestin [200]. Heparin [200] or phytic acid [211] mimic the effect of photoactivated, phosphorylated rhodopsin to some degree. A highly cationic region beginning with residue 163 was proposed to mediate the interaction with the negatively charged regions of phosphorylated rhodopsin or heparin [200]. Studies on mutated and truncated arrestins [209,212] have confirmed this hypothesis and localized another major binding site for the phosphorylated region of rhodopsin, heparin and phytic acid to the N-terminus of arrestin (residues 1–47). A sequential process was invoked from mutational and synthetic peptide approaches. Based on structural assignments, Sigler and co-workers [198,202,213] have specified a trigger mechanism, in which the phosphorylated C-terminus of the receptor interacts with the "polar core", embedded between the N- and C-terminal domains in the fulcrum of the molecule. Upon this interaction, intramolecular interactions, including a hydrogen-bonded network of buried ion pairs and salt bridges between charged side chains are disrupted, leading to structural changes, possibly involving an en bloc rearrangement of the N- and C-terminal domains [202]. It may seem intriguing that arrestin, as a "blocking" protein, goes through a conformational switch. However, this makes sense because it ensures that no interaction occurs during the amplifying phase of phototransduction. The binding sites identified are distant and do not form a flat surface. A conformational switch may thus be required to allow their simultaneous interaction with the relevant receptor loop structures.

The situation is similar for the G-protein Gt, where two distant sites (the C-termini of the α- and γ-subunits, ~45 Å apart) are involved in the signal transfer [167]. For arrestin and Gt (see [197] and [35], respectively), the simultaneous binding of two receptors at one molecule is unlikely because the titration of the complexes yields 1:1 ratios. A splice variant of arrestin, p44, in which the C-terminal residues 370–404 are replaced by a single alanine, is apparently only present in ROS [214] and has been reported to inhibit phototransduction in a similar way to the parent protein. However, it interacts not only with phosphorylated, but also with unphosphorylated rhodopsin, although with lower affinity [197]. The lack of the C-terminal sequence also lowers the activation energy of the binding reaction (70 instead of 140 kJ mol^{-1}), and it removes any specificity for the C-terminal structure of the receptor, so that even C-terminally truncated rhodopsin binds to p44. Thus it appears that, by the lack of the C-terminal structure, the conformational switch between active and inactive conformations of arrestin is absent in the splice variant [197]. This fits nicely to a proposal, based on intrinsic fluorescence and circular dichroism data, that the conformational switch involves localized movements of the N- and C-termini of arrestin; these regions may interact in the inactive conformation, and may be separated by interaction with phosphorylated rhodopsin [215].

3.7.1.3 Molecular recognition

Arrestin and Gt appear to use different mechanisms of microscopic (i.e. site to site) recognition. In contrast to the G-protein, where the Gtα and farnesylated Gtγ C-terminal sequences are identified as binding sites, all arrestin domains are intramolecular stretches. Remarkably, the Gtα and Gtγ C-termini, when prepared as synthetic peptides, have the capacity to recognize the MII species and to distinguish it from the other intermediates. None of the numerous arrestin peptides examined displayed such specificity [204]. Only the parent structure appears to provide the recognition specificity, presumably by non-linear binding domains on the side of the receptor, which form only in concert with the active arrestin conformation.

3.7.2 Rhodopsin kinase

Like G_{ι} monomeric RK (G-protein receptor kinase, GRK1) binds to cytoplasmic loops of R* forming a stable complex with a dissociation constant of 0.5 μM [196,216] (for review on properties of RK see [32]). By binding of ATP its affinity for the activated receptor is increased by a factor of ca. 10 [196] allowing phosphorylation of the rhodopsin C-terminus. Depending upon the conditions, up to 7–9 phosphates are transferred in vitro. In vivo, a single phosphate per receptor is sufficient to quench signal transduction [217,218]. Autophosphorylation of RK and phosphorylation of the receptor lowers its affinity to R* and allows dissociation of the R*·RK-complex. Like transgenic mice expressing truncated rhodopsin molecules lacking the C-terminal phosphorylation sites [219], transgenic mice lacking RK show single-photon responses which are larger and longer lasting than normal [220]. The data suggest that phosphorylation of R* by RK is solely responsible for normal rhodopsin deactivation in the dark-adapted rod.

3.7.2.1 Signaling state

RK and Gt have several important determinants of their signaling state in common: (i) rhodopsin regenerated with retinal analog 11-cis-9-desmethyl retinal (9-dm rhodopsin) is a poorer substrate for the kinase [221], as it is a poorer catalyst of Gt activation [75]; (ii) pH/rate profiles of Gt activation or phosphorylation reflect the protonation of a surface proton acceptor (presumably Glu[134]) [222]; (iii) reversal of the residues Glu[134]/Arg[135] (part of the highly conserved ERY motif) in mutant rhodopsin is deleterious to the interaction of Gt or RK with rhodopsin [176,223]; and (iv) in loop mutants of rhodopsin, the lack of loops C-II or C-III abolishs the binding of the enzyme, while mutations in loop C-III affect both binding and catalysis [223]. These results would suggest that the signaling state for both proteins coincides with the MII photoproduct, in agreement with the two-step mechanism of R* formation. Intriguingly, however, the 'extra MII' assay does not reflect any preference of RK for any of the M states [196]. When MII formation is measured in the presence of both RK and Gt, the Gt-dependent enhancement effect is reduced, consistent with the notion that the binding of RK destabilizes the MII·Gt-complex by competition. This demonstrates that RK can bind to MI and probably to all metarhodopsin forms. A possible explanation for these observations would be that kinase can bind to all M states of rhodopsin but needs the protonated form, MII$_b$, to perform the actual phosphorylation step. The idea that formation of MI, the precursor of MII, provides the trigger for the generation of the kinase substrate [224] was confirmed by studies on intact retina, in which MI but not MII was allowed to form by warming up from low temperatures, followed by subsequent photolysis. The resulting photoproduct was a substrate for the kinase. These experiments have also shown that photolysis restores the spectral identity of rhodopsin but not the conformational changes that trigger phosphorylation (in remarkable analogy to the reverted meta product discussed above; [155]). It is an open question whether constitutively active mutants of rhodopsin, which activate G$_t$ in the absence of the chromophore, can activate RK [225,226].

3.7.2.2 Interaction sites
The mapping of binding sites of RK to active rhodopsin is not yet complete but yielded the involvement of loops C-I, C-II and C-III, as concluded from partial digestion [227] and studies on mutant rhodopsin [223,228]. When studying the role of loops C-II and C-III, it turned out that loop C-II is involved in binding, whereas C-III has a role in both binding and catalysis [223].

3.7.2.3 Direct competition
In vitro data of the kinetics of R*·RK-complex formation yielded a K_D of 0.5 mM and a bimolecular rate constant of 1 μM^{-1} s^{-1} [196]. However, these data are of limited value because the necessary detergent solubilization is likely to affect the kinetic properties of RK. Kinetic constants of RK–rhodopsin interaction were derived from the behavior of the photoresponse depending on the number of R* formed per disk membrane [229]. In model calculations, the R*–RK interaction in vivo was described with two subsequent steps, binding

and phosphorylation/dissociation. Based on this model, R* binding obeys different kinetics dependent on whether the kinase is substrate saturated or not, yielding reaction times for the binding of RK and the phosphorylation step under in vivo conditions (0.25 and 1 s, respectively [230]). Such a mechanism would transcend the classical notion of a constant characteristic lifetime of activated rhodopsin [230]. Not only arrestin, but also RK competes with G_t for binding to active rhodopsin [196]. This "pre-arrestin function" has implications in understanding the shut-off mechanism of rhodopsin because, together with the above-mentioned different R* binding kinetics for saturated/subsaturated RK, it affects the lifetime of R* and thus the photoresponse.

3.8   Light-independent signaling of different forms of the apoprotein

3.8.1   Opsin

In addition to activation of Gt by light-activated rhodopsin there is, in the absence of any chromophore, light-independent activation of Gt by the apoprotein opsin. In vitro, the rate of Gt activation by opsin is in the order of 10^{-6} of MII at neutral pH [38]. Higher activities of opsin found in Gt activation assays eventually depend on residual retinal derivatives in the preparations [231]. Opsin appears also to exist in two conformational states [232], as is understood for other ligand-free GPCRs [233]. At neutral pH, opsin is in the inactive conformation which is stabilized by a salt bridge between Lys^{296} and Glu^{113} [39,120], a mechanism also used to stabilize the inactive rhodopsin ground state. Consequently, opsin mutants lacking the charge at either Lys^{296} or Glu^{113} show enhanced basal activity, denoted constitutive activity [39]. As concluded from FTIR investigations, at low pH (pH < 5, depending on type and concentration of stabilizing anions) breaking of the salt bridge due to protonation of Glu^{113} and concomitant conformational changes occur [232]. This new conformation is similar to the MII conformation, capable of binding C-terminal peptides of Gtα, and represents the active conformation of opsin. The measured opsin activity would be high enough to have a physiological role in maintaining a certain stimulation of the visual cascade, which is one of the potential explanations for "bleaching desensitization" [234]. The desensitizing activity of opsin can be distinguished from the activity expressed in "photon-like noise" which was assigned to MII formed via reversal of phosphorylation and arrestin binding [235].

3.8.2   Retinal–opsin complexes

The addition of all-*trans*-retinal to opsin enhances its activity by the formation of non-covalent complexes [120,222,236]. At a physiologically relevant 1:1

molar ratio, these ligand-like receptor•agonist complexes between opsin and all-*trans*-retinal can activate Gt in the order of 0.5 (at pH 6.5) as well as MII [237]. These non-covalent complexes can generate a conformation capable of interacting with Gt [120,236] and thus adopt a light-independent signaling state [236]. This state can also interact with RK and arrestin [222,236,238].

Besides these non-covalent complexes, a second type of opsin•all-*trans*-retinal complex exists. Schiff base formation does occur between all-*trans*-retinal and opsin, but not with the original Lys[296] [236], leading to reversible "pseudo-photoproducts" [239] which interact with arrestin and kinase, but no interaction with Gt has been measured [239]. We may summarize the results as follows:

(A)   all-*trans*-retinal + opsin $\rightleftharpoons$ opsin•all-*trans*-retinal
            $\rightarrow$ interaction with Gt, RK, arrestin
(B)   all-*trans*-retinal + opsin $\rightleftharpoons$ "pseudo-MII"
            $\rightarrow$ interaction with RK, arrestin

Starting from both products, 11-*cis*-retinal can regenerate rhodopsin by binding to Lys[296]. Moreover, the all-*trans*-retinal present does not compete with 11-*cis*-retinal, suggesting that in these products all-*trans*-retinal occupies a different binding site [236]. The level of Gt activation of opsin•all-*trans*-retinal complexes is strongly reduced when the palmitoyl groups are removed from Cys[322] and Cys[323] [237]. However, the removal of the palmitoyl anchors does not affect MII activity. This suggests that the activity of opsin•all-*trans*-retinal complexes indeed arises not from small amounts of reversibly formed MII, but from a separate form of active receptor with low intrinsic activity. Such active complexes may arise in vivo after the spontaneous decay of MII by hydrolysis of the Schiff base. Binding of 11-*cis*-retinal to opsin regenerates rhodopsin and provides a shut-off mechanism. During continuous illumination of the retina, opsin•all-*trans*-retinal may accumulate and play a role in the physiologic phenomenon of "bleaching desensitization".

3.9   Metabolism of retinal

Absorption of a photon by rhodopsin causes photoisomerization of the chromophore, 11-*cis*-retinylidene to all-*trans*-retinylidene, and concomitant changes in the apoprotein opsin. Ultimately, the Schiff base linkage is hydrolyzed and the chromophore is released from the binding pocket of opsin through a mechanism that has not been elucidated on the mechanistical level (Figure 7, reaction a). In one of the models, the chromophore is released into the intradiskal space, and then pumped out by a photoreceptor-specific ATP-dependent exchanger, ABCR protein [240]. This model may require further revisions, because the mice and humans lacking the functional ABCR protein are still able to regenerate rhodopsin at rates that are only slightly lower compared to normal controls. Therefore, the possible function of ABCR is to drive out the smallest amounts of all-*trans*-retinal to prevent formation of toxic

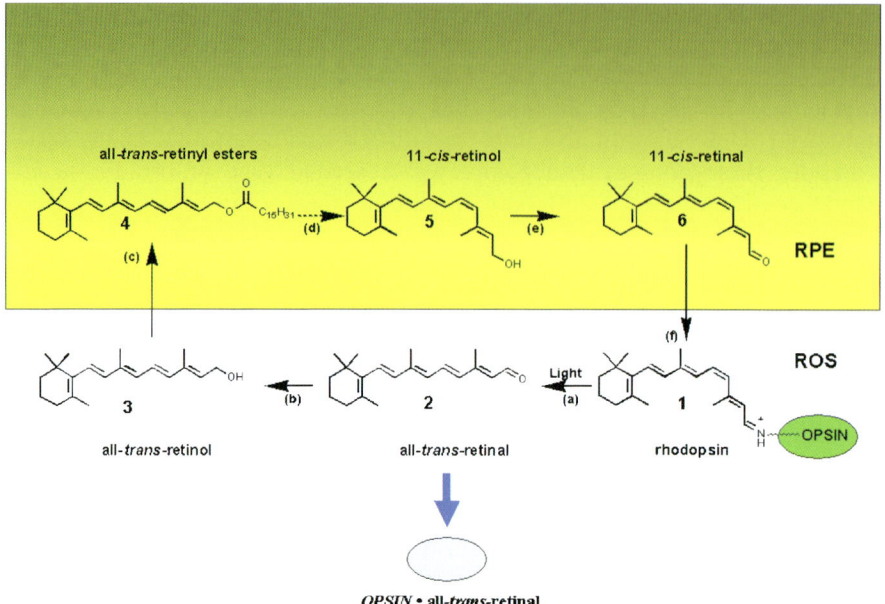

Figure 7. Flow of retinoids in the vertebrate eye. 11-*cis*-Retinal is coupled to opsin apoproteins in both rod and cone photoreceptor cells and is photoisomerized to all-*trans*-retinal by light. The regeneration of 11-*cis*-retinal, the universal chromophore of vertebrate retina, is a complex process involving photoreceptors and adjacent retinal pigment epithelial cells (RPE). For details see the text and an extensive review [241].

condensation products with phosphatidylethanolamine [241]. The cytoplasmic-assessable all-*trans*-retinal is reduced in a NADPH-dependent reaction catalyzed by retinol dehydrogenases (Figure 7, reaction b), members of the short-chain alcohol dehydrogenase super-family [242,243]. This reaction is the rate-limiting step in the retinoid cycle after intense bleaching [244]. All-*trans*-retinal also combines non-covalently with opsin (depicted by the blue arrow in Figure 7) in a different binding pocket from the original retinylidene-binding site [236,237]. As a consequence of elevated Gt stimulating activity, as compared to free opsin and lack of activity by rhodopsin, and the slow reduction process, opsin•all-*trans*-retinal complexes are possible candidates for the desensitizing form of the receptor during bleaching adaptation. This complex is also a substrate for RK, and when opsin is phosphorylated within this complex, it binds arrestin tightly [239]. Next, retinol diffuses through two cell membranes to the RPE (Figure 7, indicated by yellow), where it is trapped as insoluble fatty acid esters in a reaction catalyzed by lecithin-retinol acyl transferase (LRAT) [245] (Figure 7, reaction c). Although the isomerization process is unknown on the chemical level, retinyl esters are, most likely, converted into more reactive intermediates and isomerized to 11-*cis*-retinol (Figure 7, reaction d; see [241] for a more in-depth discussion of the potential mechanisms of isomerization).

Highly abundant RPE-specific protein RPE65 is believed to play a critical role in this process (see [241]). Finally, 11-*cis*-retinol is oxidized in a NADH/NADP-dependent reaction catalyzed by 11-*cis*-retinol dehydrogenases (Figure 7, reaction e), also like all-*trans*-retinol-specific enzymes, members of the short-chain alcohol dehydrogenases [242]. 11-*cis*-retinal diffuses back to ROS, where it irreversibly combines with opsin to form the 11-*cis*-retinylidene chromophore, and fully regenerated rhodopsin (Figure 7, reaction f). 11-*cis*-Retinal could alternatively be produced by direct photoisomerization of all-*trans*-retinal with the help of retinal G-protein-coupled receptor (RGR) protein in RPE and Müller cells [246]. Several mutations in genes involved in the retinoid cycle are associated with a subset of human retinopathies, including Stargard's fundus flavimaculatus, recessive retinitis pigmentosa (RP), Leber congenital amaurosis (LCA), or fundus albipunctatus (FA) (reviewed in [241]).

Acknowledgements

Our work has been supported by grants from the Deutsche Forschungsgemein-schaft (SFB 449), the Fonds der Chemischen Industrie, National Eye Institute (EY09339), Research to Prevent Blindness, Inc. (RPB) to the Department of Ophthalmology at the University of Washington, Foundation Fighting Blindness, Inc., the Ruth and Milton Steinbach Fund and the E.K. Bishop Foundation.

References

1. W. Kühne (1878). Über den Sehpurpur. In: *Untersuchungen aus dem Physiologischen Institut der Universität Heidelberg* (Vol. 1). Carl Winters Universitätsbuchhandlung, Heidelberg, Germany.
2. W. Kühne (1977). Chemical processes in the retina. *Vision Res.*, **17**, 1269–1316.
3. D.D. Oprian, A.B. Asenjo, N. Lee, S.L. Pelletier (1991). Design, chemical synthesis, and expression of genes for the three human color vision pigments. *Biochemistry*, **30**, 11367–11372.
4. R.W. Rodieck (1998). *The First Steps in Seeing.* Sinauer Associates, Inc., Sunderland, MA, U.S.A.
5. G. Wald (1968). The molecular basis of visual excitation. *Nature*, **219**, 800–807.
6. S. Hecht (1941). Energy, quanta, and vision. *J. Gen. Physiol.*, **25**, 819–822.
7. K.P. Hofmann (2000). Late photoproducts and signaling states of bovine rhodop-sin. In: D.G. Stavenga, W.J. DeGrip, E.N. Pugh Jr, (Eds), *Molecular Mechanisms in Visual Transduction* (Vol. 3, p. 91–142) Elsevier, Amsterdam.
8. T. Schöneberg, G. Schultz, T. Gudermann (1999). Structural basis of G protein-coupled receptor function. *Mol. Cell. Endocrinol.*, **151**, 181–193.
9. S.J. Fliesler, R.E. Anderson (1983). Chemistry and metabolism of lipids in the vertebrate retina. *Prog. Lipid Res.*, **22**, 79–131.

10. R.A. Bush, A. Malnoe, C.E. Reme, T.P. Williams (1994). Dietary deficiency of N-3 fatty acids alters rhodopsin content and function in the rat retina. *Invest. Ophthalmol. Vis. Sci.*, **35**, 91–100.

11. R.E. Anderson, M.B. Maude, K. Narfstrom, S.E. Nilsson (1997). Lipids of plasma, retina, and retinal pigment epithelium in Swedish briard dogs with a slowly progressive retinal dystrophy. *Exp. Eye Res.*, **64**, 181–187.

12. M. Heck, K.P. Hofmann (2001). Maximal rate and nucleotide dependence of rhodopsin-catalyzed transducin activation: initial rate analysis based on a double displacement mechanism. *J. Biol. Chem.*, **276**, 10000–10009.

13. H. Kühn (1981). Interaction of rod cell proteins with the disk membrane: Influence of light, ionic strength, and nucleotides. *Curr. Top. Membr. Transp.*, **15**, 171–201.

14. H.R. Seitz, M. Heck, K.P. Hofmann, T. Alt, J. Pellaud, A. Seelig (1999). Molecular determinants of the reversible membrane anchorage of the G-protein transducin. *Biochemistry*, **38**, 7950–7960.

15. T.J. Melia, J.A. Malinski, F. He, T.G. Wensel (2000). Enhancement of phototransduction protein interactions by lipid surfaces. *J. Biol. Chem.*, **275**, 3535–3542.

16. M. Heck, K.P. Hofmann (1993). G-protein-effector coupling: a real-time light-scattering assay for transducin-phosphodiesterase interaction. *Biochemistry*, **32**, 8220–8227.

17. R.S. Molday, U.B. Kaupp (2000). Ion channels of vertebrate photoreceptors. in: D.G. Stavenga, W.J. DeGrip, E.N. Pugh Jr, (Eds), *Molecular Mechanisms in Visual Transduction* (Vol. 3, pp. 143–181). Elsevier, Amsterdam.

18. P.D. Calvert, V.I. Govardovskii, N. Krasnoperova, R.E. Anderson, J. Lem, C.L. Makino (2001). Membrane protein diffusion sets the speed of rod phototransduction. *Nature*, **411**, 90–94.

19. K.C. Slep, M.A. Kercher, W. He, C.W. Cowan, T.G. Wensel, P.B. Sigler (2001). Structural determinants for regulation of phosphodiesterase by a G protein at 2.0 A. *Nature*, **409**, 1071–1077.

20. K. Palczewski, A.S. Polans, W. Baehr, J.B. Ames (2000). Ca^{2+}-binding proteins in the retina: structure, function, and the etiology of human visual diseases. *Bioessays*, **22**, 337–350.

21. A. Polans, W. Baehr, K. Palczewski (1996). Turned on by Ca^{2+} The physiology and pathology of Ca^{2+}-binding proteins in the retina *Trends Neurosci.*, **19**, 547–554.

22. K. Palczewski, T. Kumasaka, T. Hori, C.A. Behnke, H. Motoshima, B.A. Fox, I. Le Trong, D.C. Teller, T. Okada, R.E. Stenkamp M. Yamamoto, M. Miyano (2000). Crystal structure of rhodopsin: A G protein-coupled receptor. *Science*, **289**, 739–745.

23. D.C. Teller, T. Okada, C.A. Behnke, K. Palczewski, R.E. Stenkamp (2001). Advances in determination of a high-resolution three-dimensional structure of rhodopsin, a model of G-protein-coupled receptors (GPCRs). *Biochemistry*, **40**, 7761–7772.

24. D.L. Farrens, C. Altenbach, K. Yang, W.L. Hubbell, H.G. Khorana (1996). Requirement of rigid-body motion of transmembrane helices for light activation of rhodopsin. *Science*, **274**, 768–770.

25. S.P. Sheikh, T.A. Zvyaga, O. Lichtarge, T.P. Sakmar, H.R. Bourne (1996). Rhodopsin activation blocked by metal-ion-binding sites linking transmembrane helices C and F. *Nature*, **383**, 347–350.

26. K.D. Ridge, T. Ngo, S.S. Lee, N.G. Abdulaev (1999). Folding and assembly in rhodopsin. Effect of mutations in the sixth transmembrane helix on the conformation of the third cytoplasmic loop. *J. Biol. Chem.*, **274**, 21437–21442.

27. O.P. Ernst, C.K. Meyer, E.P. Marin, P. Henklein, W.Y. Fu, T.P. Sakmar, K.P. Hofmann (2000). Mutation of the fourth cytoplasmic loop of rhodopsin affects binding of transducin and peptides derived from the carboxyl-terminal sequences of transducin alpha and gamma subunits. *J. Biol. Chem.*, **275**, 1937–1943.

28. D. Deretic, S. Schmerl, P.A. Hargrave, A. Arendt, J.H. McDowell (1998). Regulation of sorting and post-Golgi trafficking of rhodopsin by its C-terminal sequence QVS(A)PA. *Proc. Natl. Acad. Sci. U.S.A.*, **95**, 10620–10625.

29. H. Ohguro, R.S. Johnson, L.H. Ericsson, K.A. Walsh, K. Palczewski (1994). Control of rhodopsin multiple phosphorylation. *Biochemistry*, **33**, 1023–1028.

30. H. Ohguro, J.P. Van Hooser, A.H. Milam, K. Palczewski (1995). Rhodopsin phosphorylation and dephosphorylation in vivo. *J. Biol. Chem.*, **270**, 14259–14262.

31. H. Ohguro, M. Rudnicka-Nawrot, J. Buczylko, X. Zhao, J.A. Taylor, K.A. Walsh, K. Palczewski (1996). Structural and enzymatic aspects of rhodopsin phosphorylation. *J. Biol. Chem.*, **271**, 5215–5224.

32. K. Palczewski (1997). GTP-binding-protein-coupled receptor kinases – two mechanistic models. *Eur. J. Biochem.*, **248**, 261–269.

33. S. AbdAlla, H. Lother, U. Quitterer (2000). AT1-receptor heterodimers show enhanced G-protein activation and altered receptor sequestration. *Nature*, **407**, 94–98.

34. N.W. Downer, R.A. Cone (1985). Transient dichroism in photoreceptor membranes indicates that stable oligomers of rhodopsin do not form during excitation. *Biophys. J.*, **47**, 277–284.

35. J.H. Parkes, S.K. Gibson, P.A. Liebman (1999). Temperature and pH dependence of the metarhodopsin I-metarhodopsin II equilibrium and the binding of metarhodopsin II to G protein in rod disk membranes. *Biochemistry*, **38**, 6862–6878.

36. G.G. Kochendoerfer, S.W. Lin, T.P. Sakmar R.A. Mathies (1999). How color visual pigments are tuned. *Trends Biochem. Sci.*, **24**, 300–305.

37. H. Imai, D. Kojima, T. Oura, S. Tachibanaki, A. Terakita, Y. Shichida (1997). Single amino acid residue as a functional determinant of rod and cone visual pigments. *Proc. Natl. Acad. Sci. U.S.A.*, **94**, 2322–2326.

38. T.J. Melia, Jr, C.W. Cowan, J.K. Angleson, T.G. Wensel (1997). A comparison of the efficiency of G protein activation by ligand-free and light-activated forms of rhodopsin. *Biophys. J.*, **73**, 3182–3191.

39. P.R. Robinson, G.B. Cohen, E.A. Zhukovsky, D.D. Oprian (1992). Constitutively active mutants of rhodopsin. *Neuron*, **9**, 719–725.

40. C.E. Elling, K. Thirstrup, B. Holst, T.W. Schwartz (1999). Conversion of agonist site to metal-ion chelator site in the beta(2)-adrenergic receptor. *Proc. Natl. Acad. Sci. U.S.A.*, **96**, 12322–12327.

41. T.G. Ebrey (2000). pKa of the protonated Schiff base of visual pigments. *Methods Enzymol.*, **315**, 196–207.

42. A. Cooper, C.A. Converse (1976). Energetics of primary processes in visual excitation: photocalorimetry of rhodopsin in rod outer segment membranes. *Biochemistry*, **15**, 2970–2978.

43. T. Okada, O.P. Ernst, K. Palczewski, K.P. Hofmann (2001). Activation of rhodopsin: new insights from structural and biochemical studies. *Trends Biochem. Sci.*, **26**, 318–324.

44. R.G. Matthews, R. Hubbard, P.K. Brown, G. Wald (1963). Tautomeric forms of metarhodopsin. *J. Gen. Physiol.*, **47**, 215–240.
45. S.E. Ostroy, F. Erhardt, E.W. Abrahamson (1966). Protein configuration changes in the photolysis of rhodopsin. II. The sequence of intermediates in thermal decay of cattle metarhodopsin in vitro. *Biochim. Biophys. Acta*, **112**, 265–277.
46. C. Baumann (1972). Kinetics of slow thermal reactions during the bleaching of rhodopsin in the perfused frog retina. *J. Physiol.*, **222**, 643–663.
47. M. Chabre, J. Breton (1979). The orientation of the chromophore of vertebrate rhodopsin in the "meta" intermediate states and the reversibility of the meta II-meta III transition. *Vision Res.*, **19**, 1005–1018.
48. J.W. Lewis, D.S. Kliger (1992). Photointermediates of visual pigments. *J. Bioenerg. Biomembr.*, **24**, 201–210.
49. T.E. Thorgeirsson, J.W. Lewis, S.E. Wallace-Williams, D.S. Kliger (1992). Photolysis of rhodopsin results in deprotonation of its retinal Schiff's base prior to formation of metarhodopsin II. *Photochem. Photobiol.*, **56**, 1135–1144.
50. R. Vogel, G.B. Fan, M. Sheves, F. Siebert (2000). The molecular origin of the inhibition of transducin activation in rhodopsin lacking the 9-methyl group of the retinal chromophore: a UV-Vis and FTIR spectroscopic study. *Biochemistry*, **39**, 8895–8908.
51. A. Cooper (1979). Energy uptake in the first step of visual excitation. *Nature*, **282**, 531–533.
52. R.W. Schoenlein, L.A. Peteanu, R.A. Mathies, C.V. Shank (1991). The first step in vision: femtosecond isomerization of rhodopsin. *Science*, **254**, 412–415.
53. Q. Wang, R.W. Schoenlein, L.A. Peteanu, R.A. Mathies, C.V. Shank (1994). Vibrationally coherent photochemistry in the femtosecond primary event of vision. *Science*, **266**, 422–424.
54. H. Kandori, H. Sasabe, K. Nakanishi, T. Yoshizawa, T. Mizukami, Y. Shichida (1996). Real-time detection of 60-femtosecond isomerization in a rhodopsin analog containing 8-membered-ring retinal. *J. Am. Chem. Soc.*, **118**, 1002–1005.
55. H. Kandori, Y. Katsuta, M. Ito, H. Sasabe (1995). Femtosecond fluorescence study of the rhodopsin chromophore in solution. *J. Am. Chem. Soc.*, **117**, 2669–2670.
56. M. Garavelli, P. Celani, F. Bernardi, M.A. Robb, M. Olivucci (1997). The $C_5H_6NH_2^+$ protonated Schiff base: An ab initio minimal model for retinal photoisomerization. *J. Am. Chem. Soc.*, **119**, 6891–6901.
57. V. Buß, O. Weingart, M. Sugihara (2000). Fast Photoisomerization of a rhodopsin model – An ab initio molecular dynamics study. *Angew. Chem. Int. Ed.*, **39**, 2784–2786.
58. B. Honig, U. Dinur, K. Nakanishi, V. Balogh-Nair, M.A. Gawinowicz, M. Arnaboldi, M. Motto (1979). An external point-charge model for wavelength regulation in visual pigments. *J. Am. Chem. Soc.*, **101**, 7084–7086.
59. A. Akita, S.P. Tanis, M. Adams, V. Balogh-Nair, K. Nakanishi (1980). Nonbleachable rhodopsins retaining the full natural chromophore. *J. Am. Chem. Soc.*, **102**, 6370–6372.
60. T. Zankel, H. Ok, R. Johnson, C.W. Chang, N. Sekiya, H. Naoki, K. Yoshihara, K. Nakanishi (1990). Bovine rhodopsin with 11-cis-locked retinal chromophore neither activates rhodopsin kinase nor undergoes conformational change upon irradiation. *J. Am. Chem. Soc.*, **112**, 5387–5388.
61. W.J. DeGrip, J. van Oostrum, P.H. Bovee-Geurts, R. van der Steen, L.J. van Amsterdam, M. Groesbeek, J. Lugtenburg (1990). 10,20-Methanorhodopsins:

(7E,9E,13E)-10,20-methanorhodopsin and (7E,9Z,13Z)-10,20-methanorhodopsin. 11-cis-locked rhodopsin analog pigments with unusual thermal and photo-stability. *Eur. J. Biochem.*, **191**, 211–220.

62. G.F. Jang, V. Kuksa, S. Filipek, F. Bartl, E. Ritter, M.H. Gelb, K.P. Hofmann, K. Palczewski (2001). Mechanism of rhodopsin activation as examined with ring-constrained retinal analogs and the crystal structure of the ground state protein. *J. Biol. Chem.*, **276**, 26148–26153.

63. S. Bhattacharya, K.D. Ridge, B.E. Knox, H.G. Khorana (1992). Light-stable rhodopsin. I. A rhodopsin analog reconstituted with a nonisomerizable 11-cis retinal derivative. *J. Biol. Chem.*, **267**, 6763–6769.

64. I. Palings, E.M. van den Berg, J. Lugtenburg, R.A. Mathies (1989). Complete assignment of the hydrogen out-of-plane wagging vibrations of bathorhodopsin: chromophore structure and energy storage in the primary photoproduct of vision. *Biochemistry*, **28**, 1498–1507.

65. U.M. Ganter, E.D. Schmid, D. Perez-Sala, R.R. Rando, F. Siebert (1989). Removal of the 9-methyl group of retinal inhibits signal transduction in the visual process. A Fourier transform infrared and biochemical investigation. *Biochemistry*, **28**, 5954–5962.

66. K. Marr, K.S. Peters (1991). Photoacoustic calorimetric study of the conversion of rhodopsin and isorhodopsin to lumirhodopsin. *Biochemistry*, **30**, 1254–1258.

67. J.M. Strassburger, W. Gärtner, S.E. Braslavsky (1997). Volume and enthalpy changes after photoexcitation of bovine rhodopsin: laser-induced optoacoustic studies. *Biophys. J.*, **72**, 2294–2303.

68. L. Ujj, F. Jäger, G.H. Atkinson (1998). Vibrational spectrum of the lumi inter-mediate in the room temperature rhodopsin photo-reaction. *Biophys. J.*, **74**, 1492–1501.

69. K. Nakanishi, R. Crouch (1995). Application of artificial pigments to structure determination and study of photoinduced transformations of retinal proteins. *Isr. J. Chem.*, **35**, 253–272.

70. B. Borhan, M.L. Souto, H. Imai, Y. Shichida, K. Nakanishi (2000). Movement of retinal along the visual transduction path. *Science*, **288**, 2209–2212.

71. F. Jager, S. Jager, O. Krutle, N. Friedman, M. Sheves, K.P. Hofmann, F. Siebert (1994). Interactions of the beta-ionone ring with the protein in the visual pigment rhodopsin control the activation mechanism. An FTIR and fluorescence study on artificial vertebrate rhodopsins. *Biochemistry*, **33**, 7389–7397.

72. H. Imai, T. Mizukami, Y. Imamoto, Y. Shichida (1994). Direct observation of the thermal equilibria among lumirhodopsin, metarhodopsin I, and metarhodopsin II in chicken rhodopsin. *Biochemistry*, **33**, 14351–14358.

73. K.P. Hofmann, S. Jäger, O.P. Ernst (1995). Structure and function of activated rhodopsin. *Isr. J. Chem.*, **35**, 339–355.

74. Y. Shichida, H. Imai (1998). Visual pigment: G-protein-coupled receptor for light signals. *Cell. Mol. Life Sci.*, **54**, 1299–1315.

75. C.K. Meyer, M. Böhme, A. Ockenfels, W. Gärtner, K.P. Hofmann, O.P. Ernst (2000). Signaling states of rhodopsin. Retinal provides a scaffold for activating proton transfer switches. *J. Biol. Chem.*, **275**, 19713–19718.

76. S. Arnis, K.P. Hofmann (1993). Two different forms of metarhodopsin II: Schiff base deprotonation precedes proton uptake and signaling state. *Proc. Natl. Acad. Sci. U.S.A.*, **90**, 7849–7853.

77. J. Kibelbek, D.C. Mitchell, J.M. Beach, B.J. Litman (1991). Functional equiva-lence of metarhodopsin II and the Gt-activating form of photolyzed bovine rhodopsin. *Biochemistry*, **30**, 6761–6768.

78. A.G. Doukas, B. Aton, R.H. Callender, T.G. Ebrey (1978). Resonance Raman studies of bovine metarhodopsin I and metarhodopsin II. *Biochemistry*, **17**, 2430–2435.

79. C. Longstaff, R.D. Calhoon, R.R. Rando (1986). Deprotonation of the Schiff base of rhodopsin is obligate in the activation of the G protein. *Proc. Natl. Acad. Sci. U.S.A.*, **83**, 4209–4213.

80. F. Jager, K. Fahmy, T.P. Sakmar, F. Siebert (1994). Identification of glutamic acid 113 as the Schiff base proton acceptor in the metarhodopsin II photointermediate of rhodopsin. *Biochemistry*, **33**, 10878–10882.

81. P.V. Attwood, H. Gutfreund (1980). The application of pressure relaxation to the study of the equilibrium between metarhodopsin I and II from bovine retinas. *FEBS Lett.*, **119**, 323–326.

82. J.H. Parkes, P.A. Liebman (1984). Temperature and pH dependence of the metarhodopsin I–metarhodopsin II kinetics and equilibria in bovine rod disk membrane suspensions. *Biochemistry*, **23**, 5054–5061.

83. M. Straume, D.C. Mitchell, J.L. Miller, B.J. Litman (1990). Interconversion of metarhodopsins I and II: a branched photointermediate decay model. *Biochemistry*, **29**, 9135–9142.

84. H. Kuhn, P.A. Hargrave (1981). Light-induced binding of guanosinetriphosphatase to bovine photoreceptor membranes: effect of limited proteolysis of the membranes. *Biochemistry*, **20**, 2410–2417.

85. U.M. Ganter, T. Charitopoulos, N. Virmaux, F. Siebert (1992). Conformational changes of cytosolic loops of bovine rhodopsin during the transition to metarhodopsin-II: an investigation by Fourier transform infrared difference spectroscopy. *Photochem. Photobiol.*, **56**, 57–62.

86. F. Siebert (1995). Application of FTIR spectroscopy to the investigation of dark structures and photoreactions of visual pigments. *Isr. J. Chem.*, **35**, 309–323.

87. A. Maeda, H. Kandori, Y. Yamazaki, S. Nishimura, M. Hatanaka, Y.S. Chon, J. Sasaki, R. Needleman, J.K. Lanyi (1997). Intramembrane signaling mediated by hydrogen-bonding of water and carboxyl groups in bacteriorhodopsin and rhodopsin. *J. Biochem. (Tokyo)*, **121**, 399–406.

88. S. Nishimura, J. Sasaki, H. Kandori, T. Matsuda, Y. Fukada, A. Maeda (1996). Structural changes in the peptide backbone in complex formation between activated rhodopsin and transducin studied by FTIR spectroscopy. *Biochemistry* **35**, 13267–13271.

89. D. Bownds (1967). Site of attachment of retinal in rhodopsin. *Nature*, **216**, 1178–1181.

90. G. Falk, P. Fatt (1968). Conductance changes produced by light in rod outer segments. *J. Physiol.*, **198**, 647–699.

91. R.H. Johnson (1970). Absence of effect of hydroxylamine upon production rates of some rhodopsin photo intermediates. *Vision Res.*, **10**, 897–900.

92. D.C. Mitchell, B.J. Litman (1999). Effect of protein hydration on receptor conformation: decreased levels of bound water promote metarhodopsin II formation. *Biochemistry*, **38**, 7617–7623.

93. C.N. Rafferty, C.G. Muellenberg, H. Shichi (1980). Tryptophan in bovine rhodopsin: its content, spectral properties and environment. *Biochemistry*, **19**, 2145–2151.

94. S.W. Lin, T.P. Sakmar (1996). Specific tryptophan UV-absorbance changes are probes of the transition of rhodopsin to its active state. *Biochemistry*, **35**, 11149–11159.

95. C.N. Rafferty, J.Y. Cassim, D.G. McConnell (1977). Circular dichroism, optical rotatory dispersion, and absorption studies on the conformation of bovine rhodopsin in situ and solubilized with detergent. *Biophys. Struct. Mech.*, **2**, 227–320.

96. P.A. Liebman, W.S. Jagger, M.W. Kaplan, F.G. Bargoot (1974). Membrane structure changes in rod outer segments associated with rhodopsin bleaching. *Nature*, **251**, 31–36.

97. M.W. Kaplan (1982). Modeling the rod outer segment birefringence change correlated with metarhodopsin II formation. *Biophys. J.*, **38**, 237–241.

98. K.P. Hofmann, R. Uhl, W. Hoffmann, W. Kreutz (1976). Measurements on fast light-induced light-scattering and -absorption changes in outer segments of vertebrate light sensitive rod cells. *Biophys. Struct. Mech.*, **2**, 61–77.

99. M. Heck, A. Pulvermüller, K.P. Hofmann (2000). Light scattering methods to monitor interactions between rhodopsin-containing membranes and soluble proteins. *Methods Enzymol.*, **315**, 329–347.

100. G.G. Kochendoerfer, S. Kaminaka, R.A. Mathies (1997). Ultraviolet resonance Raman examination of the light-induced protein structural changes in rhodopsin activation. *Biochemistry*, **36**, 13153–13159.

101. M. Chabre, J. Breton (1979). Orientation of aromatic residues in rhodopsin. Rotation of one tryptophan upon the meta I to meta II transition after illumination. *Photochem. Photobiol.*, **30**, 295–299.

102. K.T. Brown, M. Murakami (1964). A new receptor potential of the monkey retina with no detectable latency. *Nature*, **201**, 626–628.

103. R.A. Cone (1967). Early receptor potential: photoreversible charge displacement in rhodopsin. *Science*, **155**, 1128–1131.

104. M. Lindau, P. Hochstrate, H. Ruppel (1980). Two component fast photo-signals derived from rod outer segment membranes attached to porous cellulose filters. *FEBS Lett.*, **112**, 17–20.

105. M. Lindau, H. Rüppel (1983). Evidence for conformatical substates of rhodopsin from kinetics of light-induced charge displacement. *Photochem. Photobiophys.*, **5**, 219–228.

106. P.J. Bauer, E. Bamberg, A. Fahr (1984). Photoelectric signals generated by bovine rod outer segment disk membranes attached to a lecithin bilayer. *Biophys. J.*, **46**, 111–116.

107. J.M. Sullivan, P. Shukla (1999). Time-resolved rhodopsin activation currents in a unicellular expression system. *Biophys. J.*, **77**, 1333–1357.

108. T.P. Sakmar (1999). Rhodopsin early receptor potential revisited. *Biophys. J.*, **77**, 1189–1191.

109. F. Siebert, W. Mäntele, W. Kreutz (1980). Flash-induced kinetic infrared spectroscopy applied to biochemical systems. *Biophys. Struct. Mech.*, **6**, 139–146.

110. W.J. DeGrip, D. Gray, J. Gillespie, P.H. Bovee, E.M. Van den Berg, J. Lugtenburg, K.J. Rothschild (1988). Photoexcitation of rhodopsin: conformation changes in the chromophore, protein and associated lipids as determined by FTIR difference spectroscopy. *Photochem. Photobiol.*, **48**, 497–504.

111. A.L. Klinger, M.S. Braiman (1992). Structural comparison of metarhodopsin II, metarhodopsin III, and opsin based on kinetic analysis of Fourier transform infrared difference spectra. *Biophys. J.*, **63**, 1244–1255.

112. F. Siebert, W. Mäntele, K. Gerwert (1983). Fourier-transform infrared spectroscopy applied to rhodopsin. The problem of the protonation state of the retinylidene Schiff base re-investigated. *Eur. J. Biochem.*, **136**, 119–127.

113. W.J. DeGrip, J. Gillespie, K.J. Rothschild (1985). Carboxyl group involvement in the meta I and meta II stages in rhodopsin bleaching. A Fourier transform infrared spectroscopic study. *Biochim. Biophys. Acta*, **809**, 97–106.

114. Z.T. Farahbakhsh, K. Hideg, W.L. Hubbell (1993). Photoactivated conformational changes in rhodopsin: a time-resolved spin label study. *Science*, **262**, 1416–1419.

115. J.F. Resek, Z.T. Farahbakhsh, W.L. Hubbell, H.G. Khorana (1993). Formation of the meta II photointermediate is accompanied by conformational changes in the cytoplasmic surface of rhodopsin. *Biochemistry*, **32**, 12025–12032.

116. T.P. Sakmar, R.R. Franke, H.G. Khorana (1989). Glutamic acid-113 serves as the retinylidene Schiff base counterion in bovine rhodopsin. *Proc. Natl. Acad. Sci. U.S.A.*, **86**, 8309–8313.

117. T.A. Zvyaga, K. Fahmy, T.P. Sakmar (1994). Characterization of rhodopsin-transducin interaction: a mutant rhodopsin photoproduct with a protonated Schiff base activates transducin. *Biochemistry*, **33**, 9753–9761.

118. T.A. Zvyaga, K. Fahmy, F. Siebert, T.P. Sakmar (1996). Characterization of the mutant visual pigment responsible for congenital night blindness: a biochemical and Fourier-transform infrared spectroscopy study. *Biochemistry*, **35**, 7536–7545.

119. C.J. Weitz, J. Nathans (1992). Histidine residues regulate the transition of photoexcited rhodopsin to its active conformation, metarhodopsin II. *Neuron*, **8**, 465–472.

120. G.B. Cohen, D.D. Oprian, P.R. Robinson (1992). Mechanism of activation and inactivation of opsin: role of Glu113 and Lys296. *Biochemistry*, **31**, 12592–12601.

121. K. Boesze-Battaglia, S.J. Fliesler, A.D. Albert (1990). Relationship of cholesterol content to spatial distribution and age of disc membranes in retinal rod outer segments. *J. Biol. Chem.*, **265**, 18867–18870.

122. G.P. Miljanich, P.P. Nemes, D.L. White, E.A. Dratz (1981). The asymmetric transmembrane distribution of phosphatidylethanolamine, phosphatidylserine, and fatty acids of the bovine retinal rod outer segment disk membrane. *J. Membr. Biol.*, **60**, 249–255.

123. N. Wang, R.E. Anderson (1992). Enrichment of polyunsaturated fatty acids from rat retinal pigment epithelium to rod outer segments. *Curr. Eye. Res.*, **11**, 783–791.

124. D.C. Mitchell, M. Straume, J.L. Miller, B.J. Litman (1990). Modulation of metarhodopsin formation by cholesterol-induced ordering of bilayer lipids. *Biochemistry*, **29**, 9143–9149.

125. D.C. Mitchell, M. Straume, B.J. Litman (1992). Role of sn-1-saturated, sn-2-polyunsaturated phospholipids in control of membrane receptor conformational equilibrium: effects of cholesterol and acyl chain unsaturation on the metarhodopsin I in equilibrium with metarhodopsin II equilibrium. *Biochemistry*, **31**, 662–670.

126. M.F. Brown (1997). Influence of nonlamellar-forming lipids on rhodopsin. *Curr. Top. Membr.*, **44**, 285–356.

127. P.A. Baldwin, W.L. Hubbell (1985). Effects of lipid environment on the light-induced conformational changes of rhodopsin. 2. Roles of lipid chain length, unsaturation, and phase state. *Biochemistry*, **24**, 2633–2639.

128. P.A. Baldwin, W.L. Hubbell (1985). Effects of lipid environment on the light-induced conformational changes of rhodopsin. 1. Absence of metarhodopsin II production in dimyristoylphosphatidylcholine recombinant membranes. *Biochemistry*, **24**, 2624–2632.

129. D.F. O'Brien, L.F. Costa, R.A. Ott (1977). Photochemical functionality of rhodopsin-phospholipid recombinant membranes. *Biochemistry*, **16**, 1295–1303.

130. W.J. DeGrip, J. Olive, P.H.M. Bovee-Geurts (1983). Reversible modulation of rhodopsin photolysis in pure phosphatidylserine membranes. *Biochim. Biophys. Acta*, **734**, 168–179.

131. F. DeLange, M. Merkx, P.H. Bovee-Geurts, A.M. Pistorius, W.J. DeGrip (1997). Modulation of the metarhodopsin I/metarhodopsin II equilibrium of bovine rhodopsin by ionic strength – evidence for a surface-charge effect. *Eur. J. Biochem.*, **243**, 174–180.

132. D.S. Kliger, J.W. Lewis (1995). Spectral and kinetic characterization of visual pigment photointermediates. *Isr. J. Chem.*, **35**, 289–307.

133. B.J. Litman, O. Kalisky, M. Ottolenghi (1981). Rhodopsin-phospholipid interactions: dependence of rate of the meta I to meta II transition on the level of associated disc phospholipid. *Biochemistry*, **20**, 631–634.

134. M. Beck, F. Siebert, T.P. Sakmar (1998). Evidence for the specific interaction of a lipid molecule with rhodopsin which is altered in the transition to the active state metarhodopsin II. *FEBS Lett.*, **436**, 304–308.

135. E. Hessel, A. Herrmann, P. Müller, P.P. Schnetkamp, K.P. Hofmann (2000). The transbilayer distribution of phospholipids in disc membranes is a dynamic equilibrium evidence for rapid flip and flop movement. *Eur. J. Biochem.*, **267**, 1473–1483.

136. K.P. Hofmann, M. Heck (1996). Light-induced protein–protein interactions on the rod photoreceptor disc membrane. In: A.G. Lee, (ed.), *Biomembranes II*, (Vol. 2A, pp. 141–198) JAI Press, Inc., Greenwich, CT.

137. E. Hessel, P. Müller, A. Herrmann, K.P. Hofmann (2001). Light-induced reorganization of phospholipids in rod disc membranes. *J. Biol. Chem.*, **276**, 2538–2543.

138. T.P. Sakmar (1998). Rhodopsin: a prototypical G protein-coupled receptor. *Prog. Nucleic Acid Res. Mol. Biol.*, **59**, 1–34.

139. K. Fahmy, T.P. Sakmar, F. Siebert (2000). Transducin-dependent protonation of glutamic acid 134 in rhodopsin. *Biochemistry*, **39**, 10607–10612.

140. G.B. Cohen, T. Yang, P.R. Robinson, D.D. Oprian (1993). Constitutive activation of opsin: influence of charge at position 134 and size at position 296. *Biochemistry*, **32**, 6111–6115.

141. S. Arnis, K. Fahmy, K.P. Hofmann, T.P. Sakmar (1994). A conserved carboxylic acid group mediates light-dependent proton uptake and signaling by rhodopsin. *J. Biol. Chem.*, **269**, 23879–23881.

142. N.G. Abdulaev, T. Ngo, R. Chen, Z. Lu, K.D. Ridge (2000). Functionally discrete mimics of light-activated rhodopsin identified through expression of soluble cytoplasmic domains. *J. Biol. Chem.*, **275**, 39354–39363.

143. A.D. Jensen, F. Guarnieri, S.G. Rasmussen, F. Asmar, J.A. Ballesteros, U. Gether (2001). Agonist-induced conformational changes at the cytoplasmic side of transmembrane segment 6 in the beta 2 adrenergic receptor mapped by site-selective fluorescent labeling. *J. Biol. Chem.*, **276**, 9279–9290.

144. J.M. Kim, C. Altenbach, R.L. Thurmond, H.G. Khorana, W.L. Hubbell (1997). Structure and function in rhodopsin: rhodopsin mutants with a neutral amino acid at E134 have a partially activated conformation in the dark state. *Proc. Natl. Acad. Sci. U.S.A.*, **94**, 14273–14278.

145. H. Luecke, B. Schobert, H.T. Richter, J.P. Cartailler, J.K. Lanyi (1999). Structure of bacteriorhodopsin at 1.55 A resolution. *J. Mol. Biol.*, **291**, 899–911.

146. M. Kolbe, H. Besir, L.O. Essen, D. Oesterhelt (2000). Structure of the light-driven chloride pump halorhodopsin at 1.8 A resolution. *Science*, **288**, 1390–1396.

147. H. Luecke, B. Schobert, J.K. Lanyi, E.N. Spudich, J.L. Spudich (2001). Crystal structure of sensory rhodopsin II at 2.4 A: Insights into color tuning and transducer interaction. *Science*, **293**, 1499–1503.

148. H. Luecke, B. Schobert, H.T. Richter, J.P. Cartailler, J.K. Lanyi (1999). Structural changes in bacteriorhodopsin during ion transport at 2 angstrom resolution. *Science*, **286**, 255–261.

149. H.J. Sass, G. Büldt, R. Gessenich, D. Hehn, D. Neff, R. Schlesinger, J. Berendzen, P. Ormos (2000). Structural alterations for proton translocation in the M state of wild-type bacteriorhodopsin. *Nature*, **406**, 649–653.

150. J.L. Spudich (1994). Protein-protein interaction converts a proton pump into a sensory receptor. *Cell*, **79**, 747–750.

151. A.A. Wegener, I. Chizhov, M. Engelhard, H.J. Steinhoff (2000). Time-resolved detection of transient movement of helix F in spin-labelled pharaonis sensory rhodopsin II. *J. Mol. Biol.*, **301**, 881–891.

152. W.J. DeGrip, K.J. Rothschild (2000). Structure and mechanism of vertebrate visual pigments. In: D.G. Stavenga, W.J. DeGrip, E.N. Pugh Jr, (Eds), *Molecular Mechanisms in Visual Transduction*, (Vol. 3. pp. 1–54) Elsevier, Amsterdam.

153. S. Arnis, K.P. Hofmann (1995). Photoregeneration of bovine rhodopsin from its signaling state. *Biochemistry*, **34**, 9333–9340.

154. O.P. Ernst, C. Bieri, H. Vogel, K.P. Hofmann (2000). Intrinsic biophysical monitors of transducin activation: fluorescence, UV-visible spectroscopy, light scattering, and evanescent field techniques. *Methods Enzymol.*, **315**, 471–489.

155. F.J. Bartl, E. Ritter, K.P. Hofmann (2001). Signaling states of rhodopsin: Absorption of light in active metarhodopsin II generates an all-*trans*-retinal bound inactive state. *J. Biol. Chem.*, **276**, 30161–30166.

156. C. Grimm, A. Wenzel, T. Williams, P. Rol, F. Hafezi, C. Reme (2001). Rhodopsin-mediated blue-light damage to the rat retina: effect of photoreversal of bleaching. *Invest. Ophthalmol. Vis. Sci.*, **42**, 497–505.

157. D. Emeis, K.P. Hofmann (1981). Shift in the relation between flash-induced metarhodopsin I and metarhodpsin II within the first 10% rhodopsin bleaching in bovine disc membranes. *FEBS Lett.*, **136**, 201–207.

158. M. Kahlert, B. König, K.P. Hofmann (1990). Displacement of rhodopsin by GDP from three-loop interaction with transducin depends critically on the diphosphate beta-position. *J. Biol. Chem.*, **265**, 18928–18932.

159. J. Panico, J.H. Parkes, P.A. Liebman (1990). The effect of GDP on rod outer segment G-protein interactions. *J. Biol. Chem.*, **265**, 18922–18927.

160. N. Bennett, M. Michel-Villaz, H. Kühn (1982). Light-induced interaction between rhodopsin and the GTP-binding protein. Metarhodopsin II is the major photoproduct involved. *Eur. J. Biochem.*, **127**, 97–103.

161. K.P. Hofmann (1985). Effect of GTP on the rhodopsin-G-protein complex by transient formation of extra metarhodopsin II. *Biochim. Biophys. Acta*, **810**, 278–281.

162. H.E. Hamm, D. Deretic, A. Arendt, P.A. Hargrave, B. Koenig, K.P. Hofmann (1988). Site of G protein binding to rhodopsin mapped with synthetic peptides from the alpha subunit. *Science*, **241**, 832–835.

163. H.E. Hamm, H.M. Rarick (1994). Specific peptide probes for G-protein interactions with receptors. *Methods Enzymol.*, **237**, 423–436.

164. P.D. Garcia, R. Onrust, S.M. Bell, T.P. Sakmar, H.R. Bourne (1995). Transducin-alpha C-terminal mutations prevent activation by rhodopsin: a new assay using recombinant proteins expressed in cultured cells. *EMBO J.*, **14**, 4460–4469.

165. R. Onrust, P. Herzmark, P. Chi, P.D. Garcia, O. Lichtarge, C. Kingsley, H.R. Bourne (1997). Receptor and betagamma binding sites in the alpha subunit of the retinal G protein transducin. *Science*, **275**, 381–384.

166. O.G. Kisselev, M.V. Ermolaeva, N. Gautam (1994). A farnesylated domain in the G protein gamma subunit is a specific determinant of receptor coupling. *J. Biol. Chem.*, **269**, 21399–21402.

167. O.G. Kisselev, C.K. Meyer, M. Heck, O.P. Ernst, K.P. Hofmann (1999). Signal transfer from rhodopsin to the G-protein: evidence for a two-site sequential fit mechanism. *Proc. Natl. Acad. Sci. U.S.A.*, **96**, 4898–4903.

168. N. Gautam, G.B. Downes, K. Yan, O. Kisselev (1998). The G-protein betagamma complex. *Cell. Signal.*, **10**, 447–455.

169. D.G. Lambright, J. Sondek, A. Bohm, N.P. Skiba, H.E. Hamm, P.B. Sigler (1996). The 2.0 A crystal structure of a heterotrimeric G protein, *Nature.* **379**, 311–319.

170. E.J. Helmreich, K.P. Hofmann (1996). Structure and function of proteins in G-protein-coupled signal transfer. *Biochim. Biophys. Acta*, **1286**, 285–322.

171. O.G. Kisselev, J. Kao, J.W. Ponder, Y.C. Fann, N. Gautam, G.R. Marshall (1998). Light-activated rhodopsin induces structural binding motif in G protein alpha subunit. *Proc. Natl. Acad. Sci. U.S.A.*, **95**, 4270–4275.

172. L. Aris, A. Gilchrist, S. Rens-Domiano, C. Meyer, P.J. Schatz, E.A. Dratz, H.E. Hamm (2001). Structural requirements for the stabilization of metarhodopsin II by the C terminus of the alpha subunit of transducin. *J. Biol. Chem.*, **276**, 2333–2339.

173. K. Fahmy (1998). Binding of transducin and transducin-derived peptides to rhodopsin studies by attenuated total reflection–Fourier transform infrared difference spectroscopy. *Biophys. J.*, **75**, 1306–1318.

174. F. Bartl, E. Ritter, K.P. Hofmann (2000). FTIR spectroscopy of complexes formed between metarhodopsin II and C-terminal peptides from the G-protein alpha- and gamma-subunits. *FEBS Lett.*, **473**, 259–264.

175. B. König, A. Arendt, J.H. McDowell, M. Kahlert, P.A. Hargrave, K.P. Hofmann (1989). Three cytoplasmic loops of rhodopsin interact with transducin. *Proc. Natl. Acad. Sci. U.S.A.* **86**, 6878–6882.

176. R.R. Franke, B. Konig, T.P. Sakmar, H.G. Khorana, K.P. Hofmann (1990). Rhodopsin mutants that bind but fail to activate transducin. *Science*, **250**, 123–125.

177. R.R. Franke, T.P. Sakmar, R.M. Graham, H.G. Khorana (1992). Structure and function in rhodopsin. Studies of the interaction between the rhodopsin cytoplasmic domain and transducin. *J. Biol. Chem.*, **267**, 14767–14774.

178. K. Cai, J. Klein-Seetharaman, D. Farrens, C. Zhang, C. Altenbach, W.L. Hubbell, H.G. Khorana (1999). Single-cysteine substitution mutants at amino acid positions 306-321 in rhodopsin, the sequence between the cytoplasmic end of helix VII and the palmitoylation sites: sulfhydryl reactivity and transducin activation reveal a tertiary structure. *Biochemistry*, **38**, 7925–7930.

179. E.P. Marin, A.G. Krishna, T.A. Zvyaga, J. Isele, F. Siebert, T.P. Sakmar (2000). The amino terminus of the fourth cytoplasmic loop of rhodopsin modulates rhodopsin-transducin interaction. *J. Biol. Chem.*, **275**, 1930–1936.

180. O.P. Ernst, K.P. Hofmann, T.P. Sakmar (1995). Characterization of rhodopsin mutants that bind transducin but fail to induce GTP nucleotide uptake. Classification of mutant pigments by fluorescence, nucleotide release, and flash-induced light-scattering assays. *J. Biol. Chem.*, **270**, 10580–10586.

181. J. Klein-Seetharaman, J. Hwa, K. Cai, C. Altenbach, W.L. Hubbell, H.G. Khorana (1999). Single-cysteine substitution mutants at amino acid positions 55–75, the sequence connecting the cytoplasmic ends of helices I and II in rhodopsin: reactivity of the sulfhydryl groups and their derivatives identifies a tertiary structure that changes upon light-activation. *Biochemistry*, **38**, 7938–7944.

182. N.G. Abdulaev, K.D. Ridge (1998). Light-induced exposure of the cytoplasmic end of transmembrane helix seven in rhodopsin. *Proc. Natl. Acad. Sci. U.S.A.*, **95**, 12854–12859.

183. W.J. Phillips, R.A. Cerione (1992). Rhodopsin/transducin interactions. I. Characterization of the binding of the transducin-beta gamma subunit complex to rhodopsin using fluorescence spectroscopy. *J. Biol. Chem.*, **267**, 17032–17039.

184. W.J. Phillips, S.C. Wong, R.A. Cerione (1992). Rhodopsin/transducin interactions. II. Influence of the transducin-beta gamma subunit complex on the coupling of the transducin-alpha subunit to rhodopsin. *J. Biol. Chem.*, **267**, 17040–17046.

185. D.F. Morrison, P.J. O'Brien, D.R. Pepperberg (1991). Depalmitylation with hydroxylamine alters the functional properties of rhodopsin. *J. Biol. Chem.*, **266**, 20118–20123.

186. S.S. Karnik, K.D. Ridge, S. Bhattacharya, H.G. Khorana (1993). Palmitoylation of bovine opsin and its cysteine mutants in COS cells. *Proc. Natl. Acad. Sci. U.S.A.*, **90**, 40–44.

187. H.E. Hamm (2001). How activated receptors couple to G proteins. *Proc. Natl. Acad. Sci. U.S.A.*, **98**, 4819–4821.

188. W.L. Hubbell, D.S. Cafiso, C. Altenbach (2000). Identifying conformational changes with site-directed spin labelling. *Nat. Struct. Biol.*, **7**, 735–739.

189. M.C. Loewen, J. Klein-Seetharaman, E.V. Getmanova, P.J. Reeves, H. Schwalbe, H.G. Khorana (2001). Solution 19F nuclear Overhauser effects in structural studies of the cytoplasmic domain of mammalian rhodopsin. *Proc. Natl. Acad. Sci. U.S.A.*, **98**, 4888–4892.

190. Y. Itoh, K. Cai, H.G. Khorana (2001). Mapping of contact sites in complex formation between light-activated rhodopsin and transducin by covalent crosslinking: use of a chemically preactivated reagent. *Proc. Natl. Acad. Sci. U.S.A.*, **98**, 4883–4887.

191. K. Cai, Y. Itoh, H.G. Khorana (2001). Mapping of contact sites in complex formation between transducin and light-activated rhodopsin by covalent crosslinking: use of a photoactivatable reagent. *Proc. Natl. Acad. Sci. U.S.A.*, **98**, 4877–4882.

192. E.P. Marin, A.G. Krishna, T.P. Sakmar (2001). Rapid activation of transducin by mutations distant from the nucleotide-binding site: evidence for a mechanistic model of receptor-catalyzed nucleotide exchange by G proteins. *J. Biol. Chem.*, **276**, 27400–27405.

193. M. Natochin, M. Moussaif, N.O. Artemyev (2001). Probing the mechanism of rhodopsin-catalyzed transducin activation. *J. Neurochem.*, **77**, 202–210.

194. P. Rondard, T. Iiri, S. Srinivasan, E. Meng, T. Fujita, H.R. Bourne (2001). Mutant G protein alpha subunit activated by G-beta gamma: a model for receptor activation? *Proc. Natl. Acad. Sci. U.S.A.*, **98**, 6150–6155.

195. J. Xu, R.L. Dodd, C.L. Makino, M.I. Simon, D.A. Baylor, J. Chen (1997). Prolonged photoresponses in transgenic mouse rods lacking arrestin. *Nature*, **389**, 505–509.

196. A. Pulvermüller, K. Palczewski, K.P. Hofmann (1993). Interaction between photoactivated rhodopsin and its kinase: stability and kinetics of complex formation. *Biochemistry*, **32**, 14082–14088.

197. A. Pulvermüller, D. Maretzki, M. Rudnicka-Nawrot, W.C. Smith, K. Palczewski, K.P. Hofmann (1997). Functional differences in the interaction of arrestin and its splice variant, p44, with rhodopsin. *Biochemistry*, **36**, 9253–9260.

198. C. Schubert, J.A. Hirsch, V.V. Gurevich, D.M. Engelman, P.B. Sigler, K.G. Fleming (1999). Visual arrestin activity may be regulated by self-association. *J. Biol. Chem.*, **274**, 21186–21190.

199. K. Palczewski, J.P. Van Hooser, G.G. Garwin, J. Chen, G.I. Liou, J.C. Saari (1999). Kinetics of visual pigment regeneration in excised mouse eyes and in mice with a targeted disruption of the gene encoding interphotoreceptor retinoid-binding protein or arrestin. *Biochemistry*, **38**, 12012–12019.

200. K. Palczewski, A. Pulvermüller, J. Buczylko, K.P. Hofmann (1991). Phosphorylated rhodopsin and heparin induce similar conformational changes in arrestin. *J. Biol. Chem.*, **266**, 18649–18654.

201. J. Granzin, U. Wilden, H.W. Choe, J. Labahn, B. Krafft, G. Büldt (1998). X-ray crystal structure of arrestin from bovine rod outer segments. *Nature*, **391**, 918–921.

202. J.A. Hirsch, C. Schubert, V.V. Gurevich, P.B. Sigler (1999). The 2.8 A crystal structure of visual arrestin: a model for arrestin's regulation. *Cell*, **97**, 257–269.

203. W.C. Smith, J.H. McDowell, D.R. Dugger, R. Miller, A. Arendt, M.P. Popp, P.A. Hargrave (1999). Identification of regions of arrestin that bind to rhodopsin. *Biochemistry*, **38**, 2752–2761.

204. A. Pulvermüller, K. Schröder, T. Fischer, K.P. Hofmann (2000). Interactions of metarhodopsin II. Arrestin peptides compete with arrestin and transducin. *J. Biol. Chem.*, **275**, 37679–37685.

205. K. Palczewski (1994). Structure and functions of arrestins. *Protein Sci.*, **3**, 1355–1361.

206. J.G. Krupnick, V.V. Gurevich, T. Schepers, H.E. Hamm, J.L. Benovic (1994). Arrestin-rhodopsin interaction. Multi-site binding delineated by peptide inhibition. *J. Biol. Chem.*, **269**, 3226–3232.

207. D. Raman, S. Osawa, E.R. Weiss (1999). Binding of arrestin to cytoplasmic loop mutants of bovine rhodopsin. *Biochemistry*, **38**, 5117–5123.

208. A. Schleicher, H. Kühn, K.P. Hofmann (1989). Kinetics, binding constant, and activation energy of the 48-kDa protein-rhodopsin complex by extra-metarhodopsin II. *Biochemistry*, **28**, 1770–1775.

209. V.V. Gurevich, J.L. Benovic (1993). Visual arrestin interaction with rhodopsin. Sequential multisite binding ensures strict selectivity toward light-activated phosphorylated rhodopsin. *J. Biol. Chem.*, **268**, 11628–11638.

210. V.V. Gurevich, J.L. Benovic (1997). Mechanism of phosphorylation-recognition by visual arrestin and the transition of arrestin into a high affinity binding state. *Mol. Pharmacol.*, **51**, 161–169.

211. K. Palczewski, A. Pulvermüller, J. Buczylko, C. Gutmann, K.P. Hofmann (1991). Binding of inositol phosphates to arrestin. *FEBS Lett.*, **295**, 195–199.

212. V.V. Gurevich, C.Y. Chen, C.M. Kim, J.L. Benovic (1994). Visual arrestin binding to rhodopsin. Intramolecular interaction between the basic N terminus and acidic C terminus of arrestin may regulate binding selectivity. *J. Biol. Chem.*, **269**, 8721–8727.

213. S.A. Vishnivetskiy, C.L. Paz, C. Schubert, J.A. Hirsch, P.B. Sigler, V.V. Gurevich (1999). How does arrestin respond to the phosphorylated state of rhodopsin? *J. Biol. Chem.*, **274**, 11451–11454.

214. W.C. Smith, A.H. Milam, D. Dugger, A. Arendt, P.A. Hargrave, K. Palczewski (1994). A splice variant of arrestin. Molecular cloning and localization in bovine retina. *J. Biol. Chem.*, **269**, 15407–15410.

215. C.J. Wilson, R.A. Copeland (1997). Spectroscopic characterization of arrestin interactions with competitive ligands: study of heparin and phytic acid binding. *J. Protein Chem.*, **16**, 755–763.

216. K. Palczewski, J.L. Benovic (1991). G-protein-coupled receptor kinases. *Trends Biochem. Sci.*, **16**, 387–391.

217. T. Haga, K. Haga, K. Kameyama (1994). G protein-coupled receptor kinases. *J. Neurochem.*, **63**, 400–412.

218. R.T. Premont, J. Inglese, R.J. Lefkowitz (1995). Protein kinases that phosphorylate activated G protein-coupled receptors. *FASEB J.*, **9**, 175–182.

219. J. Chen, C.L. Makino, N.S. Peachey, D.A. Baylor, M.I. Simon (1995). Mechanisms of rhodopsin inactivation in vivo as revealed by a COOH-terminal truncation mutant. *Science*, **267**, 374–377.

220. C.K. Chen, M.E. Burns, M. Spencer, G.A. Niemi, J. Chen, J.B. Hurley, D.A. Baylor, M.I. Simon (1999). Abnormal photoresponses and light-induced apoptosis in rods lacking rhodopsin kinase. *Proc. Natl. Acad. Sci. U.S.A.*, **96**, 3718–3722.

221. D.F. Morrison, T.D. Ting, V. Vallury, Y.K. Ho, R.K. Crouch, D.W. Corson, N.J. Mangini, D.R. Pepperberg (1995). Reduced light-dependent phosphorylation of an analog visual pigment containing 9-demethylretinal as its chromophore. *J. Biol. Chem.*, **270**, 6718–6721.

222. J. Buczylko, J.C. Saari, R.K. Crouch, K. Palczewski (1996). Mechanisms of opsin activation. *J. Biol. Chem.*, **271**, 20621–20630.

223. R.L. Thurmond, C. Creuzenet, P.J. Reeves, H.G. Khorana (1997). Structure and function in rhodopsin: peptide sequences in the cytoplasmic loops of rhodopsin are intimately involved in interaction with rhodopsin kinase. *Proc. Natl. Acad. Sci. U.S.A.*, **94**, 1715–1720.

224. R. Paulsen, J. Bentrop (1983). Activation of rhodopsin phosphorylation is triggered by the lumirhodopsin-metarhodopsin I transition. *Nature*, **302**, 417–419.

225. P.R. Robinson, J. Buczylko, H. Ohguro, K. Palczewski (1994). Opsins with mutations at the site of chromophore attachment constitutively activate transducin but are not phosphorylated by rhodopsin kinase. *Proc. Natl. Acad. Sci. U.S.A.*, **91**, 5411–5415.

226. J. Rim, D.D. Oprian (1995). Constitutive activation of opsin: interaction of mutants with rhodopsin kinase and arrestin. *Biochemistry*, **34**, 11938–11945.

227. K. Palczewski, J. Buczylko, M.W. Kaplan, A.S. Polans, J.W. Crabb (1991). Mechanism of rhodopsin kinase activation. *J. Biol. Chem.*, **266**, 12949–12955.

228. W. Shi, S. Osawa, C.D. Dickerson, E.R. Weiss (1995). Rhodopsin mutants discriminate sites important for the activation of rhodopsin kinase and Gt. *J. Biol. Chem.*, **270**, 2112–2119.

229. D.R. Pepperberg, D.G. Birch, K.P. Hofmann, D.C. Hood (1996). Recovery kinetics of human rod phototransduction inferred from the two-branched alpha-wave saturation function. *J. Opt. Soc. Am. A*, **13**, 586–600.

230. U. Laitko, K.P. Hofmann (1998). A model for the recovery kinetics of rod phototransduction, based on the enzymatic deactivation of rhodopsin. *Biophys. J.*, **74**, 803–815.

231. A. Surya, K.W. Foster, B.E. Knox (1995). Transducin activation by the bovine opsin apoprotein. *J. Biol. Chem.*, **270**, 5024–5031.

232. R. Vogel, F. Siebert (2001). Conformations of the active and inactive states of opsin. *J. Biol. Chem.*, **276**, 38487–38493.

233. P. Samama, S. Cotecchia, T. Costa, R.J. Lefkowitz (1993). A mutation-induced activated state of the beta 2-adrenergic receptor. Extending the ternary complex model. *J. Biol. Chem.*, **268**, 4625–4636.

234. J. Jin, R.K. Crouch, D.W. Corson, B.M. Katz, E.F. MacNichol, M.C. Cornwall (1993). Noncovalent occupancy of the retinal-binding pocket of opsin diminishes bleaching adaptation of retinal cones. *Neuron*, **11**, 513–522.

235. C.S. Leibrock, T.D. Lamb (1997). Effect of hydroxylamine on photon-like events during dark adaptation in toad rod photoreceptors. *J. Physiol.*, **501**, 97–109.

236. S. Jäger, K. Palczewski, K.P. Hofmann (1996). Opsin/all-*trans*-retinal complex activates transducin by different mechanisms than photolyzed rhodopsin. *Biochemistry*, **35**, 2901–2908.

237. K. Sachs, D. Maretzki, C.K. Meyer, K.P. Hofmann (2000). Diffusible ligand all-trans-retinal activates opsin via a palmitoylation-dependent mechanism. *J. Biol. Chem.*, **275**, 6189–6194.

238. K. Palczewski, S. Jäger, J. Buczylko, R.K. Crouch, D.L. Bredberg, K.P. Hofmann, M.A. Asson-Batres, J.C. Saari (1994). Rod outer segment retinol dehydrogenase: substrate specificity and role in phototransduction. *Biochemistry*, **33**, 13741–13750.

239. K.P. Hofmann, A. Pulvermüller, J. Buczylko, P. Van Hooser, K. Palczewski (1992). The role of arrestin and retinoids in the regeneration pathway of rhodopsin. *J. Biol. Chem.*, **267**, 15701–15706.

240. J. Weng, N.L. Mata, S.M. Azarian, R.T. Tzekov, D.G. Birch, G.H. Travis (1999). Insights into the function of Rim protein in photoreceptors and etiology of Stargardt's disease from the phenotype in ABCR knockout mice. *Cell*, **98**, 13–23.

241. J.K. McBee, K. Palczewski, W. Baehr, D.R. Pepperberg (2001). Confronting complexity: the interlink of phototransduction and retinoid metabolism in the vertebrate retina. *Prog. Retin. Eye Res.*, **20**, 469–529.

242. Z. Krozowski (1994). The short-chain alcohol dehydrogenase superfamily: variations on a common theme. *J. Steroid Biochem. Mol. Biol.*, **51**, 125–130.

243. H. Jornvall, J.O. Hoog, B. Persson (1999). SDR and MDR: completed genome sequences show these protein families to be large, of old origin, and of complex nature. *FEBS Lett.*, **445**, 261–264.

244. J.C. Saari, G.G. Garwin, J.P. Van Hooser, K. Palczewski (1998). Reduction of all-*trans*-retinal limits regeneration of visual pigment in mice. *Vision Res.*, **38**, 1325–1333.

245. A. Ruiz, A. Winston, Y.H. Lim, B.A. Gilbert, R.R. Rando, D. Bok (1999). Molecular and biochemical characterization of lecithin retinol acyltransferase. *J. Biol. Chem.*, **274**, 3834–3841.

246. M. Jiang, S. Pandey, H.K. Fong (1993). An opsin homologue in the retina and pigment epithelium. *Invest. Ophthalmol. Vis. Sci.*, **34**, 3669–3678.

Chapter 4

Rhodopsin-related proteins, Cop1, Cop2 and Chop1, in *Chlamydomonas reinhardtii*

Markus Fuhrmann, Werner Deininger, Suneel Kateriya and Peter Hegemann

Table of contents

Abstract

The classical and most carefully studied biological sensory system is the visual process of higher animals. The photoreceptor is rhodopsin with retinal as the chromophoric group. In all rhodopsins known so far retinal is bound to a lysine residue of the apoprotein (opsin) via a retinylidene Schiff-base. Dark-adapted animal rhodopsins (Type II rhodopins, for a definition see [1]) contain a twisted 11-*cis* retinal that isomerises after light excitation into all-*trans*, thus triggering a conformational change, which initiates the signaling process [2]. Amino acid charges of the retinal-binding pocket modulate the retinal absorption and allow the rhodopsin absorption to cover the whole spectrum between 360 and 635 nm [3,4]. However, due to the high transmission of green light in water, most rhodopsins absorb around 500 nm. Rhodopsins are also used in the archaea and eubacteria branch (type I rhodopsins), where they serve as sensory photoreceptors for orientation of the cells in different light qualities (sensory rhodopsins) or as light-driven ion transporters (bacteriorhodopsin and halorhodopsin). These microbial rhodopsins contain all-*trans*,15-*S-anti* retinal that upon light stimulation undergoes a 13-*cis* isomerisation. The concomitant rotation of the N–H dipole induces the ion movement across the retinal barrier, resulting in pumping of a H^+ in bacteriorhodopsin or a Cl^- in halorhodopsin [5]. The function of the microbial sensory rhodopsins is surprisingly similar. A proton within the retinal binding site is displaced after retinal isomerisation. However, the proton is not released but, instead, it drives conformational changes within the rhodopsin and subsequently within the attached transducer protein (Htr, [6]). A microbial type I rhodopsin sequence has been discovered in the fungus *Neurospora crassa* [7], but its function is unknown.

In general, rhodopsins are used by motile organisms that need to respond rapidly, on a time scale of milliseconds to seconds, to changes in environmental conditions or, vice versa, to a changing positioning of the organism relative to its static surrounding. Rhodopsin-based systems are fast and the intracellular response is immediately extinguished so that the system is prepared for a new light input.

4.1 Rhodopsin as the photoreceptor for phototaxis and photophobic responses in green algae

Reconstitution of phototaxis in blind algal cells with retinal supplied convincing evidence that the photoreceptor is rhodopsin [8]. Since in the first set of experiments 11-*cis* retinal seemed to be more efficient than the all-*trans* isomer, initially algal rhodopsins were proposed to be animal rhodopsin-like [8,9]. Later, the rhodopsin chromophore was characterised in vivo more extensively by supplementation of the white cells with various retinal analogs.

The behavior was studied by several different methods [10–12], and photo-receptor currents, which are the earliest rhodopsin-triggered responses that are detectable from intact cells, were also studied in these reconstituted cells [13,14]. Finally, retinoids were extracted from wild-type cells and analysed. The clear conclusion was that the algal rhodopsins contain all-*trans* retinal, which isomerises in light to 13-*cis*, similarly to all known microbial rhodopsins (type 1 rhodopsins) but different from animal rhodopsins where light iso-merises 11-*cis* retinal to all-*trans* (type II rhodopsins). Thus, green algae are the first eucaryotes in which rhodopsins with microbial-type chromophores have been identified.

4.2   Demands on algal rhodopsins

Individual responses of freely swimming cells to flashes of different intensity were analysed using Poisson statistics, which led to the suggestion that direc-tional changes and photophobic responses are both triggered by single photons and may be mediated by different photoreceptor systems [15]. This conclusion was not considered further until Zacks et al. [16] studied phototaxis and photo-phobic responses in blind cells reconstituted with retinal analogs that cannot isomerise around certain double bonds. From competition experiments with authentic retinal the authors concluded that the chromophoric properties of the photoreceptor responsible for phototaxis are slightly different to those responsible for photophobic responses.

Electrophysiological studies, first carried out on *Haematococcus pluvialis*, revealed that the eye-specific photoreceptor current is an overlay of two components, a low-light saturating component, PC_a (or I_{P1a}) and a high-light saturating component, PC_b (I_{P1b}), [17]. In a more extended analysis on *C. rein-hardtii* Ehlenbeck et al. [18] supported this finding and presented a model with two photoreceptor systems that was able to explain quantitatively the light dependence of the photocurrent amplitudes. The fast onset of the photocurrent PC_b, within less than 30 µs after a light flash [17,19,20], led to the suggestion that in the high-light saturating photoreceptor system the rhodopsin is either directly coupled to the ion channel proteins or is itself the ion conducting protein [18,20]. In contrast, the low-light saturating current PC_a appears with a delay of many milliseconds [21], suggesting that the responsible rhodopsin activates the respective channel via amplification [18].

Recently it was found that the primary photoreceptor currents I_{P1a} and I_{P1b} are followed by H^+-carried inward currents, I_{P2a} and I_{P2b}, which are clearly vis-ible only under acidic conditions. Both I_{P1b} and I_{P2b} saturate in parallel with the rhodopsin bleaching, which suggests that they are triggered by the same recep-tor. Ehlenbeck et al. [18] presented a model of a single rhodopsin that under-goes a photocycle with two conducting intermediates, one of which mediates the fast Ca^{2+} current and the later one constitutes the proton conductance. This peculiar photoreceptor remains to be identified.

4.3 Opsin-related proteins of green microalgae

4.3.1 Cop1, the first identified opsin-related protein in green algae

Supplementation of white retinal-deficient *C. reinhardtii* cells with ^{3}H-retinal or exchanging the endogenous retinal in purified eyespot membranes with ^{3}H-retinal identified only one single retinal binding protein, which was purified and sequenced [22,23]. The amino acid sequence shows significant homology to invertebrate opsins (type 2 opsins, [23]). Based on this homology, which is described in detail elsewhere [24], and due to the retinal-binding capacity of the protein it was named chlamyopsin (Cop1). This conclusion was supported by the structure of the gene (*cop1*), i.e. the exon/intron organisation, which resembles that of animal opsin genes [25]. Immunofluorescence microscopy revealed that the Cop-protein, originally purified from eyespot membranes, is indeed located within the eyespot area. The dominance of the opsin-related Cop1 protein in *Chlamydomonas* eyes and the lack of any other identified retinal-binding protein left little doubt that Cop1 is the photoreceptor mediating the major movement responses in *C. reinhardtii* [26]. Antisera against chlamyopsin impaired the light-regulated GTPase activity of an eye-specific G-protein in the related alga *Spermatozopsis similis.* This demonstrated the existence of a rhodopsin also in this alga, but also suggested that a G-protein is involved in the transduction process of Cop1 [27]. The G-protein binding domain found in the loop between helix 4 and 5 of animal opsins is conserved in the algal Cop1 protein [24].

Over the years several arguments and experimental results have shed doubt on the hypothesis that the identified opsin-related protein is the photoreceptor for all movement responses. Most difficult to explain was how the highly charged Cop protein as well as its homolog volvoxopsin (Vop, [24,28]), which are both not compatible with a 7 transmembrane (TM)-receptor, can form functional rhodopsin photoreceptors. Second, it was always obscure as to how a protein related to animal opsins may form an archaean type all-*trans* chromophore that isomerises in light to 13-*cis*. The third difficulty occurred from experiments on *Volvox carteri.* The identified rhodopsin-related protein, Vop1, is a dominant eye protein of the somatic cells, but Vop is also expressed in eyeless gonidia and the amount increases in young embryos, long before the cells are fully differentiated and pigmented eyes are formed. Thus, it was anticipated that the Vop must have a second function besides its postulated involvement in photomovement responses [28].

Left with these inconsistent findings only a genetic analysis was expected to solve the puzzle. However, a genetic approach for the functional analysis of Cop and Vop has been unsuccessful until recently, because opsin mutants are still not available for either *C. reinhardtii* or *V. carteri*, and targeted gene disruption has yet to be established for green algae. To overcome the lack of methods for gene disruption in green algae, an antisense-RNA approach was tried first in *V. carteri* [28], in which a modified *vop*-gene with three inverted exons was prepared. Transformants with multiple copies of this antisense construct (10 to 15) showed opsin reduction down to only 10% of the wild-type level and a reduction of their phototactic rate and sensitivity. This seemed to support the claim that Vop is the phototaxis photoreceptor. Unfortunately, photocurrent measurements were not possible in these transformants due to the vast extracellular matrix material. Moreover, this type of transformant was unstable, probably due to the multicopy integration of the antisense construct.

4.3.2   Cop2, a second translation product of the primary cop-mRNA

Before an antisense strategy could be developed for *C. reinhardtii*, the *cop* mRNA had to be characterised. As seen from Figure 1, the only detectable cop mRNA is 1.7 kb long. This mRNA is significantly larger than expected from the amino acid sequence and the earlier sequenced genomic *cop1* clone. Moreover, using different primers of the original cDNA we amplified from this mRNA preparation three overlapping cDNAs that were partially identical but did not result in a contiguous open reading frame (Figure 1, right). Finally, sequencing the cDNA copy of the 1.7 kb mRNA revealed an open reading frame consisting of exons 1 to 7 and part of the earlier defined intron 7 [25], now called exon 8 (Figure 2). This exon encodes a putative retinal binding site that is more related to the putative retinal binding site of volvoxopsin, Vop,

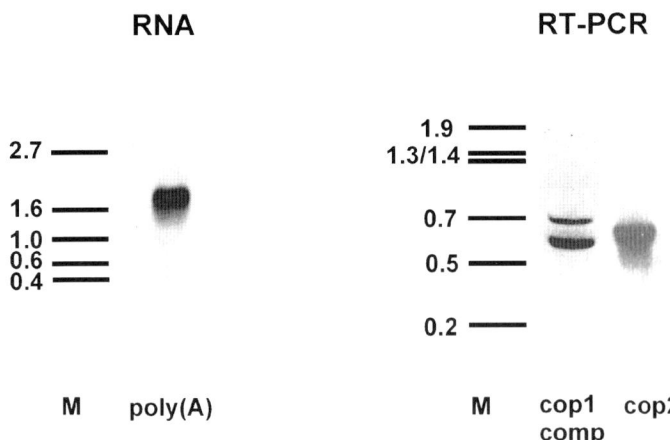

Figure 1.   Visualisation of the dominant cop-RNA (cop2) and PCR-products of all three splicing variants, *cop1*, *cop2* and *comp*. Left: RNA blot and identification of the *cop*-mRNA. As probe (E1 to E3) a digoxigenin-labeled fragment was used that identifies all three splicing variants. Right: RT-PCR from a poly(A)-RNA template originating from strain CC2454. The primer pair COPF+1/COP+708 amplified a 707 bp Cop1 fragment and a 600 bp COMP fragment, whereas COPF+1/COP2R+366 resulted in a 650 bp COP2 product. The positioning of the primers is seen in Figure 2.

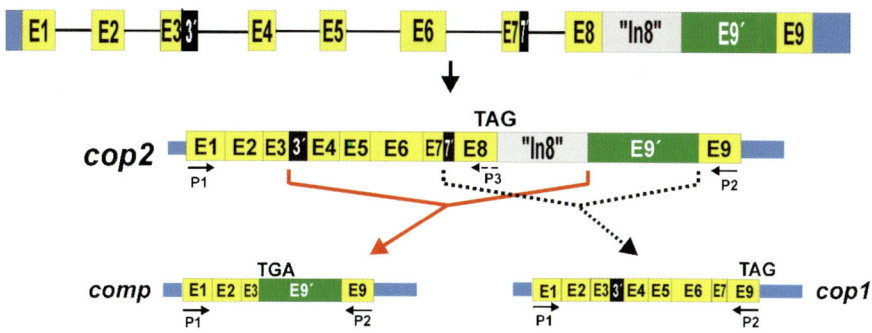

Figure 2. Genomic organisation of the opsin-related gene cop1/2. The three different mRNAs, i.e. cop1, cop2 and comp, are generated from the primary transcript by a sequential or alternative splicing at the indicated splicing sites.

than to that of Cop1. The new encoded protein is now called Cop2 (AccNo: 351823). The former E8 has been renamed E9, because it is located downstream of E8. The *cop2* mRNA may be processed in two ways. Alternative 1: the 3'end of E7, the whole E8 and E9', also part of the former I7, are spliced out, resulting in the *cop1* mRNA that encodes the Cop1 protein. Although this *cop1* message was the first identified RT-PCR product [23], the concentration is 50 times below that of the cop2-mRNA and it is not visible in Figure 1 (left). There are three reasons, why Cop2 has long been ignored. First, the molecular weights of Cop1 and Cop2 are very similar; second, all peptides generated and sequenced from the ³H-retinal-labelled eyespot protein are present in both Cop1 and Cop2 sequences; finally, PCR amplification of the whole *cop2* cDNA was never successful due to secondary structure formation in intron 8.

The 3'-end of exon E3, "E3'", exons E4 to E8, and the 5'-end of intron 8, "In8", are spliced out completely. Instead of E8, a novel exon E9', with its 5'-end upstream the formerly named exon 8, now renamed exon 9, is connected to E3 (Fig.2). This alternative splicing product, "comp", was amplified by RT-PCR and is seen in Fig. 1. The comp RNA contains two open reading frames (ORFs), spanning nucleotide 1 to 285 (comp1) and 383 to 601 (comp2). The two encoded hypothetical "opsin modified proteins" (COMPs) comprise 95 and 73 amino acids, of which 79 (N-terminus of COMP1) and 32 (C-terminus of COMP2) are identical to COP1. Both ORFs were expressed in E. coli and antibodies were raised against the purified proteins. However, we were not able to identify any of them in soluble or membrane fractions of C. reinhardtii cell extracts. Thus, these proteins are either not expressed at all, or expressed only in very low amounts, or under conditions we have not tested yet.

4.3.3   Cop1 and Cop2 are both eyespot proteins

With the identification of Cop2 it was not clear whether both Cop1 and Cop2 were expressed in the eye because the earlier used antibody identifies Cop1

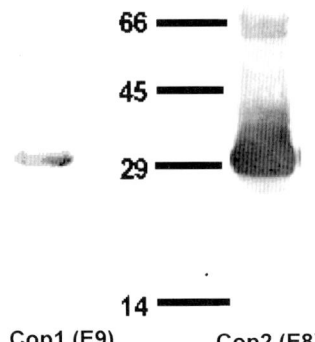

Figure 3. Specific recognition of Cop1 and Cop2. Dimers of exons 8 or 9 of the *cop1/2* gene were expressed in *E. coli* and the proteins affinity purified via their his-tag and used for antibody production. After ensuring that antisera recognized their antigens equally well, eyespot membrane fractions of strain CW15 cells were tested. The relative signal intensities indicated a Cop2:Cop1 ratio of H ≈ 50:1.

and Cop2 [23]. However, a green fluorescent protein (*gfp*)-*cop1* tandem was made from the *cop1* cDNA which led to expression of a Cop1-GFP-fusion protein [29]. The clear eyespot fluorescence demonstrates that Cop1 is directed into the eye. To evaluate the ratio between Cop1 and Cop2 in the eye, exons 8 and 9 were expressed separately in *E. coli*, purified and used for the immunisation of rabbits. The antisera were specific and recognised the respective peptides almost equally well (data not shown). Anti-Cop-2 serum labelled the 30 kD_{app} retinal protein in *C. reinhardtii* eyes about 50 times better than the anti-Cop-1 serum (Figure 3), suggesting that the ratio of Cop 1 to Cop 2 is about 1:50. The earlier presented immunolocalisation has to be reinterpreted as a localisation of Cop 1 and Cop 2, with a clear dominance of Cop2, in the eye of wild-type cells.

4.3.4   Cop1 and Cop2 are not photoreceptors for phototaxis or photophobic responses

On the basis of the known *cop* gene structure and a clear understanding of the splicing process, the antisense strategy originally developed for Cop1 also has to be reinterpreted [30]. As shown for higher plants, intron-containing gene fragments directly linked to their intron-less inverted cDNA counterparts cause more efficient post-transcriptional gene silencing (PTGS) than genes with only partially inverted segments but without inverted repeat structure [31]. A genomic *cop* fragment, consisting of exon 1 to 3, linked to the respective inverted cDNA fragment and driven by the authentic *cop* promoter reduced the Cop concentration in the cell up to 100-fold [30]. Owing to the fact that the *cop1* and *cop2* mRNA contain identical exons, 1 to 3, both protein concentrations should have been reduced similarly. In these antisense transformants the photoreceptor currents, the photophobic responses, and phototaxis were left completely unchanged compared with wild-type cells with respect to signal amplitude and light sensitivity [30]. The obligate conclusion was that the so-far identified opsin-related protein (Cop 1) is not the photo-receptor for phototaxis or photophobic responses in *C. reinhardtii*. This view has now to be extended to both type-2 opsins, namely Cop1 and Cop2, because the abundance of both proteins is reduced in the antisense transformant, as concluded after retesting the transformants with antibodies specific for the protein stretches encoded by exon 8 and 9 (Fuhrmann, unpub-lished data). The photoreceptors for phototaxis and photophobic responses are still to be identified.

4.3.5 Chop1, a microbial type algal photoreceptor

The first alga in which two classes of retinal proteins were identified is the halotolerant alga *Dunaliella salina* [32]. This was not surprising because in *Dunaliella* phototaxis and photophobic responses exhibit rhodopsin action spectra with different maxima [33]. Labelling of *Dunaliella* eyespot membranes with ^{3}H-retinal identifies a 28 kD protein, probably a homologue of Cop1/2 and Vop1. In addition, a second retinal protein with a MW of 45 kD is present in these eyes, which has not yet been purified and sequenced. However, this finding initiated an intensive search for high molecular weight rhodopsin-related proteins in the EST database of *C. reinhardtii*. Three overlapping partial cDNA sequences were identified, which together code for a 76.4 kD protein (AcNo: AF385748) that is likely to be a rhodopsin-related photo-receptor [32]. The protein contains a soluble N-terminus with 77 amino acids, of which the 22 amino acid leader peptide is likely to be cleaved off during post translational modification. The Kyte-Doolittle hydropathy plot predicts a membrane protein with 7 transmembrane helices. In this respect Chop1 is a classical rhodopsin photoreceptor. This core protein including the 7 TM-segments, reaching from amino acid 76 to 309 out of 712, shows homology (15 to 20% identity) to sensory rhodopsins from archaea, to the ion transporters bacteriorhodopsin (BR) and halorhodopsin (HR), as well as to the recently identified rhodopsin from *Neurospora* (NOP1). The homology might appear small but most amino acids that define the retinal binding site and the ion-conducting channels are conserved (Figure 4). This is most clearly documented by comparing this third opsin-like protein from *C. reinhardtii* with BR and HR. For both, the function of most amino acids is clear from countless biophysical studies and high-resolution crystal structures [5,34,35].

By comparing the primary sequences and modeling the structure on the basis of the high-resolution X-ray structure of BR of Luecke et al. [35], a func-tion for many, and probably the most essential, amino acids of Chop1 was predicted by Nagel et al. [36] and is extended here. The retinal binding Lys in type I rhodopsins is imbedded in a conserved retinal binding region LDxxx-KxxF/W^{299}, suggesting that K^{296} of Chop1 is the retinal-binding amino acid. Twenty-two amino acids are, in archaeal rhodopsins, in direct contact with

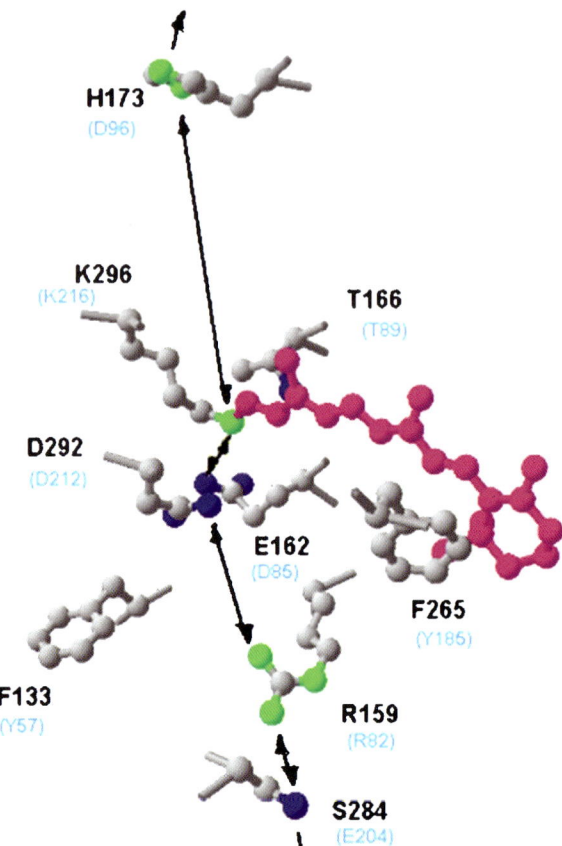

Figure 4. Scheme of the proposed H⁺-transport pathway in Chop1. The scheme includes those amino acids of Chop1 (black) that are at positions that are most critical for the proton-transporting hydrogen bonding network in BR (light blue). The all-*trans* retinal is shown in red, nitrogen atoms are in green, and oxygen atoms in dark blue. Black arrows visualise the proposed proton pathway.

the retinal moiety; nine are identical in Chop1, and four additional ones underwent conservative exchanges. The conserved ones are located in helices 3, 4, 6 and 7 and found near the more polar Schiff base side. More specifically, the 9-methyl and 13-methyl groups in BR are 3.6 to 3.7 Å from the closest heavy atom of Trp[182] and Leu[93], respectively, which is consistent with the evidence that these residues are essential for thermal reisomerisation from 13-*cis* to all-*trans* at the end of the photocycle. These residues are Trp[262] and Ile[170] in Chop1. During the BR photocycle the proton is released to Asp[85], which is Glu[162] in Chop1. In BR the H⁺ is released to the surface via Glu[204] and Glu[194], the equivalent of which in Chop1 are Glu[244] and Ser[154]. The release of the proton in BR from Asp[85] is accompanied by a new bond between Asp[85] and Arg[82]. The equivalent bonding is expected to occur in Chop1 between Glu[162] and Arg[159] after proton release from Glu[162]. This results in the proposed extracellular H⁺-transport pathway for Chop1 depicted in Figure 4.

The cytoplasmic region of BR with Asp^{96} as the proton donor of the unprotonated Schiff base is flanked by Phe^{42} and Phe^{219}. In archaeal sensors this Asp is replaced by Tyr and the reprotonation process is slowed down. In Chop1 it is His^{173}, which can be reversibly protonated and deprotonated at acidic pH. This His is expected to be in contact with Tyr^{109} and Trp^{269}. From the sequential and structural comparison we speculated that Chop1 functions as a proton transport system in an active or passive way. As a large proton current has been recorded from *C. reinhardtii* eyes at acidic pH it is not unlikely that Chop1 is the responsible photoreceptor.

4.3.6   Expression of Chop1 in Xenopus laevis oocytes

To examine the proposed ion transport function, chop1 RNA was expressed in oocytes of *Xenopus laevis* [36]. Five days after mRNA injection inward currents were recorded. These currents were monitored only upon illumination of the oocytes with green light. Red light was ineffective. At pH 6 or lower all photocurrents were inward directed between −100 and +40 mV. Replacing Cl^- by aspartate, sodium by potassium or *N*-methyl-D-glutamine, or replacing Ca^{2+} by Mg^{2+} still produced photocurrents of similar size, suggesting that neither Cl^- nor K^+ or Ca^{2+} contribute to the current. However, increasing the H^+-concentration of the bath solution to 10 μM (pH 5) led to large inward currents, between −100 and +40 mV. This indicated that Chop1 is an H^+-transport system. To discriminate between H^+-pump and H^+-channel photocurrents, the cell interior was acidified by addition of membrane-permeable butyric acid and acetic acid. Large outward currents could be observed under these conditions. The size and the direction of the current showed a nearly Nernstian behaviour, which means that the H^+-flux is a purely passive process. The authors claimed that Chop1 is the first identified light-gated ion channel and that it is not unlikely that this type of rhodopsin is widely distributed in other phototactic microalgae as well as in gametes and zoospores of macroalgae [36].

4.4   Conclusion

The 7- TM topology (Figure 2), the archaeal-type sequence and structure (Figure 2) and the intrinsic light-gated ion conductance classifies Chop1 as an excellent candidate for the photoreceptor that triggers the recently identified H^+-current I_{P2b} in the *C. reinhardtii* eye.

References

1. J.L. Spudich, C.-S. Yang, K.H., Jung, E.N. Spudich (2001). Retinylidene proteins: Structures and function from archaea to humans. *Annu. Rev. Cell. Dev. Biol.*, **16**, 393–421.

2. T. Okada, O.-P Ernst, K. Palczewski, K.-P. Hofmann (2001). Activation of rhodopsin: new insights from structural and biochemical studies. *Trends Biochem. Sci.*, **26**, 318–324.

3. J. Kleinschmidt, F.I. Harosi (1992). Anion sensitivity and spectral tuning of cone visual pigments *in situ*. *Proc. Natl. Acad. Sci. U.S.A.*, **89**, 9181–9185.

4. G. Kochendorfer, S.W. Lin, T.P. Sakmar, R.A. Mathies (1999). How color vision proteins are tuned. *Trends Biochem. Sci.*, **24**, 300–305.

5. M. Kolbe, H. Besir, L.-O. Essen, D. Oesterhelt (2000). Structure of the light-driven chloride pump Halorhodopsin at 1.8 A resolution. *Science*, **288**, 1390–1396.

6. J.L. Spudich (1994). Protein-protein interaction converts a proton pump into a sensory receptor. *Cell*, **79**, 471–474.

7. J.A Bieszke, E.L. Braun, L.E. Bean, S. Kang, D.O. Natvig, K.A. Borkovich (1999). The NOP-1 gene of *Neurospora crassa* encodes a seven transmembrane helix retinal-binding protein homologous to archeal rhodopsins. *Proc. Natl. Acad. Sci. U.S.A.* **96**, 8034–8039.

8. K.W. Foster, J. Saranak, N. Patel, G. Zarilli, M. Okabe, T. Kline, K. Nakanishi (1984). A rhodopsin is the functional phototreceptor for phototaxis in the unicellular eucaryote *Chlamydomonas*. *Nature*, **311**, 756–759.

9. H.C. Berg (1984). Bovine-like rhodopsin in algae. *Nature*, **311**, 702.

10. P. Hegemann, W. Gärtner, R. Uhl (1991). All-*trans* retinal constitutes the functional chromophore in *Chlamydomonas'* rhodopsin. *Biophys. J.*, **60**, 1477–1489.

11. M. Lawson, D.N. Zacks, F. Derguini, K. Nakanishi, J.L. Spudich (1991). Retinal analog restoration of photophobic responses in a blind *Chlamydomonas reinhardtii* mutant: Evidence for an archaebacterial-like chromophore in a eukaryotic rhodopsin. *Biophys. J.*, **60**, 1490–1498.

12. T.K. Takahashi, M. Yoshihara, M. Watanabe, M. Kubota, R. Johnson, F. Derguini, K. Nakanishi (1991). Photoisomerisation of retinal at 13-ene is important for phototaxis of *Chlamydomonas reinhardtii:* Simultaneous measurements of phototactic and photophobic responses. *Biochem. Biophys. Res. Commun.*, **178**, 1273–1279.

13. O.A. Sineshchekov, E.G. Govorunova, A. Der, L. Keszthelyi, W. Nultsch (1994). Photoinduced electric currents in carotenoid-deficient *Chlamydomonas* mutants reconstituted with retinal and its analogs. *Biophys. J.*, **66**, 2073–2084.

14. E.G. Govorunova, O.A. Sineshchekov, W. Gärtner, A.S. Chunaev, P. Hegemann (2001). Photoreceptor current and photoorientation in *Chlamydomonas* mediated by 9-demethylchlamyrhodopsin. *Biophys. J.*, **81**, 2897–2907.

15. P. Hegemann, W. Marwan (1988). Single photons are sufficient to trigger movement responses in *Chlamydomonas reinhardtii*. *Photochem. Photobiol.*, **48**, 99–106.

16. D.N. Zacks, F. Derguini K. Nakanishi, J.L. Spudich (1993). Comparative study of phototactic and photophobic receptor chromophore properties in *Chlamydomonas reinhardtii*. *Biophys. J.*, **65**, 508–518.

17. O.A. Sineshchekov, F.F. Litvin, L. Keszethely (1990). Two components of the photoreceptor potential in phototaxis of the flagellated green alga *Haematococcus pluvialis*. *Biophys. J.*, **57**, 33–39.

18. S. Ehlenbeck, D. Gradmann, F.-J. Braun, P. Hegemann (2001). Evidence for a light-induced H⁺ conductance in the eye of the green alga *Chlamydomonas reinhardtii*. *Biophys. J.*, **82**, 740–751.

19. H. Harz, C. Nonnengäßer, P. Hegemann (1992). The photoreceptor current of the green alga *Chlamydomonas*. *Phil. Trans. R. Soc. London B*, **338**, 39–52.

20. E.-M. Holland, F.-J. Braun, C. Nonnengäßer, H. Harz, P. Hegemann (1996). The nature of rhodopsin triggered photocurrents in *Chlamydomonas*. I. Kinetics and influence of divalent ions. *Biophys. J.*, **70**, 924–931.

21. F.-J. Braun, P. Hegemann (1999). Two independent photoreceptor currents in the spheroidal alga *Volvox carteri*. *Biophysical J.*, **76**, 1668–1678.

22. M. Beckmann, P. Hegemann (1991). *In vitro* identification of rhodopsin in the green alga *Chlamydomonas*. *Biochemistry*, **30**, 3692–3697.

23. W. Deininger, P. Kröger, U. Hegemann, F. Lottspeich, P. Hegemann (1995). Chlamyrhodopsin represents a new type of sensory photoreceptor. *EMBO J.*, **14**, 5849–5858.

24. P. Hegemann, W. Deininger (2001). Algal eyes and their rhodopsin photoreceptors. In: D.-P. Häder, M. Lebert (Eds), *Photomovement* (pp. 229–243). Elsevier Science B.V.

25. W. Deininger, M. Fuhrmann, P. Hegemann (2000). Opsin evolution: out of the wild green yonder. *Trends Genet.*, **16**, 158–159.

26. P. Kröger, P. Hegemann (1994). Hypothesis: Photophobic responses and phototaxis are triggered by a single rhodopsin photoreceptor. *FEBS Lett.*, **341**, 5–9.

27. M. Calenberg, U. Brohsonn, M. Zedlacher, G. Kreimer (1998). Light and Ca^{2+}-modulated GTPases in the eyespot apparatus of a flagellate green alga. *Plant Cell*, **10**, 91–103.

28. E. Ebnet, M. Fischer, W. Deininger, P. Hegemann (1999). Volvoxrhodopsin, a light-regulated sensory photoreceptor of the spheroidal alga *Volvox carteri*. *Plant Cell*, **11**, 1473–1484.

29. M. Fuhrmann, W. Oertel, P. Hegemann (1999). A synthetic gene coding for the green fluorescent protein (GFP) is a versatile reporter in *Chlamydomonas reinhardtii*. *Plant J.*, **19**, 353–361.

30. M. Fuhrmann, A. Stahlberg, S. Rank, E. Govorunova, P. Hegemann (2001). The major retinal protein of the *C. reinhardtii* eye is not the photoreceptor for phototaxis and photophobic responses. *J. Cell Sci.*, **114**, 3857–3863.

31. N.A. Smith, S.P. Singh, M.-B. Wang, P.A. Stoutsjesdijk, A.G. Green, P.M. Waterhouse (2000). Gene expression: Total silencing by intron-spliced hairpin RNAs. *Nature*, **407**, 319–320.

32. P. Hegemann, M. Fuhrmann, S. Kateriya (2001). Algal sensory photoreceptors. *J. Phycol.*, **37**, 668–676.

33. R. Wayne, A. Kadota, M. Watanabe, M. Furuya (1991). Photomovement in *Dunaliella salina*: Fluence rate-response curves and action spectra. *Planta*, **184**, 515–524.

34. Y. Kimura, D.G. Vassylyev, A. Miyazawa, A. Kidera, M. Matsushima, K. Mitsuoka, K. Murata, T. Hirai, Y. Fujiyoshi (1997). Surface of bacteriorhodopsin revealed by high resolution electron cystallography. *Nature*, **389**, 206–211.

35. H. Luecke, B. Schobert, H.-T. Richter, J.-P. Cartailler, J.K. Lanyi (1999). Structure of bacteriorhodopsin at 1.55 Å resolution. *J. Membr. Biol.*, **291**, 899–911.

36. G. Nagel, D. Ollig, M. Fuhrmann, A.-M. Musti, S. Kateriya, E. Bamberg, P. Hegemann (2002). Channelrhodopsin-1: a light-gated proton channel in green algae. *Science*, **296**, 2395–2398.

Chapter 5

The phytochromes: spectroscopy and function

Wolfgang Gärtner, Silvia E. Braslavsky

Table of contents

Abstract

Recent developments are summarized on the understanding of structure and function of the phytochromes, ubiquitous photoreceptors in higher and lower plants that have also been identified in prokaryotes. We emphasize here findings based on spectroscopic studies of native and recombinant proteins assembled with various chromophores, including *de novo* synthesized tetrapyrroles. The generation of transgenic plants, and the identification of plant phenotypes at a molecular level enlighten the involvement of the phytochromes in the light-induced signal transduction pathway. Despite the lack of a three-dimensional structure of phytochromes, the generation of recombinant proteins of various sizes and/or with a mutated sequence has allowed the identification of particular amino acids important for the correct incorporation of the chromophore and for the integrity of the spectral properties. Chemically modified tetrapyrroles, carrying modified substituents at various positions, revealed precise steric interactions between chromophore and protein binding pocket, which are of major importance for the chromoprotein assembly process, the spectroscopic properties of the chromoprotein, and the kinetics of the light-driven P_r-to-P_{fr} photoconversion. Time-resolved absorption spectroscopy has helped in elucidating the complex reaction pathway between both stable forms. This pathway implies changes in the conformations of chromophore and protein, beginning with the earliest photophysical events, femto- and picoseconds, and reaching the seconds time domain. Vibrational spectroscopy (FT-Raman and FT-infrared) has yielded a precise picture of the chromophore conformation as well as of the conformational changes of chromophore and protein upon photoexcitation. Photothermal techniques afford information on the energy content of various intermediate states in the phototransformation pathway, and have given insights into the time-resolved profile of the enthalpy and entropy changes during the photoinduced transformation. The study of plant phenotypes originating from random mutagenesis has led to the identification of protein domains involved in signal transduction.

5.1   Introduction

Understanding the sophisticated mechanisms by which plants (in most cases sessile) sense the light conditions and adapt to changes in intensity, spectral composition, duration and even polarization, is a multidisciplinary task. Plants not only respond to shading by competing with other nearby plants, they also monitor the intense irradiation which might cause deregulation of metabolic processes or even damage of the photosynthetic apparatus. The photosensors of plants act in the blue/UV- as well as in the photosynthetically relevant red/far-red wavelength range (for a description of blue light photoreceptors see Chapters 9–11 in this volume). The red/far-red range of the spectrum is scanned by the phytochromes, which are ubiquitous in higher and lower plants, and which recently have also been identified in several prokaryotic photosynthetic and even non-photosynthetic organisms (see also Chapter 7). The strong interest in the phytochromes is documented by a remarkable number of original papers and reviews [1–5]. Consequently, mainly, the most recent developments are discussed here. The reader is referred to the literature for older work.

Phytochromes constitute a relatively small protein family, with up to five variants (phyA to phyE) present in species of higher plants [6], and a variable number present in lower plants, where in some cases only one type of phytochrome has been detected. Phylogenetical alignments indicate a significant relationship between phyB and phyD (identity of 80% for *A. thaliana* proteins), suggesting a most recent gene duplication. From the phyB/D limb of the phylogenetic tree, phyE has diverged earlier, whereas phyC formed very soon after the separation between phyA and phyB/D/E which constitute a class on its own [7]. A more extended comparison showed that the recently identified bacterial phytochromes, in particular those found in cyanobacteria, represent the evolutionary ancestors of the plant phytochromes [8], whereas the so-called "phytochrome-like proteins", PLPs, e.g., RcaE from *Fremyella* [9], deviate significantly in their chromophore-binding domain and show reasonable similarity only in the C-terminal signalling domain.

Whereas in etiolated (higher) plants phyA (also called type-I phytochrome) represents the dominating protein, the complete ensemble of phytochromes is present in very similar amounts in the de-etiolated plants, with a slight dominance of the phyB-type phytochromes (green-plant derived phytochromes are historically referred to as type-II phytochromes) [10].

5.1.1   Molecular aspects

Phytochromes absorb light in the spectral range of the photosynthetic apparatus (between 600 and 750 nm), and also show a second short-wavelength absorption of lower intensity around 380 nm, which upon irradiation can induce the phytochrome-characteristic photochemistry (Figure 1).

The chromophore is an open-chain tetrapyrrole, phytochromobilin (PΦB) in higher and most lower plants, and phycocyanobilin (PCB) in the algae *Mougeotia* [11] and *Mesotaenium* [12] as well as in the cyanobacteria phytochrome, as determined for *Synechocystis* [13] (see Chapter 7). The chromophore is covalently linked to the protein through a thioether bond to a cysteine, cys321 in phyA [14–16]. Interestingly, the essential role of the covalent attachment has been documented for a mutant C321S, in which the possibility for covalent binding is absent. Expression of this mutant in a phyA-minus background caused a strong characteristic phenotype of the transgenic plant, indicating the absolute requirement of covalent chromophore attachment [17]. The sole attachment at ring A (position 3′, Figure 2) has been proven correct by NMR studies [14,15], disproving former suggestions of covalent bonds between the protein and a different or multiple positions of the chromophore.

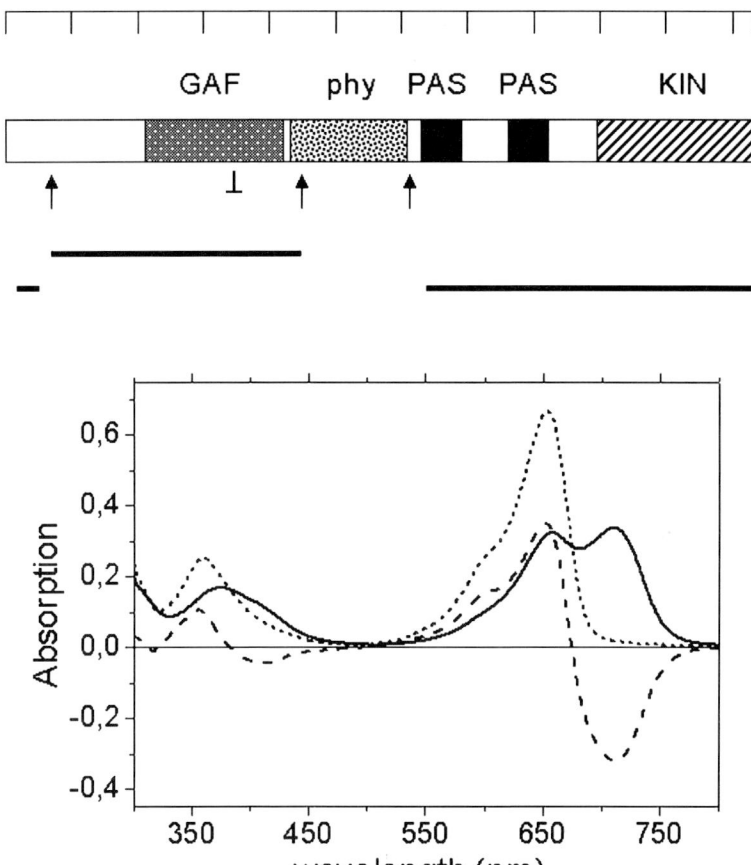

Figure 1. General features common to phytochromes. Top: Structural arrangement of the phytochrome molecules (⊥ indicates the covalently attached chromophore). The scaling at the top refers to 100 amino acids. The positions of enzymatic cleavage using trypsin are indicated by arrows. Suggested functional domains, maintenance of spectral properties (upper horizontal bar, positions 65 to 425) and regions involved in signal transduction (lower horizontal bar, utmost N-terminal part and C-terminal half) are indicated by horizontal bars. The identified dimerization domains are located within the first PAS-domain and at the end of the kinase motif (positions 1070–1128). Bottom: P_r (· ·), P_{fr} (—), and P_r – P_{fr} difference (- - -) absorption spectra of phytochrome (shown for oat phyA). Due to the spectral overlap of both spectral forms no complete conversion of P_r into P_{fr} can be accomplished.

The phytochromes are formed biosynthetically in their P_r (r = red absorbing) form with λ_{max} at ca. 665 nm (this is also the case for in vitro production [18]) and are converted upon light absorption into the P_{fr} (fr = far-red absorbing) form with λ_{max} at ca. 730 nm. (Figure 1; for a detailed discussion of the underlying light-induced molecular processes see Section 5). Together with the change of the absorption maximum in the far-red spectral region upon generation of P_{fr}, the above-mentioned absorbance around 380 nm shifts to ca.

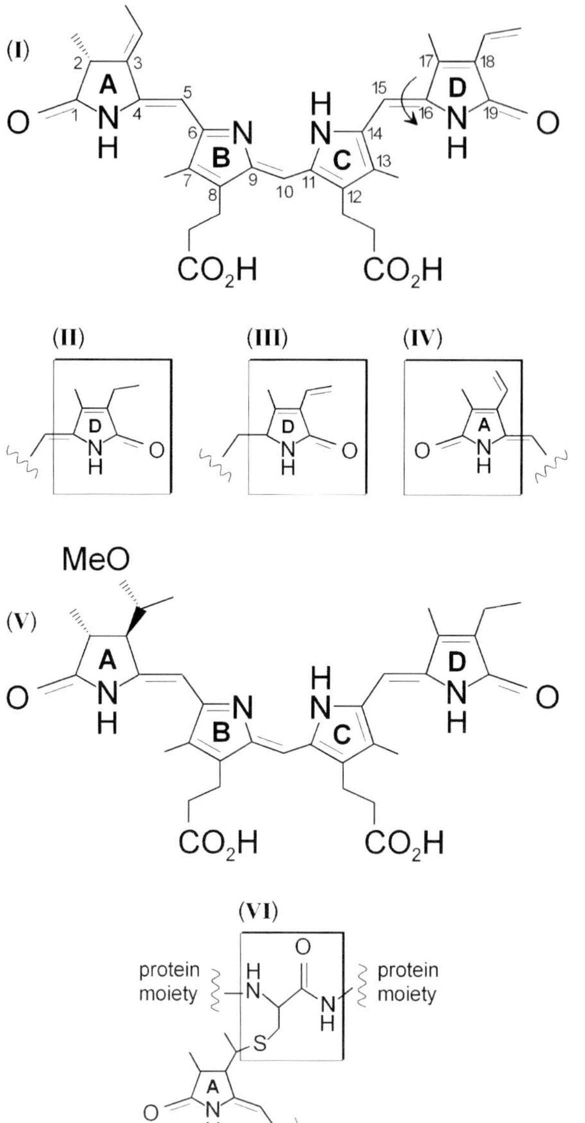

Figure 2. Structures of various open-chain tetrapyrrole chromophores, (I) PΦB, (II) PCB, (III) PEB, (IV) biliverdin IXα (BV), and (V) the 3′-methoxy adduct of PCB. For compounds (II), (III) and (IV), only those parts of the tetrapyrrole are shown that differ from PΦB, i.e. ethyl vs. vinyl substituent in PCB, 15–16 single bond vs. double bond in PEB, and 2–3 double bond and 3-vinyl group vs. 2–3 single bond and 3-ethylidene group in BV. Part (VI) depicts the type of covalent binding of compounds (I), (II), and (III) via a thiol ether bond to the protein. Note that biliverdin can not form a covalent thioether bond with the protein due to its different substituent at position 3 (vinyl instead of ethylidene). Protonation of the chromophore upon incorporation into the protein has been proven from FT-Raman experiments [155].

410 nm. Yet, due to the strong overlap of both spectral forms, no complete conversion of P_r into the P_{fr} form can be accomplished. The long wavelength tail of the P_{fr} spectrum, on the contrary, allows a 100% generation of the P_r form by far-red irradiation. Many photomorphogenical processes ascribed to phytochrome are explained by a cycling of the molecules between both states, a process resulting in varying proportions of P_r and P_{fr} depending on the spectral quality and the intensity of the incident light [19].

PhyA-P_r is remarkably stable and persists in the dark practically indefinitely. Also the P_{fr} state remains fairly unperturbed (in isolated form), with a thermal conversion into the P_r state of less than 30% of its initial amount (at ambient temperature) within three days [18,20]. This surprising stability of the P_{fr} state of phyA does not reflect the situation in the cellular milieu, where the photo-transformed phyA is rapidly degraded by proteolytic attack, following the ubiquitin pathway ([21] and references cited therein). More recent data employing modified chromophores and site-directed mutagenesis (SDM) have allowed the first insights into the mechanisms that control the thermal stability of the P_{fr} form (*vide infra*).

Conversely, *in vitro* studies of recombinant PHYB-type phytochrome have revealed a remarkably fast reversion to the P_r state [18,22]. Potato-phyB-P_{fr} rapidly thermally converts into P_r (80% of the P_{fr} form had converted into P_r within 2 h). This finding, similar to results with phyA (vide supra), apparently contradicts reports indicating that, in green plants (where phyB-type phyto-chromes prevail), light-induced photomorphogenic processes (i.e. generation of the P_{fr} form) could be reverted even after several hours [2,23]. This unexpected discrepancy has to be seen within the cellular milieu where phytochromes interact as signaling molecules with various partners that apparently extend the lifetime of the phytochrome signaling state.

The stability of the P_{fr} form has also been studied after assembling the apoprotein with various naturally occurring as well as with chemically syn-thesized chromophores, revealing a strong dependence of P_{fr} stability on the chromophore structure. This further indicates the tight interaction between chromophore and protein (*vide infra*).

5.1.2   Protein structure

Phytochromes are relatively large chromoproteins with molecular weights of ca. 125 kDa. Mature oat phyA consists of 1128 amino acids (after posttransla-tional removal of the first methionine) [24]. Phytochromes with much larger molecular weight, and consisting of more than 1300 amino acids have been reported, mainly identified in lower plants such as ferns and mosses [25,26].

Common to all phytochromes is their modular-built structure, with the chromophore-bearing domain located in the N-terminal half, whereas the C-terminal domain (together with a few amino acids at the very N-terminal part of the protein) hosts the sequences responsible for supramolecular arr-angement and signal transduction [27–29]. Although no three-dimensional structure of any phytochrome is available, several protein structural motifs have been identified from sequence alignments (*vide infra*). The physical separation of the two different functions in the protein requires precise and intimate interactions between the various protein domains to ensure an efficient light absorption in the chromophore-bearing part (= signal generation) and transmission within the protein towards the C-terminally-located output domain to allow signal transduction via protein–protein interaction. In fact, the dual functionality, i.e. sensing (in this case light sensing) and the initiation of signal transduction (generating a biological signal) is common to many biological receptors. The identification of the possibly general principle common to all biological receptors is a fascinating aspect of the research on biological photoreceptors.

The phytochrome molecules form homodimers *in vitro* with the dimerization site located in the C-terminal part of the protein. It should be mentioned that only sparse evidence has been presented indicating that also in their cellular environment phytochromes arrange as homodimers. These investigations are all based on indirect observations, including the complexity of the $P_r \rightarrow P_{fr}$ conversion (*vide infra* and [30,31]), or the mixed kinetics of the thermally driven $P_{fr} \rightarrow P_r$ back-conversion [32]. Small-angle X-ray scattering and electron microscopy studies have revealed an arrangement of the two molecules in the form of a "Y"-shaped complex with the C-terminal parts attached to each other and the N-terminal halves (forming the two arms of the "Y") separated [33,34]. The protein part responsible for homodimer formation has been narrowed by proteolytic analysis to a short domain located in the hinge region (positions 599 to 683; if not explicitly indicated, all numbering refers to mature, plant-derived oat phyA, 1128 amino acids) and in the C-terminal part of the protein (positions 1069 to 1129) [35,36]. Interestingly, two PAS domains, which are often involved in protein–protein interactions and dimerizations, are found in phytochromes located in one of the two earlier proposed interaction sites.

Whereas the isolated N-terminal half of the protein (obtainable from proteolysis or as recombinant proteins and identified to be monomeric) can undergo a photoinduced transformation between P_r and P_{fr} very similar to the full-length protein, the physiological function of phytochromes requires the presence of the C-terminus [37] (suggestive of a homodimeric arrangement *in vivo*) and also of a number of amino acids in the utmost N-terminal region of the protein. Besides truncations of C-terminal portions, SDM in that part of the protein also led to the identification of positions essential for the phytochrome function, as demonstrated for phyA [38] and for phyB [39]. This clearly indicates the essential role of the photoinduced conformational change of phytochrome that brings into proximity both ends of the protein.

Within the modular-constructed overall structure, a number of other protein motifs have been identified (based on secondary structure prediction and sequence alignment). The most important are, in addition to the above-mentioned PAS domains, those including the chromophore-binding domain (GAF-domain) and the region of the protein involved in signal transduction (vide infra). GAF domains are also found in other proteins and are related to cGMP-binding capabilities. The GAF domain of the phytochromes constitutes only part of the chromophore-binding domain (yet it contains the site for covalent binding). The chromophore-binding region is completed by a sequence motif ("phytochrome domain" in Figure 1) which is recognized in nearly all phytochromes due to its high sequence similarity. No evidence for glycosylation has been found (neither in plant-derived nor in recombinant material), whereas several putative phosphorylation sites could be identified [40–43].

The C-terminal part of phytochromes has been intensively searched for sequence motifs known from other proteins, since not only the dimerization site but also the physiological activity is located in this part of the protein, and any known signatures would indicate the site for light-induced signal transduction (for motifs involved in signal transduction see Section 5.1.4).

The identification of PCB and phycoerythrobilin (PEB) as chromophores in the antennae of cyanobacteria, which have spectral properties similar to the phytochromes (despite the lack of photochemical activity), and the knowledge of the detailed crystal structures for phycocyanin [44,45] have led to the employment of this chromoprotein as a model to calculate the phytochrome structure. Although there is good evidence, based on several structure predictions, that the amino acid residues in the direct vicinity of the chromophore binding cysteine fold into an α-helical motif, any further-reaching structure predictions have not yielded a convincing three-dimensional model for phytochrome [46,47].

5.1.3   Phytochromes as cytosolic proteins

Phytochromes are soluble proteins and few hints have been presented for a membrane-attached fraction of the phytochrome molecules, mostly in lower plants [48]. An interesting motif has been found in the phytochrome sequence of the green alga *Mougeotia scalaris* [49] with similarity to actin-binding proteins. Since this alga strongly responds by orientation of the chloroplast to irradiation with red/far-red light of various intensities and different polarization, it had been assumed that the algal phytochrome induces these macroscopically observable movements by interacting with the cytoskeleton. Besides the orientation with respect to the direction of the incident light, movements of the chloroplasts can also be observed in the plant cells of many species, although the most prominent examples are found in lower plants like ferns, mosses and algae, and evidence has not been presented for all species that a membrane-bound or -associated portion of phytochrome is involved. More recent data demonstrate that this motional activity not only responds to intensity, polarization and direction of the incident light, but also to its wavelength, such that both red and blue light can cause chloroplast movement and also stomatal cell opening, indicating the involvement of a second type of photoreceptor – putatively the phototropins [50,51].

The remarkable flexibility of phytochromes during the light-induced changes has been documented by proteolysis experiments performed with both P_r and P_{fr} states, revealing different proteolytic fragments [52], as well as by chemical assays directed towards the modification of amino acid residues, e.g. cysteine residues, which become exposed or hidden upon phototransformation [53].

5.1.4 Phytochrome domains involved in signal transduction

Light-induced, phytochrome-mediated signal transduction in plants is outlined in detail in Chapter 6. Here we restrict our discussion to aspects dealing exclusively with the phytochrome molecule and the very first steps of protein–protein interactions. Following the primary event in signal transduction, i.e. the generation of the biological signal via photoisomerization of the chromophore, a cross-talk between the N-terminal, chromophore-bearing domain and the C-terminal half is required to allow further transmission of the signal from the C-terminal part via protein–protein interactions into the cell interior. These domain interactions have been identified through chemical modification of various amino acid residues in the activated (P_{fr}) and deactivated (P_r) phytochrome forms [54].

Some search programs highlight part of the phytochromes sequence as a histidine-kinase (HK) related motif. This appeared particularly interesting, since the recently discovered prokaryotic phytochrome-like proteins were demonstrated to act as light-induced histidine kinases [55,56] (see also Chapter 7). Yet, histidine kinase activity is preferentially found in prokaryotic organisms, and is often part of a readily identifiable two-component system with response regulators that become activated via a phospho-relay mechanism. Kinase-based signal transduction in eukaryotes is more often ascribed to serine/threonine or tyrosine kinase activity. Accordingly, an SDM study has excluded the involvement of histidines of the C-terminal part of phytochrome in signal transduction, at least those which are located in common HK sequence motifs (around position 965–975) [57].

Thus, the phytochromes show histidine kinase-like sequence motifs [58], but transfer of the light-induced signal to another protein may proceed after serine or threonine phosphorylation. In fact, a number of serine residues have been identified as transiently phosphorylated in a light-dependent manner. Out of the cluster of eight serine residues at the utmost front end of the protein (positions 1–3 and 7–11) at least one has been shown to be transiently phosphorylated [40,43]. An interesting position, at which phosphorylation was also found, is serine 598, located in the hinge region between the chromophore-bearing N-terminal and the signal-transducing C-terminal part. This region of the protein is assumed to be very flexible and to undergo conformational changes during the light-induced processes (*vide infra* for spectroscopic evidence). Assuming that the flexibility is influenced by phosphorylation makes ser598 an important regulatory position in the signal transduction process.

In addition to the *in vitro* work, the analysis of transgenic plants has also revealed several positions as important for signal transduction. Most of these mutations, generated by random mutagenesis and identified after finding a strong phenotype in transgenic plants, cluster in a relatively small domain between amino acids 630 and 770 for phyA [38] and between 750 and 820 for phyB [39,59].

The utmost N-terminal portion of phytochromes also carries, besides the group of serines which are targets for phosphorylation, other positions essential for signal transduction (a less pronounced effect of several mutations and deletions is observed on the spectral properties). Studies on the involvement of various protein regions in signal transduction usually depend on *in vivo* studies and the identification of phenotypes, employing plants that either overexpress a modified protein or express a mutated protein on a deletion ("null"-mutant) background. Such approaches have identified the complete N-terminus as essential for wild-type behavior, since a strong phenotype is generated by a Δ7–69 deletion [60] and, as found by an alternative approach, by a Δ6–57 deletion [61]. This latter study has also led to the identification of another important domain between positions 652 and 712. A more detailed analysis by either deleting only parts of the 70 N-terminal amino acids, or by alanine scanning (mutating any of the investigated positions to an alanine) also revealed, in addition to the above-mentioned serine cluster, regions 25–33 and 50–62 as important parts for signal transduction [62,63]. The generation of transgenic plants overexpressing ser→ala mutants of all ten clustered serines in this protein domain led to a phenotype showing enhanced phytochrome activity [64].

5.2   Preparative aspects

5.2.1   Plant-derived material

Phytochrome preparation from plants in quantities sufficient for spectroscopic analysis is still a formidable task. During the early attempts, the extraction of material from tissue of dark-grown plants struggled with endogenous proteolytic attacks. At first, the 59 kDa fragment was isolated, later called small phytochrome. Subsequently the 118 kDa fragment, "large phytochrome" and finally the full-length 124 kDa intact protein was obtained (Figure 1) [65]. The intactness of this material was verified by the DNA sequence, which encoded a protein of matching size [66].

The first enzymatic cleavage of extracted material, supposed to be homogeneous material, generating "large phytochrome" revealed its heterogeneous character. Cleavage sites were identified at positions 33, 54, and 65, which yielded a mixture of fragments of similar size, around 114/118 kDa [52,67], some of which, fatally for functional studies, contained or lacked portions of the N-terminal domain that turned out to be important for the maintainance of the spectral and functional properties.

Studies on preferred proteolytic sites led to the identification of another fragment, spanning amino acids 66 to 425, the so-called 39 kDa fragment. This chromopeptide showed a typical P_r absorption spectrum (λ_{max} = 660 nm), but a strongly disturbed P_{fr} absorbance [68]. Irradiation of the 39 kDa P_r form reproducibly yielded an atypical photoproduct with a very broad absorbance of low intensity (λ_{max} = 660 nm) that had a reduced thermal stability and reverted back to the P_r form within several minutes. Interestingly, this chromopeptide showed absorption properties and thermal stability reminiscent of a regular P_{fr} when furnished–by molecular biology technology–with the N-terminal 6 kDa domain (amino acids 1–65, then to be called 45 kDa chromopeptide) [69]. A recombinant fragment of rice phyA with an even more extended deleted N-terminal part (Δ1–80) was reported to have lost the capability to bind the chromophore [70].

Several protocols have been worked out to prepare intact phyA from plant tissue, for which oat and pea are the most favorable sources. However, although phyA is the dominant species in etiolated (dark-grown) plants, and can be isolated following established protocols [71], it was found to be a mixture of three isogenes (AP3, 4 and 5), present in similar amounts in the plant material [66,72]. The situation in green plants is even more critical due to two obstacles: on the one hand, the amount of phyA is down-regulated upon de-etiolation to the same concentration range in which the other phytochromes (phyB to phyE) are present [73]. On the other hand, separation from the chlorophylls poses a further difficulty. To preferentially isolate type-II phytochromes, attempts have been reported utilizing a selectively raised antibody for affinity purification [74–76].

5.2.2 Recombinant approaches

The advent of molecular-biology techniques, the deciphering of phytochrome-encoding genes from various higher and lower plants, and the complete analysis of a plant genome (*Arabidopsis thaliana*) have opened up the possibility of preparing recombinant phytochromes of virtually any origin and structure, although several technical difficulties still impair the preparations. Since the various host cells utilized for protein expression are, in general, unable to provide the phytochrome chromophore, it is essential for this approach that the assembly of the chromoprotein, i.e. the formation of a covalent bond between chromophore and protein upon incubation of both components, takes place without auxiliary proteins (contrary to the case in, e.g. the phycocyanin antenna complexes of cyanobacteria, where a lyase is required for covalent attachment of the chromophore).

The unexpected discovery of the incorporation of open-chain tetrapyrroles (originating from heme degradation) into the recombinant apophytochrome during expression in the yeast *Pichia pastoris*, which yielded the complete chromoprotein [77], has allowed the identification of phycocyanobilin (PCB) as the chromophore in the green alga *Mesotaenium* [12] (see also below). This unintended chromoprotein formation can be overcome if the yeast cells are grown under strong illumination, which probably avoids the side reaction due to light-induced degradation of open-chain tetrapyrroles prior to their incorporation into the apoprotein (Remberg and Gärtner, unpublished).

Expression of apophytochrome entirely *in vitro*, making use of the rabbit reticulocyte system and binding of the chromophore without any added lyase provided the first evidence for an autocatalytic assembly process [78], although no spectroscopically detectable holophytochrome could be identified. The first attempts to express phytochrome-encoding cDNAs in bacteria (*Escherichia coli*) were successful [78–81]. However, a significant amount of recombinant apoprotein is deposited in aggregated (non-functional) form in inclusion bodies. Although the yield of the functional apoprotein from expression in *E. coli* could be improved by co-expression of phytochrome-encoding ORFs and the *E. coli* endogenous chaperons GroES and GroEL [81], a comparison of the chromophore binding activity of bacterial and yeast-derived apophytochrome indicated that, probably, the *E. coli*-derived material is not properly folded to allow the very rapid chromoprotein formation seen with the apoprotein generated in yeast (*vide infra*). More favorable hosts were found with various yeast strains: *Saccharomyces cerevisiae* [22,79,82], *Pichia pastoris*, and *Hansenula polymorpha* [18,83,84]. In particular, the latter two yeast strains show the largest yields of recombinant proteins.

Both *Pichia* and *Hansenula* are methylotrophic, i.e. they utilize methanol as the sole carbon source when grown in a minimal medium. This capability is activated by an inducible AOX (alcohol oxidase) promoter [85], and allows expression of recombinant proteins via induction with methanol, after cloning the DNA encoding the foreign protein under the control of the AOX promotor (in *H. polymorpha*, other promoters have also been utilized [86]). Transformation can be performed by electroporation as well as *via* the spheroblast generation protocol. In contrast to *S. cerevisiae*, where plasmids are autosomally replicated, *P. pastoris* and *H. polymorpha* integrate the transformed plasmid (which should be linearized prior to transformation) in several copies into the genome.

P. pastoris in general integrates between five and twelve copies, whereas *H. polymorpha* consecutively integrates up to forty copies into the genome (Piontek and Gärtner, unpublished). This phenomenon, together with the ability to grow to high cell densities in a fermenter (cell densities of up to 250 OD_{740} units mL^{-1} can be obtained) and the strong promoter acitvity, leads to massive expression of foreign proteins. For example, from a fermenter growth of a *H. polymorpha* culture, up to 250 g of cell pellet per litre can be obtained, containing up to 1 mg of recombinant N-terminal half of oat phyA per gram of cell pellet (determined after assembly of the apoprotein in the crude lysate, see Table 1) [83].

Routinely, these recombinant proteins are furnished with a tag that allows convenient affinity purification. As an alternative to the attachment of a His6 tail either at the N- or (preferentially) at the C-terminal end for purification of the protein on immobilized Ni^{2+} or Co^{2+} ions, streptavidin tags have been employed [87]. The use of larger tags, better called fusion proteins, such as the maltose-binding domain, turned out to be less advantageous. Since the addition of such large protein domains interfers with the phytochrome

Table 1. Yields of recombinant phytochrome fragments from expression in yeast and from plant extraction

Phytochrome fragment	59 kDa	65 kDa
Expression system	*Pichia pastoris*	*Hansenula polymorpha*
mg protein/g cell pellet	0.5–0.6; corresp. 100 mg (L cell culture)$^{-1}$	1.0; corresp. 250 mg (L cell culture)$^{-1}$
Purification	metal affinity + ion exchange/gel filtration column	
max. SAR (A_{654}/A_{280})	1.8	1.6
Plant derived material: native oat phytochrome	3–4 mg 124 kDa phyA/kg etiolated oat seedlings	

function and thus has to be removed before characterization of the recombinant protein, enzymatic cleavage at pre-introduced sites is essential. However, this procedure often turns out to be difficult since the cleavage sites may be buried within the fusion protein, or proteolysis can also lead to extended cleavage of the recombinant phytochrome (Hill and Gärtner, unpublished).

Although a thorough analysis of recombinant His-tagged phytochrome (59 kDa fragment) with the plant-derived tryptic fragment did not reveal strong differences in, e.g. the spectral properties and the kinetics of the P_r to P_{fr} conversion [32], removal of the tag might be advantageous for other experiments. Placement of a thrombin site between the His-tag and the last (or, in the case of N-terminal attachment, the first) genuine amino acids of phytochrome allows us to remove the polar tag and to obtain a recombinant phytochrome with only two additional amino acids from the rest of the thrombin site.

Given that the recombinant approach allows us to tailor the expressed proteins to the required size–and of course offers the chance to change its primary structure via SDM–an analysis of various parts of the protein involved in proper function can readily be performed. Whereas the generation of smaller phytochrome fragments by enzymatic cleavage has to rely on the given sites for proteolytic cleavage, the introduction of a stop codon at any desired position yields phytochromes of any anticipated size.

The advantage of the recombinant fragments became obvious during the investigation of the role of the very first amino acids in the N-terminal portion, since treatment with trypsin, which removes the complete C-terminal half of the protein, also causes cleavage after position 65 (Lys, numbering refers to oat phyA, which is cleaved at its N-terminal end, causing the removal of the first – methionine – amino acid). This position is the preferred, primarily processed site of enzymatic degradation (Figure 1). Initial spectroscopic experiments with the plant-derived, N-terminal half of the protein (59 kDa fragment) were performed without the first 6 kDa domain spanning amino acids 1–65 during the light-induced reaction, although the involvement of these first few amino acids in chromophore–protein interactions and in signal transduction has been subsequently demonstrated (*vide supra*).

Further support for the involvement of the first few amino acids in the function of phytochrome is that the spectral properties and the photoinduced $P_r \rightarrow P_{fr}$ conversion kinetics were different for the recombinant 59 and 65 kDa phytochrome fragments (spanning amino acids 66–595 and 1–595, respectively). A more detailed discussion of the spectroscopic properties of these chromoproteins is given below.

The possibility of generating recombinant phytochrome fragments, designed at the DNA level, allows us to determine molar absorption coefficients more precisely than from the same fragments produced by proteolytic cleavage. For the two recombinant N-terminal fragments of oat phyA, the 65 and 59 kDa chromoproteins (spanning amino acids 1–595 and 66–595, respectively), comparison of A_{665}/A_{280} (the so-called SAR, specific absorbance ratio) yielded values of 1.6. Based on the corresponding SAR for full-length, PΦB-containing phytochrome A of oat (ca. 1.1, and ε_{665} = 132000 M^{-1} cm^{-1} [88]) and taking into account a contribution of the chromophore absorption to the absorbance of the protein at 280 nm, it is possible to calculate for the C-terminally truncated recombinant chromopeptides ε_{665} >110000 M^{-1} cm^{-1} [89]. This estimation allows us to conclude (as suggested from other experiments) that the N-terminal part of the protein strongly affects the photochemical properties of the phytochromes.

5.3   The chromophore moiety

PΦB, the naturally occurring chromophore in all higher and most lower plants, is synthesized biochemically via the heme–biliverdin–phycocyanobilin conversion route [90,91]. The biosynthetic route was originally elucidated in plants carrying a chromophore deletion or in oat plants with a blocked tetrapyrrole biosynthesis by growing them in the presence of gabaculine [92] or 4-amino-5-hexynoic acid [93], both potent inhibitors of early steps in tetrapyrrole biosynthesis. These plants (indicated as hy1 and hy2 mutants) could be rescued to a wild-type-like behavior upon the addition of PΦB precursors. Inasmuch as these plants synthesized functional phytochrome after supplementation, they were described as chromophore-deletion mutants [94].

Recently, several genes have been identified whose products are involved in the biosynthesis of PCB or PΦB. The biosynthetic pathway of chromophore generation starts with a heme-oxygenase, which breaks heme compounds into the open-chain form, and a biliverdin reductase [95]. One of the formerly identified phytochrome-deficient mutants, *hy2*, which could be rescued by exogenous addition of biliverdin, is a loss-of-function mutant of the bilin reductase [96]. The identification of these two genes has led to a new expression system for phytochromes. The transformation of *E. coli* by the two chromophore-generating genes afforded a new host strain that upon expression of a phytochrome apoprotein yields the holoprotein already within the host cell [97,98]. This procedure yields a higher amount of photoactive phytochrome, although one has to keep in mind that expression of plant phytochrome in bacteria is difficult and of low yield and, accordingly, up to now holophytochrome formation has only been worked out with bacterial phytochrome (Cph1 from *Synechocystis*). The conversion of hemes into biliverdin – and then in plants into PCB or PΦB – should be seen in the context of the recently proposed use of biliverdin as a chromophore in several of the newly identified phytochrome-like proteins from prokaryotes (cyano- and other eubacteria [99–101], see also Chapter 7).

5.3.1   Naturally occurring open-chain tetrapyrroles as chromophores in phytochromes

Only a few naturally occurring open-chain tetrapyrroles (obtainable from plants or cyanobacteria extracts or through simple chemical treatment of other tetrapyrroles) can function as phytochrome chromophores [102]. PΦB does not occur in free form in significant amounts, but can readily be generated from phycoerythrobilin (PEB), a major component of the antennae of several cyanobacteria, by a redox reaction using mercury salts (see Figure 2) [91].

Phycocyanobilin (PCB), also extracted from cyanobacteria, is the most frequently used chromophore in phytochrome research, because the PCB-assembled chromoprotein has similar spectral and photochromic properties to those of the PΦB-assembled chromoprotein in spite of the blue shift of the P_r and the P_{fr} absorption spectra and some reproducible differences in the kinetic behavior, especially in the short microsecond time range (for a detailed description of the kinetic behavior of chromoproteins carrying one or the other chromophore, see Section 5.5).

Application of either PCB or PΦB to chromophore-deficient *Arabidopsis* plants (*hy1* and *hy2*) revealed a surprising discrimination of the phyA- and phyB binding sites especially taking into account the small differences in the structure of the two chromophores. Whereas a rescue of phyA function in these plants could only be accomplished by the external addition of PΦB (and not with PCB), the phyB function could be re-established by the addition of either chromophore [103]; apparently, the double bond of PΦB is required for a tight chromophore–protein interaction to enable an effective photochemistry. The authors claim that differences in the binding site architecture are responsible for this different effect of the two chromophores.

Phycoerythrobilin (PEB) also forms a covalently bound adduct with apophytochrome. Yet, due to its saturated 15,16-bond, it only produces a chromoprotein, with absorption maximum at 576 nm, which does not show a photochromatic behaviour [104]. Instead, the chromoprotein shows a strong fluorescence (these adducts are called phytofluorophores [105]).

Due to a modified substitution at position 3 of ring A (vinyl instead of ethylidene), biliverdin IXα (BV) cannot form a covalent bond with apophytochrome, as demonstrated by the absence of a fluorescence signal at the same position as the protein band in a zinc blot (this assay is indicative of covalent binding of the tetrapyrroles to the protein [106]). However, the embedding and the formation of non-covalent interactions within the binding site of the apophytochrome CpB from *Calothrix* PCC 7601 force BV into a phytochromelike photochemistry upon irradiation. The non-covalent incorporation becomes evident from the red-shifted absorption maxima of this chromophore–protein complex (662 and 737 nm, respectively, for the P_r and P_{fr} forms) [107]. The spectral shift indicates an unmodified, intact π-electron system in BV (both PΦB and PCB lose a double bond due to covalent attachment to the protein via thiol-ether formation).

Interestingly, incubation (in the dark) of the PHYA apoprotein with BV does not yield the P_r form, as observed with the covalently binding tetrapyrroles, but instead gives rise to an absorption band with $\lambda_{max} = 700$ nm, from which the P_r/P_{fr} photochemistry can be initiated, but into which the system relaxes upon interrupting the irradiation. This behavior is different for the cyanobacterial apophytochrome CphB from *Calothrix* PCC7601 where incubation with BV produces directly a P_r-like absorption band [107].

5.3.2   *Modified chromophores by de novo synthesis*

Very few reactions can be used to generate tetrapyrroles from naturally occurring precursors that yield phytochrome chromophores. Besides the generation of PΦB (*vide supra*) from PEB, the chemical reactivity of selected open-chain tetrapyrroles at position 10, i.e. the central position between rings B and C, can be utilized to obtain synthons for further synthesis. In particular, nucleophilic attack by reagents such as thiobarbiturate splits the central bond and yields the right and the left half as pyrromethenones. This reaction enabled the facile formation of a so-called iso-PΦB, by combination of two of these synthons. The new tetrapyrrole shows, in ring D, a reverted arrangement of the vinyl and methyl groups [108]. The use of this compound gave the first hint of a specific interaction between chromophore and protein, since the assembled chromoprotein exhibited a remarkably blue-shifted P_{fr} absorption maximum ($\lambda_{max} = 714$, compared with 730 nm for assembly with PΦB and 717 nm with PCB).

A vinyl ("iso-PΦB") or an ethyl group ("iso-PCB") at position 17, occupied by a methyl group in the native chromophore, reveals an electronic effect in the chromophore–protein interactions. These two substituents do not significantly participate in the conjugated system, and both require about the same space, although the absorbance maxima show a blue-shift for the iso-PCB-containing chromoprotein (658 and 707 nm, for P_r and P_{fr}, respectively, compared to 663 and 714 for iso-PΦB [109]).

Another tetrapyrrole derivative that can interact with the protein binding site and is prone to phytochrome-characteristic photochemistry has been found during the preparation of PCB. This tetrapyrrole, acting as a chromophore in the antenna pigments of cyanobacteria, is released from the chlorosomes by methanolysis (treatment of cyanobacteria with boiling methanol overnight). The "contaminating" derivative was a methanol adduct of PCB at its 3′,3″ position (see Figure 2) [110]. Inasmuch as this substitution inhibits the double bond between positions 3 and 3′ of PCB, and thereby eliminates the possibility of covalently binding to the protein, it served as an excellent model to study the binding pocket of the protein, and to determine to what extent a phytochrome-characteristic photochemistry could be induced.

De novo chemical synthesis of open-chain tetrapyrroles is the only adequate approach for a systematic study of the chromophore–protein interactions. Due to the demand to study the various positions of the open-chain tetrapyrrole regarding their possible contribution to the chromophore–protein interactions, each tetrapyrrole ring has to be separately synthesized and subsequently condensed with its neighbours. When the role of neither the B- nor the C-ring needs to be probed, one and the same synthon can be employed since these two rings are identically substituted and are in a mirror-image arrangement in the chromophore. A convergent synthetic route is possible, but it is laborious and time-consuming.

There is a wealth of information on the synthesis of open-chain tetrapyrroles due to their important role in many light-regulated processes, and also to their involvement in metabolic pathways of heme compounds ([102,111,112] and literature cited therein). However, since many of the formerly reported tetrapyrrole-generating synthetic approaches were designed for model compound studies, they afforded the esterified propionic side groups. To incorporate the chromophore and for a photochemistry typical of phytochromes, though, the free acids are required [113]. A detailed study showed (yet to be performed with recombinant PHYB of *A. thaliana*) that indeed monoesterified PCBs, at either of the two propionate groups, covalently bind to the apoprotein, but do not undergo P_r/P_{fr} photoreversible reactions, whereas a derivative with exchanged substituents (methyl vs. propionate group) at ring C still yields a photoactive chromoprotein [114]. Possibly, even when the propionate side chain in ring C is shifted from position 12 to 13, the residue of the protein involved in electrostatic interactions is still sufficiently close to render photochemistry possible.

Thus, to routinely generate modified chromophores, synthetic approaches leading to the acid forms were demanded. The release of the free acids could be realized by hydrolyzing the methyl esters in an ion exchange-catalyzed reaction as the final step of the reaction pathway [108]. The protection of the propionic acid groups as their allyl esters allowed mild and efficient release of the free acids [114]. The application and improvement of formerly reported synthetic routes to build open-chain tetrapyrroles has afforded several chromophore derivatives that allowed us to identify specific interactions of the protein certain positions of the chromophore. The above-mentioned work of Hanzawa *et al.* [114] represents the most comprehensive study using modified chromophores so far. Ring A is initially involved in anchoring the bilin, until a complete and covalent fixation is accomplished. When all the used derivates are incorporated into the apoprotein, they show reversible photochemistry with very similar absorption maxima for both P_r and P_{fr} forms, even those with bulkier substituents or with side groups at positions where no substituent (or only a methyl group) is present in the native chromophore. Conversely, any change of the substitution pattern at ring D has severe effects. This is clearly because this ring undergoes the greatest conformational changes upon photoisomerization. Only extensions of the chain length at positions 17 and 18 (see Figure 1 for the various positions) up to three carbon atoms (n-propyl substituent) are tolerated; further linear extension (performed up to n-octyl substituents) leads to the loss of photoreversibility, although covalent attachment within the binding site and formation of a P_r-like absorption band is still observed.

The situation at ring D was probed with a series of chromophore derivatives that do not carry linearly extended substituents, but show progressively bulkier groups at positions 17 and 18 [109]. Whereas the reduction in size of the vinyl (PΦB) and ethyl (PCB) group to a methyl group ("17,18-dimethyl-PCB") does not cause any significant effect – indicating that any anchoring at position 18 is not very important for the conformational arrangement of ring D–any increase in size of the C17- or C18-substituents causes serious changes in the photochemical behavior. Stepwise enlargement of the C18-substituent, i.e. methyl, ethyl, isopropyl, tert-butyl, still allows the formation of photoreversible chromoproteins, although with a reduced assembly rate. However, more pronounced blue-shifts of the absorption maxima are observed upon increasing the size of the substituent at position 17 from the naturally occurring methyl group to ethyl and isopropyl. This result identifies an important interaction between the protein and that part of the chromophore. Whereas a C2-substituent, vinyl or ethyl, is tolerated by the binding site and leads to photoreversible chromoproteins with only slightly blue-shifted absorption maxima, the introduction of an isopropyl group causes a selective strong hypsochromic shift of the P_r-form (λ_{max} = 550 nm), the P_{fr} absorption is only moderately blue-shifted (λ_{max} = 705 nm, compared with, e.g., 707 nm for the 17-ethyl derivative). Obviously, the chromophore experiences a strong steric hindrance in the P_r form that distorts the tetrapyrrole structure such that the full conjugation of all four rings is significantly disturbed. Such a conclusion can be justified by comparison with the absorption maxima of phytochromes assembled with PEB, showing an absorption maximum of the P_r form at 576 nm [104]. Since in PEB the saturated bond at position 15 interrupts the conjugation between the four rings which then only extends over rings A, B, and C, a blue-shifted absorbance results which is similar to that of the 17-isopropyl-PCB.

To explain the spectral properties that the bilin exhibits after binding to the protein, a number of phytochrome models have been prepared, in particular by modifying the chromophore. An interesting compound consists of the chromophore and a dipeptide, serine-cysteine (this Ser-Cys motif is also found in the sequences of phyAs). The chromophore was covalently attached in a phytochrome-like manner to the cysteine residue, but, in addition, a second bond could be formed reversibly, involving the carbonyl group of ring A and the hydroxy group of the serine generating an imino seryl ester. Whereas the non-esterified, covalently cysteine-bound chromophore absorbed at 595 nm, ester-bond formation shifted the maximum to 672 nm [115]. In an accompanying study, absorbances reminiscent of the P_r/P_{fr} absorption bands were generated by protonating/deprotonating the tetrapyrrole compound, either by the addition of strong acids or by attachment of the tetrapyrrole to a dipeptide, cysteine-glutamate, which interacted with the pyrrole rings of the chromophore via its carboxyl-/carboxylate group [116,117].

5.4 Biochemical properties of recombinant or chromophore-modified phytochromes

The availability of recombinant apoprotein allows a detailed analysis of the chromophore–protein interactions taking place during chromoprotein formation upon the incubation of the apoprotein with the chromophore (or derivatives) in the dark, and also permits to compare the effects of various point mutations or modified chromophores on the absorption spectra of the chromoproteins, their thermal stability and the kinetics of their light-induced P_{fr} formation (this latter property will be discussed in detail in Section 5.5).

5.4.1   The chromoprotein assembly

The addition of the chromophore to the apoprotein to yield the P_r form of phytochrome has been followed by spectroscopic methods. The first report made use of an enzyme-related mechanism by defining a pre-equilibrium between free and protein-associated chromophores which then becomes covalently bound in the rate-limiting step [104,118].

$$[P\Phi B] + [\text{apo-phy}] \overset{K_1}{\rightleftharpoons} [P\Phi B]{::::}\text{apo-phy} \overset{k_2}{\rightarrow} [\text{chromo-phy}]$$

From this scheme, two constants were defined: K_1, reflecting the ratio of the back- and- forward reaction rates of the equilibrium, during which a stretching of the chromophore takes place from the helical conformation that it adopts in solution [119]. The second process, k_2, is ascribed to covalent bond formation. However, a detailed analysis reveals that the "enzyme" (apo-phytochrome) is not released from the complex to continue activity, but becomes irreversibly bound and is removed from the reaction. Accordingly, a description including suicide-substrates would clearly be more appropriate.

It also turned out that observing the growth of phytochrome absorbance by a regular spectrophotometer, even via rapid scanning or by following the process at a selected wavelength, was insufficient since with yeast-derived phyA apoprotein (preferentially used as its 65 kDa N-terminal domain) the first data point – after less than 10 s – was taken when already up to 60% chromoprotein formation had occurred [18]. However, even under these circumstances, a difference in the kinetics of chromoprotein formation between PΦB and PCB could be identified. The lifetimes of these processes were 20 s for PΦB and 137 s for PCB (4°C). When the same experiment was performed with bacterial-derived apoprotein (expression in E. coli), a remarkably slower assembly was observed with a lifetime of more than 500 s [18]. This was indicative of a less prepared folding to incorporate the chromophore of the bacterial-derived apoprotein than that of the yeast-derived material.

The remarkably rapid assembly of PΦB to yield the extended and proton-ated chromophore suggests an even faster process for conformational adjust-ment of the chromophore and establishment of the necessary electrostatic interactions. Such assumptions are supported by assembly behavior in bacte-rial phytochromes, in particular in cyanobacteria, considered the evolutionary ancestors of "modern plant" phytochromes. In these proteins, the assembly initially yields a red-shifted intermediate (the formation of which again is not time-resolved with the applied methods) that converts into the typical P_r form [120]. Similar, but much more rapid kinetics, in less than 100 ms, for the conversion of a red-shifted intermediate into the P_r form, has recently been found for the assembly of plant phytochromes by stop-flow measurements (Benda, Favilla and Gärtner, unpublished).

The assembly is usually performed by incubating the apoprotein with a moderate molar excess of chromophore (no change in the kinetics was obser-ved with ratios of chromophore:apoprotein > 4:1, whereas a decrease in the assembly rate and an incomplete holoprotein formation resulted from lower ratios [18]). Under the excess conditions, the assembly kinetics can be con-sidered as a quasi-first order. Since the absorbance of phytochromes is sign-ificantly larger than that of the chromophore in aqueous solution ($\varepsilon_{667} = 132000$ M^{-1} cm^{-1} vs. ε_{600} ca. 18000 M^{-1} cm^{-1}, [121]) the formation of the chro-moprotein, generated in the P_r form, can be directly observed at 670 nm. Other detection methods identify the covalently bound chromophore through its visualization in a zinc blot [122], although this assay demands the denaturation of the protein. Alternative approaches make use of the fluorescence of a PEB-apophytochrome complex. Since formation of this complex is significantly slower than that with PΦB or PCB, a competition experiment employing vari-ous amounts of PEB allows the calculation of the assembly kinetics for PΦB- or PCB-containing phytochrome [104]. Although the absolute numbers for holophytochrome formation differ, PΦB reacts more rapidly than PCB. Only sparse information is available for rice- and tobacco-derived holophytochrome formation [104,123].

The effects of point mutations on the assembly process have been investi-gated. Mutations have preferentially been restricted to several amino acids preceding and following the chromophore binding cysteine. Positions essential for holophytochrome assembly were found to be nearly identical for phyA from pea [122] and from oat [84]. Out of the sequence Arg-Ala-Pro-His-Ser-*Cys321*-His-Leu-Gln (chromophore attachment site in bold/italics), the mutation H322L showed the strongest effect. Practically no chromophore binding was observed upon mutation. A strong effect on the stability of the recombinant protein was found for the P318K mutation, for which only small amounts of apophytochrome could be isolated. The corresponding alanine mutation (P318A) yielded a nearly unchanged, WT-like chromoprotein.

The other histidine (His319) is also strongly involved in chromophore binding since, upon mutation of this residue into leucine, the assembly time constant is nearly doubled (490 vs. 280 s, 10ºC). Conversely, the introduction of an additional positive charge (S320K) led to a faster assembly (150 s). In view of the above-mentioned suggested similarity between phytochromes and phycocyanin a double mutation was designed, i.e., L323R/Q324D. The residues at the corresponding positions in the phycocyanins are charged (arginine and aspartate) and are involved in electrostatic interactions with the chromophore [45]. The assumption that a similar protein–chromophore interaction might take place upon replacing L323/Q324 in phyA by these two charged amino acids, however, appeared to be incorrect. The double mutation had nearly no effect on the assembly kinetics or on the absorption spectra of the holoprotein.

Significant changes of the assembly kinetics could be evoked by the chemically synthesized chromophores discussed in the preceding section. An increase in substituent size at position 18 (vinyl in PΦB or ethyl in PCB) to isopropyl or tert-butyl causes slower assembly kinetics (from τ_1 and τ_2 = 1.6 and 10.8 min for the 17,18-dimethyl derivative up to 13 and 59 min for the tert-butyl derivative). More interestingly, an ethyl or isopropyl group at position 17, originally occupied by a methyl group, dramatically reduced the assembly rate: for 17-ethyl: τ_1, τ_2 are 7.6 and 44 min, for 17-isopropyl: 10.5 and 156 min, whereas for PCB: 0.9 and 10.6 min, respectively [109]. The native chromophore PΦB fits best into the binding site and assembles with times of 0.38 and 2.9 min, respectively.

5.4.2 P_r–, P_{fr}– and the P_r–P_{fr} difference spectra

Only relatively small changes of the absorption or the absorption difference spectra were observed for structurally modified phytochromes with respect to those for native phyA. Besides blue-shifts of ca. 12 nm for both the P_r and P_{fr} absorbances when using PCB instead of the native chromophore, somewhat stronger effects were found especially for P_{fr} absorbances of N- or C-terminally truncated phytochromes, finally reaching an unstructured, broad band spectrum for the thermally unstable P_{fr} in the 39 kDa tryptic peptide (see Section 2.1).

Incorporation of chemically modified chromophores causes moderate hypsochromic shifts of both absorbances, except for the above-mentioned 17-isopropyl-18-methyl derivative, which shows selectively a ca. 100 nm blue-shifted absorbance in the P_r form [109]. Exceptions are found for derivatives in which one of the two propionates remained esterified or in which the substituents at the B-ring of the chromophore were exchanged with each other [114]. Both monoesterified derivatives form only the P_r form and do not undergo light-induced conversion into P_{fr}. A tetrapyrrole carrying butyrate instead of propionate side chains forms only the P_r form. Photolytically unreactive phytochromes are also produced upon the mutual exchange of the methyl and propionate groups at ring B. This modification remarkably reduces the capability of these compounds to generate a chromoprotein. The latter study [114] was only performed with the recombinant apo-phyB protein of *Arabidopsis*. Much less information is available for mutated phytochromes.

All amino acid replacements discussed in the preceding section lead to insignificant changes in the absorption maxima [84].

5.4.3   Thermal stability of P_{fr} forms

Once formed by irradiation, the P_{fr} state of phyA-type phytochrome is nearly as stable as the P_r state, although the situation in plant cells is entirely different. In contrast, *in vitro* studies with recombinant phyB revealed a relatively unstable P_{fr} form that reverted into P_r within minutes. However, many light-induced processes in green plants can be stopped by far-red irradiation after many minutes and even some hours. Clearly, intracellular components interacting with the phytochromes modify the P_{fr} thermal stability. A possible involvement of heterodimer formation (P_r-P_{fr}) during the generation of P_{fr} was addressed by generating various amounts of P_{fr}. However, these experiments did not indicate a different spectral behavior. Conversely, two pools of P_{fr} could be identified *in vitro* (absorbing at ca. 730 and 722 nm) with different stability. A contribution of partly degraded fragments of 114/118 kDa molecular weight, known to be thermally less stable than the WT-protein, could be excluded [32]. A recent study revealed even more significant differences, since it reports that P_{fr}-P_{fr} dimers are nearly 100-fold more stable than P_{fr}-P_r dimers [124]. An *in vivo* study employing phyA-phyB chimera phytochromes demonstrated the role of the two protein parts. Clearly, the spectroscopic properties and the light responses are by the N-terminal part, whereas the signal-transduction is not as strictly localized, showing in some assays slightly overlapping functions of one or the other terminal half [125].

Attempts to more precisely identify domains or even positions in the phytochromes responsible for the different thermal stability (as well as the light-induced P_{fr} formation) have recently led to the identification of a point mutation in phyB that results in a very slow dark reversion and a hypersensitivity towards red light. This mutation also causes a strong phenotype in *Arabidopsis* [126].

Neither the C-terminal part (full-length native oat phytochrome vs. recombinant 65 kDa fragment) nor the choice of chromophore (PΦB or PCB) had any effect on the stability of the P_{fr} form in oat phyA. Only ca. 20% of the P_{fr} converted within two days into P_r in all of these chromoproteins [18]. No effect with respect to one or the other chromophore was detected in potato phyA. However, this phytochrome showed an overall lower P_{fr} stability (reversion of 50% into P_r was complete within ca. two hours) [22]. An even more rapid conversion was found for potato phyB. More than 80% of the P_{fr} form, independent of the chromophore used, thermally reverted to P_r within two hours. A difference with respect to the incorporated chromophore was found for the N-terminal halves of potato phyB. With PCB, ca. 70% of the originally formed P_{fr} converted within ca. 90 min to P_r, whereas the PΦB-containing chromoprotein reverted even more rapidly, i.e., more than 90% of the original P_{fr} material formed P_r within 150 min (all data on potato phytochromes from [18,22]). Experiments were performed to address the question as to whether the N- or the C-terminal part dominates the P_{fr} stability, based on the different rates of thermal conversion for phyA and phyB. Chimeras formed from the N-terminal part of rice phyA and the C-terminal part of tobacco phyB, named phyAB (and *vice versa*, phyBA), showed a higher stability with the phyA C-terminus ($\tau_{1/2}$ = 6 min), and a higher conversion for the phyB-derived C-terminal end ($\tau_{1/2}$ = 20 min) [127].

The thermal stability of the P_{fr} forms of some of the recombinant lower plant phytochromes was also analyzed. In addition to the chromoprotein of the green alga *Mougeotia* (see Section 5.5.7), only Cp2 from the moss *Ceratodon* has been expressed in amounts sufficient for such analysis. The P_{fr} stability of this protein is similar to that of the phyB constructs. Assembly of this Cp2 apophytochrome with either PΦB or PCB yielded a chromoprotein which decreased thermally to ca. 20% of the initially photochemically produced P_{fr} state within three hours [128].

Changes in the thermal stability of P_{fr} resulted upon incorporation of the various chemically synthesized chromophores discussed above [109]. Determination of the remaining P_{fr} content after three days revealed a higher stability for an increased size of the substituent at position 18 (tert-butyl > isopropyl > ethyl > methyl: 98, 94, 90 and 84%, respectively). However, changes at position 17 (ethyl to isopropyl) decreased P_{fr} stability with increasing substituent size (decay to 80 and 70%, respectively, after only one day).

5.5   Photophysics and photochemistry of phytochromes

5.5.1   *Chromophore structure in P_r and in P_{fr}*

The absorption spectrum of all phytochromes is very different to that of chromophore models such as the open-chain fully conjugated tetrapyrroles biliverdin and phycocyanobilin in solution. In the models, the near-UV absorption band around 370 nm is ca. 4 times more intense than the visible band at 660 nm, whereas in phytochrome (similar to the algae pigments phycocyanin and phycoerythrin) the near-UV band has a much lower intensity than the visible one [102]. In fact, absorption by phytochrome in the visible spectral region is remarkably high. PhyA shows the same maximum absorption coefficient of ε_{665} = 132000 M^{-1} cm^{-1}, as for the two monocots, i.e., for rye and oat phyA [88].

These spectroscopic features have long been taken to indicate a very different chromophore conformation in the chromoprotein [102]. An extended conformation, otherwise only present in minor amounts in the solutions of the open-chain tetrapyrroles (the esterified forms and also the acid species), in which the helical conformation dominates, should be stabilized by the protein [129]. The X-ray structure analysis of two bacterial C-phycocyanines has confirmed the stretched conformation of the chromophore in the proteins [44,45]. The latter have absorption spectra very similar to those of phytochrome, although the apoprotein structure and the function of the phycocyanines are very different from those of the phytochromes.

The strong changes induced in the absorption spectra of the chromophore by the apoprotein, including the photoreversibility, illustrates a general fundamental feature, i.e., the chromophore–protein interactions modulating the properties and function of the chromoproteins (see also [130]).

In addition to the conformational constraints, protonation of the pyrrole nitrogen of ring C in P_r explains the relatively narrow visible absorption band. In fact, the experimental resonance Raman spectra of native oat phyA P_r are compatible with the Raman spectra calculated for the protonated *ZZZasa* configuration, which has hence been suggested as the chromophore structure in P_r [131].

Most spectroscopic and photokinetic studies have been performed with phyA from several plants, preferentially from oat. Some differences have been found between samples from different species and it remains to be seen whether these differences are significant and respond to ecological constraints. For example, the maxima of the difference absorption spectra (P_r-P_{fr}) for recombinant potato PHYA-PΦB are at 660 and 712 nm [22], whereas for oat PHYA-PΦB they are at 663 and 728 nm [132] and at 666 and 730 nm [91]. The deviations in absorption probably originate from different expression systems.

The first comparative spectroscopic study of four recombinant phytochromes from the same plant, i.e., *Arabidopsis*, revealed that there is a difference in the absorption maxima of P_r and of P_{fr} of PHYA-PΦB, PHYB-PΦB, PHYC-PΦB, and PHYE-PΦB. *In vitro* assembly of the four apoproteins with phytochromobilin (PΦB) afforded difference spectra with P_r maxima at 670, 669, 661, and 670 nm and P_{fr} maxima at 737, 732, 725, and 724 nm for PHYA-PΦB, PHYB-PΦB, PHYC-PΦB, and PHYE-PΦB, respectively. Thus, the difference of the extrema (for P_r and P_{fr}) for PHYE-PΦB (54 nm) is smaller than the corresponding difference for PHYA-PΦB (67 nm), whereas the difference for PHYB-PΦB (63 nm) and for PHYC-PΦB (64 nm) are closen to those for PHYA-PΦB [133]. The authors speculate about the significance of the larger P_r-P_{fr} spectral overlap for PHYE-PΦB with respect to the role played by phyE in the shade avoidance phenomenon.

5.5.2   Excited states behavior and primary photochemical step

In view of the difficulties associated with the extraction of phytochromes other than phyA, the initial emission measurements *in vitro* were carried out with extracted full-length oat and pea phyA and their proteolytically generated fragments, [134,135] see also [136]. Song *et al.* [135] showed that no significant differences were found between the data for monocots (oat) and dicots (pea) phyA. More recently, steady-state emission data at room and low temperature have been collected for recombinant phytochromes from various sources, including moss [137], and time-resolved transient absorption spectroscopy has been performed with recombinant phytochrome from *Synechocystis* assembled with PCB (as already mentioned, the native chromophore in this type of phytochrome) and also with phycoerythrobilin [138]. The latter is the first ultrafast dynamic analysis applied to a bacterial phytochrome.

In general, the lifetimes obtained for the decay of P_r excited states do not depend on plant source and size, and very little on the environment, indicating that the excited state relaxation occurs within the protein pocket immediately surrounding the chromophore.

P_r shows a low fluorescence yield (ca. 10^{-3}), independently of apoprotein size, down to ca. 59 kDa, and of excitation wavelength [134,139]. Thus, dissipation by internal conversion and energy transfer into the protein moiety are the main deactivating pathways, competing with the photochemical primary process.

Upon excitation, a cascade of intermediates with lifetimes ranging from pico- to micro- and milliseconds is produced, eventually leading to the physiologically active far-red absorbing form P_{fr}. Upon excitation of P_{fr} a different set of intermediates is observed. The events are complex and several kinetic models have been proposed [136,140,141].

The kinetic model fitting the time-resolved fluorescence data obtained upon excitation of 124 kDa oat phytochrome involves two excited states in equilibrium, one is the initially excited P_r with a lifetime of 5–13 ps and the other is a conformationally relaxed but still electronically excited intermediate with a ca. 45 ps lifetime. The fluorescence properties are independent of the excitation wavelength [142]. Femtosecond time-resolved absorption spectra upon excitation of P_r, however, pointed to a unique excited state that decays monoexponentially with a 24 ps lifetime, which matches the rise time for the primary red-shifted photoproduct lumi-R [143], also called I_{700}. That the data could be equally fitted by a double exponential function with time constants of 13 and 44 ps the behavior was assigned to photodegradation of the sample giving rise to the second component [143]. Heyne et al. [138], however, found that the time constants (13 and 44 ps) are very similar to those derived from their own experiments with recombinant phytochrome from Synechocystis assembled with PCB, i.e., Cph1-PCB (lifetimes of 12 and 48 ps were derived from an analysis based on a sum of two exponentials), and that there were no signs of photodegradation during the measurements. Thus, the biphasic behavior found when the analysis is performed with a sum of single exponential terms in the ps time range seems to be an intrinsic property of plant and bacteria phytochromes.

Heyne et al. [138] prefer to describe the excited state decay kinetics by a distribution of rate constants. With such an analysis, after fast (ca. 150 fs) relaxation in the excited electronic state, the decay of P_r^* of 85 kDa Cph1-PCB is best described by a distribution of rate constants centered at $(16\ ps)^{-1}$. Müller and Holzwarth (personal communication), also chose a distribution of rate constants for the analysis of the transient absorption decay upon excitation of full-length oat phyA and found a ca. 100 fs decay in the excited state followed by a decay with rate constants centered at $(15\ ps)^{-1}$ and $(50\ ps)^{-1}$. These data are in good agreement with the emission data from phyA [142]. Both in the work by Heyne *et al.* and by Müller and Holzwarth an excitation wavelength dependence of the width of the transient absorption decay lifetimes distribution was observed. By femtosecond spectroscopy at a low repetition rate, lifetimes of ca. 0.4, 2, and 32 ps were observed upon excitation of phyA, with the longest correlating with the rise time of I_{700} [144]. However, fluorescence data on full-length native phyA obtained by the same group indicated emission lifetimes of 14 and 45 ps [145], which coincides with the transient absorption analysis by Müller and Holzwarth mentioned above. The lifetimes of the two fastest components observed in the transient absorption study of phyA were somehow affected by the medium viscosity, taken by the authors to indicate the major role played by the protein matrix in the relaxation of the initially excited chromophore [144]. This study also confirmed that the photoconversion of P_{fr} into P_r proceeds by a different pathway than that of P_r into P_{fr}. In particular, the excited state of P_{fr} was tentatively identified as having a 560 fs lifetime that is strongly affected by the viscosity of the medium.

Different primary reaction dynamics for excited P_r than for P_{fr} were also observed with Cph1-PCB from *Synechocystis*. Upon P_{fr} photoisomerization, two shorter lifetimes of 0.54 ps and 3.2 ps lead to the isomerized intermediate. The authors speculate that both plant phytochromes and Cph1-PCB show, at room temperature, an ultrafast P_r isomerization that is characterized (as mentioned above) by a distribution of rate constants, whereas the photoreaction of P_{fr} is pronouncedly biexponential and relatively fast [138].

The lack of a deuterium isotope effect in several of the above-mentioned time-resolved emission and absorption studies demonstrates that the primary process in the $P_r \rightarrow P_{fr}$ as well as in the $P_{fr} \rightarrow P_r$ transformation of oat phyA (and most probably of all seed plants phytochromes) does not involve a proton transfer. Thus, the primary photochemical reaction should be a $Z \rightarrow E$ isomerization. Experiments with chromopeptide fragments of phytochrome in the P_r and in the P_{fr} form already indicated that the primary photochemical process upon excitation of P_r is a Z to E photoisomerization of the 15,16 double bond of the chromophore (between rings C and D) [146]. A $Z \rightarrow E$ photoisomerization as the primary photochemical process was also confirmed by resonance Raman spectroscopy on full-length native oat phyA [131,147]. Upon excitation of P_{fr}, 15,16 double bond isomerization is also the primary photochemical step.

Thus, photoisomerization leads to the first red-shifted ground state intermediate, which grows with the longest decay lifetime of the respective excited state, i.e., ca. 30 ps for P_r and ca. 0.5 ps for P_{fr} (*vide supra*).

One hypothesis to explain the faster formation of the first ground state photoproducts upon excitation of P_{fr} than upon excitation of P_r is that in the former the chromophore is in a looser contact with the surrounding protein than in the latter [144]. This is compatible with the results of chromophore oxidation of native and degraded forms of oat phytochrome showing that the chromophore is more exposed to the medium in the P_{fr} form of the full-length form as well as of each of the fragments [148].

Two pools of phyA (subpopulations phyA′ and phyA″) have been detected by *in situ* low-temperature fluorescence spectroscopy and photochemistry [149]. The distinction between both subpopulations phyA′ and phyA″ is their different photochemical activity at low temperatures, as well as their abundance and localization patterns in plant tissue. The activation parameters for the low-temperature fluorescence are also different for phyA′ and phyA″.

Recent investigations with recombinant *Arabidopsis* and *Oryza* phyA reconstituted *in vivo* in the cellular medium used for the heterologous expression (*Saccharomyces cerevisiae*) with PCB or with PΦB chromophores, respectively, resemble, by their spectroscopic properties (low-temperature emisson and photochemistry), the minor phyA″ type (low photoconversion yield into lumi-R of < 0.1 at 85 K and absorption/emission maximum at 668/682 nm), and differ considerably from the major phyA′ type (high photoconversion yield of ca. 0.5 at 85 K and absorption/emission maximum at 673/687 nm) in plant tissues [150]. *Oryza* PHYA-PΦB is thus similar to phyA″. The authors conclude, in addition, that both phyA′ and phyA″ are full-length phytochromes encoded by the same phyA gene and that the differences are probably the result of post-translational modifications (e.g., phosphorylation), localization, or binding to other cellular components, albeit different from each other. It remains to be studied whether the two pools correspond to the two thermochromic P_r states observed for full length *Avena* phyA with absorption maxima at 657 and 672 nm [32] and postulated to give rise to the two parallel phototransformation channels with equal measured yields at room temperature [151].

The P_r excited state (in all phytochromes) decays to the first ground-state intermediate (I_{700}) with a quantum yield > 0.14 determined by laser-induced optoacoustic spectroscopy with native full-length oat phyA [152]. The 15,16 C=C bond in the chromophore is already isomerized in I_{700}. The primary quantum yield $\Phi P_r \rightarrow I_{700}$ is in the range of the overall quantum yield of P_{fr} formation, i.e., 0.16 [153]. Thus, once formed, I_{700} does not thermally return to P_r. Conversely, the photoinduced reversion of I_{700} to P_r has a quantum yield of 0.22, some 1.4 times larger than that of the forward $P_r \rightarrow I_{700}$ photoreaction. This was interpreted to indicate a similar chromophore-binding protein domain structure in P_r and I_{700} [154]. In addition, this result emphasizes the need to consider photoequilibria between parent compound and intermediates in photoreceptors, especially when working with high fluence laser pulses. The similarity between the protein structures in P_r and I_{700} is supported by the similarity of the low-temperature Fourier-transform resonance Raman spectra of these two phytochrome states [155] (Figure 3).

5.5.3 Thermal reactions after photoisomerization

The entire process of P_{fr} formation (and also the light-induced back reaction, $P_{fr} \rightarrow P_r$) is extremely complex. Initial experiments with "small phytochrome" (N-terminus truncated 59 kDa) oat phyA, have revealed four thermal (dark)

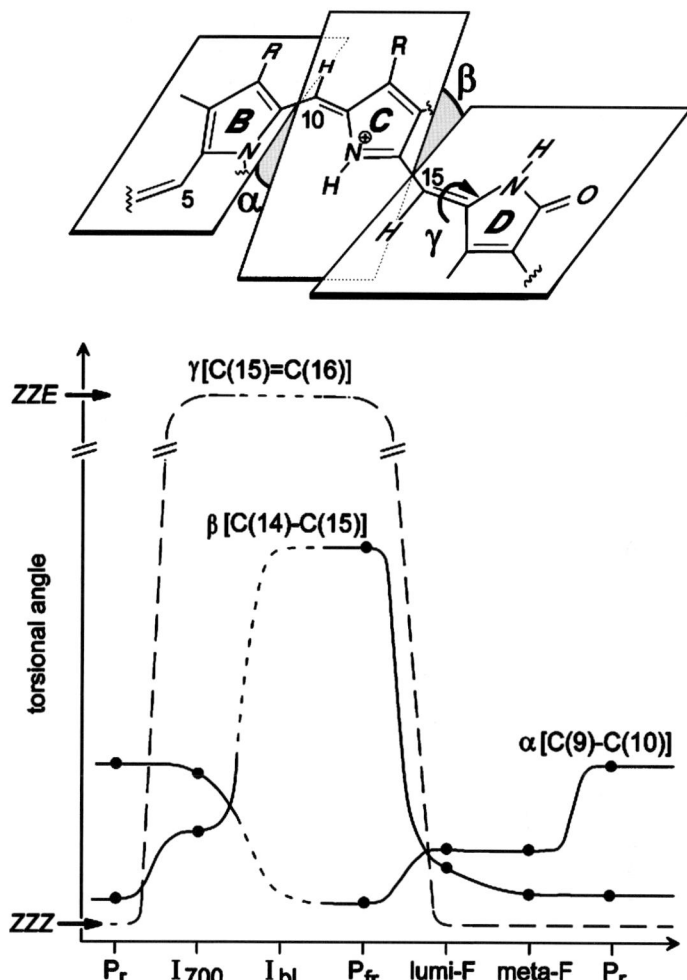

Figure 3. Schematic sketch of conformational changes of the tetrapyrrole chromophore during the light-induced $P_r \rightarrow P_{fr}$ conversion, as deduced from FT Raman measurements. Adapted from [155].

reactions between P_r and P_{fr} and two between P_{fr} and P_r [140,156]. Several studies performed with various types of phytochromes have searched for variables that would allow manipulation of the kinetics of the photoconversion to help to understand the nature of the time-resolved processes at a molecular level (see [141,157]). Models including parallel and sequential reactions as well as equilibration of intermediates have been proposed to rationalize the complex kinetics observed [158,159].

The kinetic complexity may be described in simple terms by saying that, upon pulse excitation of P_r, a sum of six single-exponential terms is needed to fit the time-resolved absorbance differences changes in the 600 to 750 nm range and in a time domain from 1 μs to ca. 3 s (final formation of P_{fr}). A global analysis is used to fit the data in the complete wavelength range. For this analysis it is assumed that the various absorbing species may have different absorption spectra, whereas for each species the lifetime should be the same over the whole wavelength range. This does not imply any specific kinetic model, since sequential, parallel, and kinetic schemes, including equilibria between transient species, all lead to a kinetic law represented by a sum of single exponential terms [158,160,161]. The wavelength-dependent amplitudes derived from this analysis are the lifetime-associated difference spectra (LADS).

After isomerization to the red-absorbing species I_{700} (lumi-R), the lifetimes associated with the six exponential terms appear in pairs, i.e., two similar LADS for the μs lifetimes (ca. 11 and 85 μs at 10°C), two similar LADS for the short ms (ca. 7 and 50 ms at 10°C), and two similar LADS in the pre-P_{fr} time range (ca. 400 ms and 3 s at 10°C); all data for full-length oat phyA [32] are very similar, albeit with different weights, to the data reported by Zhang *et al.* [158]. In particular, a very different amplitude was found in the two studies for the 7 ms lifetime component. This was attributed to differences in the preparations [32].

A central question is whether the kinetic complexity is due to heterogeneity in the phytochrome extracts, especially in view of the differences (although small) in kinetics obtained with phytochromes from different plants and extracted using different protocols. However, photoexcitation of recombinant oat PHYA assembled with PΦB yields very similar kinetics as native oat phyA, with only minor differences in the lifetimes obtained, especially in the long ms range. This means that heterogeneity based on protein sequence cannot be the origin of the multicomponent kinetics of the $P_r \rightarrow P_{fr}$ phototransformation in the nanosecond-to-second time range [32].

The P_r forms of native phyA and of a homogeneous recombinant 65 kDa fragment assembled with PCB exhibit thermochromic properties (i.e., reversible temperature-dependent absorption spectra), which are explained as arising in each case from the presence of two P_r species in thermal equilibrium. The appearance of isosbestic points when changing the temperature of the P_r solutions is more compatible with the existence of two distinct conformations (in the two P_r forms) of either the chromophore, the protein pocket surrounding the chromophore, or both [32], rather than a gradual change in the planarity of the chromophore as previously proposed to explain the P_r photochromicity [162]. The two P_r species have identical photochemical properties and their presence cannot explain the kinetic complexity upon P_r excitation [32].

A parallel mechanism by which the two microsecond lifetime I_{700} species are simultaneously produced upon phyA P_r excitation has been favoured to explain the multiexponential kinetics [32,158]. The species decaying with lifetimes of a few ms have also been called bleached intermediates (I_{bl}) due to their strongly reduced absorbance. The loss of conjugation could be explained either by a transient loss of planarity or to a loss of interaction with charged amino acids, or both. In general, the various transient species reflect interactions of various protein conformations with the chromophore which also undergoes rotations around the single bonds, following the double bond isomerization.

The Raman spectra of the trapped photoinduced intermediates at cryogenic temperatures upon excitation of full-length oat phyA in non-deuterated and in deuterated buffer solutions demonstrated that in all intermediates the chromophore is protonated at the pyrroleninic nitrogen [131]. The identification of the N–H in plane vibrations of the tetrapyrrole rings B and C demonstrated that P_{fr}, as well as the intermediates, are protonated. Thus, a previous suggestion by Mizutani *et al.* [163] that P_{fr} formation involves deprotonation of the tetrapyrrole could be ruled out [131].

That the chromophore remains protonated during the whole phototransformation has been confirmed by FTIR studies using the recombinant 65 kDa N-terminal fragment assembled with chemically modified chromophores, either at ring D or with isotopically labeled ^{18}O at the carbonyl group in ring A [164].

The kinetic analysis of recombinant phytochromes assembled with PCB, instead of PΦB, gave the first evidence for a tight chromophore–protein interaction in the photoisomerization step. PCB-assembled PHYA shows a monoexponential I_{700} decay with a lifetime of 90 μs at 10°C, i.e., it seems that the faster component (ca. 10 μs) found in the PΦB-bearing proteins is missing. This may be because this process is much faster as a result of a greater flexibility of the chromophore. The less intimate interaction between the ethyl group in ring D of PCB (instead of the vinyl group) may facilitate detachment of the chromophore from its environment, so escaping detection in the μs range. Alternatively, the parallel pathways observed in the PΦB-bearing proteins can result from very specific chromophore–protein interactions, which for PCB result in the loss of one of the two reaction channels. The PCB-assembled recombinant full-length PHYA also shows an accelerated formation of P_{fr}, again attributed to less intimate chromophore–protein interactions than in PΦB-PHYA [132].

So far, the only studies of transient absorption difference with a non-phyA phytochrome of higher plants have been performed with potato PHYB assembled with PΦB and with PCB, as well as with PHYB66 (the N-terminal apoprotein fragment, amino acids 1–596) assembled with PΦB and with PCB. The photoinduced behavior of each of these constructs was compared with those of PHA124-PΦB as well as with (1–595) PHYA65-PΦB and -PCB. Contrary to oat phyA, the I_{700} intermediate from potato PHYB, assembled with either PΦB or PCB, decayed following single exponential kinetics with a lifetime of 87 and 84 μs, respectively, at 10°C [22]. The primary photoproduct I_{700} of PHYA65-PΦB decayed biexponentially, and that of PHYA65-PCB monoexponentially, whereas I_{700} photoproduced from PHYB66 decayed monoexponentially, irrespective of the chromophore incorporated (similar to the full-length protein). Therefore, the chromophore–protein interactions in phytochrome B are such that the second path is inhibited.

These studies also showed that the formation of P_{fr} is faster with the N-terminal halves than with the full-length phytochromes, confirming an involvement of the C-terminal domain in the relatively slow protein conformational changes taking place with lifetimes in hundreds of milliseconds to seconds [18].

5.5.4   *Effect of site-directed mutagenesis on the phyA photocycle kinetics*

Only one series of laser-flash photolysis experiments with mutated oat phyA has been reported [84]. Although the mutated amino acids were all selected in the vicinity of the chromophore-binding cysteine, this study is only preliminary due to the lack of a tertiary structure of the phytochromes. The effect of site-directed mutagenesis on the $P_r \rightarrow P_{fr}$ phototransformation kinetics of PCB-assembled N-terminal (amino acid residues 1–595) recombinant oat PHYA was studied with various mutated proteins. A strong effect on the I_{700} decay was encountered when His-319 was exchanged for leucine. I_{700} produced upon excitation of H319L-65 kDa PHYA-PCB decayed with a 177 μs lifetime, in contrast to the 75 μs of I_{700} determined for 65 kDa PHYA-PCB (both lifetimes at 10°C). Loss of hydrogen-bonding ability upon exchange of His by Leu interferes not only with chromophore incorporation (the mutated protein incorporates the chromophore at a much slower rate, 490 s vs. 280 s for the WT), but also with the conformational change of the chromophore upon decay of I_{700} to I_{bl}. An effect in the opposite direction, i.e., an acceleration of the I_{700} decay from 75 to 50 μs, was found for the double mutant L323R/Q324D (the chromophore incorporation kinetics do not differ much from that of the WT). It appears that the conformational rearrangement of chromophore and protein associated with the I_{700} decay can be facilitated by electrostatic interactions. Other mutations, such as S320K (exchange of a highly conserved serine), did not influence I_{700} decay but accelerated the chromophore incorporation. P318A affected only slightly I_{700} decay but slowed down P_{fr} formation. Thus, the conformational flexibility of Pro 318 appears to be crucial for P_{fr} formation.

Interestingly, the activation energies for I_{700} decay determined for the mutants H319L and L323R/Q324D are considerably larger (E_a = 70 kJ mol^{-1}) than those for the I_{700} decay of the WT (E_a = 50 kJ mol^{-1}). Bearing in mind that the activation energies for I_{700} decay are identical for native 124 kDa phyA and PHYA-65 kDa-PCB, but higher for the mutants and for the "small" phytochrome (N-deleted 59 kDa), one should conclude that I_{700} decay is less favorable from the enthalpy factor for the N-deletion and for the two site directed mutants than for the native and N-terminus half [84]. However, the pre-exponential factors seem to compensate this effect, and for each temperature the I_{700} decay is faster for the L323R/Q324D and slower for the H319L mutant than for WT phyA.

5.5.5   *Pfr to Pr phototransformation kinetics*

In addition to the faster kinetics for the appearance of the first ground-state intermediate observed upon excitation of P_{fr} (*vide supra*), three lifetimes were detected during the $P_{fr} \rightarrow P_r$ photoreversion in the studies using full-length oat phyA [159]. Lifetimes of 320 ns, 265 μs (the species was called metaF$_a$), and 5.5 ms, similar to those obtained with truncated "small" 59 kDa phyA [140] and several other phyA samples, were observed. An unbranched sequential mechanism provides the most reasonable explanation of these data. Only one intermediate, that with a lifetime of ca. 7 ms identified in the $P_r \rightarrow P_{fr}$ photoconversion, shows spectral similarities to one of the transients identified in the back photoconversion $P_{fr} \rightarrow P_r$ [159].

5.5.6 Protein changes during phototransformation

To better understand the differences between the protein structure of the P_r and the P_{fr} forms of phyA determined by steady-state circular dichroism, i.e., that both for oat and for pea phyA an increase in the α-helix of the apoprotein polypeptide is observed upon P_r to P_{fr} phototransformation [165], time-resolved CD changes taking place upon $P_r \rightarrow P_{fr}$ and upon $P_{fr} \rightarrow P_r$ phototransformation were analyzed. Unfolding of the α-helix upon excitation of pea P_{fr} phyA occurred in 310 μs (very similar to the 265 μs lifetime for metaFa [159]), whereas folding of the N-terminal chain upon excitation of the P_r form was much slower, taking place 113 ms after excitation [166]. Folding of the N-terminal α-helix segment is considered to be a major protein structural changes upon P_r photoconversion [167].

The time-resolved enthalpy and structural volume changes after excitation of full-length oat phyA P_r were studied in the micro- to milliseconds range by photothermal beam deflection (PBD), a technique that follows the time-resolved refractive index changes after photoexcitation [151]. The first set of intermediates, I^1_{700} and I^2_{700}, stores ca. 83% of the energy of the first excited state, in agreement with previous optoacoustic data [152], whereas the second set stores only ca. 18%. The temperature dependence of the amplitudes ratio for the optical absorbances of the $(I^1_{700} + I^2_{700})$ intermediates set is explained on the basis of the previously reported thermochromic equilibrium between $P_{r,657}$ and $P_{r,672}$ [32]. The PBD data were best fitted with a parallel mechanism (with equal yield in each branch) for the production of the first set of intermediates, I^1_{700} and I^2_{700}, as well as for the second set of intermediates, I^1_{bl} and I^2_{bl}. In other words, each P_r form leads to one I_{700} transient species.

The final steps towards P_{fr} should be largely driven by positive entropic changes brought about by movements of the protein, prominently the N-terminal α-helix folding. To produce I^1_{700} and I^2_{700} an expansion of 18 ± 13 mL mol^{-1} was determined and a further expansion ≥ 7 mL mol^{-1} was estimated for the decay from I^i_{700} to the set of I_{bl} intermediates, indicating that P_{fr} has a larger volume than P_r, in agreement with chromatographic [168] and circular dichroism data [169], according to which P_{fr} shows a larger volume and the chromophore displays a higher accessibility.

5.5.7 Spectroscopic and kinetic studies with phytochromes from bacteria and lower plants

Spectral characterization of the recombinant *Synechocystis-* (Cph1) and *Calothrix*-derived chromoproteins (CphA and CphB) (see Chapter 7), exhibiting similarities to phytochromes and bacterial sensor kinases [9,170–172], underlined the proposal that these proteins are members of photoactive, bilin-binding chromoproteins. Yet, these studies revealed significant differences to the phytochromes of higher and also of lower plants. The *Synechocystis* phytochrome resembles plant phyA with regard to Raman and visible spectral properties of the P_r and P_{fr} states [173]. The FTIR-spectra also indicated similar features to those of plant phytochrome [174]. Kinetic analysis revealed a multistep photoconversion reminiscent of the phyA $P_r \rightarrow P_{fr}$ transformation but with different kinetics. Similar to phyA, the apoprotein of *Synechocystis* assembled with PΦB showed a biexponential decay of the first intermediates, whereas when assembled with PCB (the native chromophore in this case) the decay was monoexponential with a 25 μs lifetime at 10°C. With this PCB-assembled phytochrome, H/D exchange delays both growth and decay rates of the second intermediate, indicating a rate-limiting proton transfer step, perhaps an intramolecular proton release and uptake. This second intermediate (appearance with 300 μs and decay lifetime of 6–8 ms at 10°C) has no equivalence to any intermediates in the other phytochromes [173], including the closest related cyanobacterial phytochrome CphA [172]. In fact, measurements of P_{fr} formation of Cph1 revealed a pH change of the protein solution, indicating that during the photocycle reactions a proton is extruded from the protein into the bulk phase. The finding that one of the two identified bacterio-phytochromes in *Calothrix* incorporates the chromophore only non-covalently (CphB) is reflected by an entirely different photochemical pathway which lacks any of the above-described (sub-)microsecond and second processes, but consists of only two processes with 1.9 and 12.8 ms lifetimes [172].

Recombinant full-length phytochrome (CP2) from the moss *Ceratodon* assembled with PCB (CP2-PCB) yielded a holoprotein with maxima of the difference spectra at 644 (P_r) and 716 nm (P_{fr}), whereas when assembled with PΦB (CP2- PΦB) the maxima were at 659 (P_r) and 724 nm (P_{fr}), the latter in agreement with the maxima for the *Ceratodon* phytochrome extracts, implying that in this case, as with the phytochrome from seed plants, phytochromobilin is the native chromophore [128].

Flash photolysis of CP2-PCB revealed similar kinetic behavior to that for recombinant PHYA-PCB, i.e., a lifetime of ca. 110 μs at 10°C for the red-shifted intermediate and longer time kinetics, more similar to those for PHYA-PCB than those for PHYB-PCB [128]. For the similarity of CP2 and phyA from seed plants, see also Heyne *et al.* [138]. No special feature that could explain the specific action of CP2 with respect to sensing the light direction in a single moss cell was observed during these studies.

Fluorescence spectroscopy at low and room temperatures revealed a steep activation energy of the fluorescence decay upon excitation of CP2-PCB [137]. These studies revealed again a similarity to PHYA-PCB and also to the properties of the phytochromes from *Synechocystis* [175] and *Adiantum* (V.A. Sineshchekov, personal communication) with regard to the heterogeneity of the emitting species.

Only one other lower plant phytochrome has recently been expressed and characterized in detail. The apoprotein, heterologously expressed in *P. pastoris*, originated from the unicellular alga *Mougeotia scalaris* [11,49], and was assembled with PCB, which was identified to be the native chromophore. Since for this phytochrome, due to actin-binding sequence motifs in the C-terminal part, an interaction with the cell skeleton had been proposed, the full-length (FL) protein was compared to the C-terminally truncated, 66 kDa chromophore-binding half (spanning amino acids 1 to 606). In addition to slight changes in the absorption spectra, remarkable differences in the thermal stability of the P_{fr} forms of both constructs and in their photoconversion kinetics were observed [11]. The PCB-assembled chromoproteins showed 646 and 639 nm absorbances for the P_r bands (full length and 66 kDa construct, respectively), and 720 and 714 nm for the P_{fr} forms.

The thermal stability of the P_{fr} form also revealed remarkable differences between the full-length and the C-terminally truncated chromoproteins. The full-length protein arrived biexponentially (1 and 24 min) at a plateau after 2 h, and still contained <55% of the P_{fr} form, the 66 kDa construct also decayed with two time-constants, albeit both longer than the full-length protein (18 and 250 min). A stable equilibrium, still containing ca. 60% of the original P_{fr}, was reached after ca. 14 h in the latter case. In this case, the full-length protein appears to be less stable than a C-terminally truncated chromopeptide [11].

Different behavior for both constructs also became evident during the time-resolved P_{fr} formation analysis. The earliest observable process in the long-ns to μs time range is the decay of the I_{700} intermediate with (FL) 344 and (66 kDa) 590 μs. This decay is significantly longer than any other reported for this process. Further reactions proceed with 10 and 35 ms for FL, correlated with a strong loss of absorbance around 660 nm, whereas the corresponding process of the 66 kDa construct occurs quite similarly, with 10 and 32 ms. Major P_{fr} formation reactions show a biphasic behavior of 131 and 833 ms for the FL protein, and 138 and 364 ms for the 66 kDa chromopeptide. The remarkable thermal instability of the FL-P_{fr} form, discussed above, is even evident in the LADS, which show a significant P_{fr}-loss-P_r-formation feature with a lifetime of 5.7 s.

Acknowledgements

We are indebted to all co-workers (students, technical assistants and post-doctoral collaborators) and colleagues who enthusiastically participated in the "phytochrome project" in our Institute, as well as to Professor Kurt Schaffner who was a very active participant in this project.

References

1. R.E Kendrick, G.H.M. Kronenberg (1994). *Photomorphogenesis in Plants*, (2nd Edn). Kluwer Academic Publishershers, Dordrecht, The Netherlands.

2. M. Furuya, E. Schäfer (1996). Photoreception and signalling of induction reactions by different phytochromes. *Trends Plant Sci.*, **1**, 301–307.
3. A. Batschauer (1998). Photoreceptors of higher plants. *Planta*, **206**, 479–492.
4. D.R. McCarty, J. Chory (2000). Conservation and innovation in plant signaling pathways. *Cell*, **103**, 201–209.
5. P.H. Quail (2002). Phytochrome photosensory signalling networks. *Nat. Rev.*, **3**, 85–93.
6. R.A. Sharrock, P.H. Quail (1989). Novel phytochrome sequences in *Arabidopsis thaliana*: structure, evolution, and differential expression of plant regulatory photoreceptor family. *Genes Develop.*, **3**, 1745–1757.
7. T. Clack, S. Mathews, R.A. Sharrock (1994). The phytochrome apoprotein family in *Arabidopsis* is encoded by five genes: the sequences and expression of PHYD and PHYE. *Plant Mol. Biol.*, **25**, 413–427.
8. M. Herdman, T. Coursin, R. Rippka, J. Houmard, N. Tandeau de Marsac (2000). A new appraisal of the prokaryotic origin of eukaryotic phytochromes. *J. Mol. Evol.*. **51**, 205–213.
9. D.M. Kehoe, A.R. Grossmann (1996). Similarity of a chromatic adaptation sensor to phytochrome and ethylene receptors. *Science*, **273**, 1409–1412.
10. P.H. Quail (1994). Phytochrome genes and their expression. In: R.E. Kendrick, G.H.M. Kronenberg, (Eds), *Photomorphogenesis in Plants* (pp. 71–104). Kluwer Academic Publisher, Dordrecht.
11. H.J.M.M. Jorissen, S.E. Braslavsky, G. Wagner, W. Gärtner (2002). Heterologous expression and characterization of recombinant phytochrome from the green alga *Mougeotia scalaris. Photochem. Photobiol.*, **76**, 457–461.
12. S.-H. Wu, M.T. Mcdowell, J.C. Lagarias (1997). Phycocyanobilin is the natural precursor of the phytochrome chromophore in the green alga *Mesotaenium caldariorum. J. Biol. Chem.*, **272**, 25700–25705.
13. T. Hübschmann,T. Börner, E. Hartmann, T. Lamparter (2001). Characterization of the Cph1 holo-phytochrome from Synechocystis sp PCC 6803. *Eur. J. Biochem.*, **268**, 2055–2063.
14. J.C. Lagarias, H. Rapoport (1980). Chromopeptides from phytochrome. The structure and linkage of the P$_r$ form of the phytochrome chromophore. *J. Am. Chem. Soc.*, **102**, 4821–4828.
15. F. Thümmler, W. Rüdiger, E. Cmiel, S. Schneider (1983). Chromopeptides from phytochrome and phycocyanin. NMR studies of the P$_{fr}$ and P$_r$ chromophore of phytochrome and *E,Z* isomeric chromophores of phycocyanin. *Z. Naturforsch., Teil C*, **38**, 359–368.
16. J.C. Lagarias, A.V. Klotz, J.L. Dallas, A.N. Glazer, J.E. Bishop, J.F. O'Connell, H. Rapoport (1988). Exclusive A-ring linkage for singly attached phycocyanobilins and phycoerythrobilins in phycobiliproteins. *J. Biol. Chem.*, **263**, 12977–12985.
17. M.T. Boylan, P.H. Quail (1991). Phytochrome A overexpression inhibits hypocotyl elongation in transgenic *Arabidopsis. Proc. Natl. Acad. Sci. U.S.A.*, **88**, 10806–10810.
18. A. Remberg, A. Ruddat, S.E. Braslavsky, W. Gärtner, K. Schaffner (1998). Chromophore incorporation, P$_r$ to P$_{fr}$ kinetics and P$_{fr}$ thermal reversion of recombinant N-terminal fragments of phyA and phyB phytochrome chromoproteins. *Biochemistry*, **37**, 9983–9990.
19. H. Smith, G.C. Whitelam (1990). Phytochrome, a family of photoreceptors with multiple physiological roles. *Plant Cell Environ.*, **13**, 695–707.

20. R.D. Vierstra, P.H. Quail (1983). Purification and initial characterization of 124-kilodalton phytochrome from *Avena. Biochem.*, **22**, 2498–2505.
21. R.D. Vierstra (1994). Phytochrome degradation. In: R.E Kendrick, G.H.M. Kronenberg (Eds), *Photomorphogenesis in Plants* (2nd Edn, pp. 141–160). Kluwer Academic Publishers, Dordrecht.
22. A. Ruddat, P. Schmidt, C. Gatz, S.E. Braslavsky, W. Gärtner, K. Schaffner (1997). Recombinant type A and B phytochromes from potato. Transient absorption spectroscopy. *Biochemistry*, **36**, 103–111.
23. M. Furuya (1989). Molecular properties and biogenesis of phytochrome I and II. *Adv. Biophys.*, **25**, 133–167.
24. P.H. Quail, J.T. Colbert, H.P. Hershey, R.D. Vierstra (1983). Phytochrome: molecular properties and biogenesis. *Phil. Trans. R. Soc. London*, **303**, 387–402.
25. H.Ü. Kolukisaoglu, B. Braun, W.F. Martin, H.A.W. Schneider-Poetsch (1993). Mosses do express conventional, distantly B-type-related phytochromes. Phytochrome of *Physcomitrella patens* (Hedw.). *FEBS Lett.*, **334**, 95–100.
26. F. Thümmler, P. Algarra, G.M. Fobo (1995). Sequence similarities of phytochrome to protein kinases: implication for the structure, function and evolution of the phytochrome gene family. *FEBS Lett.*, **357**, 149–155.
27. H.A.W. Schneider-Poetsch, B. Braun (1991). Proposal on the nature of phytochrome action based on the C-terminal sequences of phytochrome. *J. Plant Physiol.*, **137**, 576–580.
28. R.D. Vierstra (1993). Illuminating phytochrome functions. *Plant Physiol.*, **103**, 679–684.
29. E.F. Chen, V.N. Lapko, J.W. Lewis, P.S. Song, D.S. Kliger (1996). Mechanism of native oat phytochrome photoreversion: A time-resolved absorption investigation. *Biochemistry*, **35**, 843–850.
30. W.J. van der Woude (1985). A dimeric mechanism for the action of phytochrome: evidence from photothermal interactions in lettuce seed germination. *Photochem. Photobiol.*, **42**, 655–661.
31. J. Brockmann, S. Rieble, N. Kazarinova-Fukshansky, M. Seyfried, E. Schäfer (1987). Phytochrome behaves as a dimer *in vivo. Plant Cell Environ.*, **10**, 105–111.
32. P. Schmidt, T. Gensch, A. Remberg, W. Gärtner, S.E. Braslavsky, K. Schaffner (1998). The complexity of the $P_r \rightarrow P_{fr}$ phototransformation kinetics is an intrinsic property of homogeneous native phytochrome. *Photochem. Photobiol.*, **68**, 7514–761.
33. A.M. Jones, H.P. Erickson (1989). Domain structure of phytochrome from *Avena sativa* visualized by electron microscopy. *Photochem. Photobiol.*, **49**, 479–483.
34. S. Tokutomi, M. Nakasako, J. Sakai, M. Kataoka, K.T. Yamamoto, M.Wada, F. Tokunaga, M. Furuya (1989). A model for the dimeric molecular structure of phytochrome based on small-angle X-ray scattering. *FEBS Lett.*, **247**, 139–142.
35. M.D. Edgerton, A.M. Jones (1992). Localisation of protein-protein interactions between subunits of phytochrome. *Plant Cell*, **4**, 161–171.
36. M.D. Edgerton, A.M. Jones (1994). Subunit interactions in the carboxy-terminal domain of phytochrome. *Biochemistry*, **32**, 8239–8245.
37. J.R. Cherry, D. Hondred, J.M. Walker, J.M. Keller, H.P. Hershey, R.D. Vierstra (1993). Carboxy-terminal deletion analysis of oat phytochrome A reveals the presence of separate domains required for structure and biological activity. *Plant Cell*, **5**, 565–575.
38. Y. Xu, B.M. Parks, T.W. Short, P.H. Quail (1995). Missense mutations define a restricted segment in the C-terminal domain of phytochrome A critical to its regulatory activity. *Plant Cell*, **7**, 1433–1443.

39. D. Wagner, P.H. Quail (1996). Mutational analysis of phytochrome B identifies a small COOH-terminal-domain region critical for regulatory activity. *Proc. Natl. Acad. Sci. U.S.A.*, **92**, 8596–8600.

40. R.W. McMichael, Jr. J.C. Lagarias (1990). Phosphopeptide mapping of Avena phytochrome phosphorylated by protein kinases *in vitro*. *Biochemistry*, **29**, 3872–3878.

41. V.N. Lapko, T.A. Wells, P.-S. Song (1996). Protein kinase A-catalyzed phosphorylation and its effect on conformation in phytochrome A. *Biochemistry*, **35**, 6585–6594.

42. V.N. Lapko, X.-Y. Jiang, D.L. Smith, P.-S. Song (1997). Post-translational modification of oat phytochrome A: Phosphorylation of a specific serine in a multiple serine cluster. *Biochemistry*, **36**, 10595–10599.

43. G. Choi, H. Yi, Y.K. Kwon, M.S. Soh, B.C. Shin, Z.A. Luka, T.-R. Hahn, P.-S. Song (1999). Phytochrome signalling is mediated through nucleoside diphosphate kinase 2. *Nature*, **401**, 610–613.

44. T. Schirmer, R. Huber, M. Schneider, W. Bode (1986). Crystal structure analysis and refinement at 2.5 Å of hexameric C-Phycocyanin from the cyanobacterium *Agmenellum quadruplicatum*. *J. Mol. Biol.*, **188**, 651–676.

45. T. Schirmer, W. Bode, R. Huber (1987). Refined three-dimensional structures of two cyanobacterial C-phycocyanins at 2.1 and 2.5 Å resolution. *J. Mol. Biol.*, **196**, 677–695.

46. M. Romanowski, P.-S. Song (1992). Structural domains of phytochrome deduced from homologies in amino acid sequences. *J. Protein. Chem.*, **11**, 139–155.

47. M.W. Parker, P. Goebel, C.R. Ross, P.-S. Song, J.J. Stezowski (1994). Molecular modeling of phytochrome using constitutive C-phycocyanin from *Fremyella diplosiphon* as a putative structural template. *Bioconj. Chem.*, **5**, 21–30.

48. V.A. Sineshchekov, T. Lamparter, E. Hartmann (1994). Evidence for the existence of membrane-associated phytochrome in the cell. *Photochem. Photobiol.*, **60**, 516–520.

49. A. Winands, G. Wagner (1996). Phytochrome of the green alga *Mougeotia*: cDNA sequence, autoregulation and phylogenetic position. *Plant Mol. Biol.*, **32**, 589–597.

50. T. Kagawa, M. Wada (1996). Phytochrome- and blue-light-absorbing pigment-mediated directional movement of chloroplasts in dark-adapted prothallial cells of fern Adiantum as analyzed by microbeam irradiation. *Planta*, **198**, 488–493.

51. T. Kinoshita, M. Doi, N. Suetsugu, T. Kagawa, M. Wada, K. Shimazaki (2001). Phot1 and phot2 mediate blue light regulation of stomatal opening. *Nature*, **414**, 656–660.

52. R. Grimm, C. Eckershorn, F. Lottspeich, C. Zenger, W. Rüdiger (1988). Sequence analysis of proteolytic fragments of 124-kilodalton phytochrome from etiolated *Avena sativa L.*: Conclusions on the conformation of the native protein. *Planta*, **174**, 396–401.

53. V.N. Lapko, X.-Y. Jiang, D.L. Smith, P.-S. Song (1998). Surface topography of phytochrome A deduced from specific chemical modification with iodoacetamide. *Biochemistry*, **37**, 12526–12535.

54. C.M. Park, S.H. Bhoo, P.-S. Song (2000). Inter-domain crosstalk in the phytochrome molecules. *Seminars Cell Develop. Biol.*, **11**, 449–456.

55. T. Hübschmann, H.J.M.M Jorissen, T. Börner, W. Gärtner, N. Tandeau de Marsac (2001). Phosphorylation of proteins in the light-dependent signalling pathway of a filamentous cyanobacterium. *Eur. J. Biochem.*, **268**, 3383–3389.

56. K.-C. Yeh, S.-H. Wu, J.T. Murphy, J.C. Lagarias (1997). A cyanobacterial phytochrome two-component light sensory system. *Science*, **277**, 1505–1508.

57. M.T. Boylan, P.H. Quail (1996). Are the phytochromes protein kinases? *Protoplasma*, **195**, 12–17.

58. K.C. Yeh, J.C. Lagarias (1998). Eukaryotic phytochromes: Light-regulated serine/threonine protein kinases with histidine kinase ancestry. *Proc. Natl. Acad. Sci. U.S.A.*, **95**, 13976–13981.

59. P.H. Quail, M.T. Boylan, B.M. Parks, T.W. Short, Y. Xu, D. Wagner (1995). Phytochromes: Photosensory perception and signal transduction. *Science*, **268**, 675–680.

60. J.R. Cherry, D. Hondred, J.M. Walker, R.D. Vierstra (1992). Phytochrome requires the 6-kDa N-terminal domain for full biological activity. *Proc. Natl. Acad. Sci. U.S.A.*, **89**. 5039–5043.

61. D. Wagner, M. Koloszvari, P.H. Quail (1996). Two small spatially distinct regions of phytochrome B are required for efficient signaling rates. *Plant Cell*, **8**, 859–871.

62. E.T. Jordan, J.R. Cherry, J.M. Walker, R.D. Vierstra (1995). The amino-terminus of phytochrome A contains two distinct functional domains. *Plant J.*, **9**, 243–257.

63. E.T. Jordan, J.M. Marita, R.C. Clough, R.D. Vierstra (1997). Characterization of regions within the N-terminal 6-kilodalton domain of phytochrome A that modulate its biological activity. *Plant Physiol.*, **115**, 693–704.

64. J. Stockhaus, A. Nagatani, U. Halfter, S.A. Kay, M. Furuya, N-H. Chua (1992). Serin-to-alanine substitutions at the amino-terminal region of phytochrome A result in an increase in biological activity. *Genes Develop.*, **6**, 2364–2372.

65. R.D. Vierstra, P.H. Quail (1982). Native phytochrome: Inhibition of proteolysis yields a homogeneous monomer of 124 kilodaltons from *Avena*. *Proc. Natl. Acad. Sci. U.S.A.*, **79**, 5272–5276.

66. H.P. Hershey, R.F. Barker, K.B. Idler, J.L. Lissemore, P.H. Quail (1985). Analysis of cloned cDNA and genomic sequences for phytochrome: complete amino acid sequence for two gene products expressed in etiolated *Avena*. *Nucl. Acids Res.*, **13**, 8543–8559.

67. R. Grimm, F. Lottspeich, H.A.W. Schneider-Poetsch, W. Rüdiger (1986). Investigation of the peptide chain of 124 kDa phytochrome: localization of proteolytic fragments and epitopes for monoclonal antibodies. *Z. Naturforsch., Teil C* **41**, 993–1000.

68. U. Reiff, P. Eilfeld, W. Rüdiger (1985). A photoreversible 39 kDalton fragment from the P_{fr} form of 124 kDalton oat phytochrome. *Z. Naturforsch., Teil C* **40**, 693–698.

69. W. Gärtner, C. Hill, K. Worm, S.E. Braslavsky, K. Schaffner (1996). Influence of expression system on chromophore binding and preservation of spectral properties in recombinant phytochrome. *Eur. J. Biochem.*, **236**, 978–983.

70. K.-I. Tomizawa, J. Stockhaus, N.-H. Chua, M. Furuya (1995). Spectrophotometric and molecular properties of mutated rice phytochrome A. *Plant Cell Physiol.*, **36**, 511–516.

71. R. Grimm, W. Rüdiger (1986). A simple and rapid method for isolation of 124 kDa oat phytochrome. *Z. Naturforsch., Teil C*, **41**, 988–992.

72. R. Grimm, C. Eckershorn (1991). Expression of AP 3, 4 and 5 isophytochromes in etiolated oat seedlings (*Avena sativa* L.). *Photochem. Photobiol.*, **53**, 699–700.

73. G.C. Whitelam, N.P. Harberd (1994). Action and function of phytochrome family members revealed through the study of mutant and transgenic plants. *Plant Cell Environ.*, **17**, 615–625.

74. H. Abe, K.T. Yamamoto, A. Nagatani, M. Furuya (1985). Characterization of green tissue-specific phytochrome isolated immunochemically from pea seedlings. *Plant Cell Physiol.*, **26**, 1387–1399.

75. A. Nagatani, K.T. Yamamoto, M. Furuya, T. Fukumoto, A. Yamashita (1984). Production and characterization of monoclonal antibodies which distinguish different surface structures of pea (Pisum sativum cv. Alaska) phytochrome. *Plant Cell Physiol.*, **25**, 1059–1068.

76. L.H. Pratt, S.J. Stewart, Y. Shimazaki, Y.-C. Wang, M.-M. Cordonnier (1991). Monoclonal antibodies directed to phytochrome from green leaves of *Avena sativa* L. cross-react weakly or not at all in etiolated shoots of the same species. *Planta*, **184**, 87–95.

77. S.-H. Wu, J.C. Lagarias (1996). The methylotrophic yeast *Pichia pastoris* synthesizes a functionally active chromophore precursor of the plant photoreceptor phytochrome. *Proc. Natl. Acad. Sci. U.S.A.*, **93**, 8989–8994.

78. J.C. Lagarias, D.M. Lagarias (1989). Self-assembly of synthetic phytochrome holoprotein *in vitro*. *Proc. Natl. Acad. Sci. U.S.A.*, **86**, 5778–5780.

79. J.A. Wahleithner, L. Li, J.C. Lagarias (1991). Expression and assembly of spectrally active recombinant holophytochrome. *Proc. Natl. Acad. Sci. U.S.A.*, **88**, 10387–10391.

80. T.D. Elich, J.C. Lagarias (1989). Formation of a photoreversible phycocyanobilin-apophytochrome adduct *in vitro*. *J. Biol. Chem.*, **264**, 12902–12908.

81. C. Hill, W. Gärtner, P. Towner, S.E. Braslavsky, K. Schaffner (1994). Expression of phytochrome apoprotein from *Avena sativa* in *Escherichia coli* and formation of photoactive chromoproteins by assembly with phycocyanobilin. *Eur. J. Biochem.*, **223**, 69–77.

82. L. Deforce, K.-I. Tomizawa, N. Ito, D. Farrens, P.-S. Song (1991). *In vitro* assembly of apophytochrome and apophytochrome deletion mutants expressed in yeast with phycocyanobilin. *Proc. Natl. Acad. Sci. U.S.A.*, **88**, 10392–10396.

83. D. Mozley, A. Remberg, W. Gärtner (1997). Large scale generation of affinity-purified recombinant phytochrome chromopeptide. *Photochem. Photobiol.*, **66**, 710–715.

84. A. Remberg, P. Schmidt, S.E. Braslavsky, W. Gärtner, K. Schaffner (1999). Differential effects of mutations in the chromophore pocket of recombinant phytochrome on chromoprotein assembly and P_r-to-P_{fr} photoconversion. *Eur. J. Biochem.*, **266**, 201–208.

85. G. Gellissen, M. Piontek, U. Dahlems, V. Jenzelewsi, J.E. Gavagan, R. DiCosimo, D.L. Anton, Z.A. Janowicz (1996). Recombinant *Hansenula polymorpha* as a biocatalyst - coexpression of the spinach glycolate oxidase (*GO*) and the S. cerevisiae catalase T (*CTT1*) gene. *Appl. Microbiol. Biotechnol.*, **46**, 46–54.

86. G. Gellissen, U. Weydemann, A.W.M. Strasser, M. Piontek, Z.A. Janowicz, C.P. Hollenberg (1992). Progress in developing methylotrophic yeasts as expression systems. *Trends Biotechnol.*, **10**, 413–417.

87. J.T. Murphy, J.C. Lagarias (1997). Purification and characterization of recombinant affinity peptide-tagged oat phytochrome A. *Photochem. Photobiol.*, **65**, 750–758.

88. J.C. Lagarias, J.M. Kelly, K.L. Cyr, W.O. Smith, Jr. (1987). Comparative photochemical analysis of highly purified 124 kilodalton oat and rye phytochromes *in vitro*. *Photochem. Photobiol.*, **46**, 5–13.

89. C. Benda, W. Gärtner (2002). Conformational changes of the N-terminal part of phytochromes during the light-induced P_r-to-P_{fr} transformation. *Photochem. Photobiol.*, submitted.

90. J. Matthew, J.A. Terry, J.A. Wahleithner, D.M. Lagarias (1993). Biosynthesis of the plant photoreceptor phytochrome. *Arch. Biochem. Biophys.*, **306**, 1–15.

91. J. Cornejo, S.I. Beale, M.J. Terry, J.C Lagarias (1992). Phytochrome assembly. The structure and biological activity of 2(R),3(E)-phytochromobilin derived from phycobiliproteins. *J. Biol. Chem.*, **267**, 14790–14796.

92. T.D. Elich, D.M. Lagarias (1987). Phytochrome chromophore biosynthesis. *Plant Physiol.*, **84**, 304–310.

93. T.D. Elich, J.C. Lagarias (1988). 4-Amino-5-hexynoic acid - a potent inhibitor of tetrapyrrole biosynthesis in plants. *Plant Physiol.*, **88**, 747–751.

94. B.M. Parks, P.H. Quail (1991). Phytochrome-deficient *hy1* and *hy2* long hypocotyl mutants of *Arabidopsis* are defective in phytochrome chromophore biosynthesis. *Plant Cell*, **3**, 1177–1186.

95. B.L. Montgomery, K.A. Franklin, M.J. Terry, B. Thomas, S.D. Jackson, M.W. Crepeau, J.C. Lagarias (2001). Biliverdin reductase-induced phytochrome chromophore deficiency in transgenic tobacco. *Plant Physiol.*, **125**, 266–277.

96. N. Frankenberg, K. Mukougawa, T. Kohchi, J.C. Lagarias (2001). Functional genomic analysis of the HY2 family of ferredoxin-dependent bilin reductases from oxygenic photosynthetic organisms. *Plant Cell*, **13**, 965–978.

97. G.A. Gambetta, J.C. Lagarias (2001). Genetic engineering of phytochrome biosynthesis in bacteria. *Proc. Natl. Acad. Sci. U.S.A.*, **98**, 10566–10571.

98. F.T. Landgraf, C. Forreiter, A.H. Pico, T. Lamparter, J. Hughes (2001). Recombinant holophytochrome in *Escherichia coli*. *FEBS Lett.*, **508**, 459–462.

99. Z.Y. Jiang, L.R. Swem, B.G. Rushing, S. Devanathan, G. Tollin, C.E. Bauer (1999). Bacterial photoreceptor with similarity to photoactive yellow protein and plant phytochromes. *Science*, **285**, 406–409.

100. S.J. Davis, A.V. Vener, R.D. Vierstra (1999). Bacteriophytochromes: Phytochrome-like photoreceptors from nonphotosynthetic eubacteria. *Science*, **286**, 2517–2520.

101. S.H. Bhoo, S.J. Davis, J.M. Walker, B. Karniol, R.D. Vierstra (2001). Bacteriophytochromes are photochromic histidine kinases using a biliverdin chromophore. *Nature*, **414**, 776–779.

102. H. Scheer (1981). Biliproteins. *Angew. Chem. Intl. Ed. Engl.*, **20**, 241–261.

103. H. Hanzawa, T. Shinomura, K. Inomata, T. Kakiuchi, H. Kinoshita, K. Wada, M. Furuya (2002). Structural requirement of bilin chromophore for the photosensory specificity of phytochromes A and B. *Proc. Natl. Acad. Sci. U.S.A.*, **99**, 4725–4729.

104. L. Li, J.T. Murphy, J.C. Lagarias (1995). Continuous fluorescence assay of phytochrome assembly *in vitro*. *Biochemistry*, **34**, 7923–7930.

105. J.T. Murphy, J.C. Lagarias (1997). The phytofluors: a new class of fluorescent protein probes. *Curr. Biol.*, **7**, 870–876.

106. T.R. Berkelman, J.C. Lagarias (1986). Visualization of bilin-linked peptides and proteins in polyacrylamide gels. *Anal. Biochem.*, **156**, 194–201.

107. H.J.M.M. Jorissen, B. Quest, I. Lindner, N. Tandeau de Marsac, W. Gärtner (2002). Phytochromes with noncovalently bound chromophores: The capability of apophytochromes to direct tetrapyrrole photoisomerization. *Photochem. Photobiol.*, **75**, 554–559.

108. I. Lindner, B. Knipp, S.E. Braslavsky, W. Gärtner, K. Schaffner (1998). A novel chromophore derivative alters the spectral properties of only one of the two stable states of the plant photoreceptor phytochrome. *Angew. Chem. Intl. Ed. Engl.*, **37**, 1843–1846.

109. U. Robben, I. Lindner, W. Gärtner, K. Schaffner (2001). Analysis of the topology of the chromophore binding pocket of phytochromes by variation of the chromophore substitution pattern. *Angew. Chem. Intl. Ed. Engl.*, **40**, 1048–1050.

110. I. Lindner, S.E. Braslavsky, K. Schaffner, W. Gärtner (2000). Model studies of phytochrome photochromism: Protein-mediated photoisomerization of a linear tetrapyrrole in the absence of covalent binding. *Angew. Chem. Intl. Ed. Engl.*, **39**, 3269–3271.

111. D.A. Lightner, A.F. McDonagh (1989). Phototherapy for neonatal jaundice. *The Spectrum*, **2**, 1–15.

112. H. Falk (1989). *The Chemistry of Linear Oligopyrroles and Bile Pigments.* Springer-Verlag, Wien, 1989.

113. S.H. Bhoo, T. Hirano, H.Y. Jeong, J.G. Lee, M. Furuya, P.-S. Song (1997). Phytochrome photochromism probed by site-directed mutations and chromophore esterification. *J. Am. Chem. Soc.*, **119**, 11717–11718.

114. H. Hanzawa, K. Inomata, H. Kinoshita, T. Kakiuchi, K.P. Jayasundera, D. Sawamoto, A. Ohta, K. Uchida, M. Furuya (2001). *In vitro* assembly of phytochrome B apoprotein with synthetic analogs of the phytochrome chromophore. *Proc. Natl. Acad. Sci. U.S.A.*, **98**, 3612–3617.

115. R. Micura, K. Grubmayr (1995). A phycocyanobilin serylimino-ester as a new model for the chromophore-protein interaction in phytochrome. *Angew. Chem.*, *Int. Ed. Engl.*, **34**, 1733–1735.

116. M. Stanek, K. Grubmayr (1998). Protonated 2,3-dihydrobilindiones models for the chromophores of phycocyanin and the red-absorbing form of phytochrome. *Chem. – Eur. J.*, **4**, 1653–1659.

117. M. Stanek, K. Grubmayr (1998). Deprotonated 2,3-dihydrobilindiones - Models for the chromophore of the far-red-absorbing form of phytochrome. *Chem. – Eur. J.*, **4**, 1660–1666.

118. T.D. Elich, A.F. McDonagh, L.A. Palma, J.C. Lagarias (1989). Phytochrome chromophore biosynthesis. *J. Biol. Chem.*, **264**, 183–189.

119. B. Knipp, M. Müller, N. Metzler-Nolte, T.S. Balaban, S.E. Braslavsky, K. Schaffner (1998). NMR verification of helical conformations of phycocyanobilin in organic solvents. *Helv. Chim. Acta*, **81**, 881–888.

120. T. Lamparter, F. Mittmann, W. Gärtner, T. Börner, E. Hartmann, J. Hughes (1997). Characterization of recombinant phytochrome from the cyanobacterium *Synechocystis. Proc. Natl. Acad. Sci. U.S.A.*, **94**, 11792–11797.

121. I. Lindner (1999). Neue Chromophore für den pflanzlichen Photoreceptor Phytochrom. Thesis Universität Duisburg, Germany.

122. L. Deforce, M. Furuya, P.-S. Song (1993). Mutational analysis of the pea phytochrome A chromophore pocket: chromophore assembly with apophytochrome A and photoreversibility. *Biochemistry*, **32**, 14165–14172.

123. T. Kunkel, V. Speth, C. Büche, E. Schäfer (1995). *In vivo* characterization of phytochrome-phycocyanobilin adducts in yeast. *J. Biol. Chem.*, **270**, 20193–20200.

124. L. Hennig, E. Schäfer (2001). Both subunits of the dimeric plant photoreceptor phytochrome require chromophore for stability of the far-red light-absorbing form. *J. Biol. Chem.*, **276**, 7913–7918.

125. D. Wagner, C.D. Fairchild, R.M. Kuhn, P.H. Quail (1996). Chromophore-bearing NH_2-terminal domains of phytochromes A and B determine their photosensory specificity and differential light lability. *Proc. Natl. Acad. Sci. U.S.A.*, **93**, 4011–4015.

126. T. Kretsch, C. Poppe, E. Schäfer (2002). A new type of mutation in the plant photoreceptor phytochrome B causes loss of photoreversibility and an extremely enhanced light sensitivity. *Plant J.*, **22**, 177–186.

127. K. Eichenberg, T. Kunkel, T. Kretsch, V. Speth, E. Schäfer (1999). In vivo characterization of chimeric phytochromes in yeast. *J. Biol. Chem.*, **274**, 354–359.
128. M. Zeidler, T. Lamparter, J. Hughes, E. Hartmann, A. Remberg, S.E. Braslavsky, K. Schaffner, W. Gärtner (1998). Recombinant phytochrome of the moss *Ceratodon purpureus*: heterologous expression and kinetic analysis of P_r to P_{fr} conversion. *Photochem. Photobiol.*, **68**, 857–863.
129. K. Schaffner, S.E. Braslavsky, A.R. Holzwarth (1990). Photophysics and photochemistry of phytochrome. In: D.H. Volman, G.S. Hammond, K. Gollnick (Eds), *Advances in Photochemistry* (Vol. 15, pp. 229–277). Wiley, New York.
130. R.M. Williams, S.E. Braslavsky (2001). Triggering of photomovement - molecular basis, In: D.P. Haeder, M. Lebert (Eds), *Photomovement* (pp. 16–48). Elsevier Science, Amsterdam.
131. C. Kneip, P. Hildebrandt, W. Schlamann, S.E. Braslavsky, F. Mark, K. Schaffner (1999). Protonation state and structural changes of the tetrapyrrole chromophore during the $P_r \rightarrow P_{fr}$ phototransformation of phytochrome: A resonance Raman spectroscopic study. *Biochemistry*, **38**, 15185–15192.
132. P. Schmidt, U.H. Westphal, K. Worm, S.E. Braslavsky, W. Gärtner, K. Schaffner (1996). Chromophore-protein interaction controls the complexity of the phytochrome photocycle. *J. Photochem. Photobiol.*, *B: Biol.*, **34**, 73–77.
133. K. Eichenberg, I. Bäurle, N. Paulo, R.A. Sharrock, W. Rüdiger, E. Schäfer (2000). *Arabidopsis* phytochromes C and E have different spectral characteristics from those of phytochromes A and B. *FEBS Lett.*, **470**, 107–112.
134. H. Brock, B.P. Ruzsicska, T. Arai, W. Schlamann, A.R. Holzwarth, S.E. Braslavsky, K. Schaffner (1987). Fluorescence lifetimes and relative quantum yields of 124 kDa oat phytochrome in H_2O and D_2O solutions. *Biochemistry*, **26**, 1412–1417.
135. P.-S. Song, B.R. Singh, N. Tamai, T. Yamazaki, I. Yamazaki, S. Tokutomi, M. Furuya (1989). Primary photoprocesses of phytochrome picosecond fluorescence kinetics of oat and pea phytochromes. *Biochemistry*, **28**, 3265–3271.
136. V.A. Sineshchekov (1995). Photobiophysics and photobiochemistry of the heterogeneous phytochrome system. *Biochim. Biophys. Acta*, **1228**, 125–164.
137. V. Sineshchekov, L. Koppel, J. Hughes, T. Lamparter, M. Zeidler (2000). Recombinant phytochrome of the moss *Ceratodon purpureus* (CP2): fluorescence spectroscopy and photochemistry. *J. Photochem. Photobiol. B: Biol.*, **56**, 145–153.
138. K. Heyne, J. Herbst, D. Stehlik, B. Esteban, T. Lamparter, J. Hughes, R. Diller (2002). Ultrafast dynamics of phytochrome from the cyanobacterium *Synechocystis*, reconstituted with phycocyanobilin and phycoerythrobilin. *Biophys. J.*, **82**, 1004–1016.
139. C.G. Colombano, S.E. Braslavsky, A.R. Holzwarth, K. Schaffner (1990). Fluorescence quantum yields of 124-kDa phytochrome from oat upon excitation within different absorption bands. *Photochem. Photobiol.*, **52**, 19–22.
140. H. Linschitz, V. Kasche, W.L. Butler, H.W. Siegelman (1966). The kinetics of phytochrome conversion. *J. Biol. Chem.*, **241**, 3395–3403.
141. S.E. Braslavsky, W. Gärtner, K. Schaffner (1997). Phytochrome photoconversion. *Plant Cell Environ*, **20**, 700–706.
142. A.R. Holzwarth, E. Venuti, S.E. Braslavsky, K. Schaffner (1992). The phototransformation process in phytochrome. I. Ultrafast fluorescence component and kinetic models for the initial $P_r \rightarrow P_{fr}$ transformation steps in native phytochrome. *Biochim. Biophys. Acta*, **1140**, 59–68.

143. F. Andel III, K.C. Hasson, F. Gai, P.A. Anfinrud, R.A. Mathies (1997). Femtosecond time-resolved spectroscopy of the primary photochemistry of phytochrome. *Biospectroscopy*, **3**, 421–433.

144. M. Bischoff, G. Hermann, M. Rentsch, D. Strehlow (2001). First steps in the phytochrome phototransformation: A comparative femtosecond study on the forward ($P_r \rightarrow P_{fr}$) and the back reaction ($P_{fr} \rightarrow P_r$). *Biochemistry*, **40**, 181–186.

145. G. Hermann, M. Lippitsch, H. Brunner, F. Aussenegg, E. Mueller (1990). Picosecond dynamics of the excited state relaxations in P_r. *Photochem. Photobiol.*, **52**, 13–18.

146. W. Rüdiger, F. Thümmler, E. Cmiel, S. Schneider (1983). Chromophore structure of the physiologically active form (P_{fr}) of phytochrome. *Proc. Natl. Acad. Sci. U.S.A.*, **80**, 6244–6248.

147. F. Andel III, D.M. Lagarias, R.A. Mathies (1996). Resonance Raman analysis of chromophore structure in the Lumi-R photoproduct of phytochrome. *Biochemistry*, **35**, 15997–16008.

148. D. Farrens, P.-S. Song, W. Ruediger, P. Eilfeld (1989). Site-selected chromophore oxidation of phytochrome with tetranitromethane. *J. Plant Physiol.*, **134**, 269–275.

149. V.A. Sineshchekov (1999). Phytochromes: molecular structure, photoreceptor process and physiological function. In: G.S. Singhal, G. Renger, S.K. Sopory, K.-D. Irrgang, R. Govindjee (Eds), *Concepts in Photobiology: Photosynthesis and Photomorphogenesis* (pp. 755–795). Kluwer Academic, Boston.

150. V.A. Sineshchekov, L. Hennig, T. Lamparter, J. Hughes, W. Gärtner, E. Schäfer (2001). Recombinant phytochrome A in yeast differs by its spectroscopic and photochemical properties from the major phyA' and is close to the minor phyA": evidence for posttranslational modification of the pigment in plants. *Photochem. Photobiol.*, **73**, 692–696.

151. I. Michler, S.E. Braslavsky (2001). Time-resolved thermodynamic analysis of the oat phytochrome A phototransformation. A photothermal beam deflection study. *Photochem. Photobiol.*, **74**, 624–635.

152. T. Gensch, M.S. Churio, S.E. Braslavsky, K. Schaffner (1996). Primary quantum yield and volume change of phytochrome-A phototransformation determined by laser-induced optoacoustic spectroscopy. *Photochem. Photobiol.*, **63**, 719–725.

153. J.M. Kelly, J.C. Lagarias (1985). Photochemistry of 124-kilodalton *Avena* phytochrome under constant illumination *in vitro*. *Biochemistry*, **24**, 6003–6010.

154. T. Gensch, K.J. Hellingwerf, S.E. Braslavsky, K. Schaffner (1998). Photoequilibrium in the primary steps of the photoreceptors phytochrome A and photoactive yellow protein. *J. Phys. Chem. A*, **102**, 5398–5405.

155. J. Matysik, P. Hildebrandt, W. Schlamann, S.E. Braslavsky, K. Schaffner (1995). Fourier-transform resonance Raman spectroscopy of intermediates of the phytochrome photocycle. *Biochemistry*, **34**, 10497–10507.

156. H. Linschitz, V. Kasche (1967). Kinetics of phytochrome conversion: Multiple pathways in the P_r to P_{fr} reaction, as studied by double-flash technique. *Proc. Natl. Acad. Sci. U.S.A.*, **58**, 1059–1064.

157. V.A. Sineshchekov (1995). Photobiophysics and photochemistry of the heterogeneous phytochrome system. *Biochim. Biophys. Acta*, **1228**, 125–164.

158. C.-F. Zhang, D.L. Farrens, S.C. Björling, P.-S. Song, D.S. Kliger (1992). Time-resolved absorption studies of native etiolated oat phytochrome. *J. Am. Chem. Soc.*, **114**, 4569–4580.

159. E. Chen, V.N. Lapko, J.W. Lewis, P.-S. Song, D.S. Kliger (1996). Mechanism of native oat phytochrome photoreversion: A time-resolved absorption investigation. *Biochemistry*, **35**, 843–850.

160. R.D. Scurlock, S.E. Braslavsky, K. Schaffner (1993). A phytochrome study using two-laser/two color flash photolysis: I_{700} is a mandatory intermediate in the $P_r \rightarrow P_{fr}$ phototransformation. *Photochem. Photobiol.*, **57**, 690–695.

161. R.D. Scurlock, C.H. Evans, S.E. Braslavsky, K. Schaffner (1993). A phytochrome phototransformation study using two-laser/two-color flash photolysis: analysis of the decay mechanism of I_{700}. *Photochem. Photobiol.*, **58**, 106–115.

162. N. Sasaki, O. Yasutaka, T. Yoshizawa, K.T. Yamamoto, M. Furuya (1986). Temperature dependence of absorption spectra of 114 kDa pea phytochrome and relative quantum yield of its phototransformation. *Photobiochem. Photobiophys.*, **12**, 243–251.

163. Y. Mizutani, S. Tokutomi, T. Kitagawa (1994). Resonance Raman spectra of the intermediates in phototransformation of large phytochrome: Deprotonation of the chromophore in the bleached intermediate. *Biochemistry*, **33**, 153–158.

164. H. Foerstendorf, C. Benda, W. Gärtner, M. Storf, H. Scheer, F. Siebert (2001). FTIR studies of phytochrome photoreactions reveal the C=O Bands of the chromophore: Consequences for its protonation states, conformation, and protein interaction. *Biochemistry*, **40**, 14952–14959.

165. P.-S. Song, M.H. Park, M. Furuya (1997). Chromophore:apoprotein interactions in phytochrome A. *Plant. Cell. Environ.*, **20**, 707–712.

166. E. Chen, V.N. Lapko, P.-S. Song, D.S. Kliger (1997). Dynamics of the N-terminal α-helix unfolding in the photoreversion reaction of phytochrome A. *Biochemistry*, **36**, 4903–4908.

167. Y.-G. Chai, P.-S. Song, M.-M. Cordonnier, L.H. Pratt (1987). A photoreversible circular dichroism spectral change in oat phytochrome is suppressed by a monoclonal antibody that binds near its N-terminus and by chromophore modification. *Biochemistry*, **26**, 4947–4952.

168. J.C. Lagarias, F.M. Mercurio (1985). Structure function studies on phytochrome. *J. Biol. Chem.*, **260**, 2415–2423.

169. D. Sommer, P.-S. Song (1990). Chromophore topography and secondary structure of 124-kilodalton Avena phytochrome probed by Zn^{2+}-induced chromophore modification. *Biochemistry*, **29**, 1943–1948.

170. J. Hughes, F. Mittmann, A. Wilde, W. Gärtner, T. Börner, E. Hartmann, T. Lamparter (1997). A prokaryotic phytochrome. *Nature*, **386**, 663.

171. T. Kaneko, S. Sato, H. Kotani, A. Tanaka, E. Asamizu, Y. Nakamura, N. Miyajima, M. Hirosawa, M. Sugiura, S. Sasamoto, T. Kimura, T. Hosouchi, A. Matsuno, A. Muraki, N. Nakazaki, K. Naruo, S. Okumura, S. Shimpo, C. Takeuchi, T. Wada, A. Watanabe, M. Yamada, M. Yasuda, S. Taba (1996). Sequence analysis of the genome of the unicellular cyanobacterium *Synechocystis* sp. strain PCC6803. II Sequence determination of the entire genome and assignment of potential protein-coding regions. *DNA Res.*, **3**, 109–136.

172. H.J.M.M. Jorissen, A. Remberg, T. Coursin, S.E. Braslavsky, K. Schaffner, N. Tandeau de Marsac, W. Gärtner (2002). Two independent light-induced two-component signal transduction systems in the chromatically adapting cyanobacterium *Calothrix* PCC7601. *Eur. J. Biochem.*, **269**, 2662–2671.

173. A. Remberg, I. Lindner, T. Lamparter, J. Hughes, C. Kneip, P. Hildebrandt, S.E. Braslavsky, W. Gärtner, K. Schaffner (1997). Raman spectroscopic and light-induced kinetic characterization of a recombinant phytochrome of the cyanobacterium *Synechocystis*. *Biochemistry*, **36**, 13389–13395.

174. H. Foerstendorf, T. Lamparter, J. Hughes, W. Gärtner, F. Siebert (2000). The photoreactions of recombinant phytochrome from the cyanobacterium *Synechocystis*: a low-temperature UV-vis and FT-IR spectroscopic study. *Photochem. Photobiol.*, **71**, 655–661.

175. V.A. Sineshchekov, J. Hughes, E. Hartmann, T. Lamparter (1998). Fluorescence and photochemistry of recombinant phytochrome from the cyanobacterium Synechocystis. *Photochem. Photobiol.*, **67**, 263–267.

Chapter 6

Phytochrome signal transduction

Ferenc Nagy, Eva Kevei, Klaus Harter and Eberhard Schäfer

Table of contents

Abstract

The family of phytochrome photoreceptors regulates growth and development throughout the entire life cycle of higher plants. The molecular mechanism by which the light signal is converted into a regulatory signal for gene expression is the subject of intensive research. During the past two years genetic, biochemical and cell biological studies have provided novel observations and changed our view of the phytochrome-initiated signalling cascades. It became evident that the light quality- and quantity-dependent regulation of the nucleo-cytoplasmic partitioning of phytochromes and their conformation-dependent, functional interaction with transcription factors inside the cytoplasm and the nucleus are important components of phototransduction. To understand specificity, temporal and spatial differences between the various phytochrome-controlled responses it will be essential to unravel the molecular mechanisms (import/export/degradation) that ensure controlled compartmentalization of these photoreceptors.

6.1   Introduction

6.1.1   Plants and light

Plants as sessile organisms have to cope with changing environmental conditions at the place where they grow. To regulate their fitness and adapt to unfavourable conditions, plants depend upon reliable information about environmental factors such as temperature, water, nutrient supply and light. Among these environmental parameters light is obviously the most important external factor for plants. Light not only serves as the source of energy for photosynthesis, but also functions as a morphogenic signal. Light regulates a wide range of developmental processes and adaptations during the entire life cycle. These include seed germination, the developmental switch from skotomorphogenesis (growth and development in the dark) to photomorphogenesis (growth and development in the light) of the young seedling, the detection of neighbours competing for the incident light and the onset of the reproductive phase and flowering [1,2]. To sense changes in the quality, quantity, direction and duration of light, plants have evolved at least three different photoreceptor systems. These are (i) the UV-B receptors, characterized only by action spectroscopy [3], (ii) the blue/UV-A receptors cryptochrome 1 (CRY1), cryptochrome 2 (CRY2), phototropin and NPL1 [4–9], and (iii) the red/far-red reversible phytochromes [10]. These photoreceptors monitor the entire spectrum starting from UV-B to the infrared region and control different aspects of plant growth and development. During the transition from skotomorphogenesis to photomorphogenesis the expression of more than 100 genes is altered. Modulation of gene expression is the terminal step of a complex process. It starts with the absorption of light by specialized photoreceptors, followed by the conversion of light into a gene regulatory signal. The generated biochemical signal is then transduced via complex regulatory circuits within and between the cells and results in altered gene expression, leading to light-controlled physiological and metabolic responses of higher plants. In this review we focus on the molecular aspects of light-induced signal transduction (phototransduction) events regulated by the most characterized plant photoreceptors, the phytochromes (Figure 1).

6.1.2 The phytochrome system

Phytochromes (phy) represent a group of plant photoreceptors that control a number of light-dependent processes. Small multigene families encode phytochromes and, in *Arabidopsis*, five members, *PHYA* to *PHYE*, are known [11–12]. Phytochromes exist as dimers composed of two 125 kDa polypeptides, each carrying a covalently linked tetrapyrrole chromophore in the N-terminal domain and dimerization domains in the C-terminal part of the molecule. The photosensory function of the molecule is based on its capacity for reversible interconversion between the red light absorbing P_r form and the far-red light absorbing P_{fr} form, which is mediated by sequential absorption of red (R) and far-red (FR) light. Phytochromes are synthesized in darkness in their physiological inactive red light-absorbing form (P_r). On the one hand it was shown that the five *Arabidopsis thaliana* (*AtPHY*) genes encoding phytochromes are expressed in most cell types, although their expression level seems also to be fine-tuned by a developmental program [13–15]. On the other hand it is known that light as a signal can easily penetrate through plant organs/tissues and into subcellular compartments of various cell types. Therefore, the ubiquitous expression of *PHY* genes makes feasible (i) an absorption of R and FR light in nearly any type of cell and by this (ii) a modulation of physiological responses at any stage of plant development.

Photosignal perception, absorption of a photon by the receptor is followed by conformational changes, which then activate, through an as yet poorly understood molecular mechanism, signalling pathways leading to alterations in the expression of phytochrome-responsive genes [15]. Various members of the phytochrome receptor family have specialized, yet partly overlapping, molecular functions in light-induced signalling, i.e. ultimately in light-controlled growth and development of higher plants [2].

During the past few years our knowledge about the molecular mechanisms of phytochrome-controlled responses has substantially increased. Genetic and molecular approaches have made it possible to identify novel components involved in phytochrome-mediated signal transduction. In parallel, recent studies about the subcellular localization of the photoreceptors shed light on a molecular mechanism that regulates nucleocytoplasmic partitioning of phytochromes. These data indicate that the various phytochrome-initiated phototransduction pathways, described in the following chapters in detail, are mediated by a tightly regulated interaction of molecules in the nuclear and cytoplasmic compartments (Figure 1).

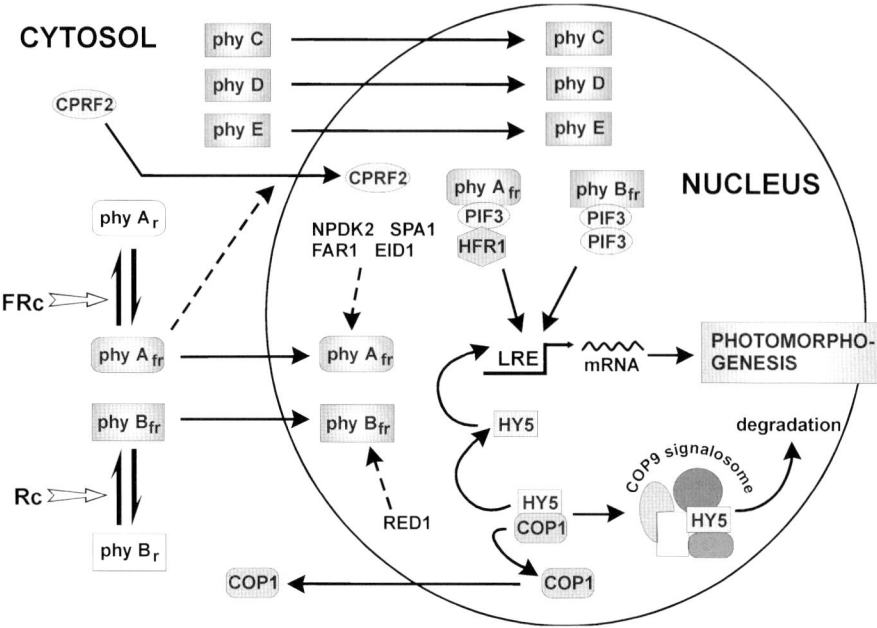

Figure 1. Light-regulated partitioning of various proteins plays a key role in photo-transduction in cells of higher plants. Phytochrome photoreceptors (phyA, phyB, phyC, phyD, phyE) are synthesized in the cytosol in their physiologically inactive form (phyA$_r$ and phyB$_r$ are shown as examples). Continuous irradiation with far-red (FRc) or red light (Rc) converts them into the physiologically active P$_{fr}$ forms (phyA$_{fr}$, phyB$_{fr}$). Photoconversion induces a yet unknown mechanism that releases phytochromes from their cytoplasmic retention. PhyA$_{fr}$ and phyB$_{fr}$ are then imported into the nuclei and interact with dimers of the PIF3/HFR1 and PIF3/PIF3 transcription factors, respectively, bound to light-responsive *cis*-regulatory elements (LRE) of various promoters. The molecular function of SPA1, FAR1, EID1 and NPDK2 affecting phyA-induced signal transduction and RED1 required for phyB-mediated signalling is not yet fully understood. Independent of their import into nuclei, phyA$_{fr}$ and phyB$_{fr}$ promote phosphorylation of other transcription factors, such as CPRF2, in the cytoplasm. Phos-phorylation then induces translocation of these proteins into the nuclei, where they contribute, probably in concert with other transcription factors, to the light-responsive transcription of subsets of genes. COP1 is imported into the nucleus in darkness, it interacts with the HY5 transcription factor and targets it for degradation and thereby represses photomorphogenesis. Light induces disassembly of the COP1/HY5 complex, export of COP1 into the cytoplasm and activation of nuclear HY5. Solid (verified) and dashed (putative) lines indicate the various protein–protein interactions and signalling steps that are thought to be involved in mediating photomorphogenesis.

6.2 Approaches to elucidate the mechanism of light-induced signal transduction

6.2.1 Photoreceptor mutants and overexpression studies

By now, single point or deletion mutations in all genes encoding photorecep-tors, except *PHYC*, have been described. The availability of these and other double/triple and quadruple mutants, together with that of transgenic plants containing overexpressed photoreceptors, made it feasible to analyse the roles of individual photoreceptors and to characterize the importance of their inter-actions in regulating light-dependent physiological responses [16]. It has become evident that cryptochromes are the dominant UVA blue-light photo-receptors that control physiological responses ranging from the inhibition of hypocotyl growth and gene expression to flowering [17]. CRY2 seems to be more effective at low blue light fluence rates, whereas CRY1 functions at higher fluence rates.

The members of the phytochrome photoreceptor family have not only dif-ferent modes of action but also different, yet partially overlapping, functions. PhyA is the most highly specialized phytochrome. It is responsible for the Very Low Fluence Rate (VLFR) and the far-red High Irradiance (HIR) Responses [18]. The extraordinary responsiveness (e.g. sensitivity to light) of this photo-receptor allows phyA to control germination of seeds buried in the soil and to induce germination when seeds are exposed to a brief light treatment, even starlight [19]. The other phytochromes control, to different extents, the classi-cal R/FR reversible induction (low fluence rate: LFR) responses and the responses to continuous R light [20].

For phyA the proteolytical degradation of the physiologically active P_{fr} form is believed to terminate signalling. As for the other phytochromes, since their P_{fr} form is relatively stable, the switch-off mechanism is still not understood, but the P_{fr} to P_r dark reversion is a possible tool to stop signalling [21]. PhyB seems to be the most prominent light-stable phytochrome. It is involved in most of the phytochrome-mediated responses, ranging from germination through inhibition of hypocotyl growth and light-quality adaptation (shade avoidance syndrome) to flowering [22–24]. PhyE has a major role in control-ling internode elongation and shade avoidance responses as well [25], whereas phyD has only a marginal role in controlling R light-mediated responses [26].

Network(s) mediating interactions between these photoreceptors seem to be extremely complex. Experimental data indicating the existence of such net-work(s), known as responsiveness amplification of certain physiological reac-tions by different light treatments, were obtained long before the complexity of the phytochrome and cryptochrome families became evident [27]. Recently, genetic and physiological studies have demonstrated that the function of CRY1 is strongly dependent on that of phyA and phyB [28,29] and a specific interaction between the individual pathways controlled by phyD and CRY1 has also been detected [30]. Furthermore, it has been shown that phyA positively controls the function of phyB [31], whereas phyB negatively effects phyA-mediated signalling. More recent identification and characterization of the SUB1 gene encoding a Ca^{2+}-binding protein indicate that SUB1 suppresses accumulation of the transcription factor HY5 and functions as a downstream component of CRY1 and CRY2 signalling pathways and as a modulator of phyA signalling cascades [32].

6.2.2   Biochemistry of phototransduction

The primary molecular function of phytochromes in mediating light signal transduction remains to be elucidated. Not long after the discovery of phytochrome a general debate started about how phytochrome regulates light responses: i.e. is phytochrome a membrane receptor, or does it function as a light-activated enzyme or does it control gene transcription directly [33]? As quite often in science, all three hypotheses may – in the end – be correct.

In lower plants like algae, mosses, and in ferns, the observed action dichroism for phytochrome-controlled orientation responses strongly favours a membrane function [34]. Recently, it was shown that the *Synechocystis* genome encodes for a phytochrome-like photoreceptor, which has all the characteristics of a prokaryotic two-component histidine kinase [35]. In higher plants, namely in oat coleoptiles and parsley cell suspension cultures, a very rapid, phytochrome-controlled protein phosphorylation was described. This observation indicates that phosphorylation cascade(s) might play a role in mediating the early steps of phytochrome-controlled signalling [36]. However, in contrast to cyanobacteria, phytochromes of higher plants are light-regulated serine/threonine protein kinases rather than two-component histidine kinase-like photoreceptors [37].

Another approach to the analysis of the mode of phytochrome-dependent signalling was microinjection of hypocotyl cells of the chromophore biosynthesis-deficient tomato *aurea* mutant [38–40]. Results obtained by this approach indicated that the phytochrome-controlled signalling cascade includes steps that affect levels of some of the well-known second messengers identified in other eukaryotic cells: namely, light absorption by phytochromes triggers activation of a trimeric GTP-binding protein that leads to induction of a bifurcated signal transduction pathway. The phyA-activated branch modulates cGMP levels and leads to the expression of *chalcone synthase* (*CHS*) gene and to the induction of anthocyanin biosynthesis. The other branch activated by both phyA and phyB regulates chloroplast development and the expression of genes encoding the chlorophyll a/b binding proteins (*CAB* genes). Additional experiments indicated a cross-talk between these branches. Although these observations have been supported by pharmacological studies, the demonstration of the modulation of internal Ca^{2+} and cGMP levels, as well as the identification of the putative target proteins, have remained elusive. However, it should be noted that mutation of the *REP1* gene selectively affects expression of genes assigned to the cGMP branch, thereby supporting the existence of the postulated bifurcating pathways [41]. Recently, an *Arabidopsis* mutant (accession number N100137) lacking detectable amounts of the alpha subunit of the only known heterotrimeric GTP-binding protein became available. It follows that the putative function of this protein in phytochrome-mediated or other signalling pathways can now be determined.

6.2.3   Genetic analysis of phototransduction

The combination of molecular and genetic approaches has been proven to be highly efficient in analysing the role of different photoreceptors controlling photomorphogenesis as well as in identifying candidate genes involved in signal transduction. In a pioneering study Koornneef screened under white light (WL) conditions for mutants resembling a dark grown phenotype [42]. This and other similar screens resulted, mainly, in the isolation and characterization of genes that either control chromophore biosynthesis or code for the photoreceptors themselves. The other type of screening, pioneered by Chory and later by Deng and their co-workers, searched for plants exhibiting light-grown phenotypes, although the plants were grown in darkness. This approach led to the isolation of the so-called COP (constitutive photomorphogenic) and DET (de-etiolated) mutants. Characterization of the COP/DET and later the FUS mutants revealed that the switch between photomorphogenesis and etiolation is regulated by a complex suppressor system that, in contrast to the photoreceptors, promotes the etiolation pathway by repressing photomorphogenesis in darkness. These two types of screens led to the isolation of dozens of mutants displaying aberrant photomorphogenic phenotypes. However, irrespectively of whether the genes underlying these aberrant phenotypes were isolated, it turned out to be unexpectedly difficult to unravel the function of these mutations at the molecular level.

The first mutants belonging to the COP/DET/FUS group, namely DET1 [43] and COP1 [44], were isolated more than a decade ago. Genetic analysis and mapping of these and other mutants identified 11 COP/DET/FUS loci in the *Arabidopsis* genome [45]. The mutants in all 11 loci are recessive. It has been shown that mutations in COP1, COP10 and DET1 loci do not affect formation of the COP9 signalosome, whereas mutations in the other 8 COP/DET loci prevent its generation [46] and are lethal at seedling stage. On the one hand this latter observation indicated that the COP9 signalosome might play a more general role in *Arabidopsis* development. On the other hand the pleiotropic nature of the *cop/det* mutants, however, made it difficult to decipher how the various *cop/det* mutants interfere with photomorphogenesis in *Arabidopsis*. At present, considering genetical, physiological and molecular aspects, the most comprehensive information is available about COP1. In recent papers it has been shown that COP1 is localized in the nucleus in darkness and functions as a repressor of photomorphogenesis by suppressing expression of light-inducible genes. The activity of COP1 is negatively controlled by the light-induced relocalization of the protein into the cytoplasm. As a result, COP1 is excluded from the nuclei of cells that are transferred to light. Moreover, localization of COP1 also has a tissue-specific component, as COP1 levels are constitutively high in the nuclei of root cells, which do not undergo photomorphogenesis [47]. The COP1 protein contains an amino-terminal Zn^{2+}-binding ring finger domain, a carboxy-terminal WD-40 repeat domain and in-between a domain with the potential to form coiled-coil structures. Detailed analysis of the localization and physiological effects of COP1 fragments revealed discrete domains that mediate light-responsive partitioning of the protein. Namely, a bipartite NLS within the core domain, including parts of the ring finger and coiled-coil domain, which is responsible for the retention in the cytoplasm [48]. In addition, a motif that mediates targeting of the COP1 protein to subnuclear foci has been described, which overlaps the cytoplasmic retention signal [49]. How the switch between the cytoplasmic and nuclear localization of the COP1 protein is accomplished at the molecular level is not yet understood. Irrespective of the mechanism that mediates subcellular partitioning of COP1 it has been documented that COP1 interacts physically with HY5 – a bZIP type transcription factor and positive regulator of photomorphogenesis [50,51]. The observations suggested that COP1 and HY5 interaction could only take place in darkness and should negatively regulate HY5 activity. In addition, Osterlund et al. [52] reported that abundance of HY5 is directly correlated with the degree of photomorphogenic development of the seedling and that proteasome inhibitors block degradation of HY5. Therefore it was concluded that (i) the level of the HY5 protein is primarily controlled by degradation, (ii) HY5 degradation is mediated most likely by the proteasome pathway. Accordingly, COP1 acts like an E3 ubiquitin-protein ligase by recruiting the ubiquitin-conjugating enzyme E2 and mediating transfer of the polyubiquitin from E2 to HY5. These steps then result in targeting of HY5 to subsequent degradation by the proteasome [53]. There is evidence that all other pleiotropic COP/DET/FUS proteins including the COP9 signalosome are required for specific degradation of HY5. The exact role of the COP9 signalosome and that of COP10 and DET1 in regulating photomorphogenesis is not yet understood. However, there is evidence that the COP9 signalosome is structurally similar to the lid subcomplex of the 19S regulatory particle of the 26S proteasome. Therefore, it seems to be plausible to assume that all COP/DET proteins act as downstream regulators of phototransduction by controlling either directly or indirectly proteasome-mediated degradation of some of the essential signalling molecules.

In recent years several laboratories have performed screens to isolate mutants, especially in *Arabidopsis*, which in contrast to the COP/DET mutants exhibit only specific light-dependent phenotypes [54,55]. These signal transduction mutations are expected to display unique phenotypes under specific light conditions as the result of modulating specific steps of signalling rather than affecting the photoreceptors themselves. Accordingly, these mutations affect (i) only the phyA-signalling pathway (either VLFR or continuous FR), (ii) only the phyB signalling (LFR or continuous R) or (iii) both phyA and phyB signalling.

For example, *RED1*, *PEF2* and *PEF3* are members of a class of genes that affect only phyB signalling. These mutations lead to reduced sensitivity to R light [56,57] and thus share some features with phyB mutants [58]. These mutants flower early in short days, have elongated petioles, and show decreased sensitivity to red light and defects in the shade avoidance response. The exact molecular nature of these mutations, however, remains to be elucidated.

In a very specific screen to obtain hypersensitive rather than hyposensitive mutants for phyB signalling only one mutant was recovered, even though more than 2 million seedlings were tested [59]. This mutant carried a point mutation within the *PHYB* gene itself and produced a phyB, which had the same photochemical properties as the wild-type phyB molecule. However, the mutant phyB exhibited three orders of magnitude higher sensitivity to continuous R light than its native counterpart [59]. We speculate that the strong responsiveness enhancement is probably due to a defect in the cytoplasmic retention of the mutated phyB protein.

The *FHY1*, *FHY3*, *FIN2*, *SPA1*, *FAR1*, *EID1* and *PAT1* gene products specifically affect phyA signalling [60–65]. Genetic analyses suggest that, except for SPA1 and EID1, they all act as positive elements in the pathway, i.e. these mutations lead to a reduced sensitivity to FR light. PAT1 was identified in a screen for mutants with long hypocotyls in FR light and survival in WL after FR pre-irradiation [65]. PAT1 is a new member of the plant specific GRAS (or VHIID) gene family [66]. In contrast to the other members of the GRAS family PAT1, tested as PAT1-GFP fusion protein in transgenic *Arabidopsis* seedlings, is localized in the cytoplasm [65]. The *spa1* mutant was identified as a suppressor of a weak phyA mutation [62]. The *SPA1* gene codes for a putative protein that shows homology to COP1, a known negative regulator of photomorphogenesis [63]. *Far1* was identified as a suppressor mutation in transgenic lines overexpressing phyA. *FAR1* is a member of a gene family specific of plants [64]. Both SPA1 and FAR1 proteins contain nuclear localization signal (NLS) motifs and are localized in the nucleus, when tested in transient expression studies using *uidA* (ß-glucuronidase, GUS) fusion constructs. The *eid* mutants were identified in a screen developed to obtain hypersensitive plants defective in phyA destruction. It was known that wild-type *Arabidopsis* seedlings show a loss of far-red HIR if continuous FR light is interrupted every 20 min by 20 min R light due to phyA degradation. The aim of the screen was to identify mutants that still show a strong HIR under these conditions. The following three different and new loci have been identified (T. Kretsch, C. Büche, M. Dieterle and E. Schäfer, unpublished). *Eid4* is a point mutation in phyA leading to altered destruction of the photoreceptor and hypersensitive responses. *Eid6* is a point mutation in the *COP1* gene having no dark phenotype but a general hypersensitivity towards all light qualities. *Eid1* has an extremely strong phenotype. In this mutation continuous FR light can be substituted by short light pulses every 30 min, whereas in the wild type the pulses must be given every 5 min [67]. The mutation even shifts the action peak from

FR to R and thereby converts the native photoreceptor into a novel system
that now exhibits maximal sensitivity to continuous R light. The *EID1* gene
codes for a nuclear F-box protein that has been shown to interact with both
ASK1 and ASK2 *in vivo* and *in vitro* and is very likely included in the formation of SCF complexes in the nucleus [68]. Thus, EID1 is believed to be an
important element involved in the turnover regulation of phyA-specific signal
transduction components [68]. The *Pef1* [57] and *psi2* [69] mutants are affected
in both phyA and phyB signalling. *Psi2* is hypersensitive to both R and FR
light, whereas *pef1* shows reduced sensitivity to light of both wavelengths.

6.2.4   Cell biological aspects of phototransduction

6.2.4.1   Phytochrome-regulated intracellular partitioning of phytochromes
Until recently the dominant view has been that plant photoreceptors are localized in the cytoplasm. Physiological studies in algae, mosses and ferns showed
action dichroism for chloroplast orientation, polarotropism and phototropism.
These observations indicated that the photoreceptors regulating these responses are localized in the cytoplasm in an oriented manner, presumably in
association with the plasmalemma or other membrane structures [34].

Computer analysis based on phytochrome sequences indicated that higher
plant phytochromes are soluble proteins and no characteristic motifs indicating
association with cell membranes were found. This hypothesis was supported
by immunocytochemical studies on the P_{fr}-dependent formation of sequestered
areas of phytochrome (SAPs) in the cytoplasm of coleoptile cells from mono-
cotyledonous seedlings [70,71], despite the demonstration of phy-dependent
transcription in nuclear run-on experiments [72]. However, results provided by
Sakamoto and Nagatani [73] further challenged the view of an exclusively
cytoplasmic localization of phytochromes. These authors could demonstrate
nuclear localization of phyB fragments fused to the GUS reporter protein in
transgenic plants, pointing to functional NLS sequences in the photoreceptor.
Additionally, in this study a substantial increase in the amount of phytochrome
in purified nuclei of plant tissues irradiated with R light was observed. More
recently, Yamaguchi et al. [74], and Kircher et al. [75] complemented phyB-
deficient *Arabidopsis* or tobacco mutants by expressing fusion proteins consisting of full-length phyB and the reporter green fluorescent protein (GFP) in
transgenic plants. The results clearly established that these types of transgenic
lines are expressing functional phytochromes and are therefore suitable to
analyse light-dependent intracellular localization of these photoreceptors. As
an outcome of these and further studies it became evident that light-dependent
nuclear import of phyA-GFP and phyB-GFP exists but is characteristically
different [75]. More recent studies, applying the same experimental approach,
extended these observations [76,77] and demonstrated that light requirements
for nuclear uptake of phyA-GFP and phyB-GFP are identical to that of
the endogenous, functionally distinct photoreceptors. The nuclear import of

phyB-GFP requires R light (high amounts of P_{fr}) and is R/FR photoreversible (LFR). Light sensitivity of this reaction is enhanced by pre-irradiation with either R or blue light but not by FR light [76]. Thus, accumulation of the phyB-GFP fusion protein in the nuclei exhibited a similar responsiveness amplification as previously described for many phyB-mediated responses [27].

The capacity to complement the corresponding photoreceptor mutant in *Arabidopsis* was also demonstrated for a phyA-GFP fusion protein [77]. However, by contrast to phyB-GFP, nuclear uptake of phyA-GFP can be initiated even by short FR pulses (VLFR, low amounts of P_{fr}) and continuous FR light (HIR) [77]. Additionally, it was shown that the kinetics of P_{fr}-dependent nuclear import of phyA is an order of magnitude faster than that of phyB. The validity of the above-described light requirements of nuclear import for the endogenous protein was also demonstrated by immunolocalization studies on the intracellular partitioning of phyA in pea seedlings [78]. These analyses provided additional evidence that the wavelength-dependence of the light-induced import of phyB-GFP and phyA-GFP into the nuclei closely resembles the reported, well-known action spectra of phyB and phyA-mediated responses.

Very recent studies on the light-regulated partitioning of phyC, phyD and phyE indicated that, similarly to phyA- and phyB-GFP fusion proteins, phyC-, phyD- and phyE-GFP are also imported into the nuclei in a light-dependent fashion [96]. Furthermore, the same authors showed that (i) import of all phytochrome photoreceptors is also regulated by a developmental program and (ii) under natural, diurnal light conditions import of all phy-GFP fusion proteins shows a diurnal oscillation. For phyA-GFP only FR/D cycles are inductive, whereas for the other phytochromes either WL/D or R/D cycles can be used. Moreover, it was also demonstrated by these authors that mutant forms of phyA and phyB, shown to be physiologically inactive in plants [15], are imported into the nuclei in a light-dependent fashion but fail to form speckles. Therefore it is postulated that accumulation of phy proteins into subnuclear speckles or foci, similarly to those formed by COP1 [49], is a characteristic feature of the physiologically active photoreceptor. Additional experiments revealed that the appearance of the phy-GFP fusion proteins in the nuclei after L/D entrainment is detectable prior to the onset of the light phase. These findings indicate that accumulation of these photoreceptors in the nucleus is regulated by the circadian clock, which indeed could be demonstrated for phyA-GFP and phyB-GFP in continuous darkness after L/D entrainment (P. Gil and E. Schäfer, unpublished).

On the one hand, these results show a strong correlation of the light requirements for physiological responses regulated by the respective phytochromes and the intracellular partitioning of these molecules. On the other hand, these observations demonstrate that import of phytochromes into the nuclei is a critical step in regulating not only phototransduction leading to photomorphogenesis but also in the input pathway for the plant circadian clock. Circadian reappearance of phytochrome indicates that interaction of phytochrome with other signalling molecules in the nuclei, including the putative clock proteins, occurs in a rhythmic fashion. It follows that studies aimed at deciphering the molecular events required for a functional circadian clock in plants should address this question.

6.2.4.2　Potential mechanisms of cytoplasmic retention and nuclear import of phytochromes

In etiolated seedlings and dark-adapted plants, where phytochromes exist in their photobiologically inactive P_r forms, phyB-GFP and phyA-GFP are localized in the cytoplasm. A mutated version of phyB-GFP, which is not able to bind its chromophore, is confined almost exclusively to the cytoplasm irrespective of light conditions [75]. These results indicate that the conformational change of P_r to the physiological active P_{fr} form is a necessary pre-requisite for the nuclear import of phytochromes. Localization experiments with truncated versions of phyB-GFP fusion proteins clearly indicate the presence of a functional NLS within the C-terminus of the photoreceptor [74,79]. The light-independent exclusive nuclear localization of the C-terminal half of phyB suggests an important role for the N-terminal part of phytochromes in cytoplasmic retention in darkness. The addition of an extra NLS to phyB-GFP does not result in the light-independent nuclear import of this modified photoreceptor protein in transgenic tobacco seedlings (S. Kircher, F. Nagy, E. Schäfer, unpublished). Taken together, these data suggest a cytoplasmic retention mechanism for phyB in its P_r form. The switch allowing the interaction of phytochrome with the nuclear import machinery could be the release of the photoreceptor from cytoplasmic retention by the light-dependent conformational change from the P_r to the P_{fr} form. It is tempting to speculate about the role of phosphorylation in regulating retention of these photoreceptors in the cytoplasm, yet there is no information available that supports this model. It is important to note that the translocation pattern of phyA and phyB into the nuclei reflects their predicted mode of action. Therefore, the specificity of phy-regulated responses should be determined, at least partially, by cytoplasmic events that occur prior to translocation into the nuclei. It follows that unravelling these cytoplasmic events will be probably one of the major targets of signal transduction research in the immediate future.

6.2.5　Phytochrome-interacting proteins

The analysis of mutants defective in phyA- and/or phyB-mediated signal transduction has identified a number of proteins involved in the regulatory process leading to the respective physiological responses. As regards to the nuclear import and function of active photoreceptors, of special interest was the identification of an *Arabidopsis* mutant that showed altered phyB-dependent light signalling [80]. The corresponding gene product, PIF3 (phytochrome-interacting factor 3), a positive acting basic helix-loop-helix (bHLH) protein of nuclear localization, was also isolated by the yeast two-hybrid approach (see below).

Recently it became possible to express different phytochrome cDNAs in yeast cells and to demonstrate an in vivo assembly of phytochrome apoproteins with chromophores [20,21,40]. Unfortunately when full-length phytochromes were fused with the GAL4 DNA binding domain the fusion protein was transactive and therefore not suitable for yeast two-hybrid screens (T. Kretsch, K. Eichenberg, E. Schäfer, unpublished data). Nevertheless, yeast two-hybrid screens with the C-terminal domains of phytochromes as "baits" were successfully used to isolate phy-interacting proteins. Three candidates have been identified so far: PIF3 [81], PKS1 [82] and NPDK2 [83]. All three proteins interact with the C-terminal part of both phyA and phyB. The function of these tentative phy-interacting partners has been analysed by using knockout mutants or antisense and overexpressor lines. Overexpression of PKS1 leads to slightly elongated hypocotyls under R light, but no effect was detectable under FR light and in antisense plants. NDPK2 loss-of-function mutants show an enhanced cotyledon opening and hook unfolding under FR light and a reduced sensitivity in R light. PIF3 antisense lines show a decreased sensitivity to R light and only a very slight effect in FR light. In vitro studies showed that the PIF3 protein physically interacts, in a R/FR reversible manner, with phyB and to a lesser extent with phyA [84]. Furthermore, recently it was also shown that PIF3 binds specifically to a G-box DNA-sequence motif present in promoters of various light-regulated genes, whereas phyB binds reversibly to G-box bound PIF3 only in its physiologically active P_{fr} form [85]. Since in PIF3 antisense plants the R light-induced activation of several light-regulated genes is reduced, it is plausible to assume that an extremely short signal transduction cascade may mediate phyB-controlled transcription of various genes. In other words, simultaneous binding of the bHLH transcription factor PIF3 to the P_{fr} forms of phytochrome and to G-box elements within the promoters of light-responsive genes could represent a straightforward mechanism for phytochrome signal transduction. Whether the characteristic speckles formed after nuclear import of phytochromes also contain PIF3 in a high-molecular weight complex has to be addressed in planta. The recent identification of another positive-acting bHLH transcription factor involved in phyA signal transduction supports the hypothesis of a direct and distinct effect on transcription of phytochromes imported into the nucleus [86]. The P_{fr} forms of phyA and phyB can physically bind only to PIF3 homodimers or PIF3/HFR1 heterodimers but not to homodimers of HFR1 [86]. Because of the low abundance of *HFR1* mRNA in plants treated with continuous R light as compared with plants irradiated with FR light it is tempting to speculate whether this factor determines the specificity of gene regulation driven by nuclear phytochromes. The nuclear function of the P_{fr} form of phyB after irradiation with R light could be mediated by binding to PIF3 homodimers, whereas the specific function of the P_{fr} form of phyA after irradiation with continuous FR could be achieved by PIF3/HFR1 heterodimer interaction. In this aspect, it would be of major interest to elucidate the binding specificities of homo- and heterodimeric bHLH transcription factors to promoter elements of individual phy-regulated genes. Besides PIF3 and HFR1, other factors involved in phytochrome signal transduction have also been shown to be nuclear proteins, which further underlines the importance of the nuclear compartment in light signalling. For phyA-dependent light-regulation, SPA1 [63], FAR1 [64], FIN219 [87], and EID1 [68] have been characterized as nuclear factors.

It was demonstrated in domain swap experiments that the C-terminal halves of phyA and phyB are interchangeable and involved in signalling, whereas the N-terminal halves are responsible for the light absorption and light quality specificity [15]. Nevertheless, no specific N-terminal interaction partner has been described so far. In prokaryotes, phytochrome is part of a phosphorelay system that, besides other constituents, also contains the so-called response regulatory proteins. In the *Arabidopsis* genome several genes encoding response regulator-like proteins (ARR) have been identified; however, the physiological role of these plant proteins is largely unclear [88]. Although several response regulators have been tested in yeast two-hybrid assays, so far only ARR4 has been shown to specifically interact with the extreme N-terminal part of phyB. In addition it was shown that ARR4 specifically and strongly inhibits dark reversion ($P_{fr} \rightarrow P_r$) of phyB in yeast and in planta and thus stabilizes phyB in its active P_{fr}-form in vivo. In accordance with this observation, overexpression of ARR4 in transgenic plants resulted in an enhanced sensitivity of all analysed responses to R light but did not affect sensitivity to FR or blue light [89]. Because response regulators are usually the final output elements of two-component systems, we suggest that the yet unidentified cognate sensor histidine kinase of ARR4 could regulate R light signalling by physical interaction with phyB.

6.2.6   *Phytochrome-regulated nuclear import of the bZIP transcription factor CPRF2*

As well as their suggested direct function in the nucleus, phytochrome photoreceptors, especially phyA, have also been shown to mediate cytoplasmic events in its active P_{fr}-form. Besides acting via trimeric G-proteins, cGMP and calcium/calmodulin on greening and anthocyanin production as revealed by pharmacological studies [38,39,90], phyA is a phosphoprotein [91]. It is considered to be a protein kinase [37]. Recently, PKS1 (phytochrome kinase substrate 1), a cytoplasmic protein identified in a yeast two-hybrid screen, was demonstrated to be phosphorylated by phyA in vitro [82].

In addition, P_{fr}-dependent phosphorylation events in the cytoplasm could lead to nuclear import of downstream regulatory proteins, which was shown to be so for a family of basic leucine-zipper motif (bZIP) transcription factors. The common plant regulatory factors (CPRF) bind to G-box and C-box promoter elements and are thought to play a role in regulating light-responsive genes in parsley [92]. In a biochemical study analysing subcellular fractions of parsley cells, a light-regulated and phosphorylation-dependent translocation of G-box binding proteins from the cytoplasm into the nucleus was demonstrated in vitro [93]. Further characterization of several members of the CPRF gene family by immunocytochemistry and transient expression of GFP fusion proteins revealed that only CPRF2 is localized in the cytoplasm in the dark. The cytoplasmic retention of this transcription factor is released by R light treatment and is, at least partially, R/FR light photoreversible. This observation points to the involvement of phytochrome in this light-regulated translocation process [94]. It is conceivable that phytochrome-dependent phosphorylation of the C-terminus of CPRF2 mediated by a cytosolic serine kinase is a prerequisite for nuclear import [95]. This study also indicates that the cytoplasmic retention of CPRF2 is achieved in a high molecular weight complex in darkness [94,95]. Localization of truncated CPRF2 fused to GFP in parsley protoplasts indicates that two structural motifs in the N-terminus, distinct from the NLS-harbouring bZIP domain, are necessary to prevent import in darkness. Additional studies show that the N-terminal domain can confer cytoplasmic retention to another nuclear bZIP factor in domain-swap experiments (S. Kircher and E. Schäfer, unpublished). It is therefore tempting to speculate that phytochrome-dependent phosphorylation of CPRF2 leads to conformational changes within the protein that releases the factor from a cytoplasmic retention complex. After release, CPRF2 could interact with the nuclear import machinery, translocate into the nucleus, and bind to light-regulated target genes. Very recently, a putative retention protein of a CPRF2 homolog from *Arabidopsis* was identified by a novel screening approach in yeast, but its function in plants has yet not been corroborated (C. Näke, E. Schäfer, K. Harter, unpublished data).

6.3   Conclusions and perspectives

Very recent progress has enabled us to develop models for the early steps of signal transduction initiated by phyA and phyB. Based on the results of mutant screens, analysis of the import of phyB into the nuclei and PIF3/phyB interactions it is predictable that a very short signal transduction chain mediates some of the phyB responses. Thus, light absorption by phyB leads to P_{fr} formation, facilitates import into the nuclei and, subsequently, results in the interaction of the photoreceptor with promoter-bound PIF3. Because phyB exhibits strong transactivation capacity when closely associated to DNA, the P_{fr}-form of the photoreceptor may function as a transactive adaptor protein directly linking PIF3 with the basal transcription machinery and, thus, induce the expression of light-regulated genes.

   The questions that await answers are the following. What mechanisms, in addition to protein import, regulate the light-induced accumulation of phyB in the nuclei? Is the import process itself light-induced or is the light-induced accumulation of phyB regulated by active retention of phyB in the cytoplasm? Does differential turnover of phyB in light and dark and in different compartments play a role in terminating signalling? Does light-induced protein phosphorylation of phyB – like for phyA – occur and is it involved in regulating retention or degradation? Are PIF3 and HFR1 the only direct targets in the nucleus and, if so, how is specificity achieved and the signalling terminated after light to dark transition?

Interestingly, a relatively large number of mutants has been isolated which affect phyA-specific signalling. This can be due either to the screening methods applied or to the fact that although phyA is a R/FR reversible pigment with photochemical properties almost identical to those of phyB, phyA must maintain its capability to respond maximally to continuous FR light. It is accepted that one of the first steps after photoconversion of phyA-P_r into P_{fr} is the formation of SAPs in the cytoplasm and the transport into the nucleus. It is tempting to predict that autophosphorylation of phyA, and its kinase activity are required for formation of SAPs and possibly for eliminating retention. It seems that P_{fr} is the primary active form of phyA, and P_{fr} is transported into the nucleus under continuous FR light. Yet even under these conditions only a minor part of phyA will exist in its P_{fr} form in the nucleus. It follows that either both the P_{fr} and P_r forms of phyA will be active in the nucleus or an unknown mechanism has to produce an HIR function not only for transport into but also for function within the nucleus. Probably, the *EID1* and *SPA1* gene products, localized in the nuclei, together with other components such as FAR1 will form the controlling network. Besides PKS1, PAT1 is so far the only known gene product of phyA signalling that is localized in the cytoplasm. It specifically affects phyA signalling in the nucleus but the mode of its action remains to be elucidated. The other candidates, namely FHY1, FHY3 and FIN219, are not yet cloned; therefore it is rather difficult to predict their functions at the molecular level.

At present, cytoplasmic functions of phyA and other phytochromes and cryptochromes symbolize the Achilles' heel of photoreceptor-mediated signal ling. Clearly, in lower plants one can expect a dominant cytoplasmic function of phytochrome and it is quite certain that, in higher plants, cytoplasmic functions like light-induced phosphorylation of PKS1, interaction with SUB1 and induction of nuclear translocation of transcription factors such as CPRF2 are important as well. In addition, data predicting involvement of G-proteins, calcium/calmodulin- and cGMP-dependent pathways still await incorporation into the network of phy-mediated signal transduction.

In general, the availability of the complete *Arabidopsis* genome will facilitate the functional characterization of nuclear transport factors in plants. Many of the factors that are known from animals and yeast have been identified in the *Arabidopsis* genome on the basis of sequence homology. However, their in vivo functions have not been investigated to date. As to the role of nucleocytoplasmic partitioning of proteins involved in the signal transduction of light, several interesting questions are waiting to be solved. To begin with the photoreceptors, the molecular mechanisms of the light-dependent relocalization of phytochromes between the nucleus and the cytoplasm are yet unknown. It is also unclear whether import of phytochromes into the nucleus depends upon importin α/β heterodimers or upon different import receptors. The mechanism that mediates cytoplasmic retention in the dark and the release of phytochrome in the light is also a crucial element of phototransduction. Yet we know nearly nothing about it. To answer these questions will be challenging but they hold the key to understanding the regulation of phytochrome-mediated signalling. In addition, apart from the nature of the speckles formed after the import of phytochromes in the nucleus and from the molecular mechanisms of the nuclear function of phytochromes, a very interesting question is whether phytochromes are degraded in the nucleus or are, at least in part, transported back to the cytoplasm. As a consequence of the latter hypothesis, at least a portion of the phytochrome pool in a cell would then show light-dependent shuttling between the nucleus and the cytoplasm. Retention mechanisms and nucleocytoplasmic shuttling are also postulated in case of the bZIP transcription factor CPRF2 and of the photomorphogenic repressor COP1, to give only two examples. The molecular mechanisms of these processes may differ from the corresponding processes of phytochromes. However, it would be interesting to know if some of the components that confer light-dependent regulation to nucleocytoplasmic partitioning of phytochromes, CPRF2 and COP1 are shared.

Acknowledgements

The work in Freiburg was supported by grants from the Graduiertenkolleg, SFB388, and DFG to K.H. and E.S., respectively, and from the Humboldt Research Award to F.N. The work in Hungary was supported by a Howard Hughes International Scholar Fellowship, HFSPO, DFG and OTKA grant T-0 32565 to F.N.

References

1. M. Neff, C. Fankhauser, J. Chory (2000). Light: an indicator of time and place. *Genes Dev.*, **14**, 257–271.
2. H. Smith (2000). Phytochromes and light signal perception by plants – an emerging synthesis. *Nature*, **407**, 585–591.
3. E. Wellmann (1983). UV irradiation in photomorphogenesis. In: H. Mohr, W. Shropshire (Eds), *Encyclopedia of Plant Physiology* (pp. 745–756). Springer Verlag, Heidelberg.
4. M. Ahmad, A.R. Cashmore (1993). HY4 gene of *A. thaliana* encodes a protein with characteristics of a blue-light photoreceptor. *Nature*, **336**, 162–166.
5. J.M. Christie, P. Reymond, G.K. Powell, P. Bernasconi, A.A. Raibekas, E. Liscum, W.R. Briggs (1998). *Arabidopsis* NPH1: A flavoprotein with the properties of a photoreceptor for phototropism. *Science*, **282**, 1698–1701.
6. C.T. Lin, H. Yang, H. Guo, T. Mockler, J. Chen, A.R. Cashmore (1998). Enhancement of blue-light sensitivity of Arabidopsis seedlings by a blue light receptor cryptochrome 2. *Proc. Natl. Acad. Sci. U.S.A.*, **95**, 2686–2690.
7. J.A. Jarillo, H. Gabrys, J. Capel, J.M. Alonso, J.R. Ecker, A.R. Cashmore (2001). Phototropin-related NPL1 controls chloroplast relocation induced by light. *Nature*, **410**, 952–954.
8. T. Kagawa, T. Sakai, N. Suetsugu, K. Oikawa, S. Ishiguro, T. Kato, S. Tabata, K. Okada, M. Wada (2001). Arabidopsis NPL1: a phototropin homolog controlling the chloroplast high-light avoidance response. *Science*, **291**, 2138–2141.

9. W.R. Briggs, C.F. Beck, A.R. Cashmore, J.M. Christie, J. Hughes, J.A. Jarillo, T. Kagawa, H. Kanegae, E. Liscum, A. Nagatani, K. Okada, M. Salomon, W. Rüdiger, T. Sakai, M. Takano, M. Wada, J.C. Watson (2001). The phototropin family of photoreceptors. *Plant Cell*, **13**, 993–997.

10. S. Mathews, R.A. Sharrock (1997). Phytochrome gene diversity. *Plant Cell Environ.*, **20**, 666–671.

11. R.A. Sharrock, P.H. Quail (1989). Novel phytochrome sequences in *Arabidopsis thaliana*: structure, evolution and differential expression of a plant regulatory photoreceptor family. *Genes Dev.*, **3**, 1745–1757.

12. T. Clack, S. Mathews, R.A. Sharrock (1994). The phytochrome apoprotein family in *Arabidopsis* is encoded by five genes - the sequences and expression of *PHYD* and *PHYE*. *Plant Mol. Biol.*, **25**, 413–427.

13. D.M. Somers, P.H. Quail (1995). Temporal and spatial expression patterns of *PHYA* and *PHYB* genes in *Arabidopsis*. *Plant J.*, **7**, 413–427.

14. L. Goosey, L. Palecanda, R.A. Sharrock (1997). Differential patterns of expression of the *Arabidopsis* PHYB, PHYD, and PHYE phytochrome genes. *Plant Phys.*, **115**, 959–969.

15. P.H. Quail, M.T. Boylan, B.M. Parks, T.W. Short, Y. Xu, D. Wagner (1995). Phytochromes: photosensory perception and signal transduction. *Science*, **268**, 675–680.

16. G.C. Whitelam, P.F. Devlin (1997). Roles of different phytochromes in *Arabidopsis* photomorphogenesis. *Plant Cell Environ.*, **20**, 752–758.

17. A.R. Cashmore, J.A. Jarillo, Y.J. Wu, D. Liu (1999). Cryptochromes: blue light receptors for plants and animals. *Science*, **284**, 760–765.

18. M. Furuya, E. Schäfer (1996). Photoperception and signaling of induction reactions by different phytochromes. *Trends Plant Sci.*, **1**, 301–307.

19. K.M. Hartmann, A. Mollwo, A. Tebbe (1998). Photocontrol of germination by moon- and starlight. *Z. Pfl. Krankh. Pfl. Schutz Sonderh.*, **XVI**, 119–127.

20. K. Eichenberg, I. Bäurle, N. Paulo, R.A. Sharrock, W. Rüdiger, E. Schäfer (2000). *Arabidopsis* phytochromes C and E have different spectral characteristics from those of phytochromes A and B. *FEBS Lett.*, **470**, 107–112.

21. K. Eichenberg, T. Kunkel, T. Kretsch, V. Speth, E. Schäfer (1999). In vivo characterization of chimeric phytochromes in yeast. *J. Biol. Chem.*, **274**, 354–359.

22. J.W. Reed, P. Nagpal, D.S. Poole, M. Furuya, J. Chory (1993). Mutations in the gene for red/far-red light receptor phytochrome B alter cell elongation and physiological responses throughout *Arabidopsis* development. *Plant Cell*, **5**, 147–157.

23. H. Smith (1994). Sensing the light environment: the function of the phytochrome family. In: R.E. Kendrick, G.M.H. Kronenberg (Eds), *Photomorphogenesis in Plants* (pp. 377–416). Kluwer Academic Publishers, Dordrecht.

24. H. Smith (1995). Physiological and ecological function within the phytochrome family. *Annu. Rev. Plant. Physiol. Plant Mol. Biol.*, **46**, 289–315.

25. M.J. Aukerman, M. Hirschfeld, L. Wester, M. Weaver, T. Clark, K. Okada, R.A. Sharrock (1997). A deletion in the *PHYD* gene of the *Arabidopis Wssilewkija* ecotype defines a role for phytochrome D in red/far-red light sensing. *Plant Cell*, **9**, 1317–1326.

26. P.F. Devlin, P.R. Robson, S.R. Patel, L. Goosey, R.A. Sharrock, G.C. Whitelam (1999). Phytochrome D acts in the shade-avoidance syndrome in *Arabidopsis* by controlling elongation growth and flowering time. *Plant Phys.*, **119**, 909–915.

27. H. Mohr (1994). Coaction between pigment systems. In: R.E. Kendrick, G.M.H. Kronenberg (Eds), *Photomorphogenesis in Plants* (pp. 353–376). Kluwer Academic Publishers, Dordrecht.

28. M. Neff, J. Chory (1998). Genetic interactions between phytochrome A, phytochrome B and cryptochrome 1 during *Arabidopsis* development. *Plant Physiol.*, **118**, 27–35.

29. L. Hennig, C. Poppe, S. Unger, E. Schäfer (1999). Control of hypocotyl elongation in *Arabidopsis thaliana* by photoreceptor interaction. *Planta*, **208**, 257–263.

30. L. Hennig, M. Funk, G.C. Whitelam, E. Schäfer (1999). Functional interaction of cryptochrome 1 and phytochrome D. *Plant J.*, **20**, 289–294.

31. D. Wagner, C. Fairchild, R. Kuhn, P.H. Quail (1996). Chromophore-bearing NH$_2$-terminal domains of phytochromes A and B determine their photosensory specificity and differential light lability. *Proc. Natl. Acad. Sci. U.S.A.*, **93**, 4011–4015.

32. H. Guo, T. Mockler, H. Duong, C.T. Lin (2001). SUB1, an *Arabidopsis* Ca^{2+}-binding protein involved in cryptochrome and phytochrome coaction. *Science*, **291**, 487–490.

33. R.E. Kendrick, G.M.H. Kronenberg (Eds) (1994). *Photomorphogenesis in Plants*. Kluwer Academic Publishers, Dordrecht.

34. M. Kraml (1994). Light direction and polarization. In: R.E. Kendrick, G.M.H. Kronenberg (Eds), *Photomorphogenesis in Plants* (pp. 417–446). Kluwer Academic Publishers, Dordrecht.

35. K.C. Yeh, S.H. Wu, J.T. Murphy, J.C. Lagarias (1997). A cyanobacterial phytochrome two-component light sensory system. *Science*, **277**, 1505–1508.

36. K. Harter, H. Frohnmeyer, S. Kircher, T. Kunkel, S. Mühlbauer, E. Schäfer (1994). Light induces rapid changes of the phosphorylation pattern in the cytosol of evacuolated parsley protoplasts. *Proc. Natl. Acad. Sci. U.S.A.*, **91**, 5038–5042.

37. K.C. Yeh, J.C. Lagarias (1998). Eukaryotic phytochromes: light-regulated serine/threonine protein kinases with histidine kinase ancestry. *Proc. Natl. Acad. Sci. U.S.A.*, **95**, 13976–13981.

38. G. Neuhaus, C. Bowler, R. Kern, N.H. Chua (1993). Calcium/calmodulin-dependent and -independent phytochrome signal transduction pathways. *Cell*, **73**, 937–952.

39. C. Bowler, G. Neuhaus, H. Yamagata, N.H. Chua (1994). Cyclic GMP and calcium mediate phytochrome phototransduction. *Cell*, **77**, 73–81.

40. T. Kunkel, G. Neuhaus, A. Batschauer, N.H. Chua, E. Schäfer (1996). Functional analysis of yeast-derived phytochrome A and B phycocyanobilin adducts. *Plant J.*, **10**, 625–636.

41. M.-S. Soh, Y.-M. Kim, S.-J. Han, P.-S. Song (2000). REP1, a basic helix-loop-helix protein, is required for a branch pathway of phytochrome A signaling in *Arabidopsis*. *Plant Cell*, **12**, 2061–2073.

42. M. Koornneef, E. Rolff, C. Spruit (1980). Genetic control of light-inhibited hypocotyl elongation in *Arabidopsis thaliana (L.)*, *Heynh. Z. Pflanzenphysiol.*, **100**, 147–160.

43. J. Chory, C. Peto, R. Feinbaum, L.H. Pratt, F. Ausubel (1989). *Arabidopsis thaliana* mutant that develops as light-grown plant in the absence of light. *Cell*, **58**, 991–999.

44. X.W. Deng, T. Caspar, P.H. Quail (1991). cop1: a regulatory locus involved in light-controlled development and gene expression in *Arabidopsis*. *Genes Dev.*, **5**, 1172–1182.

45. N. Wei, X.W. Deng (1999). Making sense of the COP9 signalosome: a regulatory protein complex conserved from *Arabidopsis* to human. *Trends Genet.*, **15**, 98–103.
46. D.A. Chamovitz, N. Wei, M.T. Osterlund, A.G. von Armin, J.M. Staub, M. Matsui, X.W. Deng (1996). The COP9 complex, a novel multisubunit nuclear regulator involved in light control of a plant developmental switch. *Cell*, **86**, 115–121.
47. A. von Armin, X.-W. Deng (1994). Light inactivation of Arabidopsis photomorphogenic repressor COP1 involves a cell type specific modulation of its nucleocytoplasmic partitioning. *Cell*, **79**, 1035–1045.
48. G. Stacey, N. Hicks, A. von Armin (1999). Discrete domains mediate the light-responsive nuclear and cytoplasmic localisation of Arabidopsis COP1. *Plant Cell*, **11**, 349–363.
49. G. Stacey, A. von Armin (1999). A novel motif mediates the targeting of the Arabidopsis COP1 protein to subnuclear foci. *J. Biol Chem.*, **274**, 27231–27236.
50. T. Oyama, Y. Shimura, K. Okada (1997). The *Arabidopsis HY5* gene encodes a bZIP protein that regulates stimulus-induced development of root and hypocotyl. *Genes Dev.*, **11**, 2983–2995.
51. L.-H. Ang, S. Chattopadhyay, N. Wei, T. Oyama, K. Okada, A. Batschauer, X.W. Deng (1998). Molecular interaction between COP1 and HY5 defines a regulatory switch for light control of *Arabidopsis* development. *Mol. Cell.*, **1**, 213–222.
52. M.T. Osterlund, C.S. Hardtke, N. Wei, X.W. Deng (2000). Targeted destabilization of HY5 during light-regulated development of *Arabidopsis*. *Nature*, **405**, 462–466.
53. C.S. Hardtke, X.W. Deng (2000). The cell biology of the COP/DET/FUS proteins. Regulating proteolysis in photomorphogenesis and beyond? *Plant Phys.*, **124**, 1548–1557.
54. X.W. Deng, P.H. Quail (1999). Signaling in light-controlled development. *Semin. Cell Dev. Biol.*, **10**, 121–129.
55. C. Fankhauser, J. Chory (1997). Light control of plant development. *Annu. Rev. Cell Dev. Biol.*, **13**, 203–229.
56. D. Wagner, U. Hoecker, P.H. Quail (1997). RED1 is necessary for phytochrome B-mediated red light-specific signal transduction in *Arabidopsis*. *Plant Cell*, **9**, 731–743.
57. M. Ahmad, A.R. Cashmore (1996). The pef mutants of *Arabidopsis thaliana* define lesions early in the phytochrome signaling pathway. *Plant J.*, **10**, 1103–1110.
58. J.W. Reed, P. Nagpal, D.S. Poole, M. Furuya, J. Chory (1993). Mutations in the gene for the red/far-red light receptor phytochrome B alter cell elongation and physiological responses throughout *Arabidopsis* development. *Plant Cell*, **5**, 147–157.
59. T. Kretsch, C. Poppe, E. Schäfer (2000). A new type of mutation in the plant photoreceptor phytochrome B causes loss of photoreversibility and an extremely enhanced light sensitivity. *Plant J.*, **22**, 177–186.
60. G.C. Whitelam, E. Johnson, J. Peng, P. Carol, M.L. Anderson, J.S. Cowl, N.P. Harberd (1993). Phytochrome A null mutants of *Arabidopsis* display a wild-type phenotype in white light. *Plant Cell*, **5**, 757–768.
61. M.S. Soh, S.H. Hong, H. Hanzawa, M. Furuya, H.G. Nam (1998). Genetic identification of FIN2, a far red light-specific signaling component of *Arabidopsis thaliana*. *Plant J.*, **16**, 411–419.
62. U. Hoecker, Y. Xu, P.H. Quail (1998). SPA1: A new genetic locus involved in phytochrome A-specific signal transduction. *Plant Cell*, **10**, 19–33.

63. U. Hoecker, J.M. Teppermann, P.H. Quail (1999). SPA1, a WD-repeat protein specific to phytochrome A signal transduction. *Science*, **284**, 496–499.

64. M. Hudson, C. Ringli, M.T. Boylan, P.H. Quail (1999). The FAR1 locus encodes a novel nuclear protein specific to phytochrome A signaling. *Genes Dev.*, **13**, 2017–2027.

65. C. Bolle, C. Koncz, N.H. Chua (2000). PAT1, a new member of the GRAS family, is involved in phytochrome A signal transduction. *Genes Dev.*, **14**, 1269–1278.

66. L.D. Pysh, J.W. Wysocka-Diller, C. Camilleri, D. Bouchez, P.N. Benfey (1999). The GRAS gene family in *Arabidopsis*: Sequence characterization and basic expression analysis of the SCARECROW-LIKE genes. *Plant J.*, **18**, 111–119.

67. C. Büche, C. Poppe, E. Schäfer, T. Kretsch (2000). *Eid1*: A new *Arabidopsis* mutant hypersensitive in phytochrome A-dependent high-irradiance responses. *Plant Cell*, **12**, 547–558.

68. M. Dieterle, Y.-C. Zhou, E. Schäfer, M. Funk, T. Kretsch (2001). *Eid1*, an F-box protein involved in phytochrome A-specific light signaling. *Genes Develop.*, **15**, 939–944.

69. T. Genoud, A.J. Millar, N. Nishizawa, S.A. Kay, E. Schäfer, A. Nagatani, N.-H. Chua (1998). An *Arabidopsis* mutant hypersensitive to red and far-red light signals. *Plant Cell*, **10**, 889–904.

70. V. Speth, V. Otto, E. Schäfer (1987). Intracellular localisation of phytochrome in oat coleoptyles by electron microscopy. *Planta*, **168**, 299–304.

71. L.H. Pratt (1994). Distribution and localization of phytochrome within the plant. In: R.E. Kendrick, G.M.H. Kronenberg (Eds), *Photomorphogenesis in Plants* (pp.163–186). Kluwer Academic Publishers, Dordrecht.

72. E. Mösinger, A. Batschauer, R.D. Vierstra, K. Apel, E. Schäfer (1987). Comparison of the effects of exogenous native phytochrome and in vivo irradiation on in vitro transcription in isolated nuclei from barley (*Hordeum vulgare*). *Planta*, **170**, 505–514.

73. K. Sakamoto, A. Nagatani (1996). Nuclear localization activity of phytochrome B. *Plant J.*, **10**, 859–868.

74. R. Yamaguchi, M. Nakamura, N. Mochizuki, S.A. Kay, A. Nagatani (1999). Light-dependent translocation of a phytochrome B-GFP fusion protein to the nucleus in transgenic *Arabidopsis*. *J. Cell Biol.*, **145**, 437–445.

75. S. Kircher, L. Kozma-Bognar, L. Kim, E. Adam, K. Harter, E. Schäfer, F. Nagy (1999). Light quality-dependent nuclear import of the plant photoreceptors phytochrome-A and B. *Plant Cell*, **11**, 1445–1456.

76. P. Gil, S. Kircher, E. Adam, E. Bury, L. Kozma-Bognar, E. Schäfer, F. Nagy (2000). Photocontrol of subcellular partitioning of phytochrome-B:GFP fusion protein in tobacco seedlings. *Plant J.*, **22**, 135–145.

77. L. Kim, S. Kircher, R. Toth, E. Adam, E. Schäfer, F. Nagy (2000). Light-induced nuclear import of phytochrome-A:GFP fusion proteins is differentially regulated in transgenic tobacco and *Arabidopsis*. *Plant J.*, **22**, 125–134.

78. A. Hisada, H. Hanzawa, J.L. Weller, A. Nagatani, J.B. Reid, M. Furuya (2000). Light-induced nuclear translocation of endogenous pea phytochrome A visualized by immunocytochemical procedures. *Plant Cell*, **12**, 1063–1078.

79. F. Nagy, S. Kircher, E. Schäfer (2000). Nucleo-cytoplasmic partitioning of the plant photoreceptors phytochromes. *Semin. Cell Dev. Biol.*, **11**, 505–510.

80. K.J. Halliday, M. Hudson, M. Ni, M. Qin, P.H. Quail (1999). *poc1*: An *Arabidopsis* mutant perturbed in phytochrome signaling because of a T DNA insertion in the promoter of PIF3, a gene encoding a phytochrome-interacting bHLH protein. *Proc. Natl. Acad. Sci. U.S.A.*, **96**, 5832–5837.

81. M. Ni, J.M. Tepperman, P.H. Quail (1998). PIF3, a phytochrome-interacting factor necessary for normal photoinduced signal transduction, is a novel basic helix-loop protein. *Cell*, **95**, 657–667.

82. C. Fankhauser, K.C. Yeh, J.C. Lagarias, H. Zhang, T.D. Elich, J. Chory (1999). PKS1, a substrate phosphorylated by phytochrome that modulates light signaling in *Arabidopsis*. *Science*, **284**, 1539–1541.

83. G. Choi, H. Yi, J. Lee, Y.K. Kwon, M.-S. Soh, B. Shin, Z. Luka, T.R. Hahn, P.-S. Song (1999). Phytochrome signaling is mediated through nucleoside diphosphate kinase 2. *Nature*, **401**, 610–613.

84. M. Ni, J.M. Tepperman, P.H. Quail (1999). Binding of phytochrome B to its nuclear signalling partner PIF3 is reversibly induced by light *Nature*, **400**, 781–784.

85. J.F. Martinez-Garcia, E. Huq, P.H. Quail (2000). Direct targeting of light signals to a promoter element-bound transcription factor. *Science*, **288**, 859–863.

86. C.D. Fairchild, M.A. Schumaker, P.H. Quail (2000). HFR1 encodes an atypical bHLH protein that acts in phytochrome A signal transduction. *Genes Dev.*, **14**, 2377–2391.

87. H.-L. Hsieh, H. Okamoto, M. Wang, L.-H. Ang, M. Matsui, H. Goodman, X. W. Deng (2000). FIN219, an auxin-regulated gene, defines a link between phytochrome A and the downstream regulator COP1 in light control of *Arabidopsis* development. *Genes Dev.*, **14**, 1958–1970.

88. I.B. D'Agostino, J.J. Kieber (1999). Posphorelay signal transduction: the emerging family of plant response regulators. *Trends Biochem. Sci.*, **24**, 452–456.

89. U. Sweere, K. Eichenberg, J. Lohrmann, V. Mira-Rodado, I. Bäuerle, J. Kudla, F. Nagy E. Schäfer, K. Harter (2001). Interaction of the response regulator ARR4 with phytochrome B in modulating red light signaling. *Science*, **294**, 1108–1111.

90. L.C. Romero, B. Biswal, P.-S. Song (1991). Protein phosphorylation in isolated nuclei from etiolated Avena seedlings, Effects of red/far-red light and cholera toxin. *FEBS Lett.*, **282**, 347–350.

91. V.N. Lapko, X.-Y. Jiang, D.L. Smith, P.-S. Song (1991). Mass spectrometric characterization of oat phytochrome A: Isoforms and posttranslational modifications. *Protein Sci.*, **8**, 1032–1044.

92. B. Weißhaar, G.A. Armstrong, A. Block, O. da Costa e Silva, K. Hahlbrock (1991). Light-inducible and constitutively expressed DNA-binding proteins recognizing a plant promotor element with functional relevance in light responsivness. *EMBO J.*, **10**, 1777–1786.

93. K. Harter, S. Kircher, H. Frohnmeyer, M. Krenz, F. Nagy, E. Schäfer (1994). Light-regulated modification and nuclear translocation of cytosolic G-box binding factors in parsley. *Plant Cell*, **6**, 545–559.

94. S. Kircher, F. Wellmer, P. Nick, A. Rügner, E. Schäfer, K. Harter (1999). Nuclear import of the parsley bZIP transcription factor CPRF2 is regulated by phytochrome photoreceptors. *J. Cell Biol.*, **144**, 201–211.

95. F. Wellmer, S. Kircher, A. Rügner, H. Frohnmeyer, E. Schäfer, K. Harter (1994). Phosphorylation of the parsley bZIP transcription factor CPRF2 is regulated by light. *J. Biol. Chem.*, **274**, 29476–29482.

96. S. Kircher, P. Gil, L. Kozma-Bognár, E. Fejes, V. Speth, T. Husselstein, E. Bury, É. Ádam, E. Schäffer, F. Nagy (2002). Nucleo-cytoplasmic partitioning of the plant photoreceptors phytochrome A, B, C, D is differentially regulated by light and exhibits a diurnal rhythm. *Plant Cell*, **74**, 1541–1544.

Chapter 7

Phytochromes and phytochrome-like proteins in cyanobacteria

Tilman Lamparter and Jon Hughes

Table of contents

Abstract

In the last few years sequencing projects have uncovered numerous prokary-otic genes which encode domains resembling the chromophore-bearing sensory module of plant phytochromes. The majority are from cyanobacteria, where up to 13 different proteins with such a domain have been found within a single species. Besides the chromophore region, many of these proteins contain PAS- and histidine kinase-like domains, both of which are also present in plant phytochromes. However, the diversity of this phytochrome-superfamily with respect to overall domain organisation and primary structure is much greater than that seen in plant phytochromes, thus offering a rich variety of strategies for research into the evolution, structure and function of the group as a whole. Based on sequence similarities and functional studies, two subgroups can be distinguished, the phytochrome-like proteins and the true phytochromes, the latter being defined by red/far-red photochromicity. To date the best analysed amongst the prokaryotic true phytochromes is Cph1 from *Synechocystis* PCC 6803. Not only the primary structure but also many spectral and biochemical properties of Cph1 are strikingly similar to those of plant phytochromes. Yet equally clear and perhaps even more interesting are the differences. Studies of more distantly-related phytochromes are also progressing rapidly, providing new vantage points for viewing structure–function relationships in the phytochrome superfamily as a whole.

7.1   Introduction

Plant phytochromes were characterised in a classically deductive manner from numerous remarkably successful physiological, biochemical and biophysical investigations. Action spectroscopy had revealed light responses in which red and far-red light acted antagonistically. These physiological observations culminated in the finding that from a sequence of alternating red/far-red pulses the last pulse, so to speak, wins. Based on these observations it was suggested that the responses were mediated by a red/far-red photochromic photorecep-tor, rather than two distinct competing photoreceptors. Even then it was speculated that the chromophore was likely to be an open-chain tetrapyrrole (bilin) [1]. The spectral identification of phytochrome by red/far-red difference spectroscopy in the late 1950s [2] was the logical consequence of these physiological studies, while this assay procedure in turn allowed biochemical characterisation and purification of the chromoprotein as biochemical meth-ods improved. A considerable amount of biochemical work confirmed the early suggestion that the chromophore was a linear tetrapyrrole (phytochro-mobilin), and showed that a Z to E isomerisation between ring C and D was an important step in the course of photoconversion from the red-absorbing Pr form into the far-red-absorbing Pfr form [3]. In the early 1980s the first phytochrome cDNA – *PHYA* from oat – was cloned in a tour de force of molecular methods [4], providing access to the other members of the family in a large number of species – data which has been valuable in establishing the phylogeny of the plant kingdom as a whole.

Cyanobacteria produce bilins in considerable quantities as chromophores for their photosynthetic antenna, so they would certainly be able to produce a holoprotein given a phytochrome gene. In contrast to the large number of conventional light responses, there have been few reports of typical red/far-red antagonistic effects in cyanobacteria. We know of only two brief papers on photoreversible spore germination that appeared in the 1970s and 1980s [5,6]; a third paper on phototaxis appeared only recently [7]. Hence it was generally assumed that phytochromes are of eukaryotic origin and do not exist in prokaryotes.

It would nevertheless be wrong at this point to ignore the remarkably percipient observation of Schneider-Poetsch and colleagues [8,9] that the C-terminal portion of plant phytochromes showed similarities to the then newly-discovered "two-component" family prokaryotic sensory histidine-kinases. This contribution might have been received more warmly by the phytochrome community had the all-important histidine target residue been as well conserved as many others clearly involved in the phosphorylation mechanism. Phytochromes, it seemed, were created by plants de novo.

The discovery of prokaryotic phytochromes awaited the era of genomic sequencing [10] and molecular genetic approaches like complementation [11], targeted knockout [12], and recombinant expression [13–15]. The sequence of the *Synechocystis* PCC6803 chromosome was an early triumph of high-throughput technology at the Kazusa Institute in Japan (http://www.kazusa.-or.jp/cyano/cyano.html). It revealed an open reading frame encoding a protein with clear similarities both to sensory histidine kinases and to primitive plant phytochromes [10,16]. Indeed, we were able to show that this apoprotein formed a typically red/far-red photochromic holophytochrome once presented with an appropriate bilin chromophore [13]. Thus, in a manner reflecting a fundamental change in our approach to biology, the history of cyanobacterial phytochrome is rather the opposite of that of plant phytochromes: first came the discovery of the gene, second came the spectral characterisation of the photoreceptor, whereas the third step, identifying its physiological role in cyanobacteria, still has to be made.

Cyanobacteria offer a wealth of phytochrome-related sequences that allow speculation regarding the possible origin of the phytochromes and deduce functions of the holoprotein molecule by comparative studies. Indeed, the discovery of a prokaryotic phytochrome caused a major paradigm shift in the field, within five years contributing substantially to our present understanding of phytochrome function. As research subjects, cyanobacteria have major advantages over plants. Their structure and organisation – both at the pheno-typic and genetic levels – are far simpler. Moreover, high rates of homologous recombination in prokaryotes make gene targeting and gene replacement straightforward. Thus we expect our knowledge of phytochrome function in prokaryotes will soon catch up with that from half a century of plant research, thereby allowing even deeper insight into phytochrome action and evolution – including that in plants.

7.2 What is a phytochrome?

Almost all known phytochromes from higher plants have a domain arrangement similar to that of phytochrome A shown in Figure 1. No aberrantly-constructed phytochromes are apparent in the *Arabidopsis* genome. This conservative pattern has already been questioned by two phytochrome sequences from the cryptogam world: an abnormal modular design was found for phytochrome 3 (also known as superchrome) from the fern *Adiantum capillusveneris* [17], and phytochrome 1 from the moss *Ceratodon purpureus* [18]. These sequences are exceptional, having arisen by gene duplication and rearrangement during the evolution of genera or species. In prokaryotes, especially cyanobacteria, the diversity of phytochrome-like proteins and of their modular design is much greater. Amongst the ~3500 genes of *Synechocystis* PCC6803 there are no less than 12 proteins with significant similarities to the chromophore-bearing sensory module of plant phytochromes, while in *Anabaena* PCC7120, 13 are apparent (both genomes having been sequenced in their entirety – see Kazusa website). This diversity makes it now important to define the term phytochrome more precisely. We propose that the name phytochrome be functionally defined for **red/far-red photochromic** holoproteins with **bilin** chromophores. The others should be termed phytochrome-like proteins. The same distinction must be made if the names "bacteriophytochromes" or "cyanochromes" are to be used, although we ourselves prefer to avoid this terminological metastasis. The division into "phytochromes" and "phytochrome-like proteins" is further supported by the evolution tree (Figure 3, below) and details discussed below.

7.3 Prokaryotic light physiology related to phytochrome

7.3.1 Photoreversible effects

Reddy and Talpasayi [6] analysed the germination of spores of *Anabaena fertilissima* and found a strong light dependence. White light induced germination to almost 100%, a ~50% induction was found with red light. The red light effect was reverted by far-red, a clear indication for a phytochrome response. Unfortunately, the work was not followed up and until recently our knowledge of red/far-red photoreversible effects was limited to that.

More attention was given to photoreversible effects that are sensitive in the green and red region of the spectrum. There are several examples where green ($\lambda_{max} \approx 520$ nm) induces and red ($\lambda_{max} \approx 650$ nm) reverts this induction [19,20].

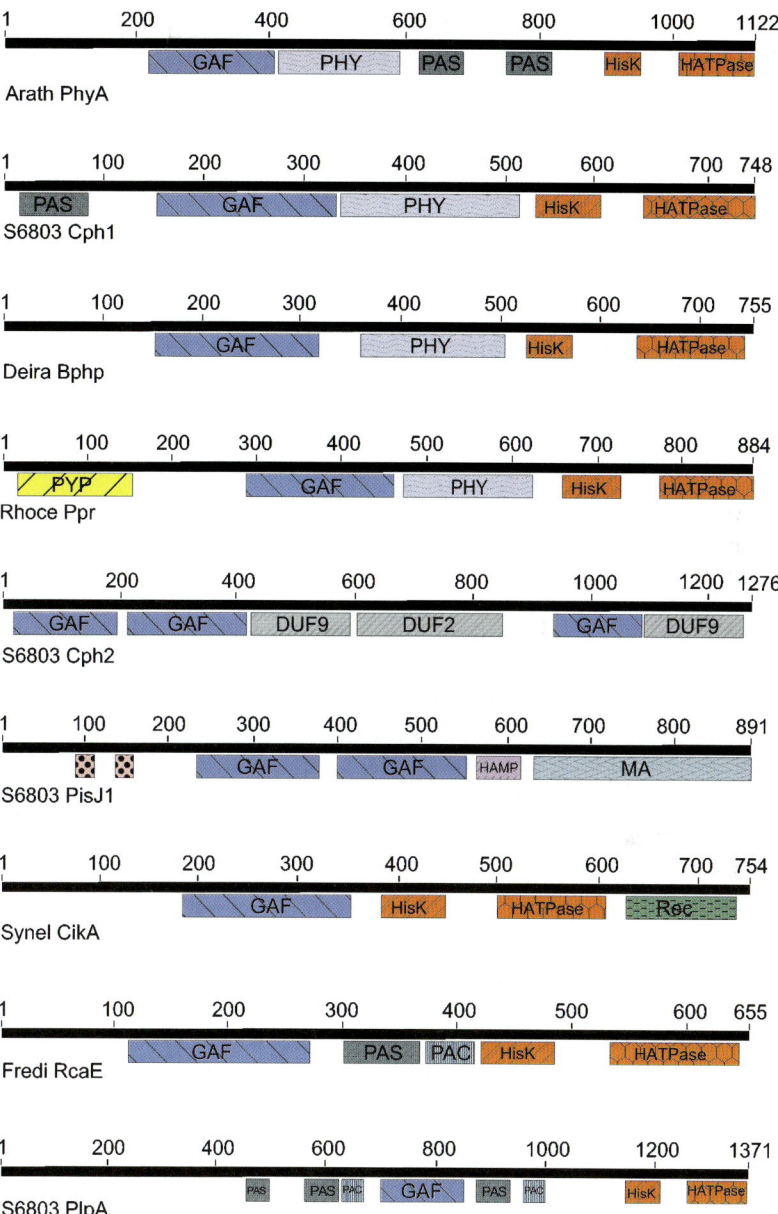

Figure 1. Domain structure of selected phytochromes and phytochrome-like proteins. The NCBI BLASTP program (http://www.ncbi.nlm.nih.gov/blast/Blast.cgi) was used for graphical display of the domain arrangement of the different proteins. The NCBI software uses SMART and PFAM algorithms/databases and displays domains from both separately. The present figure was redrawn from these displays, omitting redundant information. Species names are abbreviated by 3+2 letters or 1 letter + PCC number. The list of abbreviations is given in the legend of Figure 2, where more peptidesare shown.

The most intensely studied effect here is complementary chromatic adaptation (CCA). Unlike plants, cyanobacteria possess phycobilisome structures containing chromoprotein accessory pigments which funnel energy into the photosynthetic system. The principle chromoproteins involved are the blue–green (red-absorbing) PC and allophycocyanin and the red (blue-green-absorbing) phycoerythrin (PE). Some species are able to use CCA to adjust the ratio of these pigments according to environmental conditions: in green light PE dominates, whereas in red light PC dominates. In this way, the λ_{max} of photosynthesis is shifted to the λ_{max} of the light environment. This phenomenon was first described a century ago for *Oscillatoria* in Engelmann's Institute for Physiology in Berlin [21]. More recently, CCA has been studied in the cyanobacterium *Fremyella diplosiphon*. [According to Herdman et al. [53] *Fremyella diplosiphon* is a mutant of *Calothrix* PCC 7601.] When a series of green ($\approx$540 nm) and red ($\approx$650 nm) pulses is given to *Fremyella* and thereafter the culture kept in darkness, the last light pulse determines the dominant accessory pigment formed [20]. This kind of photoreversibility implies photochromicity, and we might thus expect a biliprotein with phytochrome-like properties to be the photoreceptor – although the $\approx$100 nm hypsochromic absorbance shift would need to be explained at the chemical level.

Phototaxis is another effect that has been studied quite extensively. It was shown by the use of electron transport inhibitors like DCMU that phototaxis is independent of photosynthesis and thus likely to be mediated by specific photoreceptors [22]. Recent physiological evidence for the involvement of a red/far-red photoreversible photoreceptor came from studies with *Synechococcus elongatus* [7]. Although the cells display phototaxis over a broad spectral range, induction by red light can be inhibited by far-red given simultaneously. Far-red can also stimulate phototaxis when applied at high fluence rates – and this in turn is inhibited by simultaneous red light irradiation. Other photoreceptors seem to be involved at shorter wavelengths, since far-red light only scarcely inhibits the effect of green or white light.

7.3.2   *Mutants*

Mutant studies provided the first tentative link between phytochrome and photocontrol in cyanobacteria. From *Fremyella* different classes of mutants with defects in chromatic adaptation were isolated based on their colour appearance. Kehoe and Grossman were able to complement some of these mutant strains and to isolate two different genes regulating CCA. These genes were called *rcaC* and *rcaE* (rca = regulator of chromatic adaptation) as they are thought to act in concert with components of the PE operon, *rcaA, rcaB* and *rcaD*. RcaE was the first prokaryotic gene product found to show similarity to portions of the phytochrome chromophore module [11]. Although there is no obvious sequence similarity to the "acid pocket" in which the plant phytochrome chromophore is held, RcaE apparently does carry a bilin chromophore and – when purified from the natural host – is photochromic with absorption maxima in the red and green region of the spectrum (David Kehoe, personal communication). It thus fulfils the above expectations for a photoreceptor for chromatic adaptation. How the red/far-red to green/red shift is achieved physico-chemically will be interesting to discover.

Wilde et al. [12] found a phytochrome-like protein, *plpA*, in *Synechocystis* PCC 6803 and created appropriate knockout mutants. Growth of these mutants was unaffected under glucose supplementation, but reduced by 25% in carbohydrate-free medium when kept in white light. Even with appropriate nutrients, the wild type refuses to grow in total darkness, the cells requiring daily pulses of dim blue light to divide. This "photomorphogenic" response is lost in the *plpA⁻* mutant. Moreover, following adaptation to such a light environment, the mutants showed an altered phycocyanin/chlorophyll ratio relative to the wild type.

Soon thereafter the complete genome sequence of this species became available [10]. As mentioned above, this reveals 12 genes with similarities to the phytochrome chromophore domain, amongst them Cph1 with the highest sequence homology to plant phytochromes. Cph1 knockout mutants were generated by several groups (e.g. A. Wilde et al., HU Berlin; D. Scanlan, Warwick University; I. Suzuki et al., NIBB Okasaki; M. Ikeuchi et al., Tokyo University and T. Ogawa et al., Nagoya University; see also the cyanobase site), yet there is no obvious phenotype correlated with this mutation. A closer examination of the mutants has shown that the content of photosynthetic pigments and their light regulation is unaltered, and the mutants show normal phototactic movements. For unknown reasons, however, their growth is considerably retarded under higher fluence rates of white light (A. Wilde et al., personal communication). This is consistent with an effect noticed earlier by Scanlan that the knockouts were more susceptible to damage from higher irradiances (personal communication). Nevertheless, the actual biochemical role of Cph1 remains unknown. In view of its close relationship to plant phytochromes and its use as a model phytochrome molecule, a physiological function of the Cph1 system is awaited almost with desperation.

Depending on light quality and fluence rates, certain *Synechocystis* PCC 6803 strains can display both positive and negative phototaxis (i.e. move towards the light or away from it). Yoshihara et al. [23] speculated that *sll0041*, a gene that encodes a hybrid protein with phytochrome-like chromophore-domain and a homology to the signalling domain of bacterial methyl-accepting chemotaxis receptors, might play a role in the tactic response. Indeed, when this gene was destroyed by targeted knocking out the cells lost their ability to move towards the light, displaying negative phototaxis instead. The same phenotype was found in *sll0038*, *sll0039* and *sll0042* knockouts, whereas positive phototaxis was weakened after inactivation of *sll0040*. These genes are arranged polycistronically together with *sll0041* and show various similarities to chemotaxis genes. Both *sll0038* and *sll0039* are similar to response regulators like CheY, *sll0043* is similar to the histidine kinase CheA and *sll0040* is similar to CheW, an adaptor of sensory histidine kinases. *sll0042* bears a signalling domain such as known bacterial chemotaxis-sensors but lacks the sensory domain. Since the gene arrangement is comparable to that of the *pil* (= pilus) genes of *Pseudomonas aeruginosa*, the homologous *Synechocystis* genes were designated *pis* (= phototaxis), so that the phytochrome-like gene *sll0041* is now *pisJ1*.

In the cyanobacterium *Synechococcus elongatus*, an elegant luciferase/gene-tagging-based approach has allowed mass screens for mutants with altered circadian rhythmicity [24,25]. Such approaches have gained detailed insight into the mechanism of the first – and as yet only – circadian clock known in a prokaryote. Complementation assays identified several novel *kai* genes involved in the physiological clock (*kai* means "cycle" in Japanese) as well as a phytochrome-homologous gene *cikA* [26]. Again, the question arises whether the protein incorporates a bilin chromophore, but sequence similarity suggests that CikA as well as the other phytochrome-like proteins might be able to attach a chromophore (see below).

Although this review focuses on cyanobacteria, based on close sequence and putative functional similarities, other prokaryotic gene products should be mentioned here. Davis et al. found phytochrome-like sequences in databases of two non-photosynthetic bacteria, *Deinococcus radiodurans* and *Pseudomonas aeruginosa* [27], DrBphP, PpBphP1 and PpBphP2, respectively (BphP stands for "bacteriophytochrome photoreceptor"). When the DrBphP gene was knocked out in *Deinococcus*, light control over synthesis of specific carotenoids was affected. The DrBphP protein expressed in *E. coli* incorporated chromophore *in vitro* and was photochromic. Thus this BphP is most likely a photoreceptor controlling carotenoid biosynthesis. Phytochrome-homologous sequences were also found in the photosynthetic bacterium *Rhodospirillum centenum* as part of a gene encoding a domain resembling photoactive yellow protein, PYP [28] (see Chapter 8 in this book). This gene was thus named *ppr* (PYP-phytochrome-related). Despite the detail in which its structure is now known, the function of PYP itself remains an enigma. Although PYP in *Ectothiorhodospira halophila* might be a photoreceptor for phototaxis [29,30], PYP knockouts in *Rhodobacter spheroides* showed normal phototaxis. Similarly, when *ppr* itself was knocked out in *Rhodospirillum centenum* the phototactic response was not affected. However, expression of chalcone synthase, an early enzyme in the flavonoid biosynthesis pathway, was negatively affected. Jiang et al. could show that the PYP domain incorporates a hydroxycinnamic acid chromophore when expressed in a heterologous system, yet speculate that the phytochrome-domain does not incorporate a chromophore [28].

7.4   Photochromic chromoproteins in cyanobacteria

Before the molecular era, cyanobacterial pigments had been screened for photochromic pigments. As outlined above, scientists did not expect to find typical phytochromes: chromatic adaptation and similar effects implied instead red/green photoreversible pigments. Such a pigment with difference maxima at 520 and 650 nm was isolated by Scheibe from *Tolopothrix tenuis* [31]. Later, various photoreversible "phycochromes" were isolated from different strains by Björn and Björn [32,33]. All these pigments absorb at lower wavelengths than phytochromes, yet the difference spectra fit very poorly to CCA action spectra [34]. Phycochrome b proved later to be phycoerythrocyanin (PEC), an integral component of the phycobilisome in some cyanobacteria. This was the last of the phycobiliproteins to be characterised [35,36]. Like the others it consists of three α/β heterodimers arranged as a triad. The β-subunit carries two PCB residues and is spectrally normal whereas PECα carries a PVB chromophore and is photochromic with absorbance difference maxima at 592 (orange) and 529 (green) nm [37,38]. However, relative to the ground state, the photochromicity in PEC is associated with an orange/green (hypsochromic) shift, whereas in the case of phytochrome a red/far-red (bathochromic) shift is seen. This implies a very different photoconversion mechanism. Furthermore, there is no significant sequence similarity between PEC and phytochrome. Thus, although at first sight PEC might seem to provide a useful model for phytochrome – and might indeed be a photoreceptor [39] – there are good reasons for doubting the analogy.

Biophysical and biochemical studies of Cph1 from recombinant *E. coli*-derived apoprotein assembled in vitro with PCB have been remarkably successful [13]. Much effort has been expended in establishing the spectral characteristics of Cph1 holoprotein in *Synechocystis* itself. Attempts to detect typical phytochrome red/far-red photoreversibility in *Synechocystis* by conventional methods failed (Wilde & Lamparter, unpublished), one reason being the high concentration of red-light-absorbing photosynthetic pigments. Even measurements with a PC⁻ mutant with low pigment content were fruitless. Success was recently achieved by modifying Cph1 in its natural host with an oligohistidine tag to allow nickel-affinity purification. With Cph1 under the control of its natural promoter, extracts from a 10 litre culture had to be concentrated to ~1 ml before difference spectroscopy became possible. The number of spectrally-active Cph1 molecules calculated on that basis is only ~25 per cell. Even under the control of the strong *psbA* promoter, this number was only about ten-fold higher [15]. These figures probably underestimate the in vivo situation considerably, but illustrate that Cph1-phytochrome concentrations are probably low and explain why spectral detection by conventional methods failed. Since difference spectra were almost indistinguishable from that of the recombinant Cph1 adduct assembled *in vitro*, it is thought that PCB is the phytochrome chromophore in the natural host.

7.5 Non-photochromic phytochrome-homologous chromoproteins

Hidden Marcov model (HMM) algorithms revealed that the chromophore domain of phytochromes overlaps at least partially with a protein domain called GAF [40]. Among the 12 phytochrome-related proteins encoded in *Synechocystis* PCC 6803, Cph2 is particularly interesting. It bears two GAF domains, which might serve as chromophore binding sites [41,42]. Both have been analysed by heterologous expression. Domain 1, which is still quite closely related to plant phytochromes and Cph1 (see below), forms a photoreversible adduct with spectral properties similar to those of plant phytochromes (thus the name cyanobacterial phytochrome 2). The second, lower-homology domain was also shown to attach a PCB chromophore auto-catalytically, *in vitro*, but is unusual in two important respects. Two absor-bance bands are seen in the visible region, that in the blue region being much stronger than another in the red, contrasting with the situation in normal phy-tochromes. In addition, it is not photochromic [42]. Cph2 might thus be able to sense both blue and red light with comparable efficiency. In this respect it resembles *Adiantum* phytochrome 3 (superchrome) mentioned above, which besides the photochromic phytochrome domain is also able to bind flavins via LOV domains similar to those of plant phototropins (see Chapter 11). Thus superchrome too would be able to sense blue and red/far-red light, albeit by a quite different photochemical process. The Cph2 domain 2 data suggest that other phytochrome-like proteins with similar primary structure might also be non-photochromic and/or show blue sensitivity. This is supported by the finding that the absorption spectrum of the heterologously expressed *slr1969* gene product is similar to that of Cph2 domain 2, as suggested by Wu and Lagarias [42].

It seems that RcaE (see above) when purified from the natural host carries a bilin chromophore, rendering it a likely candidate for the chromatic adaptation photoreceptor. The recombinant protein expressed in *E. coli* seems also able to attach bilin chromophores, although not necessarily in the same manner as in *Fremyella*. As proposed photoreceptor for chromatic adaptation, one would expect a red-green photoreversible chromoprotein. As yet, however, photoreversibility of RcaE could not be demonstrated. An essential factor, such as a missing or weakly-bound second chromophore might have been lost during purification. In any case, the sensor for chromatic adaptation need not be a photochromic photoreceptor. Instead the system might use two (or more) conventional photoreceptors.

At this point a cautionary note is appropriate. Unusual and fascinating though photochromicity in phytochromes may be, for some reason many bili-proteins display this characteristic during denaturation [43]. As cyanobacteria contain very large quantities of photosynthetic biliproteins it is always possible that a denatured – and thereby photochromic – biliprotein might contaminate an extract and lead to false-positive identification. Thus, irrespective of the extraction and purification methods used, appropriate controls should be included and documented.

7.6   Evolution of phytochromes

In the alignment shown in Figure 2, phytochrome-like proteins that are more distantly related to the plant phytochromes are placed in the lower part (S6803_PlpA to Synel_Cika). All these proteins lack several amino acids start-ing from position #350 of the alignment. [The numbers indexed with # refer to the ruler in Figure 2 which in turn corresponds to an alignment with full-length sequences of an earlier review [46] also available on the internet (http://

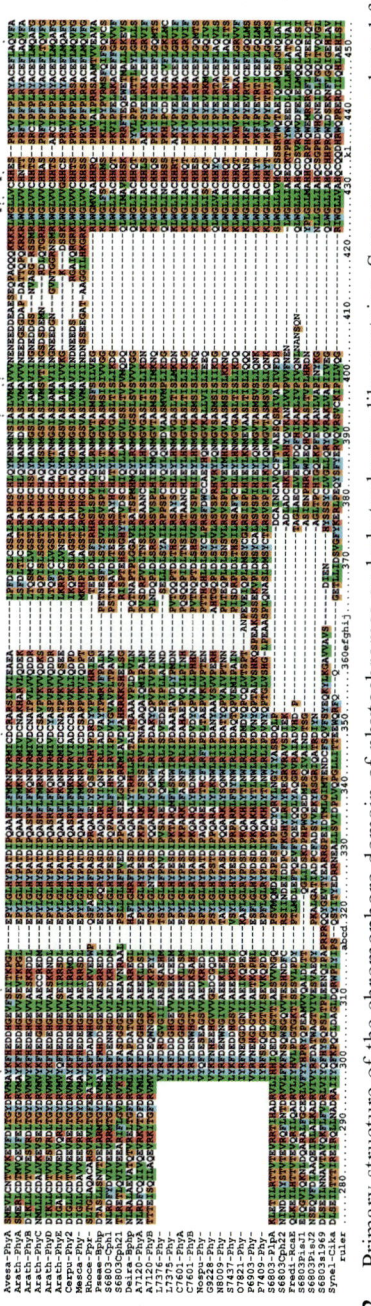

Figure 2. Primary structure of the chromophore domain of phytochromes and phytochrome-like proteins. Sequences were selected from the SWISSPROT database. The chosen region of ~180 amino acids around the (putative) chromophore binding site shows almost the entire GAF domain. Genes from a broad survey of many cyanobacterial species [53] are only available as partial sequences – in those cases the first 23 amino acids are missing. The alignment was made with ClustalX (http://www-igbmc.u-strasbg.fr/BioInfo/ClustalX/Top.html) default parameters. Color-codes were made with ClustalX (http://www-igbmc.u-strasbg.fr/BioInfo/ClustalX/Top.html) default parameters. Each peptide sequence is denominated by the species name (abbreviated by 3+2 letters or 1 letter + PCC number) followed by the name of the protein as abbreviated in published work. The numbering of the ruler refers to an earlier alignment [46] which is also available on the internet (http://www.biologie.fu-berlin.de/phytochrome/align2x.htm) The additional sequences of the present alignment incorporate further gaps indicated by letters abcd, efghij and kl. A7120 PhyA: *Anabaena* PCC 7120, phytochrome A (this species is also found under the name *Nostoc* PCC 7120); A7120 PhyB: *Anabaena* PCC 7120, phytochrome B (with lysine); Arath-PhyA .. E: *Arabidopsis thaliana*, phytochrome A .. E; Avesa-PhyA: *Avena sativa*, phytochrome A; C7601_PhyA: *Calothrix* PCC 7601; phytochrome A (with cysteine; note that the species is named *Tolopothrix* in the database); C7601 PhyB: *Calothrix* PCC 7601; phytochrome B (with lysine); Cerpu_Phy2: *Ceratodon purpureus*, phytochrome 2; Deira_Bphp: *Deinococcus radiorudans*, bacterial phytochrome homologous protein; Fredi_RcaE: *Fremyella diplosiphon*, regulator of chromatic adaptation E (note: *Fremyella* is a mutant of *Calothrix* PCC7601); G9228_Phy: *Geitlerinema* PCC 9228, phytochrome; L7375 Phy: *Leptolyngbya* PCC 7375, phytochrome; L7376 Phy: *Leptolyngbya* PCC 7376, phytochrome; Mesca_Phy: *Mesontaenium caldariorum*, phytochrome; N8009 Phy: *Nostoc* PCC 8009, phytochrome; Nospu Phy: *Nostoc punctiforme*, phytochrome; O7821 Phy: *Oscillatoria agahardii* PCC 7821, phytochrome; P6903 Phy: *Pseudanabaena* PCC 7409, phytochrome; P7409 Phy: *Pseudanabaena* PCC 7409, phytochrome; Pseae_Bphp: *Pseudmononas aeruginosa*, bacterial phytochrome homologous protein; Rhoce Ppr: *Rhodospirillum centenum*, PYP-phytochrome-like protein; S6803_Cph1: *Synechocystis* PCC 6803, cyanobacterial phytochrome 1; S6803_PisJ1: *Synechcystis* PCC 6803, phototaxis protein J domain 1; S6803_PisJ1: *Synechcystis* PCC 6803, phototaxis protein J domain 2; S6803_PlpA: *Synechocystis* PCC 6803, phytochrome-like protein A; S6803Cph21: *Synechocystis* PCC 6803, cyanobacterial phytochrome 2, domain 1; S6803Cph21: *Synechocystis* PCC 6803, cyanobacterial phytochrome 2, domain 2; S6803s1969: *Synechcystis* PCC 6803, product of gene sll1969; S7437 Phy: *Staniera* PCC 7437, phytochrome; Synel_Cika: *Synechococcus elongatus*, circadian input kinase.

www.biologie.fu-berlin.de/phytochrome/alignframes.htm). To avoid confusion, residues corresponding to three additional blocks have been allocated #a–#l. Conventional numbering (e.g."C259") refers to residues in a particular peptide sequence]. It could well be that these residues are important for the photochromic property of typical phytochromes and/or the stretched form of the chromophore that absorbs predominantly in the red [42]. In addition, the proteins listed below have two additional amino acids (positions #k and #l) absent from the other proteins. Several differences between the two groups are apparent at the single-residue level. For example, the more phytochrome-typical group members have tyrosine residues at positions #297 and #384 and a methionine at position #388; these amino acids are less well conserved in the other proteins. A phylogenetic tree constructed on the basis of this alignment (omitting the gaps) shows that the sequences of the latter group are clearly separate from the others and that they too are homologous. Further indirect evidence supports a clear distinction between phytochromes and these more distantly related proteins; based on mutant results, PlpA might be a blue light photoreceptor, while RcaE might absorb in the green and red region of the spectrum. PisJ1 seems to be a photoreceptor for phototaxis; as the action spectrum of cyanobacterial phototaxis is very broad [7], PisJ1 could well absorb in the shorter wavelength region. In conclusion, CikA, which is similar to these proteins, would also be non-photochromic and/or absorb blue light. Conversely, the phytochrome-like Ppr protein from *Rhodospirillum* – which bears a classical phytochrome domain – branches between the domain 1 of Cph2 and the other typical phytochromes. Ppr should thus incorporate a bilin chromophore to yield a photochromic pigment – despite the suggestion of the authors [44].

The evolutionary tree as presented in Figure 3 proposes the following scheme for the evolution of phytochromes: (i) most likely the first true phytochrome with red/far-red photoreversibility originated from the group of phytochrome-like proteins; (ii) one and only one of the cyanobacterial phytochromes gave rise to the plant phytochromes as we find them today, the diversity of plant phytochromes having arisen after endosymbiosis. Assumption (i) is based on the notion that phytochrome-like proteins are more diverse with respect to domain arrangement and primary structure of the chromophore domain than the true phytochromes. Even though plant physiologists associate red/far-red photochromicity with the sophisticated colour-detection involved in shade avoidance it is presently not clear how this would be useful for a cyanobacterial cell. This question might be answered once the physiological function of further cyanobacterial phytochromes is deduced. Assumption (ii) is interesting regarding the question of plant-phytochrome evolution and action. The small genome of a cyanobacterium harbours around 20 different biliproteins (phycobiliproteins, phytochromes and phytochrome-like proteins). The question arises why, from this group, only one biliprotein, namely phytochrome, has survived during the evolution of green plants. The modular design of plant phytochromes as we know it today must have brought an advantage for the green plants such that they found it worthwhile retaining the enzymes required for chromophore biosynthesis.

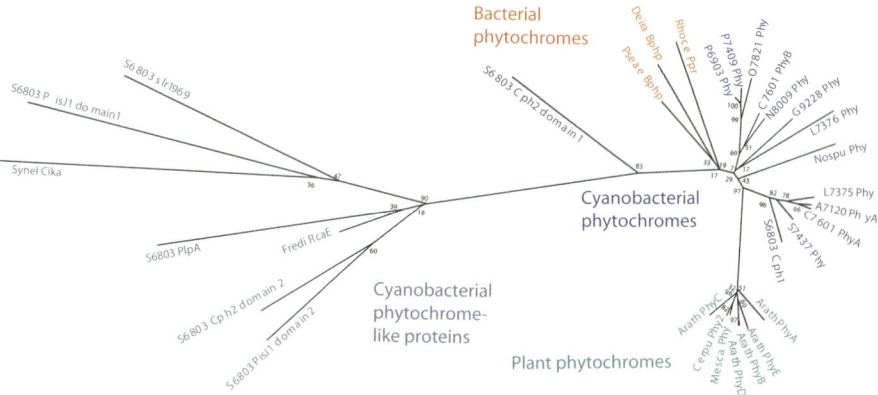

Figure 3. Evolution tree of phytochrome-like proteins and phytochromes based on the alignment of Figure 2. Only regions without gaps were used for the construction of the tree. The PHYLIP program package (http://evolution.genetics.washington.edu/phylip.html) was used for tree construction, the distance matrix was generated by the FITCH algorithm. Numbers at the branches reflect bootstrap values in %. The list of abbreviations is given in the legend of Figure 2.

Besides differences in the global domain arrangement (see Figure 1), in the chromophore-binding region the homogeneity of the plant phytochromes is reflected by the insertion #403–#421 of Figure 2. This insertion might be related to a plant-specific function, perhaps signal transduction; while it is divergent between the different plant phytochromes, flanking regions are rather similar, implying that it might be important for the specific actions of different types of plant phytochrome.

7.7   Domain arrangement and function

Figure 1 shows some examples for the arrangement of domains of phytochromes and phytochrome-like proteins as found from protein databases PFAM and SMART. The above-mentioned GAF domain [40] overlaps with the domain necessary for chromophore binding and lyase activity. Another domain found in all phytochrome proteins but with no clearly-defined function is indicated as "PHY". Many but not all of the listed proteins contain a module with homology to sensory histidine kinases. Light-dependent in vitro autophosphorylation has been demonstrated for Cph1 [14] and recently for two similar phytochromes from *Calothrix* [45]. This is the domain recognised in plant phytochromes ten years ago by Schneider-Poetsch (for discussions on its role there, see [46,47] and the signal transduction section below). Some of the proteins, again including the plant phytochromes, harbour so-called PAS-domains. Attention was drawn to these usually twinned motifs by studies of proteins involved in signal transduction and the circadian clock [48].

Importantly, it was shown that proteins can interact with each other via these domains. The exact function of the PAS domains in plant phytochromes and their allies is unclear although a cluster of residues between the two PAS domains of PHYA and PHYB in *Arabidopsis* seem to be involved in signal transduction [49]. It should be mentioned that whereas it is clear that the PAS-bearing module of plant phytochromes is absent from Cph1, HMM algorithms detect such a domain in the N-terminal part of Cph1. Other proteins with phytochrome homology have abnormal domain patterns, and include quite foreign domains (e.g. pisJ1, see above).

7.8 Biochemical and biophysical analyses on cyanobacterial phytochromes

The three different functions of phytochromes–chromophore assembly, photo-conversion and signal transduction–have extensively been studied over the last few decades with plant phytochromes. The same questions are now being addressed for prokaryotic phytochromes. Biophysical techniques which require large amounts of protein are particularly well served by Cph1 *E. coli* overexpressors. In our laboratories, the Cph1 yield per litre of pF10.his culture is about 40 mg. Since Cph1 is rather closely related to the plant phytochromes, it might well serve as a valid model for all phytochromes even at the atomic level.

7.8.1   Assembly of chromophore

In neutral aqueous buffers, PCB has a broad absorbance in the orange region (λ_{max} 610 nm). After mixing the PCB chromophore with apo-Cph1, two different species can be resolved kinetically. An intermediate with a longer-wavelength-absorption shoulder is formed rapidly, followed by the slower formation of the final Pr adduct (λ_{max} 655 nm) [50]. Equivalent spectral alterations are observed during assembly with PEB [51]. When covalent attachment of the chromophore to C259[#380] is blocked by iodoacetamide treatment, spectral analysis showed that a species resembling the intermediate was formed but that neither covalent attachment nor Pr formation was achieved. This allows the initial non-covalent step to be separated from the slower covalent linkage. Comparison with spectra of free chromophores under acidic conditions implied that the intermediate form represents a protonated form of the chromophore. It is generally accepted that in the Pr form the phytochrome chromophore is in the protonated state, which in turn implies that protonation occurs during the assembly reaction as the chromophore enters the "acid pocket". Thus protonation precedes covalent attachment. It was shown with the help of a stopped-flow machine which allows rapid mixing of apoprotein and chromophore, it was shown that the first step has a rise time of about 150 ms [82]. This might reflect the kinetics of the protonation event. Biliverdin, which is not covalently attached to the protein, undergoes spectral changes in the UV-A absorbing band (around 380 nm) but not in the red region when mixed with Cph1. This implies an early interaction between the apoprotein and biliverdin but that this does not extend to protonation. This interaction might represent a screening process in which the apoprotein "frisks" the bilin to assess its suitability as a chromophore. We thus think that the assembly process occurs in three steps: an initial screening operation, then entry into the acid pocket and associated protonation and, finally, formation of the covalent thioether link to C259[#380]. Because it is possible to distinguish two of these steps by absorbance spectroscopy, mutations affecting each can now be sought and thus the role of specific residues investigated.

A Cph2 fragment peptide consisting of 197 amino acids was capable of chromophore incorporation. The adduct was spectrally similar but not identical to typical phytochromes and showed photochromic absorbance changes [42]. This restricts the possible location of the motif necessary for covalent chromophore attachment (the lyase) and photochromicity down to ~200 amino acids. The authors have mutated three highly conserved amino acids in the corresponding domain of Cph1 and analysed these products for chromophore incorporation. D171[#292] and R172[#293] could be substituted without effect on chromophore assembly and spectral properties. Only when both were altered simultaneously (to e.g. alanine) was binding inhibited. Replacing E189[#310] also abolished chromophore binding. Whereas these results imply that this residue is important for the reaction, as always with such procedures, an indirect effect deriving from changed protein folding is also possible.

For plant phytochromes it was shown biochemically that the chromophore is attached to a particular cysteine residue via a thioether bond (C#380 in Figure 2). Indeed, this residue is conserved in all plant phytochromes. One of the interesting findings from studies of prokaryotic phytochromes and phytochrome-like proteins was that this cysteine is not conserved in many of these proteins (Figure 2). While some authors speculated that the ability for chromophore attachment might be lost [26,44], Davis and co-workers found a different mode for chromophore ligation in the case of *Deinococcus* DrBphP. Here, cysteine is replaced by a methionine, which does not, however, serve as the chromophore binding site, as shown by mutant studies. The neighbouring residue downstream is a highly-conserved histidine, also in DrBphP (Figure 2). Davis et al. showed by mass spectroscopy that this histidine serves as the attachment site for the chromophore in DrBphP. It seems that the natural chromophore of DrBphP is biliverdin instead of PCB or PΦB [68]. Formation of a thioether bond via the conventional mechanism is hindered by the vinyl side chain biliverdin. This might explain the different attachment-mechanism found for DrBphP. This is consistent with the fact that the enzymes for PCB or PΦB synthesis are lacking in *Deinococcus* and other bacterial genomes. It will thus be interesting to see whether other bacterial phytochromes also use biliverdin as their native chromophore. Herdman et al. found several "non-cysteine" phytochrome sequences in various cyanobacterial strains [53].

Gärtner and co-wokers analysed one of these, the phytochrome CphB (also referred to as PhyL or PhyB) of *Calothrix* PCC7601, which carries a lysine at #380 in place of the chromophore-binding cysteine. They could show that the chromophore binds non-covalently to the protein as it could be removed by buffer exchange. In the bound form, the chromophore undergoes photoreversible absorbance changes similar to those of typical phytochromes. We note that whereas the chromophore is assumed to bind covalently to C#380 in all phytochromes in which this residue is present, this has yet to be proven. Changes at C#380 induced either genetically [54] or biochemically (by iodoacetimide, for example) which prevent covalent attachment support the notion, but again indirect effects via protein folding are also possible.

7.8.2   *Photoconversion*

With Cph1, Fourier-transform Raman-resonance (FTRR) [55], low-temperature fluorescence [56], flash photolysis [55], Fourier-transform infrared absorbance (FTIR) [57] and circular dichroism [54] spectroscopic methods have been employed. CD, FTRR and FTIR are sensitive probes for the status of the chromophore and for changes in the protein backbone. These methods showed many similarities between Cph1 and native oat phytochrome A. FTRR revealed for both phytochromes spectral differences between the Pr and the Pfr form that reflect the $Z{\rightarrow}E$ isomerisation of the chromophore and changes in its hydrogen bonding with the protein. Moreover, as in plant phytochromes, subtle differences between the PCB and the PΦB adduct of Cph1 can be attributed to the ring D side chain (vinyl group for PΦB vs. ethyl group for PCB). FTRR studies also included intermediates of the photocycle, trapped at low temperature. The first photoproduct observed with phytochromes, *lumi*-R, was either not stable or was not spectrally detectable, coinciding with findings from low-temperature fluorescence spectroscopy [56]. The next intermediates trapped at −60 and at −30°C, *meta*-Ra and *meta*-Rc, showed considerable changes in bands assigned to the chromophore. Most likely alterations in the protein backbone take place during the *meta*-Rc → Pfr conversion, not before the formation of *meta*-Rc. The data also suggested that the chromophore is protonated both in the Pr and the Pfr form. In flash photolysis experiments, the first photoproduct *lumi*-R of Cph1 appeared substantially more quickly than that for plant phytochromes and was followed by a novel intermediate whose kinetics were delayed almost two-fold by deuterium exchange, a larger value than any reported for photoconversion in any plant phytochrome [55]. Fluorescence measurements at low temperature address the photoconversion from a different point of view. Whereas at ambient temperature phytochrome fluorescence yields are very low, these rise dramatically upon cooling; Pr → Pfr photoconversion is inhibited, although photoconversion into intermediate forms is sometimes possible. For plant PHYA at 70 K the proportion of Pr that can convert into *lumi*-R can be as high as 50%, whereas this conversion is not possible for plant PHYB and moss phytochrome. Cph1 adducts are also unable to form *lumi*-R at this temperature, implying that Cph1 is more related to PHYB than PHYA photochemically. Different activation barriers for the photoreaction are thought to explain the differences between phytochrome types.

Protonation effects during the photocycle of plant phytochromes have been analysed by various methods. The status of the chromophore is controversially discussed in this respect. Whereas protonation in the Pr form is not disputed, some authors propose that the chromophore is also protonated in the Pfr form [58]. Work on pea phytochrome implied, however, that the chromophore lost a proton upon photoconversion into Pfr [59]. In addition, deprotonation of ring A of the chromophore together with lactam–lactim tautomerization (the oxo group is transformed into a hydroxy-group and thereby the double bond of the oxygen moves into the ring) might well explain the high λ_{max} of Pfr [60]. The large deuterium effect on Cph1 awakened interest in further protonation studies on that phytochrome. UV/VIS spectra at neutral pH support the view that in both forms the chromophore is protonated. However, under more basic conditions, a slight hypsochromic shift and reduction of the extinction coefficient at λ_{max} as well as the appearance of blue-absorbing bands implies that the chromophore becomes deprotonated. The apparent pK_a of the effect in relation to the change in the extinction coefficient is close to neutrality. As the only amino acid with a pK_a around 7 is histidine it seems likely that such a residue is being titrated by the surrounding medium [61]. Thus it is thought that an imidazole side-chain from a neighbouring histidine interacts with the chromophore and thereby stabilises the protonated form. Indeed, recent results from site-directed mutants support this conclusion (Hughes *et al.*, in preparation). As with plant phytochromes, proton release from Cph1 can be observed upon Pr → Pfr conversion, a process that is fully photoreversible. Kinetic studies showed that an initial proton release within 20 ms is followed by a partial re-uptake after about 300 ms [61]. These transients could be correlated with particular intermediates of the photocycle. The observed changes might simply reflect conformational changes of the protein. That an intermediate form is more strongly deprotonated than either Pr or Pfr could as well provide a possibility for the cell to monitor Pr ↔ Pfr cycling rates.

Conformational changes during the photocycle can also be observed by size-exclusion chromatography. Cph1 holoprotein is eluted with an apparent molecular weight of a dimer, in harmony with results from plant phytochromes and sensory histidine kinases: both groups of proteins exist and act as dimers. In the Pfr form the mobility of Cph1 is slightly increased, which might indicate a change of the shape of the protein or an altered interaction with the gel matrix. The proposed dimerisation region is located in the C-terminal histidine kinase domain, in accordance with this the Pr form of the N-terminal 58 kD fragment (chromophore-binding domain) runs as an apparent monomer. Quite interestingly, Pfr elutes as an apparent dimer, pointing to conformation-specific dimerisation of the chromophore-binding domain. Whether this is of any significance in the intact Cph1 molecule is not known.

7.8.3   Signal transduction

Despite recent discoveries regarding intracellular localisation and specific interacting partners the signal transduction mechanisms of plant phytochromes are still obscure. Prokaryotic phytochromes provide interesting working models for phytochrome action. Where a histidine kinase motif is present in a prokaryotic phytochrome (see above), two-component signalling is a likely mechanism of signal transduction. Histidine kinases play a central role in prokaryotic signalling and are widespread in plants. After autophosphorylation, which is modulated by an environmental signal, the phosphate group is transferred to a conserved aspartate residue of a so-called response regulator, from which it can be further transmitted to other proteins. However, such an in vitro histidine kinase function has yet to be demonstrated for most of the prokaryotic phytochrome homologues known, and no in vivo assays have as yet been reported.

In Figure 1 the domains as found by PFAM and/or SMART algorithms are summarised for some of the phytochromes and phytochrome-like proteins. The histidine kinase consists of two parts, the subdomain containing the phospho-accepting histidine (HisK) and the region with the kinase (HATPase). In CikA the histidine kinase and response regulator domains are united on the same protein, an arrangement also seen in other sensory proteins. Recombinant holo-Cph1 was shown to have histidine-kinase activity; interestingly, this was stronger in the Pr than in the Pfr form [14] (see Figure 4), even though in plants Pfr, not Pr, is considered to be physiologically active. The cognate response regulator Rcp1 is polycistronic with Cph1. Phosphotransfer to Rcp1 is again stronger for the Pr form of Cph1 [14]. Similarly, Cph1-homologues in *Calothrix* are arranged in tandem with their response regulators, showing

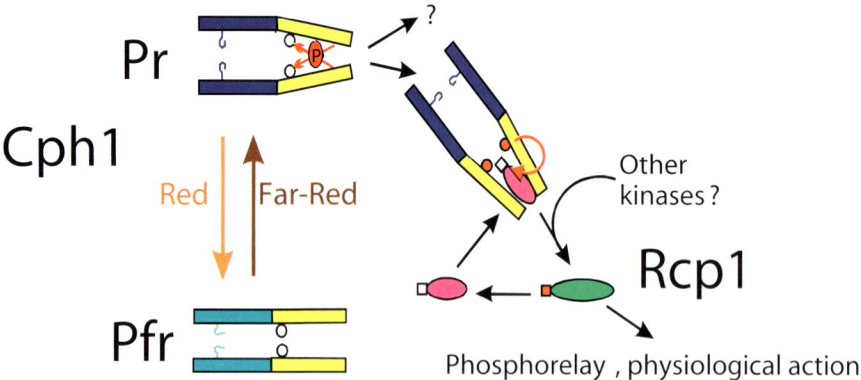

Figure 4.   Two-component signal transduction of Cph1 and Rcp1. In the first step, the histidine kinase domain of Cph1 is autophosphorylated at a conserved histidine residue. This reaction is stronger for the Pr than for the Pfr form, and thus dependent on the light conditions. The phosphate group (red) is then transferred to a conserved aspartate residue of the response regulator Rcp1.

light-regulated kinase activity and phosphotransfer [45]. In our hands auto-phosphorylation of Cph1 and phosphotransfer to Rcp1 is rather inefficient. This is consistent with the work of Yeh et al. [14], who used high protein concentrations to obtain a good phosphorylation signal. Our studies of Cph1 in *Synechocystis* [51] imply that the concentrations in vivo are many orders of magnitude lower, raising difficult questions as to how (or even if) the system operates physiologically. On the other hand, the covalent bond connecting phosphate with histidine is known to be rather labile, so it might be that most phosphate residues are lost during the electrophoretic and blotting procedures which follow the kinase reaction. The well-analysed response regulator CheY is known to dephosphorylate with a half-life of 20 s [62] – a similar mechanism in Rcp1 might be the reason for the weak signal here. Certainly, experiments are required to quantify the proportion of phosphorylated proteins and to follow the phosphorylation process kinetically.

The discovery in the Lagarias lab that Cph1 acts as a kinase prompted them to re-open the case of plant phytochrome kinase activity [63]. This by-no-means-new field of biochemical work had been thwarted by technical problems, but modern recombinant expression systems made the production of highly concentrated, undegraded, pure phytochrome holoprotein much easier. Histidine kinase action could not be demonstrated – and indeed it is unlikely that plant phytochromes can act in this manner as, even though the HisK subdomain is often recognisable, the phosphoaccepting histidine residue itself is lacking in almost all plant phytochromes (discussed as early as 1992 [64], see also [46]). Instead, serine/threonine kinase activity was observed in both algal and plant phytochromes, marking another major advance in the field. For further discussion of S/T kinase activity in plant phytochromes see Chapters 5 and 6.

This of course leaves the question of the role of the relict HisK subdomain #986–#999 in plant phytochromes unanswered. We think this plays a central role in phytochrome dimerisation, a conclusion based on four lines of evidence. Firstly, 3D-structural data (see 1B3Q, 1JOY and 1BXD – available from http://www.ncbi.nlm.nih.gov) indicate that the HisK subdomain is one of two antiparallel α-helices (residues #985–#1010 and #1023–#1044) responsible for dimerization in histidine kinases. Secondly, structural analyses *in silico* indicate that these regions are probably α-helical in plant phytochromes, too. Thirdly, deletion of the transmitter module yields monomeric Pr in the case of Cph1 (Hughes *et al.*, in preparation). Fourthly, earlier in vitro dimerisation studies implicated amongst others the #985–#1044 region [65]. Thus, while the absence of a histidine at H#995 precludes a HisK function, a conserved role for this region in dimerisation is likely.

7.9   Biosynthesis of the phytochrome chromophore

Phytochromes use either PCB or PΦB as their chromophore. The former is highly abundant in the phycobilisome antenna complexes of cyanobacteria and red algae, but is not known in green plants beyond the algae. PΦB is a very similar molecule with a single additional side-chain double bond. Studies of the phytochrome system have helped to clarify the final steps in bilin synthesis. The locus of an *Arabidopsis* mutant, *hy2*, long known to be deficient in the synthesis of the phytochrome chromophore, was recently characterised by a combination of classical genetics, *in silico* sequence comparisons and biochemical assays [66]. HY2 converts biliverdin into PΦB with ferredoxin as a cofactor. Soon thereafter cyanobacterial genes with sequence homology to HY2 and proteins encoded by them were analysed in an assay utilising recombinant expression. By this approach, an enzyme that converts biliverdin directly into PCB, PcyA, and enzymes that convert biliverdin into PEB, another chromophore in phycobilisomes of certain cyanobacteria, were identified [67]. It would appear that all enzymes of bilin synthesis have now been identified at the gene level in *Synechocystis* and *Arabidopsis*. Indeed, two groups have been able to express holo-Cph1 in *E. coli*, opening the way to several new lines of research into phytochrome structure and function [69,70].

7.10   Conclusion

The photoreceptors that mediate the various light responses in cyanobacteria and other prokaryotes have long been a mystery. Genome sequence data have now provided access to numerous putative chromoproteins, many of which include domains similar to those seen in phytochromes. These candidate photoreceptors have both provided new vantage points for viewing the action of plant phytochromes as well offering a much better understanding of light regulation in cyanobacteria. Although the molecular mechanisms of phytochrome signalling are far from completely understood, progress here is rapid. The part to which prokaryotes are contributing to this should not be underestimated. Whereas most phytochrome research is devoted to plants, numerous plant phytochrome groups now include prokaryotic phytochromes in their repertoire, and a number of cyanobacterial groups now include phytochromes in theirs. The early years of phytochrome research were concentrated on particular crop species which offered advantages for the biochemical techniques of the time. With the development of molecular-genetic approaches, *Arabidopsis* has been established as the primary subject of attention for plant science and hence phytochrome research. While prokaryotes are more distant still from crop plants and agricultural problems, as particularly convenient research subjects they will certainly continue to make important contributions to our understanding of phytochrome systems.

Since this manuscript was written, further reviews and research articles which are related to the topic of this chapter were published. For completion, these articles are listed as references [71–82].

References

1. L.C. Sage (1992). *Pigment of the Imagination – A History of Phytochrome Research.* Academic Press.
2. W.L. Butler, K.H. Norris, H.W. Siegelman, S.B. Hendricks (1959). Detection, assay, and preliminary purification of the pigment controlling photoresponsive development of plants. *Proc. Natl. Acad. Sci. U.S.A.*, **45**, 1703–1708.
3. W. Rüdiger, F. Thümmler (1994). The phytochrome chromophore. In: R.E. Kendrick, G.H.M. Kronenberg (Eds), *Photomorphogenesis in plants* (2nd Edn, pp. 51–69) Kluwer Academic Publishers, Dordrecht.
4. H.P. Hershey, J.T. Colbert, J.L. Lissemore, R.F. Barker, P.H. Quail (1984). Molecular cloning of cDNA for *Avena* phytochrome. *Proc. Natl. Acad. Sci. U.S.A*, **81**, 2332–2336.
5. P.M. Reddy, P.S.N. Rao, E.R.S. Talpasayi (1975). Effect of red and far red illumination on the germination of spores of two blue-green algae. *Curr. Sci.*, **44**, 678–679.
6. P.M. Reddy, E.R.S. Talpasayi (1981). Some observations related to red-far-red antagonism in germination of spores of the cyanobacterium *Anabaena fertilissima*. *Biochem. Physiol. Pflanzen*, **176**, 105–107.
7. Y. Kondou, M. Nakazawa, S.-I. Higashi, M. Watanabe, K. Manabe (2001). Equal-quantum action spectra indicate fluence-rate-selective action of multiple photoreceptors for photomovement of the thermophilic cyanobacterium *Synechococcus elongatus*. *Photochem. Photobiol.*, **73**, 90-95.
8. H.A. Schneider-Poetsch, B. Braun, S. Marx, A. Schaumburg (1991). Phytochromes and bacterial sensor proteins are related by structural and functional homologies. Hypothesis on phytochrome-mediated signal-transduction. *FEBS Lett.*, **281**, 245–249.
9. H.A. Schneider-Poetsch, B. Braun (1991). Proposal on the nature of phytochrome action based on the C-terminal sequences of phytochrome, *J. Plant Physiol.*, **137**, 576–580.
10. T. Kaneko, S. Sato, H. Kotani, A. Tanaka, E. Asamizu, Y. Nakamura, N. Miyajima, M. Hirosawa, M. Sugiura, S. Sasamoto, T. Kimura, T. Hosouchi, A. Matsuno, A. Muraki, N. Nakazaki, K. Naruo, S. Okumura, S. Shimpo, C. Takeuchi, T. Wada, A. Watanabe, M. Yamada, M. Yasuda, S. Taba (1996). Sequence analysis of the genome of the unicellular Cyanobacterium Synechocystis sp. strain PCC6803. II. Sequence determination of the entire genome and assignment of potential protein-coding regions. *DNA Res.*, **3**, 109–136.
11. D.M. Kehoe, R. Grossman (1996). Similarity of a chromatic adaptation sensor to phytochrome and ethylene receptors. *Science*, **273**, 1409–1412.
12. A. Wilde, Y. Churin, H. Schubert, T. Börner (1997). Disruption of a Synechocystis sp. PCC 6803 gene with partial similarity to phytochrome genes alters growth under changing light qualities. *FEBS Lett.*, **406**, 89–92.
13. J. Hughes, T. Lamparter, F. Mittmann, E. Hartmann, W. Gärtner, A. Wilde, T. Börner (1997). A prokaryotic phytochrome. *Nature*, **386**, 663–663.
14. K.C. Yeh, S.H. Wu, J.T. Murphy, J.C. Lagarias (1997). A cyanobacterial phytochrome two-component light sensory system. *Science*, **277**, 1505–1508.
15. T. Hübschmann, T. Börner, E. Hartmann, T. Lamparter (2001). Characterisation of the Cph1 holo-phytochrome from *Synechocystis* sp. PCC 6803. *Eur. J. Biochem.*, **268**, 2055–2063.

16. J. Hughes, T. Lamparter, F. Mittmann (1996). Cerpu;PHY0;2, a "normal" phytochrome in *Ceratodon. Plant Physiol.*, **112**, 446–446.

17. K. Nozue, T. Kanegae, T. Imaizumi, S. Fukuda, H. Okamoto, K.C. Yeh, J.C. Lagarias, M. Wada (1998). A phytochrome from the fern Adiantum with features of the putative photoreceptor NPH1. *Proc. Natl. Acad. Sci. U.S.A*, **95**, 15826–15830.

18. F. Thümmler, P. Algarra, G.M. Fobo (1995). Sequence similarities of phytochrome to protein kinases: implication for the structure, function and evolution of the phytochrome gene family. *FEBS Lett.*, **357**, 149–155.

19. S. Diakoff, J. Scheibe (1975). Cultivation in the dark of the blue-green alga Fremyella diplosiphon. A photoreversible effect of green and red light on growth rate. *Physiol. Plant.*, **34**, 125–128.

20. T.C. Vogelman, J. Scheibe (1978). Action spectra for chromatic adaptation in the blue-green alga *Fremyella diplosiphon. Planta*, **143**, 233–239.

21. N. Gaidukov (1902). Über den einfluß farbigen lichts auf die färbung lebender oscillarien. *Abh. Preuss. Akad. Wiss.*, **5**, 1–36.

22. W. Nultsch (1969). Effect of desaspidin and DCMU on photokinesis of blue-green algae. *Photochem. Photobiol.*, **10**, 119–123.

23. S. Yoshihara, F. Suzuki, H. Fujita, X.X. Geng, M. Ikeuchi (2000). Novel putative photoreceptor and regulatory genes required for the phototactic movement of the unicellular motile cyanobacterium Synechocystis PCC 6803. *Plant Cell Physiol.*, **41**, 1299–1304.

24. T. Kondo, T. Mori, N.V. Lebedeva, S. Aoki, M. Ishiura, S.S. Golden (1997). Circadian rhythms in rapidly dividing cyanobacteria. *Science*, **275**, 224–227.

25. T. Kondo, M. Ishiura (1999). The circadian clocks of plants and cyanobacteria. *Trends Plant Sci.*, **4**, 171–176.

26. O. Schmitz, M. Katayama, S.B. Williams, T. Kondo, S.S. Golden (2000). CikA, a bacteriophytochrome that resets the cyanobacterial circadian clock. *Science*, **289**, 765–768.

27. S.J. Davis, A.V. Vener, R.D. Vierstra (1999). Bacteriophytochromes: Phytochrome-like photoreceptors from nonphotosynthetic eubacteria. *Science*, **286**, 2517–2520.

28. Z.Y. Jiang, C.E. Bauer (1997). Analysis of a chemotaxis operon from *Rhodospirillum centenum. J. Bacteriol.*, **179**, 5712–5719.

29. W.D. Hoff, H.C.P. Matthijs, H. Schubert, W. Crielaard, K.J. Hellingwerf (1995). Rhodopsin(s) in eubacteria. *Biophys. Chem.*, **56**, 193–199.

30. W.W. Sprenger, W.D. Hoff, J.P. Armitage, K.J. Hellingwerf (1993). The eubacterium *Ectothiorhodospira halophila* is negatively phototactic, with a wavelength dependence that fits the absorption spectrum of the photoactive yellow protein. *J. Bacteriol.*, **175**, 3096–3104.

31. J. Scheibe (1972). Photoreversible pigment: occurrence in a blue-green alga. *Science*, **176**, 1037–1039.

32. G.S. Björn, L.O. Björn (1976). Photochromic pigments from blue-green algae: phycochromes a,b, and c. *Physiol. Plant.*, **36**, 297–304.

33. L.O. Björn (1979). Photoreversibly photochromic pigments in organisms: properties and role in biological light perception. *Q. Rev. Biophys.*, **12**, 1–23.

34. G.S. Björn (1979). Action spectra for in vivo conversion of phycochrome B, a reversibly photochromic pigment in a blue-green alga, and its separation from other pigments. *Physiol. Plant.*, **46**, 281–286.

35. D.A. Bryant, A.N. Glazer, F.A. Eiserling (1976). Characterization and structural properties of the major biliproteins of Anabaena sp. *Arch. Microbiol.*, **110**, 61–75.
36. R.V. Swanson, J. Zhou, J.A. Leary, T. Williams, R. de Lorimier, D.A. Bryant, A.N. Glazer (1992). Characterization of phycocyanin produced by cpcE and cpcF mutants and identification of an intergenic suppressor of the defect in bilin attachment. *J. Biol. Chem.*, **267**, 16146–16154.
37. K.H. Zhao, R. Haessner, E. Cmiel, H. Scheer (1995). Type I reversible photochemistry of phycoerythrocyanin involves Z/E-isomerization of alpha-84 phycoviolobilin chromophore. *Biochim. Biophys. Acta*, **1228**, 235–243.
38. K.H. Zhao, H. Scheer (1995). Type I and type II reversible photochemistry of phycoerythrocyanin alpha-subunit from *Mastigocladus laminosus* both involve Z, E isomerization of phycoviolobilin chromophore and are controlled by sulfhydryls in apoprotein. *Biochim. Biophys. Acta*, **1228**, 244–253.
39. W. Braune, T. Wilczok, R. Waclawek (1988). Indications for photoreversible reactions in the range of phycochrome b absorption obtained by automated microscopic image analysis of germinating *Anabaena* akinetes. *Cytobios*, **54**, 39–48.
40. L. Aravind, C.P. Ponting (1997). The GAF domain: an evolutionary link between diverse phototransducing proteins. *Trends. Biochem. Sci.*, **22**, 458–459.
41. C.M. Park, J.I. Kim, S.S. Yang, J.G. Kang, J.H. Kang, J.Y. Shim, Y.H. Chung, Y.M. Park, P.S. Song (2000). A second photochromic bacteriophytochrome from Synechocystis sp. PCC 6803: spectral analysis and down-regulation by light. *Biochemistry*, **39**, 10840–10847.
42. S.H. Wu, J.C. Lagarias (2000). Defining the bilin lyase domain: lessons from the extended phytochrome superfamily. *Biochemistry*, **39**, 13487–13495.
43. H. Scheer (1987). Photochemistry and photophysics of C-phycocyanin. In: J. Biggins (Ed.), *Progress in Photosynthesis Research: Proceedings of the VIIth International Congress on Photosynthesis, 1986* (pp. 143–149). M. Nijhoff Publishers, Dordrecht.
44. Z. Y. Jiang, L. R. Swem, B. G. Rushing, S. Devanathan, G. Tollin, C. E. Bauer (1999). Bacterial photoreceptor with similarity to photoactive yellow protein and plant phytochromes. *Science*, **285**, 406–409.
45. T. Hübschmann, H. J. M. M. Jorissen, T. Börner, W. Gärtner, N. Tandeau de Marsac (2001). Phosphorylation of proteins in the light-dependent signalling pathway of a filamentous cyanobacterium. *Eur. J. Biochem.*, **268**, 3383–3389.
46. M.T. Boylan, P.H. Quail (1996). Are the phytochromes protein kinases? *Protoplasma*, **195**, 12–17.
47. J. Hughes, T. Lamparter (1999). Prokaryotes and phytochrome - the connection to chromophores and signaling. *Plant Physiol.*, **121**, 1059–1068.
48. C.P. Ponting, L. Aravind (1997). PAS: a multifunctional domain family comes to light. *Curr. Biol.*, **7**, R674–R777.
49. P.H. Quail, M.T. Boylan, B.M. Parks, T.W. Short, Y. Xu, D. Wagner (1995). Phytochromes: Photosensory perception and signal transduction. *Science*, **268**, 675–680.
50. T. Lamparter, F. Mittmann, W. Gärtner, T. Börner, E. Hartmann, J. Hughes (1997). Characterization of recombinant phytochrome from the cyanobacterium Synechocystis. *Proc. Natl. Acad. Sci. U.S.A*, **94**, 11792–11797.
51. T. Lamparter, B. Esteban, J. Hughes (2001). Phytochrome Cph1 from the cyanobacterium *Synechocystis* PCC6803: purification, assembly, and quarternary structure. *Eur. J. Biochem.*, **268**, 4720–4730.

52. R.D. Vierstra, S.J. Davis (2000). Bacteriophytochromes: new tools for understanding phytochrome signal transduction. *Semin. Cell Dev. Biol.*, **11**, 511–521.

53. M. Herdman, T. Coursin, R. Rippka, J. Houmard, N. Tandeau de Marsac (2000). A new appraisal of the prokaryotic origin of eukaryotic phytochromes. *J. Mol. Evol.*, **51**, 205–213.

54. C.M. Park, J.Y. Shim, S.S. Yang, J.G. Kang, J.I. Kim, Z. Luka, P.S. Song (2000). Chromophore-apoprotein interactions in Synechocystis sp. PCC6803 phytochrome Cph1. *Biochemistry*, **30**, 6349–6356.

55. A. Remberg, I. Lindner, T. Lamparter, J. Hughes, K. Kneip, P. Hildebrandt, S.E. Braslavsky, W. Gärtner, K. Schaffner (1997). Raman spectroscopic and light-induced-kinetic characterization of a recombinant phytochrome of the cyanobacterium *Synechocystis*. *Biochemistry*, **36**, 13389–13395.

56. V. A. Sineschekov, J. Hughes, E. Hartmann, T. Lamparter (1998). Fluorescence and photochemistry of recombinant phytochrome from the cyanobacterium Synechocystis. *Photochem. Photobiol.*, **67**, 263–267.

57. H. Foerstendorf, T. Lamparter, J. Hughes, W. Gärtner, F. Siebert (2000). The photoreactions of recombinant phytochrome from the cyanobacterium *Synechocystis*: A low-temperature UV-Vis and FT-IR spectroscopic study. *Photochem. Photobiol.*, **71**, 655–661.

58. C. Kneip, P. Hildebrandt, W. Schlamann, S.E. Braslavsky (1999). Protonation state and structural changes of the tetrapyrrole chromophore. *Biochemistry*, **16**, 15185–15192.

59. Y. Mizutani, S. Tokutomi, T. Kitagawa (1994). Resonance Raman spectra of the intermediates in phototransformation of large phytochrome: deprotonation of the chromophore in the bleached intermediate. *Biochemistry*, **33**, 153–158.

60. M. Stanek, K. Grubmayr (1998). Deprotonated 2,3-Dihydrobilindiones - Models for the chromophore of the far-red-absorbing form of phytochrome. *Chem. Eur. J.*, **4**, 1660–1666.

61. J.J. van Thor, B. Borucki, W. Crielaard, H. Otto, T. Lamparter, J. Hughes, K.J. Hellingwerf, M.P. Heyn (2001). Light-induced proton release and proton uptake reactions in the cyanobacterial phytochrome Cph1. *Biochemistry,* **40**, 11460–11471.

62. A.M. Stock, V.L. Robinson, P.N. Goudreau (2000). Two-component signal transduction. *Annu. Rev. Biochem.*, **69**, 183–215.

63. K.C. Yeh, J.C. Lagarias (1998). Eukaryotic phytochromes: Light-regulated serine/threonine protein kinases with histidine kinase ancestry. *Proc. Natl. Acad. Sci. U.S.A.*, **95**, 13976–13981.

64. S. Hanelt, B. Braun, S. Marx, H.A. Schneider-Poetsch (1992). Phytochrome evolution: a phylogenetic tree with the first complete sequence of phytochrome from a cryptogamic plant (Selaginella martensii spring). *Photochem. Photobiol.*, **56**, 751–758.

65. P.H. Quail (1997). An emerging molecular map of the phytochromes. *Plant Cell Environ.*, **20**, 657–665.

66. T. Kohchi, K. Mukougawa, N. Frankenberg, M. Masuda, A. Yokota, J.C. Lagarias (2001). The arabidopsis hy2 gene encodes phytochromobilin synthase, a ferredoxin-dependent biliverdin reductase. *Plant Cell*, **13**, 425–436.

67. N. Frankenberg, K. Mukougawa, T. Kohchi, J.C. Lagarias (2001). Functional genomic analysis of the hy2 family of ferredoxin-dependent bilin reductases from oxygenic photosynthetic organisms. *Plant Cell*, **13**, 965–978.

68. S.H. Bhoo, S.J. Davis, J. Walker, B. Karniol, R.D. Vierstra (2001). Bacterio-phytochromes are photochromic histidine kinases using a biliverdin chromophore. *Nature*, **414**, 776–779.
69. G.A. Gambetta, J.C. Lagarias (2001). Genetic engineering of phytochrome biosynthesis in bacteria. *Proc. Natl. Acad. Sci. U. S. A.*, **98**, 10566–10571.
70. F.T. Landgraf, C. Forreiter, P.A. Hurtado, T. Lamparter, J. Hughes (2001). Recombinant holophytochrome in *Escherichia coli*. *FEBS Lett.*, **508**, 459–462.
71. A. Wilde, B. Fiedler, T. Börner (2002). The cyanobacterial phytochrome Cph2 inhibits phototaxis towards blue light. *Mol. Microbiol.*, **44**, 981–988.
72. H.J. Jorissen, B. Quest, I. Lindner, N. Tandeau de Marsac, W. Gärtner (2002). Phytochromes with noncovalently bound chromophores: the ability of apophy-tochromes to direct tetrapyrrole photoisomerization. *Photochem. Photobiol.*, **75**, 554–559.
73. H.J. Jorissen, B. Quest, A. Remberg, T. Coursin, S.E. Braslavsky, K. Schaffner, N. Tandeau de Marsac, W. Gärtner (2002). Two independent, light-sensing two-component systems in a filamentous cyanobacterium. *Eur. J. Biochem.*, **269**, 2662–2671.
74. K. Heyne, J. Herbst, D. Stehlik, B. Esteban, T. Lamparter, J. Hughes, R. Diller (2002). Ultrafast dynamics of phytochrome from the cyanobacterium synecho-cystis, reconstituted with phycocyanobilin and phycoerythrobilin. *Biophys. J.*, **82**, 1004–1016.
75. T. Lamparter, N. Michael, F. Mittmann, B. Esteban (2002). Phytochrome from *Agrobacterium tumefaciens* has unusual spectral properties and reveals an N-terminal chromophore attachment site . *Proc. Natl. Acad. Sci. U. S. A.*, **99**, 11628–11633.
76. Y.J. Im, S.H. Rho, C.M. Park, S.S. Yang, J.G. Kang, J.Y. Lee, P.S. Song, S.H. Eom (2002). Crystal structure of a cyanobacterial phytochrome response regulator. *Protein Sci.*, **11**, 614–624.
77. E. Giraud, J. Fardoux, N. Fourrier, L. Hannibal, B. Genty, P. Bouyer, B. Dreyfus, A.Vermeglio (2002). Bacteriophytochrome controls photosystem synthesis in anoxygenic bacteria. *Nature*, **417**, 202–205.
78. B.L. Montgomery, J.C. Lagarias (2002). Phytochrome ancestry: sensors of bilins and light. *Trends Plant Sci.*, **7**, 357–366.
79. H. Otto, T. Lamparter, B. Borucki, J. Hughes, M.P. Heyn (2003). Dimerization and inter-chromophore distance of Cph1 phytochrome from Synechocystis, as monitored by fluorescence homo and hetero energy transfer. *Biochemistry*, **42**, 5885–5895.
80. N. Frankenberg, J.C. Lagarias (2003). Phycocyanobilin:ferredoxin oxidoreductase of Anabaena sp. PCC 7120. Biochemical and spectroscopic. *J. Biol. Chem.*, **278**, 9219–9226.
81. T. Lamparter, N. Michael, O. Caspani, T. Miyata, K. Shirai, K. Inomata (2003). Biliverdin binds covalently to *Agrobacterium* phytochrome Agp1 via its ring A vinyl side chain. *J. Biol. Chem.*, *in press*.
82. B. Borucki, H. Otto, G. Rottwinkel, J. Hughes, M.P. Heyn, T. Lamparter (2003). Mechanism of Cph1 Phytochrome Assembly from Stopped-Flow Kinetics and Circular Dichroism. *Biochemistry*, **submitted**.

Chapter 8

The family of photoactive yellow proteins, the xanthopsins: from structure and mechanism of photoactivation to biological function

Klaas J. Hellingwerf, Johnny Hendriks, Michael van der Horst, Andrea Haker, Wim Crielaard and Thomas Gensch

Table of contents

Abstract

Several small, water-soluble, yellow-coloured proteins, all containing thiol-ester linked 4-OH-cinnamic acid as their chromophore, have been discovered since 1985, in a range of proteobacteria. This family of photoreceptor proteins, the xanthopsins, has a function in a wide range of processes, from genetic regulation of chalcone synthesis to the tactic migration of bacteria.

Photoactive yellow protein (PYP), from the purple-sulfur bacterium *Ectothiorhodospira halophila* is the archetype of this family. It is an α/β-fold protein that functions as the photoreceptor in a photophobic response. Light absorption initiates isomerisation of its anionic chromophore, from the 7-*trans*, 9-*S-cis-* to the 7-*cis*, 9-*S-trans* configuration. This leads to the formation of several transient intermediates, initially red-shifted compared with the ground state (pG_{446}), of which the most stable (pR_{465}) decays bi-exponentially to a blue-shifted state (pB_{355}), the tentative signalling state of PYP. pG_{446} recovers within a few hundred ms.

This change in configuration of the buried chromophore is relayed to the surface of the protein through a concerted conformational transition, allowing activation of a downstream signal transduction partner(s). Upon formation of pB_{355} the chromophore is protonated by nearby Glu-46, to which it initially was hydrogen-bonded. This creates a de-stabilising buried negative charge in the hydrophobic interior of the protein, which then transiently changes its conformation. The extent of this change is sensitive to conditions in the mesoscopic surroundings of the photoreceptor protein.

8.1   Introduction

In 1985, in a project aimed at making an inventory of all coloured proteins present in an anoxygenic phototrophic and extremophilic bacterium, *Ectothiorhodospira halophila* (or *Halorhodospira halophila*), a small yellow protein was detected that differed from the well-known flavoproteins [1]. Soon afterwards the protein was shown to be photoactive and hence it was named photoactive yellow protein (PYP). Similar proteins were found in a number of related phototrophic bacteria [2,3]. These proteins are yellow because of their unique chromophore, an anionic cinnamin derivative – a chemical structure not previously shown to play a role in photobiological signal transduction processes.

Subsequent to its discovery, evidence was presented that PYP has a role in bacterial phototaxis. Furthermore, similar proteins, and/or the genes encoding such proteins, were found in several bacterial species belonging to the family of the proteobacteria. This has led to the proposal to group these proteins in the xanthopsins family [4]. This then is a group of blue-light photoreceptor proteins that contain 4-hydroxycinnamic acid as their photoactive chromophore. The xanthopsin from *Ectothiorhodospira halophila* will be referred to as E-PYP.

Photoactive yellow protein from *E. halophila* displays excellent stability, both chemically and photochemically. Furthermore, the possibility to heterologously express the apo-protein in *Escherichia coli* and reconstitute fully functional holo-protein in vitro has made it possible to produce large quantities of PYP. This led to an avalanche of biophysical studies on the mechanism of the intramolecular signal transduction in the PYP protein, which is initiated by photoactivation.

These experiments have revealed many details of the mechanism of signalling state formation in PYP, and have made this protein a model system for studies in photochemistry and in protein folding. These studies will be reviewed below.

8.2   Biological function of the members of the xanthopsin family

The first photoactive yellow protein (PYP) was identified in *E. halophila* [1]. Subsequently, strikingly similar photoactive yellow proteins were isolated and purified from *Rhodospirillum salexigens* and *Chromatium salexigens*, which are also halophilic phototrophic purple bacteria [2,3]. All these proteins are present in very low copy number (approximately 500 molecules per cell [1]). This observation and their similarity to the sensory rhodopsins from archaebacteria suggested that they might function in a photosensory process. Like many other anoxygenic phototrophic bacteria, *E. halophila* also shows a positive phototactic response to (infra)red light. However, additionally, this organism is repelled by light of shorter wavelength and/or higher intensity. The wavelength-dependence of this light-induced repellent response of *E. halophila* follows the absorption spectrum of PYP [5]. Therefore, the current consensus on the function of PYP in extremophilic eubacteria is that of a photoreceptor in a light-induced behavioural response that allows the bacterium to avoid regions with (high intensities of) blue light.

Genes encoding a xanthopsin have also been identified in several additional purple bacteria [4,6,7] (see also Figure 1). For the xanthopsins of *Rhodobacter sphaeroides* [8] and *Rhodospirillum centenum* [7] it has also been possible to heterologously express and isolate the encoded (hybrid) photoactive yellow protein (see below and Section 8.6.3 for a more detailed description of these two photoreceptor proteins).

In part because the application of genetic techniques in extremophilic bacteria, such as *E. halophila*, is not well developed, a genetic proof for the function of PYP in the repellent photoresponse has not been provided. In contrast, *Rb. sphaeroides*, another phototrophic member of the proteobacteria that also contains a *pyp* gene, is, genetically, very well accessible. Computer-assisted motion analysis of this species clearly demonstrated a blue-light-induced repellent response analogous to that in *E. halophila* (see above), including a demethylation response, characteristic of behavioural adaptation [9]. In contrast to the previously reported motility response in these bacteria to a decrease in infrared light, this blue-light response does not depend on the number of photosynthetic pigments per cell, suggesting that it is mediated by a

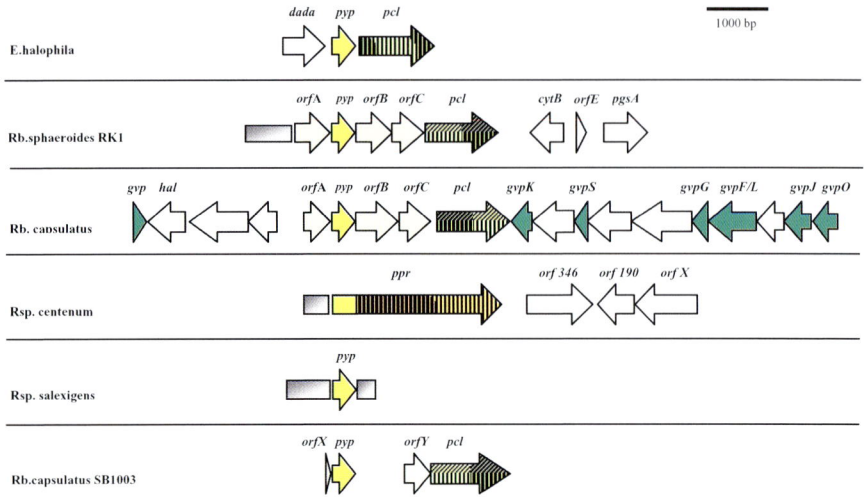

Figure 1. Structure of all known *pyp* operons. Sequences and *orf* assignment were taken from generally accessible databases: *orf*-organisation for *R. capsulatus* from the genome sequence project. *Orf*X and *orf*Y from *R. capsulatus* SB1003 are homologous to *orf*A and *orf*C, respectively. *pyp*–Photoactive yellow protein, *pcl*–putative CoA-ligase, *dada*–putative oxidoreductase, *cytb*–putative cytochrome b [12], *pgsA*–homologous to sll1522 (Cyanobase) and *pgsA* from *Bacillus subtilis*, *gvp*–gas vesicle protein, *hal*–putative histidine ammonia-lyase, *ppr*–phytochrome and pyp related (hybrid of pyp and cyanobacterial phytochrome-like protein). Grey boxes represent available DNA sequence in which significant Orfs could not be detected.

separate sensing system. Surprisingly, however, *pyp* knockout strains, derived from the 2.4.1 wild type, in which the *pyp*-gene was either interrupted by an antibiotic-resistance cassette or completely removed from the chromosome, did not show any impairment in their repellent response [9]. *Rhodobacter* also shows a phototactic response towards *low* intensities of blue light, but this stimulus then functions as an attractant, with the components of the photosynthetic apparatus as the sensor (A. Haker and J.P. Armitage, unpublished results). Also, for this latter response no involvement of PYP could be shown, nor did knockout of other genes in the *pyp* region of the chromosome of *Rb. sphaeroides* have this effect (A. Haker, unpublished results and [9]). Further molecular genetic experiments with *E. halophila*, as well as two-hybrid analyses with *Rb. sphaeroides* are in progress to obtain genetic proof for the function of these members of the xanthopsin family.

In *Rh. centenum*, however, a very explicit physiological function has been identified for its xanthopsin, i.e. the (light-induced) regulation of the expression of the enzyme chalcone synthase (an enzyme which plays a role, in plants, in flavonoid biosynthesis). A most noteworthy aspect of the xanthopsin moiety in *Rh. centenum* is that it is translationally fused to a phytochrome-like protein, Ppr [7]. Ppr, however, has an amino-terminal domain that is homologous to other members of the xanthopsin family (see Figure 2).

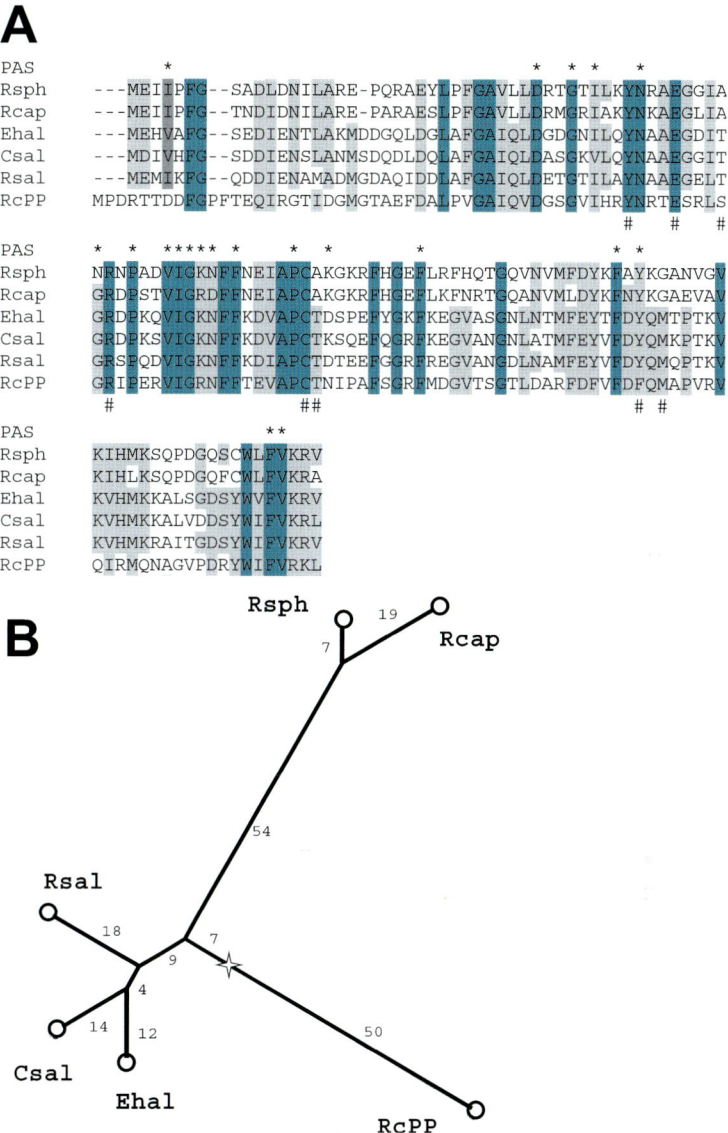

Figure 2. Sequence alignment (A) and phylogenetic tree (B) of all known PYP's. Rsph –*Rhodobacter sphaeroides* (EMBL AJ002398), Rcap–*Rhodobacter capsulatus* (EMBL AF064095), Ehal–*Ectothiorhodospira halophila* (EMBL X98887), Csal–*Chromatium salexigens* (EMBL X98888), Rsal–*Rhodospirillum salexigens* (EMBL X98888), RcPP–*Rhodosprillum centenum* (PYP domain) (AF064527). Sequence alignment was performed using CLUSTAL W. Conserved residues are indicated in green (in all six sequences) or in grey (homologous amino acids or residues conserved in at least four sequences). The symbol * indicates residues conserved in PAS domains [19], and # refers to key residues for chromophore binding in PYP (see Section 8.3). The phylogenetic tree was constructed using the program AllAll: Related peptide (http://cbrg.inf.ethz.ch), numbers represent calculated PAM distances, the cross indicates the weighted centroid of the tree.

8.3   Gene and protein structure in the xanthopsin family

8.3.1   Primary sequence of pyp genes

The primary sequence of six members of the xanthopsin family is known. All the organisms in which PYP has been identified are phototrophic purple bacteria, belonging either to the α- or the γ-subgroup of the proteobacteria. Comparison of these sequences shows the conservation of the characteristic amino acids involved in binding and tuning the chromophore (i.e. Y42, E46, R52, C69, Y98 (F for *Rs. centenum*); see also Figure 2 and Sections 8.3.5 and 8.3.6). However, other key residues for the functioning of E-PYP are not conserved throughout all members of the xanthopsins, i.e. T50, T70 and M100. These residues, also located in the chromophore binding pocket, are not present in both *Rhodobacter* proteins, and T50 is exclusively present in the xanthopsins from *E. halophila, C. salexigens and Rs. salexigens.* Moreover, most residues conserved in the PAS protein family are also conserved in the xanthopsins (see Section 8.3.7).

In addition, comparison of the amino acid sequence of the different xanthopsins reveals the existence of possible sub-groups in this protein family (Figure 2B). The proteins from *E. halophila, Rs. salexigens* and *C. salexigens* share a very high sequence similarity (between 69% and 77% sequence identity), as well as, mutually, the xanthopsins from the two purple non-sulphur bacteria, *Rhodobacter sphaeroides* and *R. capsulatus.* The PYP-domain of the hybrid protein Ppr from *Rhodospirillum centenum* takes a somewhat intermediate position with no strong sequence similarity with either of the two other groups. The sequence identity between these sub-groups is rather low (32–49%). The bio- and photochemical properties of these proteins correlate with the group assignment (see Section 8.6).

No new members of the xanthopsin family have yet been discovered through sequence analyses of the large amount of genome sequence information that is accumulating in the public domain (A. Haker and W. Crielaard, unpublished results). This may seem surprising as many organisms show a single cross-reacting band in a Western analysis with a specific antiserum against E-PYP [10]. The identification of *pyp* homologues, however, is complicated because many sequences obtained from the public database have sequence similarity with E-PYP, presumably because they are related to the family of PAS proteins. However, in none of these is a residue corresponding to the crucial side chain C69 present.

8.3.2   Genetic structure of the operons encoding a pyp gene

So far, the DNA-sequences of *pyp* and its flanking regions from five different organisms have been elucidated (see Figure 1). Also, the hypothetical operon organisation (only preliminary results are available with respect to the expressed transcripts) correlates with the assignment of sub-groups within the xanthopsins (see also Section 8.3.1). Whereas both *Rhodobacter* species share a similar operon structure (over five orfs), no similarity with the flanking regions of the other *pyp* genes can be identified. The only additional common gene present in three of the five organisms is a downstream gene encoding a CoA ligase homologue. This gene product is most likely involved in the conversion of the 4-OH-cinnamyl-chromophore into a CoA derivative. This reaction is well known in plants, where 4-OH-cinnamyl CoA is a key intermediate in phenylpropanoid metabolism (for review see [11]). Therefore, in these purple bacteria an activated CoA chromophore is presumably also formed and covalently linked to apo-PYP by either autocatalysis or an additional transferase, which remains to be identified.

In the upstream region of *pyp* from *E. halophila* an *orf* has been identified with homology to an oxidoreductase belonging to the Dada (D-alanine dehydrogenase) family. Whether this gene is connected to the functioning of PYP in this organism cannot be decided yet, in part because a homologous gene has not been found in the flanking region of any other *pyp* cluster.

Striking is the appearance of three additional *orfs* (A–C; *orf*X and *orf*Y from *R. capsulatus* SB1003 are homologous to *orf*A and *orf*C, respectively) in the two *Rhodobacter* species, which form a sub-group within the xanthopsin family. The identity of the products of these open reading frames is not known, since no significant sequence similarity with any other gene in the database could be found. The hypothetical operons are surrounded by genes involved in the synthesis of gas vesicles in *Rb. capsulatus* and in *Rb. sphaeroides* by a putative cytochrome (*orf*D–[12]) gene and an open reading frame encoding a protein possibly involved in the synthesis of acidic phospholipids (*orf*F –homologous to *pgs*A from *Bacillus subtilis*; A. Haker, unpublished results).

8.3.3   Heterologous expression of pyp genes

As already mentioned, genes encoding a xanthopsin can be heterologously expressed in, e.g. *Escherichia coli*. Constructed poly-histidine tagged versions of *E. halophila* PYP can be overproduced up to 2500-fold in *E. coli* in apo-protein form, which then allows a straightforward purification of the protein. A similar procedure has successfully been followed for the xanthopsins from *Rs. centennum* and *R. sphaeroides* [7,8]. In all cases holo-PYP has to, and can be, reconstituted by the addition of an anhydride derivative of 4-OH-cinnamic acid, to yield holo-PYP, as was first shown by Imamoto and co-workers [13]. As an alternative, activated cinnamyl-derivatives of *N*, *N*-carbonyl-di-imidazole may be used for this reaction. The availability of this reconstitution pathway has allowed detailed analyses of the effect of varying the chemical structure of the chromophore (see also Table 1): (i) Three separate contributions to tuning of the chromophore (to the visible wavelength region) were quantified. (ii) The chromophore-binding pocket was explored with respect to additional groups that can be accommodated. (iii) The effect of substitution was tested for the double bond that, supposedly, is sensitive to

Table 1. List of all hybrid PYP proteins, reconstituted with a non-physiological chromophore

No.	Name	Structure	λ_{max}	Photoactive?	Ref.*
I	4-Hydroxycinnamic acid		446		
II	3,4-Dihydroxycinnamic acid		457	yes	a, b
III	3-Methoxy-4-hydroxycinnamic acid		460	yes	a, c
IV	3,5-Dimethoxy-4-hydroxycinnamic acid		488	no	a
V	4-Aminocinnamic acid		353	no	a
VI	4-Dimethylaminocinnamic acid		436	no	a
VII	4-Methoxycinnamic acid		355	no	a
VIII	4-Fluorocinnamic acid		317	no	d
IX	Cinnamic acid		317	no	d
X	7-Hydroxy-coumarin-3-carboxylic acid		443	yes	c, e
XI	4-Hydroxyphenylpropiolic acid		460	yes	d
XII	4-Hydroxy-α-bromocinnamic acid		447	yes	d, f
XIII	4-Hydroxy-α,β-dideuterocinnamic acid		446	yes	d, f
XIV	Imidazole-4-acrylic acid		343	no	d

The physiological chromophore of PYP is shown in line 1 for comparison.* a, [26]; b, [99]; c, [51]; d, [107]; e, [14]; f, M.J. van der Meer, PhD Thesis, University of Amsterdam.

Table 2. List of all known PYP mutant proteins, generated through site-directed mutagenesis

Mutation	λ_{max} (absorption)	Most typical characteristics
Y42F	458 (~385)	Has been crystallised. Y42 hydrogen bonds to chromophore. Second chromophore conformation generates 391/380 nm shoulder
E46Q	462	E46 hydrogen bonds to chromophore and is proton donor for chromophore protonation
E46D	444 (345)	Mutation tunes chromophore pK, resulting in chromophore protonation already at neutral pH
E46A	465 (365)	See E46D
T50V	457	Has been crystallised. Weakened hydrogen bonding of chromophore leaves more electron density on the latter and causes a red shift
R52A	452	Shields chromophore from solution; not involved in chromophore charge stabilisation
R52Q	447	See R52A
C69S	–	No pigment binding
M100A	446	Hydrogen bonds to R52 and strongly affects the recovery rate. Causes shoulder at 355 nm
H108F	446	Hydrogen bonds to water-200
G47S, G51S, G47S/G51S	446	Have been crystallised. Hinge-bending mutants that show faster pB formation and slower pG recovery; additive effect in double mutant
D25, D27	444	Deletion of the first 25 (or 27) N-terminal residues. Slows down pG recovery and decreases the non-linearity of the Arrhenius plot of pG recovery

photoisomerisation, with a covalently closed six-membered hetero-aromatic ring, or even by substitution with a triple bond [14]. The availability of this convenient expression system has greatly advanced the biophysical experiments that have elucidated the detailed properties of the xanthopsins.

Site-directed mutations have been introduced in the *pyp* gene to change most of the key residues important for functioning of the xanthopsins (see Table 2), using standard techniques for site-directed mutagenesis in *E. coli*. The corresponding proteins have been used in many detailed biophysical studies to improve our understanding of PYP function (see further below).

8.3.4   Spatial structure of PYP

Detailed information on the 3D structure of the best-studied xanthopsin, PYP from *E. halophila* (i.e. E-PYP), has become available via X-ray crystallography and multi-nuclear NMR spectroscopy [15,16]. Whereas, initially, its structure was resolved to 1.4 Å resolution, this has been improved to 0.82 Å [17]. E-PYP is a relatively small protein containing 125 amino acids, with a relatively high abundance of negatively charged residues, as is often seen in proteins from halophiles. PYP displays a typical α/β fold, with an open, twisted, six-stranded, antiparallel β-sheet, flanked by four α-helices and a long, well-defined loop containing C69 (Figure 3A). This part forms the major hydrophobic core of the protein. Two of the α-helical segments (Asp-10 to Leu-15 and Asp-19 to Leu-23) are folded independently at the back of the central β-sheet and cover a second, minor, hydrophobic core. In the solution (i.e. NMR) structure this latter helix is disordered. E-PYP contains only one tryptophan (W119), which is clamped between the central β-sheet and the N-terminal α-helices in the minor hydrophobic core. This secondary structure confirms an early CD spectrum of the protein [18], which predicted this α-helical content. In terms of the "PAS fold" (see further below; [19]), the major hydrophobic core is composed of (i) the β-scaffold, (ii) the helical connector (i.e. the longest α-helix of PYP (α-5)) and the PAS core. The last contains a π-helix (flanking C69). This helix may be crucial for the flexibility of the protein backbone required for signalling state formation [20,21].

In addition, PYP carries a light-absorbing chromophore. From initial confusion [22] it took a rather long time to discover the precise molecular structure of PYP. Initially it was presumed that it would be a retinal protein. The first evidence showing its chromophore could not be a Schiff-base-linked retinal derivative was that the chromophore was linked to C69; a year later it was shown by NMR and various additional techniques to be identical to *trans* 4-OH-cinnamic acid [23,24]. This coumaryl chromophore is present in the anionic (i.e. phenolate) form in the ground state (pG_{446}) of the protein [25]. It is buried within the major hydrophobic core of the protein, where it is stabilised via a hydrogen-bonding network, involving the amino acids Y42, T50 and (protonated) E46 [15]. With site-directed mutagenesis the role of most of the amino acid side chains lining the chromophore-binding pocket has been investigated. In particular, the E46Q mutant has been instrumental in resolving proton transfer within the protein upon photoactivation.

Based on the (electrostatics of the) surface structure of PYP detailed speculations about a possible specific surface region that would be involved in signal transfer to a downstream partner were published [15]. Analyses of the structure of photocycle intermediates and the molecular dynamics analysis of E-PYP reveal, however, that such conclusions must still be considered as very preliminary.

8.3.5   *The chromophore-binding pocket*

In Figure 3B the chromophore-binding pocket of E-PYP is represented in its ground state configuration, as measured with X-ray crystallography [15]. Hydrogen bonds are represented as green dashed lines. The phenolate moiety of the chromophore is involved in a hydrogen-bonding network, in which E46 and Y42 directly interact with the negatively charged oxygen atom of the chromophore and T50 is hydrogen-bonded to Y42. On the opposite side of the chromophore the acyl-group oxygen hydrogen bonds to the backbone nitrogen of C69. Note that the chromophore, in solution with a pK_a of 8.9 [26], is

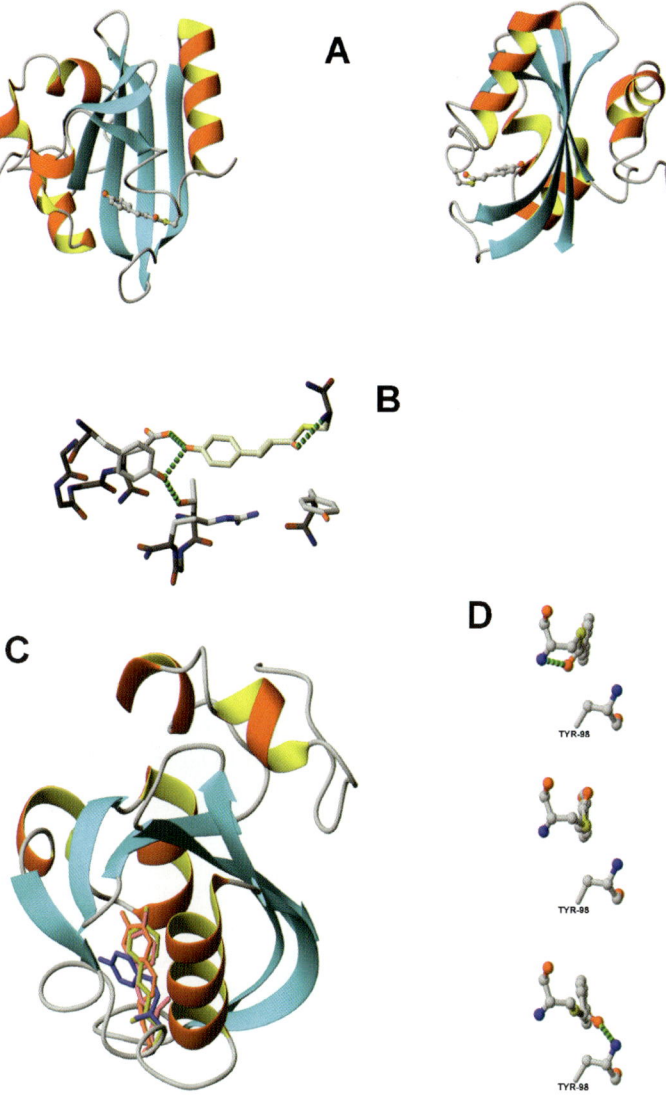

Figure 3. Spatial structure of the various states of E-PYP. (A) Two views of the backbone structure of PYP. Left: Front-view showing the major hydrophobic core containing the chromophore-binding pocket. Right: View highlighting the minor hydrophobic core (for further details: see text). (B) Detailed view of the chromophore-binding pocket of PYP. Colour coding has been selected according to standard rules. The dotted green lines represent hydrogen bonds. (C) Structural transitions upon photoactivation of crystalline PYP as determined by time-resolved Laue diffraction and low-temperature trapping. Colour coding of the configuration of the chromophore in the various intermediate states characterised: yellow: pG [15]; pink: PYPBL [17]; red: pR [20]; blue: pB [47]. (D) View along the long axis of the chromophore in pG (top), PYPBL (middle) and pR (bottom).

deprotonated and the E46, typically with a pK_a around 4.4, is protonated. Thus not only does E-PYP show spectral tuning, the pK_as of the chromophore and Glu-46 are also significantly tuned.

There is one major difference between this active site structure and that obtained with NMR spectroscopy. In the NMR structure R52 has two possible orientations, both different from that found with X-ray crystallography. Also, the Y98 side chain has a different orientation. In one orientation of R52 in the NMR (solution) structure, its two free amino groups are about 4 Å from the aromatic ring of the chromophore. In the other orientation they are about 4 Å from the aromatic ring of Y98. This is in line with the observation that positively charged amino groups like to pack within 3.4 to 6 Å of the centroids of aromatic rings [27]. That the R52A mutation has no influence on the absorption maximum of PYP [28] is in line with the NMR data.

8.3.6 Chromophore tuning

The absorption maximum of the free chromophore, *trans*-4-hydroxycinnamic acid, is 284 nm [29], whereas in the holo-protein the absorption maximum is at 446 nm [1]. This large red-shift of 162 nm (12790 cm^{-1}) is caused by the interaction of the chromophore with the protein. Three factors contribute to this shift [26]: (i) The thiolester linkage of the chromophore with C69 causes a shift from 284 to 338 nm, i.e. a shift of 5625 cm^{-1}. (ii) Deprotonation of the hydroxy group of the chromophore [25] causes a further shift to 410 nm, i.c. 5196 cm^{-1}. (iii) Various interactions of the phenolate chromophore with residues of the protein, which also help stabilise the negative charge on the chromophore, induce a final shift to 446 nm, i.e. 1969 cm^{-1}. The latter is a combination of several (possible) interactions, of which only the formation of a hydrogen-bonding network is undisputed. Other possible contributions are strain induced in the chromophore by the surrounding protein (this was shown to contribute to retinal tuning in bacteriorhodopsin [30]) and ionic and/or dipolar interactions between charged and/or dipolar residues and the anionic chromophore [31]. With this knowledge about the various contributions to the tuning of the chromophore it is possible to deduce that the large blue-shift observed upon formation of the pB intermediate (i.e. the signaling state) is at least in part caused by the protonation of the hydroxy group of the chromophore. Even in this state, however, the chromophore is still tuned by the protein, since the absorption maximum of pB is 355 nm. If the thiolester linkage alone tuned the chromophore an absorption maximum of 338 nm would be predicted. The chromophore-binding pocket of E-PYP allows considerable modifications of the endogenous chromophore, without significant impairment of yellow pigment formation [26]. Up to two bulky methoxy groups can be added. As expected, elimination of the 4-hydroxy group is not permissible in this respect.

8.3.7   The PAS domain: A key element in biological signal transfer

The PAS domain [32] is a key element in many signal transduction pathways, including many in eukaryotic cells [19,33]. It was first noted by Lagarias that there was identifiable sequence conservation between the PAS consensus sequence and PYP [34]. Subsequently the spatial structure has been resolved for four members of the PAS-domain family (including PYP) [33], and they all indeed show a strongly similar fold of their backbone. PYP has been nominated as *the* structural prototype for the complete (not only PAS-core) three-dimensional fold of the PAS domain superfamily [19]. It displays all the structural and functional features characteristic of the PAS domain superfamily (i.e. N-terminal cap; PAS core; helical connector and β-scaffold). In PYP, the essential PAS-core region extends from residues 29 to 69 and contains a number of conserved residues, including three glycines (G47, G51 and G59). Of these three, G47 is conserved only in the xanthopsins, whereas G51 and G59 are conserved in the entire family of PAS domains. Strikingly these are the same (conserved) glycines that are of fundamental importance in the concerted motions of the protein [35]. A mechanistic function often ascribed to PAS domains is a role as a domain for (hetero)dimerization. Surprisingly, no significant indications for such a function in E-PYP have yet been obtained.

8.4   Dynamical changes in the configuration of the chromophore of PYP from *E. halophila* upon photoactivation

8.4.1   Introduction

The functional activity of almost any protein, including all enzymes, requires rapid dynamical fluctuations in its structure. Signal-receptor proteins are ideal model systems to study such dynamic transitions, because of their intrinsic capacity to transiently form a signalling state, which has to have a long enough lifetime to be recognised by the downstream signal transduction chain. In a photoreceptor protein these dynamical changes are initiated by photon absorption, and therefore can be resolved – through the use of pulsed lasers – with up to fs time resolution. In most cases these dynamical transitions in structure are reflected in the colour of the pigment(s) in the photosensor protein. Nevertheless, many additional techniques are available to characterise the structural rearrangements. Below we will discuss these studies as performed on the xanthopsins, in particular on E-PYP. We will discuss first the changes in the configuration in the chromophore of PYP and, subsequently, discuss the conformational transition triggered by chromophore isomerisation in the surrounding protein

In photoactive proteins the chromophore is at the heart of the changes that occur inside the protein as a result of photon absorption. This chromophore is usually – the green fluorescent protein is an exception–a prosthetic group that is bound to the apo-protein to form the holo-protein. Since the usual function of a chromophore is to catch light from the visible region of the electromagnetic spectrum, UV/Vis absorption spectroscopy is the first obvious technique to study such a protein. PYP displays a photocycle, i.e. absorption of a photon induces temporary changes that eventually make the holo-protein return to its ground state (or dark adapted state). This greatly facilitates kinetic studies because of the possibility to apply signal-averaging techniques. Where there is absorption of photons, there is also a non-zero probability for emission of photons. Therefore fluorescence spectroscopy can report on the initial events following photon absorption in the holo-protein. Nevertheless, many additional techniques can give us information on the state of the chromophore during the photocycle, such as Raman spectroscopy, Fourier-transform infrared spectroscopy (FTIR), photoacoustic spectroscopy (PAS), nuclear magnetic resonance spectroscopy (NMR), and X-Ray diffraction crystallography. All of these have been used to characterise the events elicited in E-PYP by photon absorption both in the chromophore and in the surrounding protein.

8.4.2   Photocycle of E-PYP

Models of the photocycle of E-PYP have become ever more complex. Nevertheless, it can still be depicted simply with only three basic types of species (Figure 4). The (most) stable state is the ground state (or dark adapted state) that is ready for photon absorption and in which the chromophore is deprotonated and in the *trans*-configuration. The second species is spectrally red-shifted with respect to the ground state and is formed on the ns timescale. In this state the chomophore is still deprotonated, but is isomerised to the *cis*-configuration. The third (type of) species is spectrally blue-shifted with respect to the ground state, is formed on a μs timescale and lives long enough to transmit a signal to a downstream partner. In this state the chromophore has become protonated and is still in the *cis*-configuration. The photoreceptor protein spontaneously recovers to its ground state on a sub-second timescale.

The naming of transient- or intermediate states in photoreceptor proteins, unfortunately, is not subject to strict rules; it is often quite arbitrary. One uses characters, numbers and abbreviations derived from the name of a property specific for a certain intermediate, etc. For some photoreceptors (such as phytochrome [36,37]) different nomenclatures have coexisted for decades. Due to the application of a growing number of spectroscopic techniques, with capabilities for time-resolved detection other than visible color, additional intermediate states with structural differences but identical absorption properties are observed and have to fit into the existing photocycle schemes. E-PYP also suffers from the annoying situation of the use of more than one nomenclature. Three different ways have been proposed to name the basic photocycle intermediates, as differentiated by UV/Vis absorption spectroscopy (Figure 4A). As a result these different nomenclatures will have to be used here. However, the different nomenclatures can be overlaid and compared, using the three basic

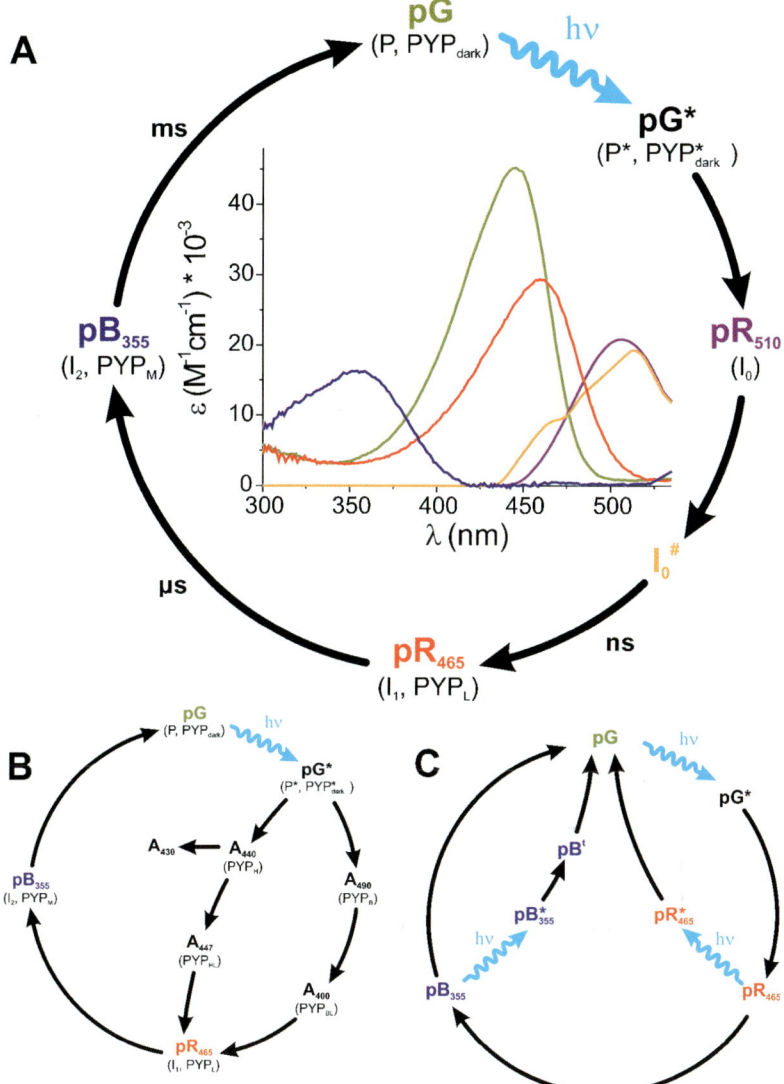

Figure 4. Photocycle of PYP at room and cryogenic temperatures. (A) Photocycle showing all known intermediates (with all published nomenclature) and approximate rate constants for E-PYP, incubated at neutral pH and ambient temperature. For further explanation and references. see the text. (B) Branched photocycle of E-PYP at cryogenic temperatures [40]. (C) Light-dependent branching reactions detected in the photocycle of E-PYP [71] (see text for further details).

types of photocycle intermediates described above. Initially, the ground, the red-shifted, and the blue-shifted states were called P, I_1 and I_2, respectively, by Meyer et al. [38]; there is uncertainty about the involvement of an I_3 intermediate [18]. Subsequently, the names pG, pR and pB were introduced by Hoff et al. [39]. Imamoto et al. introduced yet another nomenclature [40] with PYP (or PYP_{dark}), PYP_L and PYP_M, which was based on similar nomenclature to that introduced for the rhodopsins. In addition, absorption spectroscopy at cryogenic temperatures has added five intermediate states between pG and pR. Only one of them has been tentatively assigned to a room temperature intermediate found in ultrafast absorption spectroscopy. For the low temperature states two nomenclatures exist (Figure 4B). This situation is very confusing, and with new results from time-resolved FTIR, photoacoustic spectroscopy etc. the number of intermediates will grow even further. Also, the use of a subscript stating the absorption maximum can be confusing, since the same intermediate can have a different absorption maximum, depending e.g. on pH or temperature. A unified naming scheme is urgently needed but is outside the scope of this review. In the following we will use the names given first in Figures 4A and 4B.

8.4.3   Primary photoreaction(s) of PYP as studied with transient absorption detection

It was shown with fs transient absorption spectroscopy ($\lambda_{exc} = 400$, 460 nm), observing the first 10 ps of the photocycle, that an intermediate with a red-shifted absorption spectrum compared with pG is formed with a time constant of 2.8×10^{11} s^{-1} [41]. This process is preceded by a relaxation process with a rate constant of 1.4×10^{12} s^{-1} on the excited state surface, producing an excited state intermediate in which the *trans–cis* isomerisation may have already occurred. The intermediate formed in several picoseconds was tentatively assigned to pR_{465}, the first intermediate found in transient absorption measurements with nanosecond time-resolution. A subsequent experiment with ps time resolution and a ns observation time window ($\lambda_{exc} = 452$ nm), however, established the appearance of an intermediate preceding the formation of pR_{465} (I_0; from now on: pR_{510}) and formed in less then 3 ps [42]. The time constant of pR_{465} formation was determined to be 3 ns. In the same report another intermediate ($I_0^\#$) – either a real state with a broader absorption spectrum, very similar to pR_{510}, or a thermal equilibrium of pR_{465} and pR_{510} – was detected in between pR_{510} and pR_{465}, appearing with a rate constant of 4.55×10^9 s^{-1}. More recently, Devanathan et al. found a wavelength dependence of PYP excitation in a transient-absorption study with fs-excitation and -time resolution [43]. According to their report excitation at 395 nm leads to a much more red-shifted stimulated emission signal than excitation at 460 nm. The existence of an energetically higher-lying excited state was concluded from this difference, which – when excited – would form pR_{500} via an excited state (pR_{510}^*). In their interpretation [43] excitation at 400 nm leads to the formation of a higher lying pG** which can relax to pG* or cross to pR_{510}^*. The pR_{510}^*–pR_{510} energy difference is assumed to be smaller than that of pG*–pG which would lead to the red-shifted stimulated emission. In other words, it is proposed that isomerisation takes place on the excited state surface, if the isomerisation is the primary reaction. The time-dependence of the difference absorption signal within the first 7 ps was described in a global fit with a mono-exponential function, and a characteristic time of 1.1 ps was derived. From this they estimated the rate constant of pR_{510} formation as 4.4×10^{11} s^{-1} [43,44]. In a similar investigation by the same authors of the E46Q mutant, the same sequence of appearance of intermediates (pG $\rightarrow$ pR_{510} $\rightarrow$ $I_0^{\#}$ $\rightarrow$ pR_{465}) was observed although spectra were not determined [44,45]. E46 forms a strong hydrogen bond with the negatively charged phenolic oxygen in the ground state of PYP. Replacement of E46 by a glutamine is supposed to weaken this hydrogen bond substantially [28,44]. While the rate constant of pR_{510} formation in the mutant protein was very similar to the corresponding one in wild-type PYP, the two slower transitions were significantly faster in the two studies (although they mutually differ by almost a factor of two, i.e. 8 ps [44] / 14 ps [45] vs. 220 ps [42] and 0.7 ns [45] / 1.7 ns [42] vs. 3 ns [42]). These differences are assigned to a significant difference in the signal-to-noise ratio in the two studies (with this ratio being superior for Zhou et al. [45]). In both studies it is proposed, on the basis of the larger rate constants for $pR_{510} \rightarrow I_0^{\#}$ and $I_0^{\#} \rightarrow pR_{465}$, that these two processes involve movements of the phenolate ring of the chromophore. The larger rate constants, however, could also be because mutant protein movements of this phenolate ring are possible, which would be largely blocked in wild-type E-PYP, due to stronger hydrogen bonding. The fact that the rate of the primary reaction (pG $\rightarrow$ pR_{510}) is not altered is interpreted by Devanathan et al. [44] as support for the rotation of the chromophore carbonyl group as the primary photoreaction (as proposed by Xie et al. [46]; see also Genick et al. [17]) rather than a single bond photoisomerisation, which would require release of the phenolate ring from its binding pocket (see also Genick et al. [47]). However, Zhou et al. [45] derive, from the differences between wt-PYP and the E46Q mutant, with respect to the kinetics of the pG $\rightarrow$ pR_{465} transition as well as the vibrational spectra (measured by ps resonance coherent anti-Stokes Raman spectroscopy PR/CARS), a hypothetical model where the C=C isomerisation takes place after formation of $I_0^{\#}$. Both interpretations are rather speculative and need further support by, e.g., ultrafast vibrational spectroscopies. Since there is evidence for chromophore isomerisation in a low temperature intermediate, we consider it most plausible that the isomerisation takes place before the formation of pR_{465}.

The photoisomerisation of the free, deprotonated chromophore in the *trans*-configuration has also been investigated by ultrafast transient absorption spectroscopy [48]. The *trans*/*cis* isomerisation occurs with a single rate of 7.7×10^{10} s^{-1}. No evidence for intermediate states was obtained. This is in contrast to the complex multi-exponential pG $\rightarrow$ pR_{465} transition which lasts for nanoseconds and most likely represents the full isomerisation of the chromophore in the protein matrix. If the pG $\rightarrow$ pR_{510} transition contains the photoisomerisation, then the protein matrix speeds up this first event as has been shown for other photoreceptors with isomerisable chromophores (such as in the rhodopsins [49]).

8.4.4   Primary photoreaction(s) of PYP as studied with time-resolved fluorescence

At room temperature PYP is only weakly fluorescent ($\Phi_F \approx 0.0015$) [50]. Nevertheless, the time course of the fluorescence decay after excitation can help to elucidate the photoreaction by giving information about the excited state surface. A multi-exponential decay of PYP fluorescence has been observed in time-resolved fluorescence spectroscopy studies, based on the fluorescence up-conversion technique [50,51]. Its two fastest components account for 80% of the total fluorescence and have characteristic times of 700 fs and 3–4 ps. From this, one can conclude that the excited state manifold is depopulated within 10 ps and to a large extent already in the first picosecond. This also sets the timeframe for the primary photoreaction event. Two or three minor components in fluorescence decay have a time constant in the ns time range, as measured with time-correlated single photon counting with low amplitude, but no functional role has been assigned to them [51]. In a recent more detailed study, the temperature dependence of the fluorescence decay of PYP, and some site-directed mutants with substitutions in the chromophore binding pocket, has been investigated [52]. Interestingly, the sub-ps component is not temperature dependent (reflecting an activation free process), whereas the next two ps components show normal Arrhenius behaviour, with activation energies of 8 and 30 kJ mol^{-1}. It is suggested that the sub-ps process reflects a barrierless or coherent process from the non-relaxed excited state to a twisted state in the course of the *trans*-to-*cis* photoisomerisation. All the mutants lacked the fast component while the other processes were slower compared with wild-type E-PYP. This is explained by a looser structure of the chromophore-binding pocket in the mutants. Interestingly, fluorescence decay from the chromophore in the denatured protein is much slower and shows mono-exponential behaviour [52]. Under these conditions the time course of the fluorescence is very similar to that of the *trans/cis* isomerisation of the deprotonated free chromophore [48]. This can be taken as a proof for the involvement of the protein matrix in the increased rates of the primary photochemistry of the chromophore in PYP.

8.4.5   Photocycle of E-PYP at low temperatures

Low temperature spectroscopy allows trapping of intermediates after photoactivation and in this way an investigation of the early steps in a photoreaction cycle. Evidence suggesting a rather complex photophysical scheme of the first part of the PYP photocycle has been obtained in spectroscopic studies at low temperatures, in particular through the excitation wavelength dependence of its primary photochemistry. Hoff et al. [53] reported the existence of at least three species upon prolonged illumination (λ_{exc} = 400 and 447 nm) at 77 K. Two of them are primary photoproducts of the ground state pG (A_{490} and A_{440}, named after their absorption maxima). The third is a secondary photoproduct from A_{440}, which is highly fluorescent (A_{430}). When using excitation in the red edge of the absorption spectrum (λ_{exc} = 475 nm) the rate of production of A_{490} as well as of A_{430} decreases substantially, while A_{440} becomes the major photoproduct. In a hole-burning study at 10 K, Masciangliolo et al. have observed a similar difference in the composition of the photoproducts produced upon illumination at 431 and 475 nm [54]. Imamoto et al. investigated this unusual excitation wavelength dependent photochemistry by testing seven excitation wavelengths in the blue- and red-edge of the pG spectrum in steady-state absorption measurements at 83 K [40]. They report the formation of different amounts of A_{490} (PYP_B) and A_{440} (named PYP_H and PYP_B), depending on the excitation wavelength. The largest difference was found for the step from 460 to 480 nm. Both states transform upon heating on separate pathways, first via yet another intermediate [PYP_{HL} (λ_{abs} = 447 nm) and PYP_{BL} (λ_{abs} = 400 nm), respectively] into the intermediate pR_{465} and subsequently pB_{355} and pG (see Figure 4B). In all three studies, however the overlapping absorption spectra of ground state and trapped photointermediates, and the prolonged illumination times mean that photoback- and side-reactions can occur. The A_{430} species may be the result of such behaviour. For excitation wavelengths shorter than 460 nm, A_{490} and A_{440} are approximately equally (i.e. within 20%) populated. Only excitation at 475 or 480 nm shows mainly A_{440} formation. It is necessary to further test the existence of two distinctively different primary photoproducts with pulsed excitation, under conditions where no secondary photoreactions can occur.

It is very surprising that both primary photoproducts at low temperature form the same pR_{465} [40]. Definite proof for this behaviour – that is an excitation wavelength independent photocycle from ns onwards – at room temperature is not yet available. So far, it has only been shown that the yield of pB formation is independent of excitation wavelength from 408 to 458 nm [55].

8.4.6   Formation of pR as studied by photoacoustic and photothermal methods

Absorption of a photon drives a molecule into an excited state. Typically, several non-radiative relaxation processes will take place within picoseconds on the excited state surface, before the molecule returns to the ground state. This return can occur by emitting a photon, the formation of a long-living excited state (triplet state), reversible- (isomerisation, charge transfer state) and irreversible photochemical reactions or non-radiative deactivation. Most excited state relaxations and photochemical reactions can be time-resolved via transient absorption spectroscopy in a time range starting from femtoseconds. The same is true for fluorescence detection, with respect to the fluorescence deactivation channel. But the non-radiative deactivations are generally not accessible to optical spectroscopy. They can be followed, however, in a time-resolved manner by photoacoustic and photothermal methods [56–58], which record among other phenomena the amount of heat released, i.e. the change in enthalpy. The time resolution in this technique spans from ps to seconds, depending on the particular method selected. These methods can be very sensitive since the energy released in non-radiative deactivation processes can easily amount to 50–95% of the absorbed energy. Although this principle has long been known only a few experimental set-ups have been established in the past thirty years. Two of them have been used to characterise the enthalpy changes during the PYP photocycle, in the pG-to-pR$_{465}$ transition, i.e. photoacoustic spectroscopy [PAS, also named light-induced optoacoustic spectroscopy (LIOAS)] and thermal grating (TG). Both methods suffer because processes other than heat release contribute to the signal generated, such as structural volume changes (PAS, TG) and absorption changes (TG). One can, however, separate the different contributions to the signal and obtain additional information from the structural volume changes. They reflect alterations in bond length, solvation, protonation state and interactions of the chromophore with surrounding amino acid side chains. The structural volume change can be related–under certain circumstances – to the difference in entropy between two photocycle intermediates. [59,60]

In the first photoacoustic investigation [55] the energy content of the pR$_{465}$ intermediate was determined as 120 kJ mol^{-1}. The formation of pR$_{465}$ is accompanied by a large negative volume change (–23 Å^3). To estimate the two values it was assumed that the photocycle quantum yield and the enthalpy and structural volume changes are temperature independent. This is standard procedure in PAS studies [57] and was found to be valid for a vast amount of molecules. Terazima and co-workers applied PAS in combination with TG to PYP [61]. With TG it is possible to separate the heat dissipation in the first step of a reaction (pG $\rightarrow$ pR$_{465}$ for PYP) from the other contributions to the TG signal, due to the orders of magnitude faster heat diffusion compared with the sample molecule diffusion [62]. In this way the structural volume change could be determined without assuming its temperature independence. A value of –12 Å^3 was found at 293 K, which is also a contraction, but only half the value calculated from PAS [55]. An energy content of 160 kJ mol^{-1} was obtained for pR$_{465}$ at this temperature. Furthermore, the structural volume change of the pR$_{465}$ formation was found to be temperature dependent. With decreasing temperature the magnitude of the contraction increases. At 273 K it amounts to –25 Å^3, a value very similar to the one measured with PAS only [55]. The TG signal was not large enough at lower than 293 K to decide whether the energy content of pR$_{465}$ is also temperature dependent. The good agreement between the values of the structural volume change at 273 K from TG and PAS [61] and that from PAS alone [55] can be rationalized by an inherent property of PAS studies in aqueous solution, then by high weighting of the experiments at temperatures near the temperature with zero thermal expansion coefficient (277 K in pure water). The unusual and strong temperature dependence of the structural volume change is attributed to changes in the void volume of PYP or a change in interactions of certain amino acid residues with the chromophore and/or the solvent in the pG-to-pR_{465} transition [61].

TG and PAS measurements have also revealed the enthalpy change and the structural volume change of the *trans/cis* photoisomerisation of the bare chromophore *p*-coumaric acid [63]. Interestingly the photoisomerisation yield of the chromophore in aqueous solution is high compared with that in the protein. In the retinal proteins this is the other way around; the protein increases the yield of the isomerisation of the chromophore. The *cis* isomer has increased energy content (about 50 kJ mol^{-1}) and a small contraction (-1.2 Å^3) occurs during the isomerisation reaction, much smaller, however, compared to contraction during photoisomerisation in PYP. Direct comparisons of these values with the ones in the protein are not possible for two reasons. First, the chromophore is deprotonated in pG (while it was protonated in the study of Takeshita et al. [63]). Second, the chromophore has very specific interactions with its protein environment, which most probably change the energetic and structural features of the photoisomerisation dramatically compared with the situation in free solution.

It is interesting to compare the amount of stored energy and the structural volume change with the values obtained for other photoreceptors. The energy content in the first intermediate (accessible with the timescale of the photo-thermal methods used) of the PYP photocycle is about 60%. Similarly high values have been found for the membrane-bound photoreceptors bovine rhod-opsin (bathorhodopsin stores 54% [64]), sensory rhodopsin I from *Halobac-terium salinarium* (62% stored in K_{610} intermediate [65]) and sensory rhodopsin II from *Natronobacterium pharaonis* (40% stored in K_{510} [66]) as well as for the soluble phytochrome (82% stored in I_{700} [67]). In contrast, the K intermediate of bacteriorhodopsin from *Halobacterium salinarium* has a very low energy content of only 14% [68,69]. Formation of the first intermediate of all these photoreceptors involves the photoisomerisation of the chromophore. The high energy content of the signal-transducing photoreceptors, as compared with the proton pump bacteriorhodopsin, could be required to drive larger conforma-tional changes in the later steps of the photocycle of the sensory proteins, which presumably trigger signal transduction. The absolute value of the struc-tural volume change of pR_{465} formation is one of the largest among the studied photoreceptors. Due to the many processes contributing to the structural volume change and the complex superposition of their effects, however, a comparison of the structural volume changes of the photoreceptors is not very substantial. One should also bear in mind that slight modifications of the buffer (such as the ionic strength) as well as of the protein (His-tag, presence of transducer protein) can lead to significant changes in the energy content and structural volume of particular photocycle intermediates as has been recently reported for the K-to-L transition of sensory rhodopsin II from *Natronobacterium pharaonis* [60].

So far, only results on the pG-to-pR_{465} transition as studied with photo-thermal methods have been published. New, as yet unpublished, results on the timescale of pR_{465} formation have been obtained recently. While pR_{465}

formation is completed within 10 ns [42], when measured with transient optical spectroscopy, it extends to several microseconds as measured with opto-acoustic methods (unpublished experiments T. Gensch and others). More detailed insight into the energetics and dynamics of the PYP photocycle is also expected to result from application of photoacoustic and photothermal methods when applied to the later photocycle steps.

8.4.7 Transient absorption spectroscopy of E-PYP in the μs to ms time domain

Many studies have been performed to characterize the conversion of pR into pB, and the subsequent reaction, pG recovery, using UV/Vis transient spectroscopy in the μs to ms time range. Various models can fit these transitions, from single exponential fits, via multi-exponential deconvolution, up to distributed kinetics (i.e. a Gaussian distribution of the rate constants), depending on the signal-to-noise ratio present in the available data. For instance, for the pH-dependence of the recovery kinetics the best fit was obtained with a mono-exponential fit at acidic and alkaline pH values, but with a bi-exponential fit for the neutral pH range [108]. Also, when a sample is irradiated for an extended period, an additional slow component develops in the recovery reaction (T. Gensch et al., unpublished observations). Generally, however, in a fresh sample, the pR to pB transition is bi-exponential (with roughly equal amplitudes) and the recovery reaction mono-exponential.

The pH is probably the most widely varied parameter to modulate photocycle kinetics. Such experiments show that the formation of pB from pR is fastest in an acidic environment and slows down in a neutral to basic environment, via a transition with a pK_a of 5.7. Recovery from pB to pG shows very different pH dependence. Here, the rate has a bell-shape dependence on pH, with a maximum rate observed at pH 8 and apparent pK_as at 6.4 and 9.4. This pH dependence must be taken into account when comparing different sets of data. In addition, there is also evidence that the spectral characteristics of E-PYP intermediates are pH dependent. For instance the maximum of pB shifts from 365 nm (at low pH), via 355 nm (at neutral pH), to ~430 nm at high pH [70,71]. The latter can only be explained by assuming deprotonation of the chromophore in the pB state. Low recovery rates can be exploited to accumulate the signalling state of PYP (i.e. pB; see e.g. [70,72]). At the alkaline side of the optimal pH for recovery, very low rates are not achievable, however, because of the limited stability of the thiolester linkage between the chromophore and C69. At acidic pH values, the recovery rates can be slowed down to 10^{-3} s^{-1} at pH 2.5 [73]. Measurements at this very low pH, however, are complicated by acid denaturation of PYP into pB$_{dark}$, which occurs with a pK_a of 2.7 [73]. Besides information on the immediate environment of the chromophore, kinetic UV/Vis studies can also provide information on the behaviour of the entire protein, e.g. via the dependence of the recovery rate as a function of temperature. This will be discussed in Section 8.5.

8.4.8 Light-induced branching reactions

After photoisomerisation, the chromophore of PYP is re-isomerised in the dark during the recovery from pB to pG by the PYP protein. There is no reason to believe that this cannot also be achieved with the aid of photons. Indeed it was initially shown by Miller et al. that illumination of PYP with light of 366 nm speeds up the recovery from pB to pG. We [71,74,75] later showed that upon photoisomerisation of the chromophore in pB, a light-dependent shortcut reaction to pG is initiated that is about 1000 times faster than the normal dark recovery reaction. In this branching reaction an intermediate pB^t is formed instantaneously, on the ns timescale that is slightly blue-shifted with respect to pB. This slight blue-shift can be explained by the isomerisation of the chromophore from *cis* to *trans*. No additional intermediates are observed with UV/Vis spectroscopy in going from pB^t to pG.

The acceleration of the cyclic photoreaction by light-induced back-isomerisation is a general phenomenon for photoreceptors that contain a photoisomerisable chromophore, such as bacteriorhodopsin, phytochromes and sensory rhodopsin I. In the last protein it even has a physiological function. Such light-induced back-isomerisations are not only observed for the late intermediates [such as pB_{355} (PYP) [71,75], M (bacteriorhodopsin) [76], I_{bl} (phytochrome) [77], SR_{373} [78]], but also for intermediates formed on the ps time scale [K (bacteriorhodopsin) [79,80], I_{700} (phytochrome) [81]]. The quantum yield for light-induced back-isomerisations for the early intermediates is very high and usually substantially larger (1.5 and 2 times in the case of bacteriorhodopsin [80] and phytochrome [82], respectively) than the quantum yield of the forward isomerisation, i.e. the primary and biologically most relevant photoreaction. This is in line with the general picture of the photoreaction of photoreceptors with photoisomerisation as the primary step. The chromophore experiences, after the photoisomerisation, a strained configuration and unfavourable interactions with some nearby amino acid residues, which are still in a position that stabilises the ground-state configuration. The probability for a re-isomerisation is therefore high when a second photon is absorbed by the chromophore in such a non-equilibrium state.

For PYP such an analysis has been performed for the pR_{465} state by using photoacoustic measurements with a laser pulse with 10 ns FWHM. In contrast to the other photoreceptors the probability for the back-isomerisation from pR_{465} to pG is very low, less than one-fifth of the forward isomerisation ($\Phi_{pR465 \rightarrow pG} < 0.07$). When this study was made the pR_{465} precursors pR_{510} and $I_0^\#$ were unknown and it was believed that pR_{465} would be formed in picoseconds [41]. As shown later, pR_{465} is formed only within several nanoseconds [42]. Based on this it is likely that pR_{510} and $I_0^\#$ have a back-isomerisation quantum yield which is at least as high as the primary photoreaction. This implies that the formation of pR_{465} involves changes of the chromophore configuration and/or its protein environment, which stabilise the isomerised chromophore so that the subsequent step in the photocycle, i.e. the proton transfer from E46 to the chromophore, can occur. In the first intermediate pR_{510} these changes have not yet taken place.

When E-PYP is illuminated for prolonged times, particularly in relatively high concentrations of urea [18], but also in buffers with low ionic strength (T. Gensch et al., unpublished experiments), recovery of the ground state of PYP slows down considerably. The nature of these "slow-forms" of E-PYP remains to be determined, but is seems likely that secondary photoreactions of the intermediates of PYP may be involved in their formation.

8.4.9 Excitation wavelength dependence of the quantum yield

The quantum yield values reported for the PYP photocycle are surprisingly variable. Three studies determined the quantum yield of pB-formation and obtained values of 0.64 (λ_{exc} = 446 nm) [38] and 0.35 (λ_{exc} = 408, 440, 458, 480 nm) [55,82]. The numbers also differ in fs transient absorption spectroscopy, varying from 0.5 [43,44] and 0.35 [41] to 0.2 [42]. Of course, the accuracy of the quantum yield determination in the latter experiments is much lower. A thorough analysis of the quantum yield in the photocycle of E-PYP – also with respect to the excitation wavelength dependence–is still missing.

8.5   Dynamical alterations in the conformation of the PYP protein during signalling state formation

8.5.1 Intrinsic dynamics of the ground state of E-PYP

The intrinsic dynamics of the ground state of a protein is reflected in the correlated motions of groups of atoms, around their equilibrium position, as determined by, e.g., X-ray diffraction and/or NMR spectroscopy. These dynamics can be analysed by molecular dynamics modelling. In this approach the trajectories of each atom of the molecule are simulated on the basis of classical rate laws. Given current limitations in computer capacity these calculations can realistically be extended only up to the ns timescale for a single protein molecule in a box of water molecules.

From the correlated motion of different atoms, the eigenvectors of the intrinsic dynamics in a protein can be calculated. For PYP this analysis reveals several striking features. The protein shows correlated motion of its chromophore and the surrounding amino acid side chains; this dynamic flexibility is largely described by the three main eigenvectors from the essential dynamics analysis, and in each eigenvector the majority of the atoms in the protein are involved [35].

For several proteins it has been shown that the eigenvectors from the essential dynamics analysis describe the initial part of the trajectory of the conformational change that is important for functional activity of the protein [83,84]. It is a major challenge to relate the eigenvectors from the essential dynamics analysis of PYP to the conformational transitions that occur in the photocycle of this signal transduction protein.

8.5.2   Structural transitions during initiation of the photocycle of E-PYP

The changes in the spatial structure of PYP that occur upon initiation of photocycling have been studied with X-ray crystallography in various forms, e.g. with continuous laser illumination (for pB; [47]), cryo-trapping (for a very early intermediate; [17]), and time-resolved X-ray diffraction at ambient temperatures (for pR, [20] and unpublished results). The cryo-trapped intermediate was identified as A_{400} but, considering the temperature and illumination conditions used, could also be composed of a mixture of intermediates. The initial alterations of the structure of PYP, elicited by light, can be described as a flip of the carbonyl group around the long axis of the chromophore, with a 3.4 Å displacement of the oxygen atom and rearrangement of the dihedral angles around the sulfur atom (leading to a 1.4 Å displacement).

The chromophore itself has a stretched *cis*-configuration, showing that isomerisation has already taken place, and the movement of the thiolester carbonyl oxygen, as proposed from FTIR data [46], is evident from the structure of this cryo-trapped intermediate. The chromophore's aromatic ring moves slightly (0.4 Å) because of the "contraction" of the chromophore, due to the isomerisation [17]. To accommodate these changes, the most notable alteration in the residues lining the chromophore is a 0.5 Å shift in the position of the aromatic ring of Phe-96. Together with the small packing gaps that flank this residue, this allows the carbonyl "flip", without significant collisions.

No hydrogen bond is yet obtained with Y98, but the thiolester carbonyl oxygen is rotating *clockwise* towards the backbone nitrogen of Y98, when viewed from the C69. In the cryo-trapped intermediate the degree of rotation is 166° of the 282° needed to reach its position in the pR intermediate (see Figure 3D). The time-resolved crystallographic data [20] also initially shows only changes around the chromophore, but now with the orientation of the carbonyl oxygen in a position such that it can hydrogen bond to Y98. In addition, the *cis*-configuration of the chromophore is less stretched. In the cryotrapped intermediate the hydrogen bonding between the chromophore and E46 and Y42 is still intact. For the pR structure obtained at room temperature initially it was concluded that the hydrogen bond between the chromophore and E46 had been disrupted (because their mutual distance is >5 Å; [20]). More recent refinements, however, show that E46 and the chromophore move in concert, so that this hydrogen bond presumably will remain intact during pR formation, which is in full agreement with FTIR results [46,85].

The transition to the pB intermediate is next initiated by transfer of the proton from E46 to the chromophore. The buried charge that is generated in that process destabilises the PYP protein to such an extent that a partially unfolded protein with a protonated chromophore results (see also [86]). The resulting structure of the pB intermediate is, however, dependent on the mesoscopic context in which the protein functions (see further below).

Re-isomerisation of the chromophore to the *trans*-configuration (which also proceeds at high rates in crystalline PYP) has to be initiated from the pB structure. In this conformation the carbonyl oxygen of the thiol-ester group is again hydrogen bonded to the backbone nitrogen of C69. However, no detailed rationalisation as to how the PYP protein may facilitate this re-isomerisation has been given.

8.5.3   Conformational transition underlying signalling state formation in E-PYP

The crucial activity of a photosensor protein is to transduce photon absorption into a change in conformation that can be detected by a downstream partner. To achieve this the photoreceptor protein has to enter a so-called signalling state. Information about the nature of transient signalling states in photoreceptor proteins is very scarce. Clearly identified signalling states at the functional level have only been described for the bacterial rhodopsins, initially through the use of retinal analogues (e.g. [87]) and predominantly with respect to their spectral properties. These studies have revealed that the long-living blue-shifted intermediate functions as the signalling state for the archaeal sensory rhodopsins. Based on their mutual similarities it has been presumed that the blue-shifted intermediate pB is the signalling state of PYP. Formation of this state shows characteristics typical for a (partial) protein unfolding reaction (see further below); its rate (i.e. 10^4 s^{-1} [88]) is compatible with this interpretation [89].

It is therefore very exciting that the spatial structure of transient intermediates of E-PYP, with lifetimes ranging from ns to s have become available through the application of time-resolved X-ray diffraction (or Laue diffraction) analysis, applied to crystalline PYP [20,47]. This approach allows Ångstrom resolution at the ns time scale and so may even come close to revealing the nature of the transition state in conformational transitions in proteins. Figure 3C summarises results so far obtained. Upon excitation, initially the carbonyl group rotates around the long axis of the coumaryl chromophore (Figure 3B); subsequently the phenolate moiety is gradually (in the sub-ms timescale) exposed to solvent (Figure 3C). This structure relaxes to the ground state in the s timescale. Surprisingly, the backbone structure of PYP is hardly affected by these transitions, except for some very small displacements near the chromophore-binding site. Some side chains, however, in particular R52 and R124, do significantly alter their orientation during pB formation, leading to a structure with a chromophore fully exposed to the surrounding solvent.

8.5.4   Mesoscopic dependence of the conformation of PYP in the signalling state pB

NMR studies of the structure of the pB intermediate of PYP, however, have resulted in a very different picture of the characteristics of this intermediate. Changes in the HSQC spectrum of PYP [72] suggest that a large part of the secondary structure of the protein has been altered in the presumed signalling state pB, which actually is better described as a collection of rapidly (ms) exchanging states. Because of this rapid exchange it has not yet been possible to determine the spatial structure of the pB intermediate in solution. Nevertheless, fully in line with this supposed significant change in secondary structure in the pB state, it was subsequently shown from amide H/D exchange data that although the protein has a solid exchange-protected core, in two other regions the chromophore loop and the N-terminal domain, the secondary structure of E-PYP in the pB state alters significantly [90].

We interpret this difference between the results of Laue diffraction and NMR experiments as evidence that the precise nature of the pB intermediate formed strongly depends on the mesoscopic context of PYP (i.e. whether the protein is dissolved as a monomer in aqueous solution, or present in a crystalline lattice). This view has now been fully confirmed by results of FTIR analyses, which show that the extent of the conformational transition upon pB formation is far larger in solution than in crystalline PYP [86]. In agreement with this we have observed that, although entrapment of E-PYP in a crystalline lattice does not affect its spectral characteristics, with respect to the kinetics of its photocycle transitions significant differences exist (J. Hendriks et al., unpublished observations).

In the intact cell, PYP presumably functions in a (large) signalling complex. Therefore, in vivo, PYP may have characteristics in-between those of the crystalline and the dissolved state. We consider it a major challenge to resolve these in vivo characteristics of PYP and its intermediates, and to elucidate their role(s) in xanthopsin-mediated signal transduction. Such studies have additional relevance because of the prototype function of PYP for all PAS proteins, and therefore may elucidate key features of the regulation of (eukaryotic) signal transfer in general.

8.5.5   Thermodynamics support for partial unfolding of E-PYP in the signalling state

The first indication to suggest a large structural change in E-PYP during pB formation was obtained from quantitative studies of the temperature dependence of photocycle transitions [91]. When the rate of the pG-recovery reaction is plotted against reciprocal temperature, strongly non-linear Arrhenius curves are obtained. Whereas simple (gas-phase) chemical reactions usually show a linear Arrhenius curve (i.e. a plot of $\ln k$ versus $1/T$), complex protein-folding reactions typically show (strongly) curved plots. This is usually explained by invoking a change in heat capacity (ΔC_p) in the transition under study [92], which arises from the exposure of hydrophobic surface area to water, in the denatured state; the hydrophobic side chains are surrounded by "icebergs" of water that melt with increasing temperature, thus making a large contribution to the heat capacity of the denatured state and a smaller one to the compact transition state for folding [93]. In this interpretation the thermodynamic characteristics of the transition are assumed to be independent of temperature; as an alternative one may also assume that ΔS and/or ΔH are strongly temperature dependent [94].

Whereas in PYP the recovery of pG shows a strongly curved Arrhenius plot, the corresponding plot for the pR to pB transition is only slightly curved, and not convex but concave. These results are fully in line with the results of protein folding studies, in which protein unfolding also shows a slightly concave Arrhenius plot. If the transitions in the photocycle are considered as a pseudo-equilibrium transition with the transition state, thermodynamic activation parameters can be calculated, from which the area of the exposed hydrophobic contact surface can be derived. For the ground-state recovery in E-PYP this results in ΔC_{p} values that agree with the renaturation of proteins of a comparable size.

To pinpoint the domain involved in the unfolding event of signalling state formation of E-PYP, genetic truncations of PYP were constructed that lack (part of) their N-terminal domain [95]. Both N-terminal α-helices can be deleted without abolishing photocycling and signalling state formation by the protein, although the kinetics of the recovery reaction in the photocycle of these truncated variants is strongly decelerated. Surprisingly, in the Arrhenius plot of the recovery reaction of these truncated proteins the extent of curvature has significantly decreased, up to the point that for the protein with the first 25 amino acids deleted an essentially linear Arrhenius plot is obtained [95]. This has been interpreted as evidence that the transient functional unfolding of PYP in its signalling state predominantly takes place in the N-terminal domain of the protein, possibly due to the low intrinsic stability of this part of the protein [90]. Apparently, the rearrangements that also take place in these truncated derivatives in the region surrounding the chromophore do not display the thermodynamic characteristics of an unfolding reaction.

8.5.6   *Various biophysical techniques confirm the occurrence of a significant conformational transition upon signalling state formation in E-PYP*

Because of the seemingly different results obtained with (transient) crystallography and photocycle kinetics as to the extent of unfolding during signalling state formation in PYP, a series of biophysical techniques were used to characterise the signalling state (i.e. pB) of E-PYP in solution. The results confirm the significant change in conformation of PYP in the pB state, which has the characteristics of a partial protein unfolding. This is the first example of a very clear *functional* protein unfolding process. The most important of these biophysical studies will be briefly discussed below:

(I) Surface plasmon resonance spectroscopy (SPR) was used to show [96] that PYP can bind to positively-charged, and negatively-charged, as well as to neutral bilayers; with the highest extent of binding to the positively-charged bilayers. Irradiation with white or blue light results in a relatively large increase in the extent of PYP binding to the bilayer interphase. These experiments suggest that in the pB state E-PYP has a much higher affinity for the lipid bilayer than in the pG_{446} state. This is consistent with an increase in the hydrophobic character of the surface of PYP in the pB state. The fact that these measurements were interpreted with an incorrect spatial structure of PYP does not weaken this conclusion.

(II) Proton uptake: During its photocycle, the PYP chromophore is transiently protonated [24,47,97]. This protonation was shown to take place during pB formation, with E46 as the proton donor [46,98]. In the E46Q mutant protein the proton may be taken up directly from the buffer [28]. Accordingly, one would not expect PYP to take up protons from, or release protons into, solution, upon transfer into the pB state.

Analysis of steady-state pH changes in an unbuffered solution of PYP [70] indeed confirmed this at neutral to slightly alkaline pH. Nevertheless, it was further revealed that PYP in the pB state can show both reversible proton uptake from, and proton release to, the solvent, depending on the ambient pH. This net change in protonation state is most probably due to the burial/exposure of other residues than E46. The apparent pKs involved suggest that not only carboxylic and amino groups are involved but possibly also a histidine side chain [70].

(III) Nile Red probe binding: Nile Red is a small organic fluorescent probe, the emission characteristics of which depend on the hydrophobicity of its molecular environment. In water, and in the presence of the ground state of PYP, this emission is at the same λ_{max}. However, on additional blue-shifted emission is observed after blue-light-induced accumulation of pB. This implies that a hydrophobic binding site is exposed during the conformational changes that accompany pB formation. It was tested whether changes in Nile Red emission would occur because this fluorophore is expected to probe the unfolding hydrophobic surface of the protein in the signalling state that also gives rise to the strongly curved Arrhenius plots (see above). It was, therefore, surprising to detect that Nile Red binding to the truncated variants of E-PYP was indistinguishable from its binding to wild-type PYP (except that the pB recovery rates were dramatically decreased in the truncated variants [95]). These results have been interpreted as evidence that Nile Red binds to the region of the PYP protein where the chromophore is exposed to solvent in the pB state [109].

(IV) Several additional techniques are now being used to further characterise the spatial structure of PYP in its signalling state pB. Among these are: NMR spectroscopy, tryptophan fluorescence spectroscopy (of the unique W119 of PYP), small-angle X-ray and neutron scattering experiments, time-resolved optical rotary dispersion, etc. These techniques may contribute to a detailed understanding of the spatial structure of this transient signalling state.

8.6 Comparison of variant forms of xanthopsins

8.6.1 Chromophore hybrids of E-PYP

Besides mutating the protein, it is also possible to engineer its chromophore. In Table 1 an overview is given of all chromophore analogues that have been used to reconstitute PYP hybrids [14,26,99]. Of these 14 chromophore analogues, only five provide PYP with measurable photoactivity (chromophores I–III and XII–XIII), this includes the native 4-hydroxycinnamic acid chromophore. The key to a successful chromophore is the negatively charged 4-hydroxy group of the phenolate moiety of the chromophore in the ground state of the protein and the possibility to isomerise and/or display "carbonyl rotation". Besides that, the chromophore of course must also be able to fit into the chromophore-binding pocket. For the two non-native chromophores II and III, the photocycle is slower than in wild-type PYP. For chromophore II, the ground-state recovery rate was also determined as a function of pH [99]. Surprisingly, it did not show the characteristic bell shape. The rate constant remains constant in the range between pH 5 and 10. Formation of pB_{dark} has a characteristic pK_a of 3.8 for chromophore II and 3.5 for chromophore III.

Two hybrids have been made using chromophores that cannot give rise to *trans* to *cis* isomerisation, i.e. chromophores X and XI. Surprisingly, both turned out to show (partial) photoactivity [14].

8.6.2   Comparison of mutant forms of E-PYP

The detailed structural information available for E-PYP and the sequence conservation within the Xanthopsin family reveals several residues that are important for PYP function. Here we will only consider mutations of residues that can interact with the chromophore. Of these, the following residues have been mutated: Y42, E46, T50, R52 and C69. As described above, Y42 and E46 both form a hydrogen bond with the negatively-charged phenolic oxygen of the chromophore; in addition T50 hydrogen-bonds to Y42 to complete the hydrogen-bonding network surrounding the aromatic ring of the chromophore. In the following, we discuss each residue separately with respect to its effect on spectral tuning of the holo-protein, its place in the active site of the protein and its effect on the photocycle kinetics.

Tyr42 has been mutated to Phe [100,101], effectively removing its hydrogen bond with the chromophore. Also, the hydrogen bond between T50 and Y42 can no longer be formed. This mutation causes the absorption maximum to shift 12 nm (587 cm^{-1}), from 446 to 458 nm. Additionally, a second absorption peak is formed with a maximum of 391 nm. This shows that Tyr42 indeed has an important role in the tuning of E-PYP. Subsequent studies to investigate the effects of denaturants, pH, kosmotropic agents (molecules that strongly order water molecules and help proteins fold), and chaotropic agents (molecules that do the opposite, i.e. interact weakly with water and don't help proteins fold) show that the equilibrium between the 458 and 391 nm peak can be shifted towards the 391 nm peak by adding small amounts of denaturant (large amounts will denature the protein), adding a chaotropic agent, or by lowering the pH (lowering the pH too much will cause the formation of pB_{dark}, pK_a 4.4). Adding a kosmotropic agent or increasing the pH has the opposite effect and shifts the equilibrium towards the 458 nm peak. This suggests that the 391 nm species is probably due to a less ordered protein structure, probably caused by a less efficient stabilisation of the negative charge of the chromophore. FT-Raman spectroscopy revealed that in the 391 nm species the chromophore is still deprotonated. This implies that tuning of the chromophore due to the protein environment is 19 nm (1185 cm^{-1}) to the blue, rather than 36 nm (1969 cm^{-1}) to the red in E-PYP. The crystal structure of the 458 nm species shows that removal of the hydroxy group of Y42 leads to a rearrangement that allows T50 to hydrogen bond directly to the chromophore. With respect to photocycle kinetics, formation of the pB state is faster and recovery to the ground state is slower in Y42F as compared with E-PYP. The pH dependence of the recovery rate still has the same bell shape with the same apparent pK_as.

E46 has been mutated to Q, D and A [28,74,100]. In E46Q the direct proton donor to the chromophore has been eliminated but hydrogen bonding between the residue at this position and the chromophore is still possible. In E46D the length of the side chain of the hydrogen bond/proton donor is effectively shortened by one carbon atom, which leads to an increased distance between the carboxyl group and the chromophore. In E46A all possibilities for polar interactions between the side chain of residue 46 and the chromophore have been eliminated. The absorption maximum of these three proteins is at 460, 444 and 465 nm, respectively, and the pK_a of pB$_{dark}$ formation is 4.5, 8.6 and 7.9, respectively. Especially, the last two values are very close to the pK_a of 8.9 for 4-OH-cinnamyl esters in solution, suggesting that in these proteins there is very little stabilisation of the phenolate form of the chromophore by the surrounding protein. All three mutant proteins no longer have the bell-shaped pH dependence for their recovery reaction, but instead show an increase in this rate, on going to more alkaline pH, with characteristic pK_as of 8.2, 9.6 and 9.0, respectively. This suggests that E46 is not only important for spectral tuning of E-PYP, but also has a profound influence on the pH dependence of the recovery rate of these proteins.

T50 has been mutated to V [100,101], which prevents the possibility of this residue being part of the hydrogen bonding network, stabilising the negative charge of the chromophore. The absorption maximum of T50V is 457 nm, thus T50 also contributes to tuning E-PYP. In T50V pB$_{dark}$ is formed with a characteristic pK_a of 3.6, which is closer to the wild-type value than when either of the other two residues that are part of the hydrogen-bonding network are mutated. The pH dependence of the rate of photocycle recovery is similar to wild-type PYP. Also in T50V, formation of pB is faster and the rate of recovery slower, compared with wild-type E-PYP. The most important role of T50 therefore seems to be the spectral and kinetic fine-tuning of E-PYP.

R52 has been mutated to Q and A [28,100]. Particularly in R52A, the residue at this position can no longer have a role in shielding the chromophore from the solvent. The absorption maximum of these proteins is at 447 and 452 nm, respectively. This shows that the shielding function is not crucial and that ionic interaction between R52 and the chromophore has no, or very little, influence on spectral tuning.

C69 is of course the most important residue in the list since without it the chromophore cannot be attached to the protein [100] and thus the protein can no longer be photoactive. Of the others, E46 influences the kinetics of the photocycle most profoundly. Additional mutants have been constructed, such as a series of G to S substitutions of glycine residues strongly conserved within the sequence of the PAS family, but discussion of these is beyond the scope of this chapter. There is, however, one additional mutation that has a very profound and surprising effect on PYP function: M100A slows down the rate of PYP recovery by three orders of magnitude [74]. It is therefore possible that M100 has a key role in the re-isomerisation step of the chromophore.

8.6.3   Comparison of xanthopsins from different bacterial genera

In addition to PYP from *E. halophila*, the biochemical and photophysical properties of three other members of the xanthopsin family have been described. The photoactive yellow protein isolated from *Rhodospirillum salexigens* [3], which shares 71% amino acid sequence identity with E-PYP, has virtually the same absorption spectrum, the absorption maximum of PYP from *Rs. salexigens* is just slightly blue-shifted (λ_{max} 445 nm). Also the kinetics of photobleaching and recovery are very similar, with rate constants for *Rs. salexigens* PYP of $k_{pR>pB}$ 1.2×10^4 s^{-1} and $k_{recovery}$ 4.7 s^{-1}.

Another member of the xanthopsin family that has been characterised in some detail is R-PYP from the phototrophic purple non-sulfur bacterium *Rhodobacter sphaeroides*. R-PYP was heterologously expressed in *E. coli* and reconstituted in vitro with 4-OH-cinnamic acid. This resulted in the formation of a yellow-coloured protein with the characteristic xanthopsin absorption in the visible part of the spectrum (λ_{max} 446 nm) [8]. This absorbance peak can be reversibly bleached by irradiation with blue light. Subsequently, a blue-shifted intermediate is formed with a difference-absorbance maximum at 350 nm. This pB form of R-PYP relaxes to the ground-state pG with a rate constant of 500 s^{-1}. This recovery process is about 100-fold faster than in E-PYP. The absorption maximum at 446 nm and the ability of the protein to undergo a photocycle indicate that R-PYP also shows the typical features of a xanthopsin.

However, the UV-Vis absorption spectrum of R-PYP shows an additional peak, positioned at 360 nm and which can be reversibly converted into the 446 nm form by adjusting the temperature. Lowering the temperature leads to the (reversible) accumulation of the 360 nm form. Titration of the ground state of R-PYP in the pH range 1.5 to 9 revealed two separate transitions, with pK values of 3.8 and 6.5, respectively. Below pH 9 the absorbance at 446 nm decreases, whereas that at 360 nm increases. At pH < 5, yet another spectral intermediate is formed, with a clearly blue-shifted absorbance maximum (at 345 nm), which is probably a pB$_{dark}$ form of R-PYP. Combined, these observations suggest that a dark equilibrium between the *cis* (λ_{max} = 360) and *trans* (λ_{max} = 446) form of the chromophore exists that is pH and temperature dependent. Therefore R-PYP may well be an important tool in understanding dark (re)isomerisation in the xanthopsins.

In *Rhodospirillum centenum* a gene encoding a fusion protein was discovered, which was called Ppr (from: PYP-phytochrome related). Besides a central phytochrome-like domain and a C-terminal histidine kinase domain [7] this protein contains an N-terminal domain that shows striking similarity with PYP. This Ppr protein, consisting of 884 amino acids, was heterologously expressed in *E. coli* and subsequently reconstituted with 4-OH-cinnamic acid. The reconstituted protein turns out to have an absorbance maximum at 434 nm. Furthermore, it displays a photocycle upon illumination with a blue or white flash. The bleach at 434 nm is accompanied by the initial formation of a red-shifted intermediate with a difference-absorbance maximum at about 470 nm, followed by the formation of a blue-shifted intermediate with a difference absorbance maximum at about 330 nm. The subsequent recovery to the ground state is biphasic with a fast and a very slow component (τs of 0.21 ms and 46 s, respectively). This dark recovery rate of Ppr is ~300-fold slower than in E-PYP and ~23.000-fold slower than in R-PYP.

8.7   Prospects and Conclusions

Because of its extraordinary physicochemical- and photo-stability and the ease with which it can be made to form crystals that yield high-resolution structures, E-PYP has almost become a photochemistry laboratory in itself. Accordingly, PYP has become a very attractive model system for studies of the primary photochemistry of photosensing and for studies of (functional) protein folding processes. In addition, intriguing questions as to the biological function of many members of the xanthopsin family lie ahead (see also [39]). This function as a model system also results in ever more theoretical and experimental approaches and techniques being applied to and tried out in PYP, as a test case for their applicability. This, in turn, has provided a wealth of information regarding the physicochemical basis of PYP functioning. Examples are the (quantum) chemical calculation of energy levels in the UV/Vis spectrum of PYP and the development of time-resolved diffraction analysis of biological function at ambient temperature. Below we will briefly discuss these developments, illustrated via some selected techniques.

8.7.1   Quantum chemical calculations

Quantum chemical calculations are increasingly used to predict the crucial characteristics of photoactive proteins such as absorption spectra and the trajectory of the photo-isomerisation process (e.g. [102–104]). In most cases a combined quantum chemical/molecular mechanics approach is used, initially applied mostly to the isolated chromophore. However, with the increasing power of computation, with this approach even entire protein molecules come within reach [31]. Extension of this approach must lead to a complete description of the activation of photosensor proteins, based on first principles.

8.7.2   Ultrafast spectroscopy

Further application of fs spectroscopy will be a powerful tool to unravel the primary photochemical events that are initiated by blue photons in PYP. This applies not only to studies in the visible, but particularly also to studies in the mid-infrared region of the spectrum, a region in which many functional groups of the chromophore strongly absorb (e.g. the phenolate ring). Such studies can be performed in a visible pump–IR probe mode, but several alternative modes can be thought of as well. In addition, ultrafast time-resolved resonance Raman studies of photocycle intermediates will provide crucial insight into the correlation between PYP colour (i.e. spectra) and the chemical structure of the chromophore during the primary photochemical transitions.

Important remaining questions to be answered for PYP include the involvement of a bi-radical pair in the isomerisation of the cinnamic acid chromophore of PYP, the chemical basis of the ultrafast colour changes during the initial stages of the photocycle and the dependence of the quantum yield of photochemistry in PYP on the excitation wavelength.

More detailed structural information can further be obtained by making use of polarised spectroscopy in random or ordered samples. In this respect the ease with which (crunched) PYP crystals can be made is important.

8.7.3   Time-resolved Laue diffraction experiments

Using time-resolved Laue diffraction analysis the structures of the PYP-intermediates, which dominate the sample 10 ms and 1 ns after photo-stimulation, have been determined. These structures correspond to the pB [47] and pR [20] intermediates, respectively. In the latter study, however, a very large data set was obtained, spanning the entire time range that is relevant for photocycle transitions from 1 ns onwards. The time resolution could in principle be even further improved, because the length of the X-ray pulses is approximately 100 ps. However, so far no successful fs laser experiments have been reported in combination with diffraction analysis.

This very large data set of diffraction data against time has been evaluated with respect to the number of exponents that are required to describe the changes in the concentration of its most significant features. This information has subsequently been used to time-smooth the diffraction data. From this a much more detailed "film" of the events occurring upon photostimulation of PYP is emerging [110]. From the amplitudes of the significant exponentials in this data set, the structure of relevant intermediates can be calculated.

Although most of the backbone structure of PYP is hardly affected by the transitions in the photocycle, significant displacements of side chains and the backbone near the chromophore-binding site occur. The most important side chains in this respect are R52 and R124. The availability of these structures will provide the opportunity for future tests as to whether available programs for modelling protein dynamics can reliably predict the structure of these transient intermediates. Accordingly, these Laue diffraction experiments may open a route to bridge the gap between understanding functional dynamics of proteins (and enzymes) at the seconds timescale (i.e. the timescale relevant for their function) and the modelling of protein dynamics. This will also impact greatly on further understanding of the electrostatics in proteins, which is in turn of great importance in understanding protein/protein interactions.

8.7.4   Time-resolved FTIR measurements

Fourier-transform infrared spectroscopy (FTIR) is a very powerful technique for the study of photocycle events in PYP. The largest problem with this technique is the assignment the spectral features observed to specific changes in the structure of the protein. Nonetheless several spectral features have been assigned, using model compounds, mutants, isotopic labelling, and deductive reasoning.

Using time-resolved FTIR, the global conformational changes during the PYP photocycle can be probed by studying the amide A and amide I bands in infrared difference spectra. Amide A, centred at 3500–3000 cm^{-1} and arising from the N–H stretching mode, and amide I, at 1624 cm^{-1}, which arises predominantly from the C=O stretching of protein backbone [105], showed large changes in pB-pG difference spectra measured in hydrated films, indicating that large structural changes occur during pB formation. This study also showed that reduced hydration in the PYP films strongly reduces the amide I signals, indicating that in these conditions a pB state is formed without the occurrence of large structural changes in the protein backbone.

Using FTIR, structural changes of water can also be investigated. The PYP crystal structure shows that no water molecules are present inside PYP; they are all found near the protein surface. Upon light-induced pB formation, a negative band is seen at 3658 cm^{-1}, characteristic of the stretch vibration of water O–H (and confirmed using $H_2^{18}O$; [106]). Using a H108F mutation, water-200 was assigned as the responsible molecule. This water is located in a binding pocket between F6 and F121, and is within hydrogen-bonding distance with H108 and G7. The distance between this water molecule and the chromophore is larger than 15 Å. The fact that this water signal is modulated in the pB state emphasises the fact that the conformational transition that occurs in PYP upon pB formation affects large parts of the molecule, in agreement with the essential dynamics results.

Recently two ns time-resolved FTIR studies were published in parallel [85,86]. In the first it was observed that the hydrogen-bonding network surrounding the phenolate moiety of the chromophore is not disrupted until the proton is transferred from E46 to the chromophore in the transition from pR to pB, whereas Xie et al. [86] compared the conformational changes occurring in aqueous solution and in crystals. They observed that the structure of the pB intermediate is far less well developed in $P6_3$ crystals, as judged by the much smaller change in the amide I region of difference spectra. They furthermore showed that the large structural change during pB formation takes place after proton transfer from E46 to the chromophore (these two events can be separately recorded at 1726 and 1498 cm^{-1}, respectively). These observations suggest that the conformational changes are caused by the ionisation of E46, which induces a buried negative charge in a highly hydrophobic pocket.

8.7.5 Scattering experiments

The observed dependence of the extent of unfolding of PYP in the pB state upon the mesoscopic context of the protein has sparked intense interest in the conformation of the protein under various conditions. Because of this, a number of groups have initiated the application of small-angle X-ray scattering and small-angle neutron scattering analyses of PYP under various conditions. These types of measurements may provide valuable additions to the toolbox of researchers interested in the changes in protein structure that accompany signalling state formation.

Acknowledgements

This work has been supported by the Netherlands Foundation for Chemical Research and for Earth and Life Sciences Research, with financial assistance from the Netherlands Organisation for Scientific Research (NWO). R. Cordfunke is gratefully acknowledged for expert technical assistance.

References

1. T.E. Meyer (1985). Isolation and characterization of soluble cytochromes, ferredoxins and other chromophoric proteins from the halophilic phototrophic bacterium *Ectothiorhodospira halophila*. *Biochim. Biophys. Acta*, **806**, 175–183.
2. M. Koh, G. Van Driessche, B. Samyn, W.D. Hoff, T.E. Meyer, M.A. Cusanovich, J.J. Van Beeumen (1996). Sequence evidence for strong conservation of the photoactive yellow proteins from the halophilic phototrophic bacteria *Chromatium salexigens* and *Rhodospirillum salexigens*. *Biochemistry*, **35**, 2526–2534.
3. T.E. Meyer, J.C. Fitch, R.G. Bartsch, G. Tollin, M.A. Cusanovich (1990). Soluble cytochromes and a photoactive yellow protein isolated from the moderately halophilic purple phototrophic bacterium, *Rhodospirillum salexigens*. *Biochim. Biophys. Acta*, **1016**, 364–370.

4. R. Kort, W.D. Hoff, M. Van West, A.R. Kroon, S.M. Hoffer, K.H. Vlieg, W. Crielaand, J.J. Van Beeumen, K.J. Hellingwerf (1996). The xanthopsins: a new family of eubacterial blue-light photoreceptors. *EMBO J.*, **15**, 3209–3218.

5. W.W. Sprenger, W.D. Hoff, J.P. Armitage, K.J. Hellingwerf (1993). The eubacterium *Ectothiorhodospira halophila* is negatively phototactic, with a wavelength dependence that fits the absorption spectrum of the photoactive yellow protein. *J. Bacteriol.*, **175**, 3096–3104.

6. R. Kort, M.K. Phillips-Jones, D.M. van Aalten, A. Haker, S.M. Hoffer, K.J. Hellingwerf, W. Crielaard (1998). Sequence, chromophore extraction and 3-D model of the photoactive yellow protein from *Rhodobacter sphaeroides*. *Biochim. Biophys. Acta*, **1385**, 1–6.

7. Z. Jiang, L.R. Swem, B.G. Rushing, S. Devanathan, G. Tollin, C.E. Bauer (1999). Bacterial photoreceptor with similarity to photoactive yellow protein and plant phytochromes. *Science*, **285**, 406–409.

8. A. Haker, J. Hendriks, T. Gensch, K. Hellingwerf, W. Crielaard (2000). Isolation, reconstitution and functional characterisation of the *Rhodobacter sphaeroides* photoactive yellow protein. *FEBS Lett.*, **486**, 52–56.

9. R. Kort, W. Crielaard, J.L. Spudich, K.J. Hellingwerf (2000). Color-sensitive motility and methanol release responses in *Rhodobacter sphaeroides*. *J. Bacteriol.*, **182**, 3017–3021.

10. W.D. Hoff, W.W. Sprenger, P.W. Postma, T.E. Meyer, M. Veenhuis, T. Leguijt, K.J. Hellingwerf (1994). The photoactive yellow protein from *Ectothiorhodospira halophila* as studied with a highly specific polyclonal antiserum: (intra)cellular localization, regulation of expression, and taxonomic distribution of cross-reacting proteins. *J. Bacteriol.*, **176**, 3920–3927.

11. K. Hahlbrock, D. Scheel (1989). Physiology and molecular-biology of phenylpropanoid metabolism. *Annu. Rev. Plant Physiol. Plant Mol. Biol.*, **40**, 347–369.

12. P.S. Duggan, S.D. Parker, M.K. Phillips-Jones (2000). Characterisation of a *Rhodobacter sphaeroides* gene that encodes a product resembling *Escherichia coli* cytochrome b(561) and *R. sphaeroides* cytochrome b(562). *FEMS Microbiol. Lett.*, **189**, 239–246.

13. Y. Imamoto, T. Ito, M. Kataoka, F. Tokunaga (1995). Reconstitution of photoactive yellow protein from apoprotein and p-coumaric acid derivatives. *FEBS Lett.*, **374**, 157–160.

14. R. Cordfunke, R. Kort, A. Pierik, B. Gobets, G.J. Koomen, J.W. Verhoeven, K.J. Hellingwerf (1998). Trans/cis (Z/E) photoisomerization of the chromophore of photoactive yellow protein is not a prerequisite for the initiation of the photocycle of this photoreceptor protein. *Proc. Natl. Acad. Sci. U.S.A.*, **95**, 7396–7401.

15. G.E. Borgstahl, D.R. Williams, E.D. Getzoff (1995). 1.4 A structure of photoactive yellow protein, a cytosolic photoreceptor: unusual fold, active site, and chromophore. *Biochemistry*, **34**, 6278–6287.

16. P. Dux, G. Rubinstenn, G.W. Vuister, R. Boelens, F.A. Mulder, K. Hard, W.D. Hoff, A.R. Kroon, W. Crielaard, K.J. Hellingwerf, R. Kaptein (1998). Solution structure and backbone dynamics of the photoactive yellow protein. *Biochemistry*, **37**, 12689–12699.

17. U.K. Genick, S.M. Soltis, P. Kuhn, I.L. Canestrelli, E.D. Getzoff (1998). Structure at 0.85 A resolution of an early protein photocycle intermediate. *Nature*, **392**, 206–209.

18. T.E. Meyer, E. Yakali, M.A. Cusanovich, G. Tollin (1987). Properties of a water-soluble, yellow protein isolated from a halophilic phototrophic bacterium

that has photochemical activity analogous to sensory rhodopsin. *Biochemistry*, **26**, 418–423.

19. J.L. Pellequer, K.A. Wager-Smith, S.A. Kay, E.D. Getzoff (1998). Photoactive yellow protein: a structural prototype for the three-dimensional fold of the PAS domain superfamily. *Proc. Natl. Acad. Sci. U.S.A.*, **95**, 5884–5890.

20. B. Perman, V. Srajer, Z. Ren, T. Teng, C. Pradervand, T. Ursby, D. Bourgeois, F. Schotte, M. Wulff, R. Kort, K. Hellingwerf, K. Moffat (1998). Energy transduction on the nanosecond timescale: early structural events in a xanthopsin photocycle. *Science*, **279**, 1946–1950.

21. T.M. Weaver (2000). The pi-helix translates structure into function. *Protein Sci.*, **9**, 201–206.

22. D.E. McRee, J.A. Tainer, T.E. Meyer, J. Van Beeumen, M.A. Cusanovich, E.D. Getzoff (1989). Crystallographic structure of a photoreceptor protein at 2.4 A resolution. *Proc. Natl. Acad. Sci. U.S.A.*, **86**, 6533–6537.

23. M. Baca, G.E. Borgstahl, M. Boissinot, P.M. Burke, D.R. Williams, K.A. Slater, E.D. Getzoff (1994). Complete chemical structure of photoactive yellow protein: novel thioester-linked 4-hydroxycinnamyl chromophore and photocycle chemistry. *Biochemistry*, **33**, 14369–14377.

24. W.D. Hoff, P. Dux, K. Hard, B. Devreese, I.M. Nugteren-Roodzant, W. Crielaard, R. Boelens, R. Kaptein, J. van Beeumen, K.J. Hellingwerf (1994). Thiol ester-linked p-coumaric acid as a new photoactive prosthetic group in a protein with rhodopsin-like photochemistry. *Biochemistry*, **33**, 13959–13962.

25. M. Kim, R.A. Mathies, W.D. Hoff, K.J. Hellingwerf (1995). Resonance Raman evidence that the thioester-linked 4-hydroxycinnamyl chromophore of photoactive yellow protein is deprotonated. *Biochemistry*, **34**, 12669–12672.

26. A.R. Kroon, W.D. Hoff, H.P. Fennema, J. Gijzen, G.J. Koomen, J.W. Verhoeven, W. Crielaard, K.J. Hellingwerf (1996). Spectral tuning, fluorescence, and photoactivity in hybrids of photoactive yellow protein, reconstituted with native or modified chromophores. *J. Biol. Chem.*, **271**, 31949–31956.

27. N.S. Scrutton, A.R. Raine (1996). Cation-pi bonding and amino-aromatic interactions in the biomolecular recognition of substituted ammonium ligands. *Biochem. J.*, **319**, 1–8.

28. U.K. Genick, S. Devanathan, T.E. Meyer, I.L. Canestrelli, E. Williams, M.A. Cusanovich, G. Tollin, E.D. Getzoff (1997). Active site mutants implicate key residues for control of color and light cycle kinetics of photoactive yellow protein. *Biochemistry*, **36**, 8–14.

29. G. Aulin-Erdtman, R. Sanden (1968). Absorption properties of some 4-hydroxyphenyl, guaiacyl, and 4-hydroxy-3,5-dimethoxyphenyl type model compounds for hardwood lignins. *Acta Chem. Scand.*, **22**, 1187–1209.

30. G.S. Harbison, S.O. Smith, J.A. Pardoen, J.M. Courtin, J. Lugtenburg, J. Herzfeld, R.A. Mathies, R.G. Griffin (1985). Solid-state 13C NMR detection of a perturbed 6-s-trans chromophore in bacteriorhodopsin. *Biochemistry*, **24**, 6955–6962.

31. H. Houjou, Y. Inoue, M. Sakurai (2001). Study of the opsin shift of bacteriorhodopsin: Insight from QM/MM calculations with electronic polarization effects of the protein environment. *J. Phys. Chem. B*, **105**, 867–879.

32. J.R. Nambu, J.O. Lewis, K.A. Wharton, S.T. Crews (1991). The Drosophila single-minded gene encodes a helix-loop-helix protein that acts as a master regulator of Cns midline development. *Cell*, **67**, 1157–1167.

33. B.L. Taylor, I.B. Zhulin (1999). PAS domains: Internal sensors of oxygen, redox potential, and light. *Microbiol. Mol. Biol. Rev.*, **63**, 479–506.

34. D.M. Lagarias, S.H. Wu, J.C. Lagarias (1995). Atypical phytochrome gene structure in the green alga *Mesotaenium caldariorum*. *Plant Mol. Biol.*, **29**, 1127–1142.

35. D.M. van Aalten, W.D. Hoff, J.B. Findlay, W. Crielaard, K.J. Hellingwerf (1998). Concerted motions in the photoactive yellow protein. *Protein Eng.*, **11**, 873–879.

36. W. Rüdiger, F. Thümmler (1991). Phytochrom, das sehpigment der pflanzen. *Angew. Chem.*, **103**, 1242–1254.

37. K. Schaffner, S.E. Braslavsky, A.R. Holzwarth (1990). Photophysics and Photo-chemistry of Phytochrome. In: D.H. Volman, G. Hammond and K. Gollnick (Eds), *Advances in Photochemistry* (pp. 229–277). John Wiley & Sons, Ltd., New York.

38. T.E. Meyer, G. Tollin, J.H. Hazzard, M.A. Cusanovich (1989). Photoactive yellow protein from the purple phototrophic bacterium, *Ectothiorhodospira halophila*. Quantum yield of photobleaching and effects of temperature, alcohols, glycerol, and sucrose on kinetics of photobleaching and recovery. *Biophys. J.*, **56**, 559–564.

39. W.D. Hoff, H.C.P. Matthijs, H. Schubert, W. Crielaard, K. J. Hellingwerf (1995). Rhodopsin(s) in eubacteria. *Biophys. Chem.*, **56**, 193–199.

40. Y. Imamoto, M. Kataoka, F. Tokunaga (1996). Photoreaction cycle of photoac-tive yellow protein from *Ectothiorhodospira halophila* studied by low-temperature spectroscopy. *Biochemistry*, **35**, 14047–14053.

41. A. Baltuška, I.H.M. van Stokkum, A. Kroon, R. Monshouwer, K.J. Hellingwerf, R. van Grondelle (1997). The primary events in the photoactivation of yellow protein. *Chem. Phys. Lett.*, **270**, 263–266.

42. L. Ujj, S. Devanathan, T.E. Meyer, M.A. Cusanovich, G. Tollin, G.H. Atkinson (1998). New photocycle intermediates in the photoactive yellow protein from *Ectothiorhodospira halophila*: picosecond transient absorption spectroscopy. *Biophys. J.*, **75**, 406–412.

43. S. Devanathan, A. Pacheco, L. Ujj, M. Cusanovich, G. Tollin, S. Lin, N. Woodbury (1999). Femtosecond spectroscopic observations of initial intermedi-ates in the photocycle of the photoactive yellow protein from *Ectothiorhodospira halophila*. *Biophys. J.*, **77**, 1017–1023.

44. S. Devanathan, S. Lin, M.A. Cusanovich, N. Woodbury, G. Tollin (2000). Early intermediates in the photocycle of the Glu46Gln mutant of photoactive yellow protein: femtosecond spectroscopy. *Biophys. J.*, **81**, 2132–2137.

45. Y. Zhou, L. Ujj, T.E. Meyer, M.A. Cussanovich, G.H. Atkinson (2001). Photocycle dynamics and vibrational spectroscopy of the E46Q mutant of photoactive yellow protein. *J. Phys. Chem. A*, **105**, 5719–5726.

46. A. Xie, W.D. Hoff, A.R. Kroon, K.J. Hellingwerf (1996). Glu46 donates a proton to the 4-hydroxycinnamate anion chromophore during the photocycle of photoactive yellow protein. *Biochemistry*, **35**, 14671–14678.

47. U.K. Genick, G.E. Borgstahl, K. Ng, Z. Ren, C. Pradervand, P.M. Burke, V. Srajer, T.Y. Teng, W. Schildkamp, D.E. McRee, K. Moffat, E.D. Getzoff (1997). Structure of a protein photocycle intermediate by millisecond time-resolved crystallography. *Science*, **275**, 1471–1475.

48. P. Changenet-Barret, P. Plaza, M.M. Martin (2001). Primary events in the photoactive yellow protein chromophore in solution. *Chem. Phys. Lett.*, **336**, 439–444.

49. G.G. Kochendoerfer, R. A. Mathies (1995). Ultrafast spectroscopy of rhodopsins – photochemistry at its best!. *Isr. J. Chem.*, **35**, 211–226.

50. H. Chosrowjan, N. Mataga, N. Nakashima, I. Yasushi, F. Tokunaga (1997). Femtosecond–picosecond fluorescence studies on excited state dynamics of photoactive yellow protein from *Ectothiorhodospira halophila*. *Chem. Physics Lett.*, **270**, 267–272.

51. P. Changenet, H. Zhang, M.J. van der Meer, K.J. Hellingwerf, M. Glasbeek (1998). Subpicosecond fluorescence upconversion measurements of primary events in yellow proteins. *Chem. Phys. Lett.*, **282**, 276–282.

52. N. Mataga, H. Chosrowjan, Y. Shibata, Y. Imamoto, F. Tokunaga (2000). Effects of modification of protein nanospace structure and change of temperature on the femtosecond to picosecond fluorescence dynamics of photoactive yellow protein. *J. Phys. Chem. B*, **104**, 5191–5199.

53. W.D. Hoff, S.L.S. Kwa, R. van Grondelle, K.J. Hellingwerf (1992). Low temperature absorbance and fluorescence spectroscopy of the photoactive yellow protein from *Ectothiorhodospira halophila*. *Photochem. Photobiol.*, **56**, 529–539.

54. T. Masciangioli, S. Devanathan, M.A. Cusanovich, G. Tollin, M.A. el-Sayed (2000). Probing the primary event in the photocycle of photoactive yellow protein using photochemical hole-burning technique. *Photochem Photobiol*, **72**, 639–644.

55. M.E. van Brederode, T. Gensch, W.D. Hoff, K.J. Hellingwerf, S.E. Braslavsky (1995). Photoinduced volume change and energy storage associated with the early transformations of the photoactive yellow protein from *Ectothiorhodospira halophila*. *Biophys. J.*, **68**, 1101–1109.

56. P. Schulenberg, S.E. Braslavsky (2001). Time-resolved photothermal studies with biological supramolecular systems. In: A. Mandelis, P. Hess (Eds), *Progress in Photothermal and Photoacoustic Science and Technology, Volume III: Life and Earth Sciences* (pp. 58–81). SPIE Press, Washington.

57. S.E. Braslavsky, G.E. Heibel (1992). Time-resolved photothermal and photoacoustic methods applied to photoinduced processes in solution. *Chem. Rev.*, **92**, 1381–1410.

58. T. Gensch, C. Viappiani, S.E. Braslavsky (1999). Laser induced optoacoustic spectroscopy. In: G.E. Tranter, J.L. Holmes (Eds), *Encyclopedia of Spectroscopy and Spectrometry* (pp. 1124–1132). Academic Press Ltd., London.

59. C.D. Borsarelli, S.E. Braslavsky (1998). Volume changes correlate with enthalpy changes during the photoinduced formation of the (MLCT)-M-3 state of ruthenium(II) bipyridine cyano complexes in the presence of salts. A case of the entropy-enthalpy compensation effect. *J. Phys. Chem. B*, **102**, 6231–6238.

60. A. Losi, A.A. Wegener, M. Engelhard, S.E. Braslavsky (2001). Enthalpy-entropy compensation in a photocycle: The K-to-L transition in sensory rhodopsin II from *Natronobacterium pharaonis*. *J. Am. Chem. Soc.*, **123**, 1766–1767.

61. K. Takeshita, N. Hirota, Y. Imamoto, M. Kataoka, F. Tokunaga, M. Terazima (2000). Temperature-dependent volume change of the initial step of the photoreaction of photoactive yellow protein (PYP) studied by transient grating. *J. Am. Chem. Soc.*, **122**, 8524–8528.

62. T. Hara, N. Hirota, M. Terazima (1996). New application of the transient grating method to a photochemical reaction: The enthalpy, reaction volume change, and partial molar volume measurements. *J. Phys. Chem.*, **100**, 10194–10200.

63. K. Takeshita, N. Hirota, M. Terazima (2000). Enthalpy changes and reaction volumes of photoisomerization reactions in solution: azobenzene and p-coumaric acid. *J. Photochem. Photobiol. A-Chem.*, **134**, 103–109.

64. T. Gensch, J.M. Strassburger, W. Gartner, S.E. Braslavsky (1998). Volume and enthalpy changes upon photoexcitation of bovine rhodopsin derived from

optoacoustic studies by using an equilibrium between bathorhodopsin and blue-shifted intermediate. *Isr. J. Chem.*, **38**, 231–236.

65. A. Losi, S.E. Braslavsky, W. Gartner, J.L. Spudich (1999). Time-resolved absorption and photothermal measurements with sensory rhodopsin I from *Halobacterium salinarum*. *Biophys. J.*, **76**, 2183–2191.

66. A. Losi, A.A. Wegener, M. Engelhard, W. Gartner, S.E. Braslavsky (1999). Time-resolved absorption and photothermal measurements with recombinant sensory rhodopsin II from *Natronobacterium pharaonis*. *Biophys. J.*, **77**, 3277–3286.

67. T. Gensch, M.S. Churio, S.E. Braslavsky, K. Schaffner (1996). Primary quantum yield and volume change of phytochrome-A phototransformation determined by laser-induced optoacoustic spectroscopy. *Photochem. Photobiol.*, **63**, 719–725.

68. A. Losi, I. Michler, W. Gartner, S.E. Braslavsky (2000). Time-resolved thermodynamic changes photoinduced in 5,12-trans-locked bacteriorhodopsin. Evidence that retinal isomerization is required for protein activation. *Photochem. Photobiol.*, **72**, 590–597.

69. D. Zhang, D. Mauzerall (1996). Volume and enthalpy changes in the early steps of bacteriorhodopsin photocycle studied by time-resolved photoacoustics. *Biophys. J.*, **71**, 381–388.

70. J. Hendriks, W.D. Hoff, W. Crielaard, K.J. Hellingwerf (1999). Protonation/deprotonation reactions triggered by photoactivation of photoactive yellow protein from *Ectothiorhodospira halophila*. *J. Biol. Chem.*, **274**, 17655–17660.

71. J. Hendriks, I.H. van Stokkum, W. Crielaard, K.J. Hellingwerf (1999). Kinetics of and intermediates in a photocycle branching reaction of the photoactive yellow protein from *Ectothiorhodospira halophila*. *FEBS Lett.*, **458**, 252–256.

72. G. Rubinstenn, G.W. Vuister, F.A. Mulder, P.E. Dux, R. Boelens, K.J. Hellingwerf, R. Kaptein (1998). Structural and dynamic changes of photoactive yellow protein during its photocycle in solution. *Nat. Struct. Biol.*, **5**, 568–570.

73. W.D. Hoff, I.H.M. Van Stokkum, J. Gural, K.J. Hellingwerf (1997). Comparison of acid denaturation and light activation in the eubacterial blue-light receptor photoactive yellow protein. *Biochim. Biophys. Acta-Bioenerg.*, **1322**, 151–162.

74. S. Devanathan, U.K. Genick, I.L. Canestrelli, T.E. Meyer, M.A. Cusanovich, E.D. Getzoff, G. Tollin (1998). New insights into the photocycle of *Ectothiorhodospira halophila* photoactive yellow protein: photorecovery of the long-lived photobleached intermediate in the Met100Ala mutant. *Biochemistry*, **37**, 11563–11568.

75. A. Miller, H. Leigeber, W.D. Hoff, K.J. Hellingwerf (1993). A light-dependent branching-reaction in the photocycle of the yellow protein from *Ectothiorhodospira halophila*. *Biochim. Biophys. Acta*, **1141**, 190–196.

76. S.P. Balashov (1995). Photoreactions of the photointermediates of bacteriorhodopsin. *Isr. J. Chem.*, **35**, 415–428.

77. L.H. Pratt, Y. Inoue, M. Furuya (1984). Photoactivity of transient intermediates in the pathway from the red-absorbing to the far-red-absorbing form of *Avena* phytochrome as observed by a double-flash transient-spectrum analyzer. *Photochem. Photobiol.*, **39**, 241–246.

78. T.E. Swartz, I. Szundi, J.L. Spudich, R.A. Bogomolni (2000). New photointermediates in the two photon signaling pathway of sensory rhodopsin-I. *Biochemistry*, **39**, 15101–15109.

79. J.K. Delaney, P.K. Schmidt, T.L. Brack, G.H. Atkinson (2000). Photochemistry of K-590 in the room-temperature bacteriorhodopsin photocycle. *J. Phys. Chem. B*, **104**, 10827–10834.

80. V. Bazhenov, P. Schmidt, G.H. Atkinson (1992). Nanosecond photolytic interruption of bacteriorhodopsin photocycle - K-590-Br-570 reaction. *Biophys. J.*, **61**, 1630–1637.

81. R.D. Scurlock, C.H. Evans, S.E. Braslavsky, K. Schaffner (1993). A Phytochrome phototransformation study using 2-laser 2-color flash-photolysis – analysis of the decay mechanism of I(700). *Photochem. Photobiol.*, **58**, 106–115.

82. T. Gensch, K.J. Hellingwerf, S.E. Braslavsky, K. Schaffner (1998). Photoequilibrium in the primary steps of the photoreceptors phytochrome A and photoactive yellow protein. *J. Phys. Chem. A*, **102**, 5398–5405.

83. A. Amadei, A.B. Linssen, H.J. Berendsen (1993). Essential dynamics of proteins. *Proteins*, **17**, 412–425.

84. D.M. van Aalten, J.B. Findlay, A. Amadei, H.J. Berendsen (1995). Essential dynamics of the cellular retinol-binding protein–evidence for ligand-induced conformational changes. *Protein Eng.*, **8**, 1129–1135.

85. R. Brudler, R. Rammelsberg, T.T. Woo, E.D. Getzoff, K. Gerwert (2001). Structure of the I1 early intermediate of photoactive yellow protein by FTIR spectroscopy. *Nat. Struct. Biol.*, **8**, 265–270.

86. A. Xie, L. Kelemen, J. Hendriks, B.J. White, K.J. Hellingwerf, W.D. Hoff (2001). Formation of a new buried charge drives a large-amplitude protein quake in photoreceptor activation. *Biochemistry*, **40**, 1510–1517.

87. D.A. McCain, L.A. Amici, J.L. Spudich (1987). Kinetically resolved states of the *Halobacterium halobium* flagellar motor switch and modulation of the switch by sensory rhodopsin I. *J. Bacteriol.*, **169**, 4750–4758.

88. W.D. Hoff, I.H. van Stokkum, H.J. van Ramesdonk, M.E. van Brederode, A.M. Brouwer, J.C. Fitch, T.E. Meyer, R. van Grondelle, K.J. Hellingwerf (1994). Measurement and global analysis of the absorbance changes in the photocycle of the photoactive yellow protein from *Ectothiorhodospira halophila*. *Biophys. J.*, **67**, 1691–1705.

89. U. Mayor, C.M. Johnson, V. Daggett, A.R. Fersht (2000). Protein folding and unfolding in microseconds to nanoseconds by experiment and simulation. *Proc. Natl. Acad. Sci. U.S.A.*, **97**, 13518–13522.

90. C.J. Craven, N.M. Derix, J. Hendriks, R. Boelens, K.J. Hellingwerf, R. Kaptein (2000). Probing the nature of the blue-shifted intermediate of photoactive yellow protein in isolation by NMR: Hydrogen-deuterium exchange data and pH studies. *Biochemistry*, **39**, 14392–14399.

91. M.E. Van Brederode, W.D. Hoff, I.H. Van Stokkum, M.L. Groot, K.J. Hellingwerf (1996). Protein folding thermodynamics applied to the photocycle of the photoactive yellow protein. *Biophys. J.*, **71**, 365–380.

92. P.L. Privalov, G.I. Makhatadze (1990). Heat capacity of proteins. II. Partial molar heat capacity of the unfolded polypeptide chain of proteins: protein unfolding effects. *J. Mol. Biol.*, **213**, 385–391.

93. M. Oliveberg, Y.J. Tan, A.R. Fersht (1995). Negative activation enthalpies in the kinetics of protein folding. *Proc. Natl. Acad. Sci. U.S.A.*, **92**, 8926–8929.

94. M. Karplus (2000). Aspects of protein reaction dynamics: Deviations from simple behavior. *J. Phys. Chem. B*, **104**, 11–27.

95. M. van der Horst, I.H. van Stokkum, W. Crielaard, K.J. Hellingwerf (2001). The role of the N-terminal domain of photoactive yellow protein in the transient partial unfolding during signalling state formation. *FEBS Lett.*, **497**, 26–30.

96. Z. Salamon, T.E. Meyer, G. Tollin (1995). Photobleaching of the photoactive yellow protein from Ectothiorhodospira halophila promotes binding to lipid bilayers: evidence from surface plasmon resonance spectroscopy. *Biophys. J.*, **68**, 648–654.

97. T.E. Meyer, M.A. Cusanovich, G. Tollin (1993). Transient proton uptake and release is associated with the photocycle of the photoactive yellow protein from the purple phototrophic bacterium *Ectothiorhodospira halophila*. *Arch. Biochem. Biophys.*, **366**, 515–517.

98. Y. Imamoto, K. Mihara, O. Hisatomi, M. Kataoka, F. Tokunaga, N. Bojkova, K. Yoshihara (1997). Evidence for proton transfer from Glu-46 to the chromophore during the photocycle of photoactive yellow protein. *J. Biol. Chem.*, **272**, 12905–12908.

99. S. Devanathan, U.K. Genick, E.D. Getzoff, T.E. Meyer, M.A. Cusanovich, G. Tollin (1997). Preparation and properties of a 3,4-dihydroxycinnamic acid chromophore variant of the photoactive yellow protein. *Arch. Biochem. Biophys.*, **340**, 83–89.

100. K. Mihara, O. Hisatomi, Y. Imamoto, M. Kataoka, F. Tokunaga (1997). Functional expression and site-directed mutagenesis of photoactive yellow protein. *J. Biochem. (Tokyo)*, **12**, 876–880.

101. R. Brudler, T.E. Meyer, U.K. Genick, S. Devanathan, T.T. Woo, D.P. Millar, K. Gerwert, M.A. Cusanovich, G. Tollin, E.D. Getzoff (2000). Coupling of hydrogen bonding to chromophore conformation and function in photoactive yellow protein. *Biochemistry*, **39**, 13478–13486.

102. H.Y. Yoo, J.A. Boatz, V. Helms, J.A. McCammon, P.W. Langhoff (2001). Chromophore protonation states and the proton shuttle mechanism in green fluorescent protein: Inferences drawn from ab initio theoretical studies of chemical structures and vibrational spectra. *J. Phys. Chem. B*, **105**, 2850–2857.

103. A. Sergi, M. Grüning, M. Ferrario, F. Buda (2001). Density functional study of the photoactive yellow protein's chromophore. *J. Phys. Chem. B*, **105**, 4386–4391.

104. V. Molina, M. Merchan (2001). On the absorbance changes in the photocycle of the photoactive yellow protein: A quantum-chemical analysis. *Proc. Natl. Acad. Sci. U.S.A.*, **98**, 4299–4304.

105. W.D. Hoff, A. Xie, I.H. Van Stokkum, X.J. Tang, J. Gural, A.R. Kroon, K.J. Hellingwerf (1999). Global conformational changes upon receptor stimulation in photoactive yellow protein. *Biochemistry*, **38**, 1009–1017.

106. H. Kandori, T. Iwata, J. Hendriks, A. Maeda, K.J. Hellingwerf (2000). Water structural changes involved in the activation process of photoactive yellow protein. *Biochemistry*, **39**, 7902–7909.

107. J. Hendriks (2002). Shining light on Photoactive Yellow Protein from *Halorhodospira halophila*. PhD thesis, University of Amsterdam, PrintPartners, Ipskamp, Enschede, The Netherlands.

108. J. Hendriks, I.H.M. van Stokkum, K.J. Hellingwerf (2002). Deuterium isotope effects in the photocycle transitions of the Photoactive Yellow Protein. *Biophys. J.*, **84**, 1180–1191.

109. J. Hendriks, T. Gensch, K.J. Hellingwerf, J.J. van Thor (2002). Transient exposure of hydrophobic surface in the photoactive yellow protein monitored with Nile Red. *Biophys. J.*, **82**, 1632–1643.

110. Z. Ren, B. Perman, V. Srajer, T.Y. Teng, C. Pradervand, D. Bourgeois, F. Schotte, T. Ursby, R. Kort, M. Wulff, K. Moffat (2001). A molecular movie at 1.8 Å resolution displays the photocycle of photoactive yellow protein, a eubacterial blue-light receptor, from nanoseconds to seconds. *Biochemistry*, **40**, 13788–13801.

Chapter 9

Higher plant phototropins, photoreceptors not only for phototropism

Michael Salomon

Table of contents

Abstract

Phototropins, a new family of UV-A/blue light receptors of higher plants, are light-activated, autophosphorylating serine/threonine kinases that function as primary photoreceptors for phototropism, which is the directional growth of plant organs toward or away from the light. Since 1988, when light-mediated phosphorylation of phototropin was first described, significant advances have been made in our understanding of the molecular mechanisms by which light signals are processed by these photoreceptors. This chapter is aimed at giving a detailed overview and discussion of all of the relevant experimental data that meanwhile have been brought together. It will basically follow the sequence of events leading from the initial photophysiological and biochemical analysis of the light-induced phosphorylation reaction and its correlation with phototropism to the isolation of the phototropin gene and the subsequent characterization of the structural, biochemical and, in particular, the photochemical properties of the encoded protein. Most recent genetic evidence for the involvement of at least a second member of the phototropin family in the regulation of phototropic responses, the participation of both phototropins in chloroplast movement, another well-known blue light response of higher plants, as well as currently existing hypotheses concerning possible early events in downstream signaling are further focal points.

9.1   Introduction

Photosensory systems control a variety of morphogenic and growth processes in higher plants. This includes chloroplast biogenesis, the development of leaves or flowers, hypocotyl growth inhibition, pigment synthesis, circadian timing and phototropism [1,2]. All of the response-mediating photoreceptors known to date are chromoproteins. Depending on the wavelength range of solar light they absorb, plant photoreceptors basically can be divided into three distinct classes, namely UV-B receptors, UV-A/blue light receptors and the red/far-red light-absorbing family of the phytochromes. The latter are still by far the best-characterized plant light-sensors [reviewed in 3–7]. Nevertheless, over the past decade exciting progress has also been made in the identification of UV-A/blue light receptors and in our understanding of their mode of action, a fact corroborated by the multitude of recently published reviews [8–15].

In the course of evolution plants have also evolved the ability to react in a spontaneous and reversible way to changes in the ambient light environment. For example, growing plants that receive light only from one direction are able to sense the lateral differences in light intensity and quality and typically respond to these illumination conditions by curvature of the shoot and orientation of the leaves toward and curvature of the root away from the light source. This differential growth reaction, commonly known as phototropism, is specifically induced by UV-A and blue light. Phototropism was studied long before the photomorphogenic effects of red light were even mentioned and research in the field traces back to Charles Darwin. In 1881, he carried out the first elaborate experiments to investigate the phototropic behavior of grass coleoptiles [16], plant organs that were to become the classical objects for physiological studies on phototropism. Two years later, Julius Sachs presented the first direct evidence that wavelengths from the blue region of the electromagnetic spectrum can induce the curving reaction [17], but it would be 1934 before Johnston unequivocally confirmed that phototropism is a genuine blue-light response when he obtained the first reliable action spectrum, this time for oat coleoptiles [18]. Despite this early knowledge of the wavelength dependency of phototropism the nature of the acting chromophore was to remain elusive for more than 60 years. During this long period a series of different putative chromophores were discussed, including carotenoids [19–21] flavins [22,23], pterins [24] and retinal [25]. At least temporarily, all these pigments found strong supporters. However, those being most favored throughout the years certainly were carotenoids and flavins, for two major reasons. First, both pigments were found to be present in substantial amounts in the oat coleoptile, initially demonstrated in 1936 for carotenoids [21] and 13 years later for ribo-flavin [22]. Second and more importantly, refined action spectra for phototropism of the *Avena* coleoptile [26,27] resembled from their shape the combined absorption spectra of free flavins and carotenoids. In the blue, the spectra exhibited the typical carotenoid-like fine structure whereas in the UV-A there was a broad action peak near 370 nm characteristic for flavins. A corresponding action spectrum subsequently determined for alfalfa (*Medicago sativa*) [28] confirmed the seemingly hybrid spectral properties of the photoreceptor for phototropism in higher plants. These findings led to a passionate controversy as to whether flavins, carotenoids or both pigments could mediate the response. However, the flavin supporters gained ground when phototropic responsiveness was shown to be normal in *w3*, a maize mutant lacking phytoene desaturase, a lesion that results in the disruption of carotenoid biosynthesis [29].

Another major problem facing researchers in these years was the accurate interpretation of the unexpectedly complex fluence–response curves of the phototropic bending reactions. As exemplified for the coleoptiles of monocot grass seedlings, but being valid for dicot plants as well, short pulses of blue light at low fluences are effective in inducing first positive phototropic curvature. This response occurs within the tip cm of the coleoptile in a fluence-dependent manner and obeys the Bunsen–Roscoe reciprocity law. By contrast, second positive curvature, a less photosensitive response occurring along the entire length of the coleoptile, is a time-dependent reaction as curvature is the more pronounced as the same fluence of blue light is administered at increasing irradiation times. Many groups were occupied with the analysis of the phototropic fluence response relationships under various conditions using different plants [30–34]. However, all hypotheses proposed on the basis of the data obtained either on assuming the action of a single [33] or the coaction of at least a second photoreceptor [31,34] were unsuccessful in providing an accurate explanation for the striking differences between first and second positive phototropism. Thus, physiological and biochemical approaches clearly failed to resolve both the chromophore question and that of the number of photoreceptors participating in phototropism.

In this respect the genetic characterization of photomorphogenic mutants from *Arabidopsis thaliana* turned out to be much more straightforward. Working with *hy4*, an *Arabidopsis* mutant impaired in blue-light-controlled hypocotyl growth inhibition, Ahmad and Cashmore in 1993 identified the first UV-A/blue light receptor [35], a flavoprotein closely related to DNA-photolyases that they designated as cryptochrome 1 (CRY1). Four years later the Briggs lab successfully cloned the gene defective in the phototropic null mutant *nph1-5* [36] and demonstrated subsequently that this gene encodes a blue-light-regulated, flavin-binding, autophosphorylating photoreceptor kinase. According to its function as a photoreceptor for phototropism, they named it phototropin.

This chapter aims to cover recent advances in the molecular characterization of phototropins. It consists of a detailed summary of the properties of the light-driven autophosphorylation reaction and of those of the reverse reaction in the dark, followed by a description of the relationships between phototropin structure and function with particular focus on the biochemical, spectral and photochemical properties of the chromophore binding domains and will finally outline recent evidence for another member of this new family of UV-A/blue light receptors in higher plants as well as possible early steps in postperception signaling.

9.2   Blue-light-driven autophosphorylation of phototropin

9.2.1   Getting a first glimpse of the photoreceptor for phototropism

In 1988 Gallagher et al. [37] made the pioneering discovery that phosphorylation of a protein (about 120 kDa) by [γ-^{32}P] ATP was specifically induced upon blue-light irradiation of microsomal membranes prepared from elongating stem tissues of dark grown pea seedlings (Figure 1). By contrast, red light was ineffective in stimulating this response. Subsequently homologous phosphoproteins with comparable characteristics were reported for various other monocot as well as dicot plant species, including maize [38,39] sunflower [40], tomato [40], *Arabidopsis thaliana* [41], zucchini [40], wheat [40], barley [40], sorghum [40] and oat [40,42], indicating that the protein is likely to be found in all higher plants. In all cases the phosphoprotein from grass coleoptiles (105–115 kDa) turned out to exhibit a somewhat lower apparent molecular mass when compared with that from dicot plants (120–130 kDa). Almost ten years later it eventually turned out that this ubiquitous plasma membrane-associated phosphoprotein of higher plants was the UV-A/blue light receptor phototropin, the photoreceptor for phototropism.

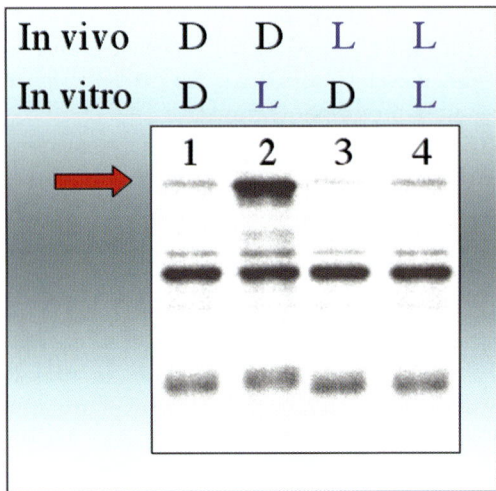

Figure 1. Phosphorylation of mock-irradiated (D, lanes 1 and 3) and blue-light-irradiated (L, lanes 2 and 4) cell extracts prepared from coleoptile tips of etiolated oat seedlings. Strong light-induced autophosphorylation of phototropin (marked by a red arrow) is only found for plants kept in the dark (lanes 1 and 2), whereas the reaction in vitro can not be stimulated when the seedlings have received a saturating pulse of blue light prior to extract preparation (lanes 3 and 4).

9.2.2   Basic properties of the phosphorylation reaction

The light-induced phosphorylation reaction is highest in tissues of etiolated plants. In all subcellular localization studies published up to now phototropin was found to be almost exclusively present in microsomal membranes. For several plants further fractionation of the microsomal membranes was carried out either on sucrose gradients or by partitioning in a polyethyleneglycol/dextran-based aqueous two-phase system [43]. In each case, the phospho-protein was highly enriched in those fractions that were high in plasma membranes [37,44 (pea); 38,39 (maize); 42 (oat); 45 (wheat)]. The protein is a peripheral membrane-associated rather than an integral component of the plasma-membrane since treatment with high salt disrupts the interactions with the membranes (E. Knieb, M. Salomon, unpublished results).

As demonstrated first for pea [46] and in subsequent work with maize [38] and oat [47,48], exposure of etiolated seedlings to blue light prior to membrane isolation in each case led to a dramatic reduction in the capacity of subsequent in vitro phosphorylation of phototropin (see Figure 1). Furthermore, in etiolated tissues that were preincubated in $^{32}P_i$ to enable synthesis of endogenous ^{32}P-ATP, in vivo illumination resulted in strong labelling of the phosphopro-tein [38,39,49]. From these results it was concluded that the light-mediated phosphorylation reaction can be elicited both *in vitro* and *in vivo*. The dramatic reduction in ^{32}P-labelling of the protein, found for pre-illuminated plants, can be reasonably explained on assuming that the majority of the reacting phosphorylation sites were already filled with phosphate derived from intracellular non-labelled ATP as a result of the in vivo light treatment. Hence, the amount of protein that becomes phosphorylated in vivo in response to exposure of the plants to blue light of a particular quantity or quality, can be calculated indirectly by subtracting the in vitro phosphorylation values obtained for the pre-irradiated plants (CP_L) from those determined for non-irradiated plants (CP_D). Thus, in vivo phosphorylation equals to CP_D minus CP_L. This relationship should be kept in mind, as it will be relevant for several studies described below.

9.2.3 Correlations between phototropin phosphorylation and phototropism

With this relatively simple in vitro system at hand the way was paved for a detailed analysis of this higher plant-specific blue-light effect at the cellular, biochemical and molecular levels. Consequently, the emphasis of most studies in this field published up to 1998 was on the further characterization of the phosphorylation reaction. One of the first questions that had to be resolved was to assign this blue-light effect to a particular physiological plant response. Over the years a body of correlative evidence has been brought together that strongly suggests this reaction is involved in the regulation of phototropic responses of higher plants, most likely representing a very early step within the signaling cascade. The most striking correlations derived from these early studies are outlined below:

(i) *Tissue distribution.* In all plants investigated to date, phototropin phosphorylation is highest in those tissues that are most sensitive for stimulation of phototropic curvature by a given, effective, directional light pulse. In etiolated seedlings of dicot plants the corresponding tissue is located just below the hypocotyl or epicotyl hook [46, pea] and in those of monocot grass seedlings in the coleoptile tip [38,50 (maize), 47 (oat), 45 (wheat)]. For the coleoptiles of maize [50] and oat [47] the response, even though being most prominent in the tip region and declining rapidly towards the node, is detectable along the entire length of the coleoptile. The same is valid for the hypocotyls of cress, mustard and soybean and the pea epicotyl. Western analysis with an anti-phototropin antibody [51] revealed that in all cases the basipetal decrease in the amount of light-inducible phototropin phosphorylation reflected the actual expression levels of the photoreceptor protein in the respective tissues (E. Knieb, M. Salomon, unpublished results).

(ii) *Fluence response relationship.* Both first positive phototropic curvature [31,34] and blue-light-potentiated phosphorylation of phototropin [46] adhere to the Bunson–Roscoe reciprocity law since the magnitude of each response is solely determined by the sum of photons applied, independent of whether they are administered as a short pulse at a higher fluence rate or vice versa. Therefore, both responses obviously follow first-order photochemistry.

(iii) *Action spectra.* The action spectrum for blue-light-activated phototropin phosphorylation in vivo [38] and in vitro [50], i.e. the magnitude of the response versus the wavelength at constant fluences, closely matches the action spectrum for phototropism [26–28], i.e. the measured curvature angles versus the wavelength at constant fluences.

(iv) *Dark recovery kinetics.* In the regulation of phototropic responses adaptive processes play an important role. For example, when plants are exposed to directional light conditions above those inducing first positive curvature but below those inducing second positive curvature, a consequent desensitization of the photosensory system is observed and the plants have to pass through a period of regeneration in the dark before a normally effective second light stimulus can elicit a bending reaction [30]. Depending on the plant species, the refractory period required to fully restore phototropic sensitivity can last up to 30 min. Likewise, in membranes prepared from maize coleoptile tips at different times following a preceding blue light treatment, at intensities sufficiently high to saturate phototropin phosphorylation in vivo, the phosphorylation capacity in vitro gradually returns in the dark to the level of non-irradiated plants within 20–30 min [50]. The recovery kinetics almost perfectly match those for dark recovery of phototropic responsiveness. However, in microsomal membranes from pea upper epicotyls and in crude cell extracts from oat coleoptile tips the reaction appears to proceed at much slower rates and total regeneration of non-phosphorylated phototropin lasts between 60 [46, pea] and 90 min [52, oat]. We will discuss possible reasons for this discrepancy in Section 9.3.1.

(v) *Unilateral light induces a lateral gradient of phototropin phosphorylation.* To better understand what will be outlined below a short methodological excursion needs to be inserted here. Because phototropin is localized in the plasmalemma membrane, all of the initial studies aimed at characterizing light-potentiated phosphorylation were either carried out with microsomal or plasma membranes. However, working with membrane preparations to quantitate phototropin phosphorylation in vivo using the aforementioned indirect method bears two major disadvantages. First, relatively large amounts of plant material are needed. Second, during the relatively long-lasting preparation procedure partial dephosphorylation of the in vivo phosphorylated phototropin is likely to occur.

Salomon et al. [47,48] resolved these problems by working with crude cell extracts either prepared from a single 5 mm section of an oat coleoptile or from two corresponding longitudinal half sections. Using this cell extract-based in vitro system, they demonstrated that exposure of the seedlings to a unilateral blue light generates, at least for certain fluences, a directional gradient of phototropin phosphorylation across the coleoptile. In coleoptile tips, a plot of the measured differences in phosphorylation between the irradiated and the shaded side versus the log of fluence gives a bell-shaped curve very similar to the fluence–response curve for first positive phototropic curvature. However, there is one striking inconsistency, the resulting curve, and thus the threshold of the reaction, is shifted by about two orders of magnitude towards higher fluences.

It must be noted here that this discrepancy is generally valid for blue-light-potentiated phototropin phosphorylation, as for plants so far tested, i.e. pea, maize, oat, and *Arabidopsis,* the threshold of the reaction in vivo [40,44,46,52] as well as in vitro [41,42,44,49,50,53] is between 1 and 10 μmol m^{-2} of blue light, a fluence range at which first positive curvature has already reached its maximum. This result suggests that the phosphorylation event is probably not an absolute prerequisite for the signaling process. However, the real reasons for the strikingly different photosensitivities of phosphorylation and phototropic bending are currently far from being fully understood. The easiest explanation may be that there are limitations in detectability of phosphorylated phototropin both in vivo and in vitro by the methods used. However, alternative possibilities cannot be ruled out, and we will discuss an existing hypothesis based upon the assumption of a hierarchical nature of phototropin phosphorylation at the end of this chapter.

In agreement with the photophysiological properties of second positive phototropic curvature, which is a time-dependent rather than a fluence-dependent process, longer irradiation times in the more basal regions of the coleoptile at a constant fluence produced a steeper phosphorylation gradient [47]. This indicates that the adjustment of the gradient within these tissues also appears to involve a time-dependent reaction and is not determined solely by the fluence. Together, these studies provided the first direct demonstration for a biochemical gradient brought about by a corresponding blue light gradient within phototropically sensitive plant tissues and its correlation with first and second positive phototropism.

9.2.4 *Early evidence for an autophosphorylating photoreceptor kinase*

Prior to cloning of the phototropin gene and characterization of its gene product, several biochemical observations already provided a body of indirect evidence that the protein that became phosphorylated in response to blue light might combine three distinct functions within the same molecule, that of a photoreceptor, a kinase, and a substrate for the kinase. First, for a membrane associated photoreceptor kinase system composed of separate proteins one would expect that detergent solubilization of the membranes would lead at least in part to a disruption of the functional complex by dissociation of its components. These predicted properties are in apparent contrast to the observation made by several authors that the addition of the non-ionic detergent Triton X-100 to membranes has a strong stimulatory rather than an inhibitory effect on light-activated phosphorylation [38,42,49]. Moreover, Triton-solublization does not affect the photosensitivity of the system nor does it alter the kinetics of the phosphorylation reaction [44], results that are also in agreement with the assumption of a single tri-functional protein. Second, following native electrophorectic separation of solubilized membranes, photo-tropin migrated as a 330 kDa complex that retained its capacity for auto-phosphorylation on incubation of the gel in a $(\gamma\text{-}^{32}\text{P})$-ATP-containing buffer under blue light [54]. The migration behavior of phototropin in native gels further suggests that the light-responsive, functional protein could be a homo-dimer. Third, the ATP antagonist 5'-p-fluorosulfonylbenzoyladenosine, which because it covalently binds to ATP-binding sites, is a potent inhibitor of ATP-binding enzymes, including kinases [55], also inhibits blue-light-activated phos-phorylation of phototropin [39]. Furthermore, an antibody against this reagent utilized in Western analysis of inhibitor-treated microsomal membrane pro-teins from pea specifically recognized a protein with an electrophorectic mobil-ity (120 kDa) indistinguishable from that of the phosphoprotein [39,44]. Based on these results, it was concluded that the 120 kDa protein at least functions as a kinase and is its own substrate, implying that we are dealing with an autophosphorylation.

9.2.5   Biochemical properties of phototropin phosphorylation

Similar kinetics for light-driven phototropin phosphorylation have been reported for pea [44,49], maize [39,47] and *Arabidopsis* [41]. In membranes pre-irradiated with blue light, strong phosphorylation is detectable within a few seconds on addition of ATP and saturation is reached after about 2 min. Some authors observed a continuous disappearance of the phosphoprotein to near dark levels within 20 min [39,49], whereas others found that the maximum phosphorylation level remained almost constant upon prolonged phosphoryla-tion times [38,51]. We will come back to this phenomenon in Section 9.3.2. The reaction requires Mg^{2+} but no Ca^{2+} [38,39,56]. The kinase is highly specific for ATP [56,57] and shows a much lower turnover of GTP [57], while CTP and UTP completely fail to mediate autophosphorylation [56]. The reaction has a pH optimum between 7 and 8 [38,56], is much more sensitive toward more acidic than more alkaline pH, and is inhibited in the presence of nanomolar concentrations of the protein kinase C inhibitor staurosporine [38,45].

Hager et al. [53] investigated the effects of several reducing agents on phototropin phosphorylation and found that NADH, NADPH, ascorbate and dithiothreitol (DTT) at concentrations ranging between 2 and 20 mM strongly enhanced the capacity for phosphorylation, particularly in older, frozen and thawed membrane preparations. They concluded that the photoactivation mechanism may involve the transfer of an electron from the excited chro-mophore to a target molecule via a light-driven redox reaction, and that exter-nally added electron donors may facilitate the recovery of the light-responsive, reduced chromophore. However, in a later study, Rüdiger and Briggs [58] reported that reagents reacting with sulfhydryl groups such as iodoacetamide, N-ethylmaleimide (NEM), and N-phenylmaleimide (NPM) inhibited photo-tropin phosphorylation and, hence, provided some evidence that the reaction may require a free cysteine-SH, which could be a target site for reducing agents as well.

9.2.6 *Phosphorylation sites of phototropin*

Phosphoamino acid analysis of in vitro phosphorylated phototropin from pea [56] and maize [39] revealed that autophosphorylation occurs almost exclusively on serine residues. Upon protease treatment of the phosphoprotein one obtains a variety of distinctly strongly labelled phosphopeptides of different size [49,42,56], a finding consistent with phosphorylation on multiple sites. Indeed, preliminary mapping of the autophosphorylation sites in the oat photoreceptor indicates that 4–6 different serine residues are the targets of the kinase, and that all of these residues, presumably, are located in the N-terminal region of the protein [59,103].

9.3 The reverse reaction: Regeneration of non-phosphorylated phototropin in the dark

9.3.1 *Dark recovery in vivo*

In contrast to light-driven phosphorylation of phototropin, which occurs in a matter of seconds, the reverse reaction in the dark is very slow. In maize, pea, and oat dark recovery has been investigated in detail by monitoring the return of the in vitro phosphorylation capacity to the dark level following a saturating blue light pulse *in vivo*. The resulting regeneration times of 30 min for maize [38,39], 60 min for pea [49] and even 90 min for oat [52] suggest that relatively broad variations between plant species might exist. The same, however, is not valid for the respective refractory periods for these plants to regain photosensitivity following a preceding phototropic stimulus. One possible explanation for this discrepancy may be by differences in the experimental conditions. For example, in both studies with maize membranes, blue-light pre-treatment of coleoptile tips could reduce subsequent phototropin phosphorylation *in vitro* to only 70–75% whereas with cell extracts from oat the corresponding value was greater than 90% (see also Figure 1). If one considers that the time required to prepare membranes (maize) is much longer than that for cell extracts (oat), the above differences may result from different degrees of dark recovery (dephosphorylation) between *in vivo* light treatment and subsequent in vitro phosphorylation. Since the kinetic data reported for maize do not take sample preparation into account, they possibly do not cover the entire process. Thus, the effective time required for quantitative regeneration of non-phosphorylated phototropin in maize may exceed 30 min.

For the basal regions of oat coleoptiles the kinetics for *in vivo* dark recovery of phosphorylation differed strikingly from those found for the tip. Pre-irradiation of these tissues led to a dramatic overshoot of the reaction. The maximum phosphorylation after 90 min in the dark, typically, were 3 to 4-fold higher than those of the dark controls [52]. Since the observed effect was found to be associated with an increase in photosensitivity of the pre-irradiated seedlings for second positive phototropism, it was reasoned that the amplified phosphorylation may reflect a sensitization mechanism which most likely is brought about by a blue-light-regulated increase in the expression levels of phototropin. However, more recent antibody studies carried out to monitor the actual amounts of phototropin expressed during the recovery process did not confirm this hypothesis (Knieb, Rüdiger, Salomon, unpublished results). Though not as pronounced as in oat, an overshoot in the phosphorylation capacity has also been reported to occur in pea in response to subsaturating and supersaturating blue-light pulses [46], but was not found in maize or *Arabidopsis*.

9.3.2   Dark recovery in vitro

In microsomal membranes the level of phosphorylated phototropin remained almost constant for up to 10 min in maize [38] and even 40 min in *Arabidopsis* [51] after the phosphorylation reaction reached its maximum, indicating that no dark recovery occurs in isolated membranes. The absence of any phosphate turnover and hence phosphatase activity in membranes finds further support in that no decline in the amount of ^{32}P-labelled phototropin was observed upon addition of cold ATP even at high concentrations [38,49]. Conflicting results have been reported by Palmer et al. [39] for the phosphorylation reaction in membranes prepared from maize coleoptiles. They observed a very rapid and almost quantitative disappearance of the phosphoprotein within 20 min of the reaction starting (see Section 9.2.5), even in the presence of the phosphatase inhibitor, sodium fluoride, and the protease inhibitors PMSF and leupeptin. The only apparent difference between these studies was the presence of the detergent Triton X-100 in the work of Palmer et al. For membrane-associated phototropin from oat it has been demonstrated that Triton X-100 has adverse effects on the stability and/or the activity of the protein [42]. Furthermore, since Palmer et al. did not test whether a second pulse of blue light, given after 20 min together with fresh (γ-^{32}P)-ATP, could restore phototropin phosphorylation, it is possible that the observed rapid decline in radio-labelled phototropin may have resulted from proteolytic degradation or from intrinsic instability of the protein caused by the detergent, rather than reflecting dark recovery as it exists in vivo.

The molecular mechanism underlying in vivo dark recovery of phototropin is still unclear. To date, there is no direct experimental evidence for a phototropin-dephosphorylating phosphatase. On postulating a dephosphorylation reaction, the required phosphatase might be localized in the cytoplasm since microsomal and plasmalemma membranes appear to lack a phosphatase that specifically acts on phosphorylated phototropin in vitro. However, if one considers the long-lasting restoration of the ground state in the dark, it could be that the slow recovery rates may be the consequence of proteolytic degradation of phosphorylated phototropin and subsequent de novo synthesis of the non-phosphorylated protein.

9.3.3   Relaxation of the activated state in the dark

Even though the membrane-based in vitro system apparently lacks components required for the regeneration of the dephosphorylated form of phototropin, another phenomenon, often designated as the in vitro memory for a blue-light pulse, can be readily investigated by this method. Since membranes do not contain cellular ATP, irradiation will lead to the activation of the phototropin kinase, but in contrast to the in vivo situation, no phosphorylation will occur as long as external ATP is not added. This allows one to quantitate the time that the activity of the kinase and hence the activated state of the photoreceptor is maintained in the dark independent of phosphorylation following a saturating blue light treatment. Surprisingly, the dark decay in kinase activity followed kinetics as slow as those determined for the regeneration of photosensitive, non-phosphorylated phototropin in vivo. In membranes from oat [42] and maize [53] the phosphorylation levels were still above those of the dark controls when the ATP was added as much as 60 min after irradiation. About 80% (oat) and 95% (maize) of the initial phosphorylation activity could be restored by a second light pulse administered after a dark period of 70 and 60 min, respectively. Therefore, at least for the majority of the membrane-bound phototropin, the long-term incubation obviously did not cause a loss of function, and the kinetics obtained reflect rather the slow relaxation of the activated state to the ground state. Corresponding relaxation kinetics reported by Palmer et al. for maize membranes solubilized in Triton X-100 are not consistent with those from the aforementioned studies as they found that light activation of the kinase only persisted for 20 min in the dark [39]. However, a second light treatment almost completely failed to reactivate the phosphorylation reaction. Again, the most probable reason for this difference is the presence of detergent and its effect on phototropin stability and/or function.

In conclusion, the strikingly similar rates of in vivo dark recovery, which involves disappearance of the phosphorylated and reappearance of the non-phosphorylated forms of the photoreceptor, and the relaxation of the activated state in vitro following a light pulse are unlikely to be the result of a mere coincidence, but rather suggest that we might be dealing here with interdependent processes. One possible way by which both reactions could depend on each other is as follows. Given that phototropin retains its kinase activity after autophosphorylation has occurred–to date still not known–and further given that dark recovery results from the action of a phosphatase, the slow return to the ground state could be explained by mutual phosphorylation and dephosphorylation, whereby the equilibrium of the overall reaction will be shifted towards dephosphorylation at the same rate as relaxation of photoactivated phototropin and, hence, suppression of kinase activity proceeds. Alternatively, blue-light activation and subsequent autophosphorylation could induce a change in photoreceptor conformation such that the phosphorylated sites are protected against the action of a phosphatase. As the protein folds back to its ground state conformation, these sites become accessible for the phosphatase, enabling dephosphorylation of the photoreceptor.

9.4   Cloning and characterization of the NPH1 gene

9.4.1   Generation and analysis of mutants impaired in phototropism

The final breakthrough towards the identification of the photoreceptor photo-tropin came from analysis of mutants from *Arabidopsis thaliana* that were affected in their phototropic responses. Among the first set of such mutants characterized in 1989, the mutant *JK224* exhibited both reduced blue-light-induced phosphorylation of the 120 kDa protein and a lower photosensitivity, as the fluence response curve was significantly shifted towards higher fluences while the magnitude of the response was normal [60,61]. From these photo-physiological properties it was argued that such a sensitivity shift can be best explained by a mutation directly affecting photoreceptor function. The *JK224* mutant, however, was not further characterized at the molecular level at that time.

Six years later, the Briggs laboratory isolated and characterized four fast neutron-generated non-allelic *Arabidopsis* mutants, designated *nph1* through *nph4* (nph is the acronym for non-phototropic-hypocotyl) that were impaired in their phototropic responses [62]. One of these mutants, *nph1*, failed to show any blue-light-potentiated phosphorylation and also lacked a 120 kDa protein band detectable in microsomal membranes of wild-type plants [63]. Since *nph2, nph3* and *nph4* exhibited normal levels of blue-light-induced phosphorylation with respect to the wild-type it was concluded that their gene products act downstream from *nph1* within the signaling pathway for phototropism. Consequently, it was tempting to surmise that the gene affected in *nph1* encodes the photoreceptor itself. Huala et al. [36] finally identified the *NPH1* gene by sequence analysis of three allelic *nph1* mutants that turned out to bear distinct lesions within the same gene. These results, along with the fact that phototropic sensitivity in *nph1–5* plants transformed with the corresponding wild-type gene was restored, unequivocally demonstrated that the lesions present in the *nph1* genes were responsible for the non-phototropic phenotype of *nph1* plants.

9.4.2   Sequence analysis of the NPH1 gene and primary structure of
        phototropin

The *NPH1* gene from *Arabidopsis* encodes for a protein of 996 amino acids with a calculated molecular mass of 112 kDa [36]. In regard to the still unan-swered question whether the *NPH1* gene product could function as substrate, kinase, and photoreceptor for phototropism, at least one answer was immedi-ately provided from the sequence analysis. The encoded protein is a protein kinase closely related to the PVPK1 family within the protein kinase C group and contains, within its C-terminal third, all eleven signature motifs typically found in serine/threonine kinases [64]. The aforementioned *JK224* mutant turned out to be allelic to the *NPH1* locus and bears a single point mutation near the C-terminal end of the kinase resulting in a substitution of a lysine for an arginine. JK224 therefore obtained the new designation *nph1-2*. The replaced arginine residue appears to be critical for maintaining the correct folding properties of the carboxy-terminal region of the kinase. Thus, the lower phototropic sensitivity measured for *nph1-2* plants most probably reflects adverse effects of the substituted amino acid on the catalytic activity of the kinase [36].

The second striking feature of *NPH1* are two PAS-domains in the N-terminal region that share about 40% identity and consist of 110 amino acids each. The acronym PAS is derived from their occurrence in *Drosophila* period (PER), the vertebrate aryl hydrocarbon receptor nuclear translocator (ARNT) and *Drosophila* single minded (SIM) [65,66]. The subsequent isolation of homologous cDNAs from other plant species including oat [67] and maize [68] and somewhat later also from rice [69] and the fern *Adiantum capillus veneris* [70] demonstrated that the described domain structure is common to all phototropins. One characteristic feature of the *NPH1* homologues from monocot plants is the somewhat shorter N-terminus when compared with that of the *Arabidopsis* protein, resulting in an averaged calculated molecular mass of only 100 kDa. If one recalls the aforementioned size differences between the phosphoprotein of dicot and monocot plant species (see Section 9.2.1), the results obtained from the sequence analysis of *NPH1* genes were clearly consistent with the hypothesis that nph1 might be the phosphorylated protein.

9.4.3   Phototropin is a flavoprotein

However, the most important question that still had to be answered was to find out whether the nph1 protein can bind a chromophore, a prerequisite for its proposed function as a photoreceptor. Since at that time, and even up to now, all efforts aimed at purifying phototropin from plants by classical bio-chemical approaches have failed, Christie et al. [51] struck a new and more successful path to resolve the photoreceptor question. They expressed NPH1 from *Arabidopsis* heterologously in insect cells (baculovirus system) and found that a yellow protein (BAC-NPH1) was produced by the cells. The yellow color resulted from flavin mononucleotide (FMN) that was non-covalently attached to BAC-NPH1. No other chromophores were found to be associated with the protein in detectable amounts. Furthermore, Bac-NPH1 in crude cell extracts exhibited essentially the same kinetics for blue-light-potentiated phos-phorylation as the plasma membrane associated phosphoprotein of higher plants. In the UV-A- and blue region of the electromagnetic spectrum the fluo-rescence excitation spectrum of cell extracts containing Bac-NPH1 was very similar to the action spectrum for phototropism with two major peaks occur-ring at 370 and 450 nm and two shoulders near 420 and 470 nm. In contrast, the FMN-chromophore when released from the apoprotein by acid den-aturation exhibited the typical flavin-like excitation spectrum and entirely lacked the above-mentioned fine structure in the blue. In summary, both the properties of *NPH1* as derived from its sequence and those of the Bac-NPH1 protein provided clear and convincing evidence that nph1 functions as an autophosphorylating photoreceptor regulated by UV-A/blue light. According to its evident role as a photoreceptor for phototropism the protein was thereafter designated as phototropin (meanwhile named phot1, see Section 9.6).

9.5   Characterization of the chromophore-binding domains LOV1 and LOV2

9.5.1   LOV1 and LOV2 function as FMN-binding sites

As mentioned above, in addition to the serine/threonine kinase two PAS domains are the only regions in *NPH1* that exhibit significant homologies to known proteins. Therefore, after the Briggs lab had successfully demonstrated that phototropin is a flavoprotein, it made intuitive sense to hypothesize that the most obvious and likely role of the PAS domains is that they function as flavin binding sites.

The major function originally ascribed to PAS domains was to mediate protein/protein interactions. Meanwhile, the remarkable versatility of PAS domains with regard to the diverse functions they can exhibit in various functionally unrelated proteins is well documented by the increasing number of publications that have become available over the past decade. For example, PAS domains have been reported as binding sites for different cofactors. For the oxygen sensor protein FixL it is a heme [71], in NIFL from *Azotobacter vinelandii* [72] a protein controlling the expression of genes involved in nitrogen fixation and in the *E. coli* aerotaxis chemoreceptor AER [73,74] it is the flavin FAD, while photoactive yellow protein (PYP), a classical PAS domain, binds the unusual chromophore 4-hydroxycinnamic acid [65,75]. Since the phototropin PAS domains are mostly closely related to a subfamily of PAS domains found in those proteins that are either regulated by light, oxygen or voltage they were named LOV1 and LOV2, respectively [36].

Christie et al. [76] showed unambiguously that these domains represent the flavin-binding sites in phototropin. They expressed only those regions of the *NPH1* cDNAs from *Arabidopsis*, oat and the *PHY3* cDNA from the fern *Adiantum capillus veneris* that contained the coding information for either LOV1, LOV2 or both domains in *E.coli*, and obtained exclusively recombinant flavin-binding polypeptides. As for BAC-NPH1 the bound chromophore was a non-covalently attached FMN in each case. The FMN versus protein ratio was approximately 1 for LOV1 and LOV2 and about 2 for constructs that contained both domains, indicating that one molecule of FMN is bound to each LOV domain. Biochemical characterization of the singly expressed domains showed that LOV1 represents a homodimer whereas LOV2 is a monomer [8]. Given the experimental evidence that native phototropin may be a homodimer as well (see Section 9.2.4), the role of LOV1 could be dual: that of a blue-light sensor and a mediator of protein/protein interactions. In contrast to full-length

Arabidopsis phototropin expressed heterologously in the baculovirus system, where the amount of soluble protein turned out to be extremely low, the solubility of the singly expressed LOV proteins was sufficiently high to allow purification to near homogeneity at protein concentrations suitable for spectroscopic analysis.

9.5.2 Spectral properties of LOV1 and LOV2

Either of the two LOV holoproteins generated absorption spectra that, like the action spectrum for phototropism, were characterized by the typical carotenoid-like fine structure in the blue (two peaks near 450 and 470 nm and the shoulder at 420 nm) and a second major absorption band at 370 nm [76,77] (see Figure 2).

From these results two principal conclusions can be drawn. First, the LOV domains are the chromophore-binding sites of phototropin, which is therefore a dual chromophore photoreceptor. Second, the high degree of agreement between the absorption spectra of the separately expressed chromophore domains and that of the full-length photoreceptor (Bac-NPH1) allow only one interpretation. The 110 amino acids of LOV1 and LOV2 must contain all of the structural information necessary for proper folding of the polypeptide

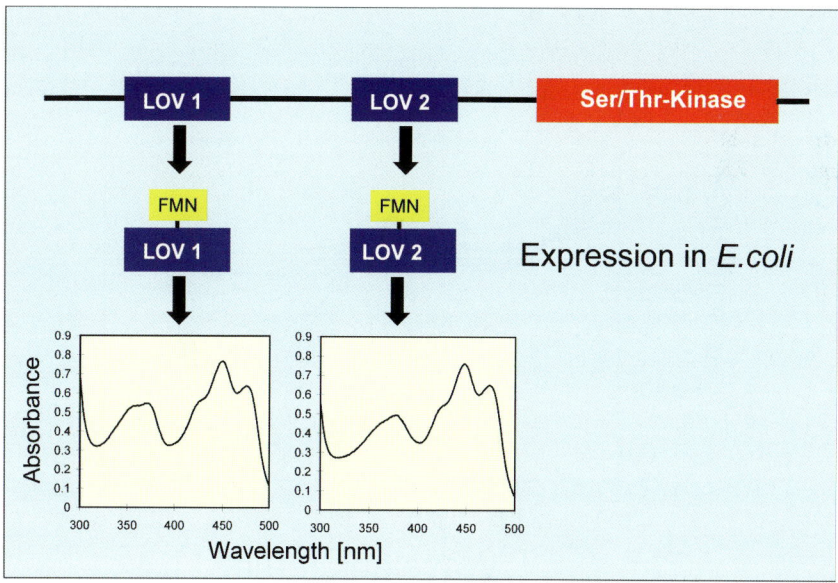

Figure 2. Schematic illustration of the experimental strategy by which the phototropin PAS domains, LOV1 and LOV2, were shown to function as chromophore-binding sites. Both LOV domains non-covalently bind one molecule of FMN when singly expressed in *E. coli*. The diagrams show the absorption spectra of the purified chromopeptides, which are very similar to the action spectrum for phototropism.

chain in a way that ensures the correct interactions between the FMN and the apoprotein. Regions in phototropin that flank the LOV domains do not appear to be involved in these processes. Therefore, both LOV1 and LOV2 can be regarded as self-contained functional modules.

9.5.3   The LOV photocycle

Both LOV domains are photochemically active chromoproteins that undergo a classical photocycle upon light exposure. For both FMN-binding domains of phototropin a rapid loss of absorbance in the spectral region between 410 and 500 nm is observed in response to irradiation with either white or blue light. The final product generated in this photoreaction exhibits maximum absorption at 380 nm and is stable just as long as illumination persists (see Figure 3). In the dark the initial state, i.e. the FMN in its oxidized form, is fully restored within 2–4 min at room temperature. The light-driven forward reaction as well as the reverse reaction seem to be monomolecular processes since they both follow first order kinetics. However, the resultant rate constants are strikingly different between LOV1 and LOV2. While photoproduct formation proceeds more than twice as fast for LOV2 as for LOV1, the opposite is the case for the dark reaction. The relative quantum efficiency of the photoreaction is 0.45 for LOV2 and 0.05 for LOV1. Based on these data the equilibrium for the overall reaction for LOV1 is expected to lie much more on the side of the reactant

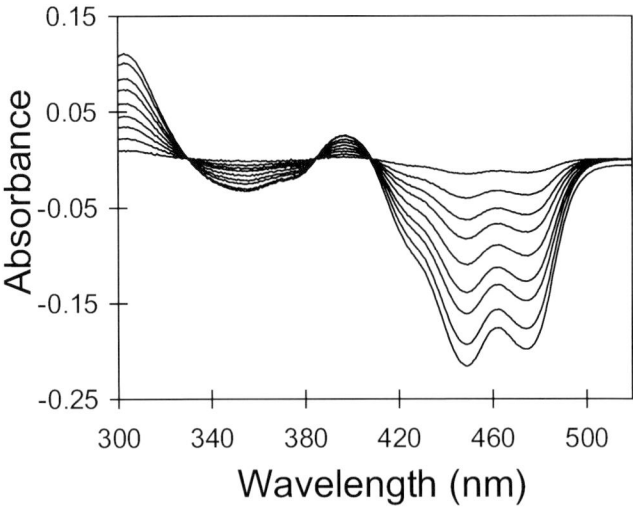

Figure 3.   Differential light minus dark absorption spectrum of the LOV2 holoprotein from oat phototropin. Spectra were recorded at 200 ms intervals after the onset of blue light illumination. Prominent light-induced absorbance changes between 300 and 500 nm only occur in the blue region of the electromagnetic spectrum – spectral properties consistent with those of flavin-cysteinyl C(4a) adducts.

than is the case for LOV2. Accordingly, for LOV1 the yields in photoproduct generated by a given blue light fluence were roughly half of those obtained for LOV2.

Thus, the isolated LOV proteins exhibit strikingly different photosensitivities. Whether LOV1 and LOV2 have similar properties in the native plant photoreceptor is not known to date. However, since many other photoreceptors, e.g. the phytochromes and rhodopsins, typically possess only a single chromophore domain it is reasonable to assume that in the course of evolution the presence of a second light sensor in phototropins must have been advantageous for plants in processing directional light stimuli. Therefore, it is tempting to speculate that two differentially light-sensitive domains may execute distinct functions in the signaling process.

The relatively fast relaxation kinetics of LOV1 and LOV2 in the dark are not consistent with the slow decline in kinase activity determined for the native plant photoreceptor in response to a preceding blue-light pulse (see Section 9.3.3). At first sight, this result might suggest that the so-called memory for a blue-light pulse cannot be explained by the LOV photocycle itself. However, when both LOV domains of oat phototropin were expressed *in tandem* the decay to the ground state occurred at about a ten-fold slower rate, resulting in an overall recovery time of 30 min [8, M. Salomon, unpublished results]. Even though these kinetics are still almost twice as fast as those measured for in vivo dark recovery (see Section 9.3.1) and in vitro memory in oat, they approach the time required to restore phototropic sensitivity. Assuming that this double domain construct, like LOV1, forms a homodimer, the kinetic properties suggest that in the resultant quaternary structure the LOV1 and LOV2 domains might interact in a way that stabilizes the photoproduct.

9.5.4   Blue-light activation of phototropin occurs via a novel photochemical reaction

We will now turn to the photochemical reaction itself. The light reaction associated with the LOV domains is unrelated to typical photoreductions of flavins or flavoproteins for the following reasons. First, the reaction occurs equally effectively in the presence or absence of oxygen. Second, reducing agents as electron donors are not required. Third, the photoproduct generated in the LOV photocycle exhibits a prominent absorption peak at 380 nm while photoreduced flavins almost entirely fail to absorb light in the UV-A [78,79]. The spectral properties of this UV-A-absorbing flavin species rather resemble those of a flavin-cysteinyl C(4a) adduct [80], a designation that stands for the covalent bonding of a sulfur atom of a cysteine residue to the carbon 4a atom of the flavin isoalloxazine ring (see Figure 4). These adducts are known reaction intermediates in several flavoprotein reductases, typically formed during the reduction of protein disulfides.

For such a reaction to take place a cysteine residue is necessarily required. The most likely candidate is a cysteine residue within the motif GRNCRFLQ

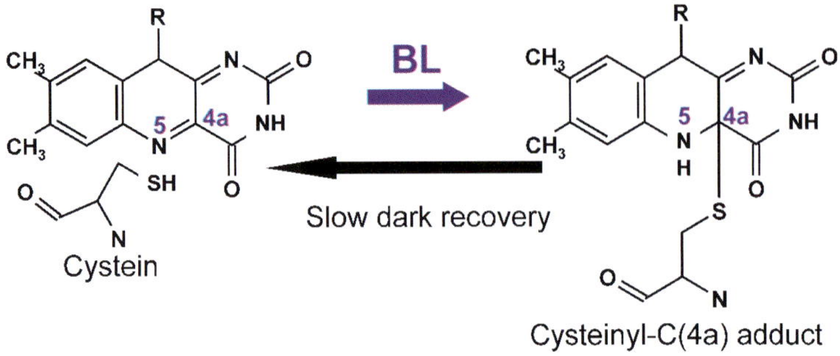

Figure 4. Proposed photoactivation mechanism for higher plant phototropins. A conserved cysteine within the FMN-binding pocket of the LOV domains becomes covalently bonded to the carbon 4a atom of the FMN isoalloxazine ring as a result of the photochemical reaction. In the dark, the ground state, i.e. the fully oxidized flavin, is slowly recovered.

that is highly conserved in the LOV1 and LOV2 domains of all known phototropins and phototropin-related proteins. Moreover, it is the only cysteine in LOV2, whereas a second cysteine residue is present in LOV1. When this residue was replaced in either domain with an alanine, the resulting recombinant proteins (LOV1-C39A and LOV2-C39A) still bound FMN as tightly as the wild-type flavopeptides and also exhibited very similar absorption characteristics, but they failed to show any photoreactions, even under conditions of prolonged blue-light exposure to very high fluences. Thus, on eliminating this residue one generates knock-out mutants of the photocycle, an effect that is best explained by a mutation that directly affects one of the reacting partners of the photochemical reaction.

Another experimental approach aimed at gaining more detailed information on the structural changes that occur in the flavin isoalloxazine in response to irradiation is based on NMR spectroscopy carried out on the LOV2 protein of oat phototropin. Following release of the bound chromophore and subsequent reconstitution of the apoprotein with FMN chromophores bearing ^{13}C and/or ^{15}N labels at various positions within the isoalloxazine ring, NMR spectra were recorded in the dark and under continuous, saturating blue light. For all atoms of the isoalloxazine ring prominent light-induced chemical upfield shifts consistent with a sp^2–sp^3 transition of these atoms were observed only for the resonances of carbon 4a [C(4a)] and nitrogen 5 (N5) [104]. As formation of a flavin C(4a)-cysteinyl adduct involves the breaking of the double bond between C4a and N5 (see Figure 4), the above results clearly confirm the proposed reaction mechanism for phototropin photoactivation.

Although a detailed photochemical analysis has been carried out so far only for LOV1 and LOV2 of the oat photoreceptor, this novel photochemistry seems to be shared by the domains of all phototropins investigated to date and even those of the more distantly related phy3 protein from the fern *Adiantum capillus- veneris* (J.M. Christie, M. Kasahara, T. Swartz, W.R. Briggs, unpublished results; see also Chapter 11).

The adduct-forming cysteine in the LOV2 domain is apparently highly protected against access from small molecules from the outside of the holoprotein. Long-term incubation of LOV2 in the presence of 10 mM NPM affected neither the spectral nor the photochemical properties of this domain. By contrast, LOV1 along with a mutant chromopeptide lacking the second cysteine in LOV1 responded to a corresponding treatment by a slow release (lasting several hours) of the FMN from the apoprotein, indicating that covalent bonding of NPM to the reacting cysteine disrupts the chromophore/apoprotein interactions [77]. Nevertheless, from these results this cysteine is an unlikely candidate target to explain the adverse effect of NPM on light-induced phototropin phosphorylation reported by Rüdiger and Briggs [58]. In their study the inhibitory effect was the same independent of whether NPM was added before or after photoexcitation.

Working with three-dimensional models for LOV1 and LOV2 based on the crystal structure of the PAS domain of the human potassium channel protein HERG [81], and underlying the above-mentioned results, Salomon et al. [77] predicted that cysteine 39 is located within the central region of the proteins being part of a putative FMN-binding pocket. More recently, the crystal structure of the LOV2 domain of phy3 (Phy3-LOV2) from *Adiantum capillus veneris* has been obtained at a resolution of 2.7 Å by Crosson and Moffat [84]. Both the composition of the secondary structure elements (helix-turn-helix αA/αB, five-stranded β-scaffold and helical connector αC) [65] and their overall three-dimensional folding in Phy3-LOV2 are strikingly similar to those properties determined for other PAS domains [75,82 (PYP); 71,83 (FixL), 81 (HERG)]. Moreover, and consistent with the above prediction, cysteine 39 is located within the hydrophobic FMN-binding pocket of the protein with the sulfur atom being 4.2 Å from carbon 4a of the FMN isoalloxazine.

Even though, from the now available data, the general principle of phototropin photochemistry seems to be understood, our current knowledge of the exact reaction mechanism by which the cysteine becomes covalently bound to the flavin is rather poor. In this regard it should be noted that the differential dark/light absorption spectra of both LOV holoproteins exhibit three isosbestic regions between 300 and 500 nm. The fact that we, thereby, do not obtain clearly defined isosbestic points can be indicative of a spectrally different flavin species that is generated prior to adduct formation.

9.5.5   From photochemistry to conformational changes

In the wavelength range of FMN absorption (300–500 nm) circular dichroism spectra recorded for LOV1 and LOV2 from oat phototropin revealed a high degree of divergence between the non-irradiated and the irradiated forms of the proteins [77]. These prominent light-induced changes in absorption of the circularly polarized light seem to be in direct connection to the adduct formation as they were not detectable for the photochemically deficient mutant proteins LOV1-C39A and LOV2-C39A. Thus, at least the FMN appears to undergo notable changes of its conformation when becoming covalently bound to the apoprotein by the photoreaction. That such conformational changes are not unexpected becomes evident on noting that the fully oxidized isoalloxazine, as it exists in the ground state of the LOV domains, is almost planar. However, formation of the flavin C(4a)-cysteinyl adduct will disrupt the conjugated electron system due to the aforementioned sp^2–sp^3 transition at positions C4a and N5. As a consequence, the isoalloxazine ring will become more angular.

Although further experimental work is required, preliminary results derived from Fourier-transform infrared spectroscopy (FTIR) analysis of the LOV domains indicate that at least some of the light-induced vibrational changes most probably have to be assigned to the apoprotein portion [8]. Thus, if one remembers that phototropin represents a blue-light-regulated kinase, whose activity is repressed in the dark and switched on in the light, a change of protein conformation as a direct consequence of the photochemical reactions is the most likely mechanism by which the activity of the carboxy-terminal kinase may be controlled.

9.6   Evidence for a second phototropin in higher plants

Phototropin from *Arabidopsis thaliana* shares roughly 65% identity and 74% similarity with homologous proteins from other higher plants, e.g. maize, oat, rice and pea. In 1998, a gene encoding a second, more distantly related member (55% identity/68% similarity) of the phototropin family of UV-A/blue light receptors was identified in *Arabidopsis* and was designated as NPL1 (**NP**H1-**L**ike) [85]. The npl1 protein possesses the same structural elements as nph1, i.e. two N-terminal LOV domains and a C-terminal serine/threonine kinase but has, due to a shorter N-terminus, a somewhat lower calculated molecular mass than nph1 (102 kDa instead of 112 kDa).

Sakai et al. [86] recently demonstrated that the LOV domains of npl1, like those of nph1, function as FMN-binding sites and undergo the same light-induced absorbance changes, indicating that photochemistry of npl1 also might involve the formation of a flavin C(4a)-cysteinyl adduct. Even though light-induced autophosphorylation has been successfully demonstrated for npl1 expressed in insect cells [86], it has not yet been observed in *Arabidopsis*. Nevertheless, npl1 meets all the criteria of a second higher plant blue-light receptor within the phototropin family.

Recent genetic analysis of *npl1* and *nph1* mutants further corroborate this assumption. In their original characterization of *nph1* mutants Liscum and Briggs failed to observe any phototropic responses for fluences that normally induce first or second positive curvature [63]. This prompted them to conclude that all differential growth reactions toward or away from a directional light source are abolished in the absence of the *NPH1*-encoded protein. When the protein encoded by the *NPH1* locus was identified to be a flavoprotein with the characteristic features of a UV-A/blue light receptor kinase, it became an

accepted view that, at least in *Arabidopsis*, the various and quite complex phototropic reactions were regulated by a single photoreceptor. However, Sakai
et al. recently examined the phototropic responsiveness of the *A. thaliana nph1-
101* mutant to continuous unilateral illumination at fluence rates ranging
between 1 and 100 μmol m^{-2} s^{-1} and found strong second positive curvature
[87] under these conditions. Since this response was almost absent in a *nph1
npl1* double mutant [86], the npl1 protein appears to function as a second
photoreceptor for phototropism that apparently mediates directional growth
of the plants in response to fluence rates of continuous blue light greater than
1 μmol m^{-2} s^{-1}. By contrast, nph1 mediates phototropic curvature at both low
(0.01–1 μmol m^{-2} s^{-1}) and high fluence rates (1–100 μmol m^{-2} s^{-1}) of continuous
light, but also in response to short low intensity light pulses.

In addition to their light-sensing role in phototropism there is now growing
evidence that the two phototropins are probably also implicated in the regulation of another blue light response of higher plants, namely the movements of
chloroplasts within a cell. As with the diverse phototropic bending reactions,
one can also distinguish between low fluence and high fluence reactions for
chloroplast movement. Under low light conditions the chloroplasts accumulate
perpendicular to the direction of light to trap the maximum amount of available photons they need for photosynthesis (accumulation response). By contrast, to prevent photodamage when the light intensities become too high the
orientation of the chloroplast will be parallel to the incoming light (avoidance
response).

It was recently found that a *npl1* mutant lacked the avoidance response
but still showed a normal accumulation response [88]. However, a *nph1 npl1*
double mutant turned out to be deficient in both of the blue-light-dependent
orientation reactions of chloroplasts [86]. These results provide compelling
evidence that, as with phototropism, nph1 mediates the more photosensitive
and npl1 the less photosensitive responses in the regulation of chloroplast
movement as well.

To date, it is not yet clear whether two distinct light-sensitive phototropins
exist in all higher plants. However, the two phototropin cDNA sequences
NPH1a and *NPH1b* that have been reported for rice share about the same low
degree of homology as do nph1 and npl1 [69]. Moreover, expression of these
genes and tissue distribution of the resulting gene products are differently
regulated. While the *NPH1a*-encoded protein, the nph1 homologue in rice, is
predominantly present in coleoptiles, the *NPH1b*-encoded protein is mainly
found in leaves. When etiolated rice seedlings are exposed to light, *NPH1a*
mRNA is rapidly down-regulated while at the same time the expression levels
of *NPH1b* mRNA continually increase. It now remains to be determined if the
photophysiological functions of nph1a and nph1b in rice are parallel to those
of nph1 and npl1 in *Arabidopsis*.

From the results discussed in this section, we clearly have to modify our
current understanding about phototropin(s) in two ways. First, there are
at least two phylogenetically and functionally distinct members of the
phototropin family in higher plants. Second, phototropism can no longer be

Table 1. Old and new designations of higher plant phototropins

Plant species	Old gene name	New gene name	Ref.
Oryza sativa	Os*NPH*1a	*PHOT1*	[69]
Oryza sativa	OsNPH1b	*PHOT2*	[69]
Arabidopsis thaliana	*NPH1*	*PHOT1*	[36]
A. thaliana	*NPL1*	*PHOT2*	[85]
Avena sativa	*NPH1-1*	*PHOT1a*	[67]
Avena sativa	*NPH1-2*	*PHOT1b*	[67]
Pisum sativum	*PsPK4*	*PHOT1*	[102]
Zea mays	*NPH1*	*PHOT1*	[68]

regarded as the sole plant response controlled by these photoreceptors. The following new nomenclature was recently proposed: all phototropins that fall into the same homology group as *Arabidopsis* nph1 are now designated as phot1, while those more closely related to *Arabidopsis* npl1 are named phot2 (see Table 1).

9.7   Early events in downstream signaling

In contrast to nph1, the three remaining phototropic mutants *nph2–nph4* described by Liscum and Briggs [62] exhibited normal blue-light-induced phototropin autophosphorylation. Therefore, their gene products very likely act downstream from nph1. The protein encoded by the *NPH4* locus has been demonstrated to be involved in the regulation of auxin-mediated gene expression and appears to function as a conditional regulator of auxin-dependent differential growth responses [89]. Since *nph4* plants were also found to be severely affected in their gravitropic responses, the nph4 protein apparently is an essential component of the signal transduction chains for both phototropism and gravitropism, indicating that both plant sensory systems utilize the same signal output pathway.

The NPH3 gene encodes for a protein that is characterized by two putative protein/protein interaction domains [90], a BTB (broad complex, tramtrack, bric a brac)/POZ (poxvirus and zinc finger) domain [91,92] in the amino-terminal and a coiled-coil [93,94] in the carboxyl terminal portion of the protein. More recently, identical motifs have been reported in RPT2 (root phototropism 2), a protein closely related to NPH3, which appears to be an essential component of the signaling pathway for phototropic curvature in response to high fluences of blue light [95]. Plants lacking RPT2 also exhibit impaired root phototropism. At least, NPH3 seems to act very close to the photoreceptor and may represent an early participant within the signaling cascade for phototropism. Direct evidence for this assumption is derived from two independent experimental approaches (yeast two-hybrid and in vitro co-immunoprecipitation studies), both of which demonstrate a direct interaction between the chromophore-binding region of *Arabidopsis* phot1 and a carboxyl terminal NPH3-construct that contains the complete coiled-coil motif

[90]. Based on these results the authors hypothesized that NPH3 and phot1 may form a functional signaling complex, possibly together with additional proteins required for the transduction of the light signal, whereby the role of NPH3 could be as a scaffold protein coordinating the correct assembly of such a multimolecular complex [90,96]. The observation that NPH3 is a plasma-membrane-associated protein as is phot1 is also consistent with such a postulate. Moreover, in Western blots of cell extracts prepared from dark-grown *nph1* seedlings, NPH3 exhibits an increased electrophoretic mobility when compared with the protein from etiolated wild-type plants [90]. In the latter plants a correspondingly faster migrating protein band is only detectable after pre-irradiation with blue light, but not with red light. Since both NPH3 and RPT2 possess several potential phosphorylation sites it has been argued that the above differences in electrophoretic mobility could be the result of NPH3 phosphorylation in the dark and its dephosphorylation in the light. One possible mechanism by which dephosphorylation of NPH3 could take place is by means of a light-induced release of NPH3 from the signaling complex and hence from the plasmalemma membrane followed by dephosphorylation of the protein by a cytosolic phosphatase. On the other hand, the lower apparent molecular mass of NPH3 in wild-type seedlings pre-illuminated with blue light and in *nph1* plants need not necessarily reflect the non-phosphorylated form of the protein but could be the consequence of limited proteolysis caused by an increased instability of the protein when phototropin is either lacking or is phosphorylated. However, either possibility is consistent with the postulated light-mediated release of NPH3 from the complex with phot1. Whether RPT2 can also interact with either phot1 or phot2 is currently unknown.

If one considers that multiple autophosphorylation of phototropin occurs in the N-terminal, light-sensing portion of the protein for which, in turn, specific binding of NPH3 has been demonstrated, it appears reasonable to speculate that phosphorylation within this region of phototropin may lead to a disruption of these protein/protein interactions. Under those circumstances, light-potentiated phototropin phosphorylation could fulfil two distinct tasks. First, it might provoke the release of downstream signaling components from the putative complex and, second, it might serve as a desensitization mechanism since reassembly of the complex might be prevented as long as the non-phosphorylated form of phototropin is not restored in the dark.

9.8   Conclusions and future perspectives

Phototropins belong to a new family of UV-A/blue light receptors that are characterized by their unique domain structure consisting of two N-terminal light-sensing and FMN-binding PAS domains, designated as LOV1 and LOV2, and a C-terminal serine/threonine kinase. Recent advances in the elucidation of the biochemistry and photochemistry of the LOV domains meanwhile provide a compelling body of evidence that the photoactivation mechanism of phototropin involves the covalent binding of a cysteine residue to the isoalloxazine C(4a) position of FMN. Even though we currently do not understand the very early photochemical events in full detail, this reaction is a nucleophilic attack of a cysteine thiolate anion on a carbon atom. Thus, phototropin photochemistry is unrelated to a redox reaction, a mechanism that has been generally predicted for flavin-photoreceptors, among them the cryptochromes [97]. Hence, the transduction of the light signal obviously does not occur via the transfer of an electron to a signaling component, but likely is mediated by conformational changes of the phot1 holoprotein, a recurring principle also realized by other photoreceptors such as the phytochromes [98], rhodopsins [99] and PYP [82]. In all of these cases we are dealing with light-induced molecular motions occurring within the chromophore (here the result of a *cis–trans* isomerization) that trigger structural changes of the apoprotein.

Assuming a related mode of action, the activation of the C-terminal kinase in the light is very likely the direct consequence of such structural rearrangements. To date, the molecular basis underlying dark repression of kinase activity is still unknown. A possible control mechanism could be the binding of an autoinhibitory domain (pseudo-substrate) [100] to the catalytic center of the kinase in the dark and its subsequent release in the light, brought about by the proposed conformational changes associated with the photochemical process.

Following kinase activation phototropin becomes autophosphorylated at multiple sites. From our current knowledge, this occurs almost exclusively at serine residues located in the N-terminal region of the protein. The true function of this autophosphorylation reaction with regard to phototropic signaling is still obscure. At present, there are no indications for a phototropin-activated phosphorelay or a phosphorylation cascade. Reports of proteins other than phototropin being phosphorylated in response to illumination with blue light probably just represent stable degradation products of the photoreceptor itself [42,45]. The slow in vivo regeneration of non-phosphorylated phototropin in the dark is consistent with the hypothesis that phosphorylation may be a mechanism for photoreceptor desensitization. Nevertheless, the reduced photosensitivity of the *nph1-2* (JK224) mutant is almost certainly due to a partially impaired kinase function, and one must conclude that the phosphorylation response also plays an important role in signal transduction. However, the strikingly different photosensitivities of first positive phototropic curvature and phototropin phosphorylation are in apparent contrast to this. Based on results from earlier studies, Briggs and co-workers [101] presented an attractive hypothesis to overcome this apparent discrepancy. They proposed that at least one phosphorylation site already becomes phosphorylated at very low fluences to mediate transfer of the perceived light information to downstream transduction chain components, while the remaining sites are phosphorylated only at higher fluences and are involved in desensitization. Experiments are now being performed aimed at testing this possibility that we hope will clarify this important topic. Further questions that await an answer in the near future are: Do NPH3 and RPT2 interact with phot1 and/or phot2 in vivo and what is the exact composition of the proposed signaling complex with phototropin? Are there two or even more distinctly photosensitive phototropins in all higher plants? Will a more detailed physiological analysis of the phototropic curvature responses in *phot1* and *phot2* mutants lead to a model that can finally provide a satisfying explanation of the complex phototropic fluence response relationships?

Acknowledgements

I thank Winslow R. Briggs, Wolfhart Rüdiger and John M. Christie for critical reading of the manuscript and for helpful comments.

References

1. J. Chory, E.R. Susek (1994). *Light Signal Transduction and the Control of Seedling Development*. Cold Spring Harbor Press, New York.
2. R.E. Kendrick, G.H.M. Kronenberg (Eds) (1994). *Photomorphogenesis In Plants*. Kluwer Academic Publishers, Dordrecht.
3. R. Sharma (2001). Phytochrome: A serine kinase illuminates the nucleus! *Curr. Sci.*, **80**, 178–188.
4. M.E. Hudson (2000). The genetics of phytochrome signaling in Arabidopsis. *Semin. Cell Dev. Biol.*, **11**, 475–483.
5. C. Fankhauser (2000). Phytochromes as light-modulated protein kinases. *Semin. Cell Dev. Biol.*, **11**, 467–473.
6. P.H. Quail (2000). Phytochrome-interacting factors. *Semin. Cell Dev. Biol.*, **11**, 457–466.
7. H. Smith (2000). Phytochromes and light signal perception by plants – an emerging synthesis. *Nature*, **407**, 585–591.
8. W.R. Briggs, J.M Christie, M. Salomon (2001). Phototropins: A new family of flavin-binding blue light receptors in plants. *Antioxid. Redox Signal.*, **3**, 775–788.
9. C. Lin (2000). Plant blue-light receptors. *Trends Plant Sci.*, **5**, 337–342.
10. J.J. Casal (2000). Phytochromes, cryptochromes, phototropin: photoreceptor interactions in plants. *Photochem. Photobiol.*, **71**, 1–11.
11. M. Ahmad (1999). Seeing the world in red and blue: insight into plant vision and photoreceptors. *Curr. Opin. Biol.*, **2**, 230–235.
12. W.R. Briggs, J.M. Christie, E. Knieb, M. Salomon (1999). Phototropin (nph1), a photoreceptor for phototropism, is a FMN-binding chromoprotein. In: S. Gishla, P. Kroneck, P. Macheroux, H. Sund (Eds), *Flavins and Flavoproteins, Proceedings of the Thirteenth International Symposium, Konstanz, Germany*. (pp. 299–308). Rudolf Weber, Agency for Scientific Publications, Berlin, Germany.
13. A. Batschauer (1999). Light perception in higher plants, *Cell. Mol. Life Sci.*, **55**, 153–166.
14. W.R. Briggs, E. Huala (1999). Blue-light photoreceptors in higher plants. *Annu. Rev. Cell Dev. Biol.*, **15**, 33–62.
15. A.R. Cashmore, J.A. Jarillo, Y.-J. Wu, D. Liu (1999). Cryptochromes: blue light receptors for plants and animals. *Science*, **284**, 760–765.
16. C. Darwin (1881). *The Power of Movement in Plants*. Da Capo Press Reprint Ed. (1966). New York.

17. J. von Sachs (1887). In (Eng. edn. H.M. Ward translator) *Lectures on the Physiology of Plants*, (p. 696). Clarendon, Oxford.

18. E.S. Johnston (1934). *Phototropic Sensitivity in Relation to Wavelength*. (Vol. 92, pp. 1–17). Smithsonian Miscellaneous Collections Smithsonian Institution, Washington DC.

19. M.A. Quinones, Z. Lu, E. Zeiger (1996). Close correspondence between the action spectra for the blue light responses of the guard cell and coleoptile chloroplasts, and the spectra for blue light-dependent stomatal opening and coleoptile phototropism. *Proc. Natl. Acad. Sci. U.S.A.*, **93**, 2224–2228.

20. M.A Quinones, E. Zeiger (1994). A putative role of the xanthophyll, zeaxanthin, in the blue light photoperception of corn coleoptiles. *Science*, **264**, 558–561.

21. G. Wald, H.G. DuBuy (1936). Pigments of the oat coleoptile. *Science*, **84**, 237.

22. A.W. Galston (1949). Riboflavin-sensitized photo-oxidation of indoleacetic acid and related compounds. *Proc. Natl. Acad. Sci. U.S.A.*, **35**, 10–17.

23. A.W. Galston (1950). Riboflavin, light, and the growth of plants. *Science*, **111**, 619–624.

24. P. Galland, H. Senger (1988). The role of pterins in the photoreception and metabolism of plants. *Photochem. Photobiol.*, **48**, 811–820.

25. R. Lorenzi, N. Cerccarelli, B. Lercari, P. Gaultieri (1994). Identification of retinol in higher plants: is a rhodopsin-like protein the blue light receptor? *Phytochemistry* **36**, 599–601.

26. W. Shropshire Jr., R.B. Withrow (1958). Action spectrum of phototropic tip-curvature of Avena. *Plant Physiol.*, **33**, 360–365.

27. K.V. Thimann, G.M. Curry (1960). Phototropism and phototaxis. In: M. Florkin, H. Mason (Eds), *Comparative Biochemistry* (Vol. 1, pp. 243–309). Academic Press, New York.

28. T.I. Baskin, M. Iino (1987). An action spectrum in the blue and ultraviolet for phototropism in *alfalfa*. *Photochem. Photobiol.*, **46**, 127–136.

29. J.M. Palmer, K.M.F. Warpeha, W.R. Briggs (1996). Evidence that zeaxanthin is not the photoreceptor for phototropism in maize coleoptiles. *Plant Physiol.*, **110**, 1323–1328.

30. W.R. Briggs (1960). Light dosage and phototropic responses of corn and oat coleoptiles. *Plant Physiol.*, **35**, 951–962.

31. B.K. Zimmermann, W.R. Briggs (1963). Phototropic dosage-response curves for oat coleoptiles, *Plant Physiol.*, **38**, 248–253.

32. M. Everett, K.V. Thimann (1968). Second positive phototropism in the *Avena* coleoptile. *Plant Physiol.*, **13**, 1786–1792.

33. M. Iino (1987). Kinetic modeling of phototropism in maize coleoptiles. *Planta*, **171**, 110–126.

34. R. Konjevic, B. Steinitz, K.L. Poff (1989). Dependence of the phototropic response of Arabidopsis thaliana on fluence rate and wavelength. *Proc. Natl. Acad. Sci. U.S.A.*, **86**, 9876–9880.

35. M. Ahmad, A.R Cashmore (1993). HY4 gene of *A. thaliana* encodes a protein with the characteristic features of a blue-light photoreceptor. *Nature*, **366**, 162–166.

36. E. Huala, P.W. Oeller, E. Liscum, I.-S. Han, E. Larsen, W.R. Briggs (1997). *Arabidopsis* NPH1: A protein kinase with a putative redox-sensing domain. *Science*, **278**, 2121–2123.

37. S. Gallagher, T.W. Short, P.M. Ray, L.H. Pratt, W.R. Briggs (1988). Light-mediated changes in two proteins found associated with plasma membrane fractions from pea stem sections. *Proc. Natl. Acad. Sci. U.S.A.*, **85**, 8003–8007.

38. A. Hager, M. Brich (1993). Blue light-induced phosphorylation of a plasma-membrane protein from phototropically sensitive tips of maize coleoptiles. *Planta*, **189**, 567–576.

39. J.M. Palmer, T.W. Short, S. Gallagher, W.R. Briggs (1993). Blue light-induced phosphorylation of a plasma membrane-associated protein in *Zea mays* L. *Plant Physiol.*, **102**, 1211–1218.

40. P. Reymond, T.W. Short, W.R. Briggs (1992). Blue light activates a specific kinase in higher plants. *Plant Physiol.*, **100**, 655–661.

41. P. Reymond, T.W. Short, W.R. Briggs, K.L. Poff (1992). Light-induced phosphorylation of a membrane protein plays an early role in signal transduction for phototropism in *Arabidopsis thaliana*. *Proc. Natl. Acad. Sci. U.S.A.*, **89**, 4718–4721.

42. M. Salomon, M. Zacherl, W. Rüdiger (1996). Changes in blue-light-dependent protein phosphorylation during early development of etiolated oat seedlings. *Planta*, **199**, 336–342.

43. S. Widell, C. Larsson (1987). Plasma membrane purification. In: H. Senger (Eds), *Blue Light Responses: Phenomena and Occurrence in Plants and Microorganisms*, (pp. 99–107). CRC Press, Boca Raton, FL.

44. T.W. Short, P. Reymond, W.R. Briggs (1993). A pea plasma membrane protein exhibiting blue light-induced phosphorylation retains photosensitivity following triton solubilization. *Plant Physiol.*, **101**, 647–655.

45. V.K. Sharma, P.K. Jain, S.C. Maheshwari, J.P. Khurana (1997). Rapid blue-light-induced phosphorylation of plasma-membrane-associated proteins in wheat. *Phytochemistry*, **44**, 775–780.

46. T.W. Short, W.R. Briggs (1990). Characterization of a rapid blue light-mediated change in detectable phosphorylation of a plasma membrane protein from etiolated pea (*Pisum sativum* L.) seedlings. *Plant Physiol.*, **92**, 179–185.

47. M. Salomon, M. Zacherl, W. Rüdiger (1997). Asymmetric, blue-light-dependent phosphorylation of a 116-kilodalton plasma membrane protein can be correlated with the first- and second-positive phototropic curvature of oat coleoptiles. *Plant Physiol.*, **115**, 485–491.

48. M. Salomon, M. Zacherl, W. Rüdiger (1997). Phototropism and protein phosphorylation in higher plants: unilateral blue light irradiation generates a directional gradient of protein phosphorylation across oat coleoptiles. *Bot. Acta*, **110**, 214–216.

49. T.W. Short, M. Porst, W.R. Briggs (1992). A photoreceptor system regulating *in vivo* and *in vitro* phosphorylation of a pea plasma membrane protein. *Photochem. Photobiol.*, **55**, 773–381.

50. J.M. Palmer, T.W. Short, W.R. Briggs (1993). Correlation of blue light-induced phosphorylation to phototropism in *Zea mays* L.. *Plant Physiol.*, **102**, 1219–1225.

51. J.M. Christie, P. Reymond, G.K. Powell, P. Bernasconi, A. Raibekas, E. Liscum, W.R. Briggs (1998). *Arabidopsis* NPH1: A flavoprotein with the properties of a photoreceptor for phototropism. *Science*, **282**, 1698–1701.

52. M. Salomon, M. Zacherl, L. Luff, W. Rüdiger (1997). Exposure of oat seedlings to blue light results in amplified phosphorylation of a putative photoreceptor for phototropism and higher sensitivity of the plants to phototropic stimulation. *Plant Physiol.*, **115**, 493–500.

53. A. Hager, M. Brich, I. Bazlen (1993). Redox dependence of the blue-light-induced phosphorylation of a 100-kDa protein on isolated plasma membranes from tips of coleoptiles. *Planta*, **190**, 120–126.

54. K.M.F. Warpeha, W.R. Briggs (1993). Blue light-induced phosphorylation of a plasma membrane protein in pea: a step in the signal transduction chain for phototropism. *Austr. J. Plant Physiol.*, **20**, 393–403.

55. J.L. Wyatt, R.F. Coleman (1977). Affinity labelling of rabbit muscle pyruvate kinase by 5'-p-fluorosulfonylbenzoyladenosine. *Biochemistry*, **17**, 1333–1342.

56. T.W. Short, M. Porst, J. Palmer, E. Fernbach, W.R. Briggs (1994). Blue light induces phosphorylation of seryl residues on a pea (*Pisum sativum* L.) plasma membrane protein. *Plant Physiol.*, **104**, 1317–1324.

57. A. Hager (1996). Properties of a blue-light-absorbing photoreceptor kinase localized in the plasma membrane of the coleoptile tip region. *Planta*, **198**, 294–299.

58. W. Rüdiger, W.R. Briggs (1995). Involvement of thiol groups in blue light-induced phosphorylation of a plasma-membrane-associated protein from coleoptile tips of *Zea mays* L. *Z. Naturforsch.*, *Teil C*, **51**, 231–234.

59. E. Knieb, W. Rüdiger, M. Salomon (2000). Identifizierung der autophosphorylierungsstellen des blaulichtrezeptors phototropin. In: *Abstract Book Botanikertagung*, (Abstract P04-17, p. 75). Jena, Germany.

60. J.P. Khurana, K.L Poff (1989). Mutants of *Arabidopsis thaliana* with altered phototropism. *Planta*, **178**, 400–406.

61. J.P. Khurana, Z. Ren, B. Steinitz, B. Park, T.K. Best, K.L. Poff (1989). Mutants of Arabidopsis thaliana with decreased amplitude in their phototropic response. *Plant Physiol.*, **91**, 685–689.

62. E. Liscum, W.R. Briggs (1996). Mutations of Arabidopsis in potential transduction and response components of the phototropic signaling pathway. *Plant Physiol.*, **112**, 291–296.

63. E. Liscum, W.R. Briggs (1995). Mutations in the *NPH1* locus of Arabidopsis disrupt the perception of phototropic stimuli. *Plant Cell*, **7**, 473–485.

64. S.K. Hanks and T. Hunter (1995). The eukaryotic protein kinase superfamily: kinase (catalytic) domain structure and classification. *FASEB J.*, **9**, 576–596.

65. J.L. Pellequer, K.A Wager-Smith, S.A. Kay, E.D. Getzoff (1998). Photoactive yellow protein: A structural prototype for the three-dimensional folding of the PAS domain superfamily. *Proc. Natl. Acad. Sci. U.S.A.*, **95**, 5884–5890.

66. B.L. Taylor, I.B. Zhulin (1999). PAS domains: internal sensors of oxygen, redox potential, and light, *Microbiol. Mol. Biol. Rev.*, **22**, 479–506.

67. M. Zacherl, E. Huala, W. Rüdiger, W.R. Briggs, M. Salomon (1998). Isolation and characterization of cDNAs from oat encoding a serine/threonine kinase: An early component in signal transduction for phototropism. *Plant Physiol.*, **116**, 869.

68. Gene Bank accession number: AF033263.

69. H. Kanagae, M. Tahir, F. Savazzini, K. Yamamoto, M. Yano, T. Sasaki, T. Kanagae, M. Wada, M. Takano (2000). Rice NPH1 homologues, OsNPH1a and OsNPH1b are differently regulated. *Plant Cell Physiol.*, **41**, 415–423.

70. K. Nozue, T. Kanagae, T. Imaizumi, S. Fukuda, H. Okamoto, K.-C. Yeh, J.C. Lagarias, M. Wada (1998). A phytochrome from the fern Adiantum with features of the putative photoreceptor *NPH1*. *Proc. Natl. Acad. Sci. U.S.A.*, **95**, 15826–15830.

71. W. Gong, B. Hao, S.S. Mansy, G. Gonzalez, M.A. Gilles-Gonzales, W.R. Chan (1998). Structure of a biological oxygen sensor; a new mechanism for heme-driven signal transduction. *Proc. Natl. Acad. Sci. U.S.A.*, **95**, 15177–15182.

72. S. Hill, S. Austin, T. Eydmann, T. Jones, R. Dixon (1996). *Acetobacter vinelandii* NIFL is a flavoprotein that modulates transcriptional activation of nitrogen-fixation genes via a redox-sensitive switch. *Proc. Natl. Acad. Sci. U.S.A.*, **93**, 2143–2148.

73. S.I. Bibikov, R. Biran, K.E. Rudd, J.S. Parkinson (1997). A signal transducer for aerotaxis in *Escherichia coli*. *J. Bacteriol.*, **179**, 4075–4079.

74. R.N. Grishanin, S.I Bibikov (1997). Mechanism of oxygen taxis in bacteria. *Biosci. Rep.*, **17**, 77–83.

75. G.E.O Borgstahl, D.R. Williams, E.D. Getzoff (1995). A 1.4 Å structure of photoactive yellow protein, a cytosolic photoreceptor: Unusual fold, active site, and chromophore. *Biochemistry*, **34**, 6278–6287.

76. J.M. Christie, M. Salomon, K. Nozue, M. Wada, W.R. Briggs (1999). LOV (light, oxygen, or voltage) domains of the blue light photoreceptor phototropin (nph1): binding sites for the chromophore flavin mononucleotide. *Proc. Natl. Acad. Sci. U.S.A.*, **96**, 8779–8783.

77. M. Salomon, J.M. Christie, E. Knieb, U. Lempert, W.R. Briggs (2000). Photochemical and mutational analysis of the FMN-binding domains of the plant blue light receptor, phototropin. *Biochemistry*, **39**, 9401–9410.

78. V. Massey, M. Stankovic, P. Hemmerich (1978). Light-mediated reduction of flavoproteins with flavins as catalysts. *Biochemistry*, **17**, 1–8.

79. V. Massey and P. Hemmerich (1978). Photoreduction of flavoproteins and other biological compounds catalyzed by deazaflavins. *Biochemistry*, **17**, 9–16.

80. S.M. Miller, V. Massey, D. Ballou, C.H. Williams Jr., M.D. Distefano, M.J. Moore, C.T. Walsh (1990). Use of a site-directed triple mutant to trap intermediates: Demonstration that the flavin C(4a)-thiol adduct and reduced flavin are kinetically competent intermediates in mercuric ion reductase. *Biochemistry*, **29**, 2831–2841.

81. J.H. Cabral, A. Lee, S.L. Cohen, B.T. Chait, M. Li, R. Mackinnnon (1998). Crystal structure and functional analysis of the HERG Potassium channel N terminus: a Eukaryotic PAS domain. *Cell*, **95**, 649–655.

82. U.K. Genick, G.E. Borgstahl, K. Ng, Z. Ren, C. Pradervand, P.M. Burke, V. Srajer, T.-Y. Teng, W. Schildkamp, D.E. McRee, et al. (1997). Structure of protein photocycle intermediate by millisecond time-resolved crystallography. *Science*, **275**, 1471–1475.

83. H. Miyatake, M. Mukai, S.-Y. Park, S.-L. Adachi, K. Tamura, K. Nakamura, T. Tsuchiya, T. Iizuka, Y. Shiro (2000). Sensory mechanism of oxygen sensor FixL from *Rhizobium melioti*: crystallographic, mutagenesis and resonance Raman spectroscopic studies. *J. Mol. Biol.*, **301**, 415–431.

84. S. Crosson, K. Moffat (2001). Structure of a flavin-binding domain, LOV2, from the chimeric phytochrome/phototropin photoreceptor, phy3. *Proc. Natl. Acad. Sci. U.S.A.*, **98**, 2995–3000.

85. J.A Jarillo, M. Ahmad, A.R. Cashmore (1998). NPL1 (accession No. AF053941): a second member of the NPH serine/threonine kinase family of Arabidopsis. *Plant Physiol.*, **117**, 719.

86. T. Sakai, T. Kagawa, M. Kasahara, T.E. Swartz, J.M. Christie, W.R. Briggs, M. Wada, K. Okada (2001). Arabidopsis nph1 and npl1: blue-light receptors that mediate both phototropism and chloroplast relocation. *Proc. Natl. Acad. Sci. U.S.A.*, **98**, 6969–6974.

87. T. Sakai, T. Wada, S. Ishiguro, K. Okada (2000). RPT2: a signal transducer of the phototropic response in Arabidopsis. *Plant Cell*, **12**, 225–236.

88. T. Kagawa, T. Sakai, N. Suetsugu, K. Oikawa, S. Ishiguro, T. Kato, S. Tabata, K. Okada, M. Wada (2001). NPL1, a phototropin homologue controlling the chloroplast high-light avoidance response. *Science*, **291**, 2138–2141.

89. E.L. Stowe-Evans, R.M. Harper, A.V. Motchoulski, E. Liscum (1998). NPH4, a conditional modulator of auxin-dependent differential growth responses in Arabidopsis. *Plant Physiol.*, **118**, 1265–1275.

90. A. Motchoulski, E. Liscum (1999). Arabidopsis NPH3: a NPH1 photoreceptor-interacting protein essential for phototropism. *Science*, **286**, 961–964.

91. O. Albagli, P. Dhordain, C. Deweindt, G. Lecocq, D. Leprince (1995). The BTB/POZ domain: a new protein-protein interaction motif common to DNA- and actin-binding proteins. *Cell Growth Differ.*, **6**, 1193–1198.

92. L. Aravind, E.V. Koonin (1999). Fold prediction and evolutionary relationship with the potassium channel tetramerization domain. *J. Mol. Biol.*, **285**, 1353–1361.

93. C. Coen, D.A.D. Parry (1996). α-Helical coiled coils and bundles: how to design an α-helical protein. *Proteins*, **7**, 1–15.

94. A. Lupas (1996). Coiled coils: new structures and new functions. *Trends Biochem. Sci.*, **21**, 375–382.

95. T. Sakai, T. Wada, S. Ishiguro, K. Okada (2000). RPT2: A signal transducer of the phototropic response in Arabidopsis. *Plant Cell*, **12**, 225–236.

96. E. Liscum, E.L. Stowe-Evans (2000). A "simple" physiological response modulated by multiple interacting photosensory-response pathways. *Photochem. Photobiol.*, **72**, 273–282.

97. A.R. Cashmore (1997). The cryptochrome family of photoreceptors. *Plant Cell Environ.*, **20**, 764–767.

98. F. Thümmler, W. Rüdiger, E. Cmiel, S. Schneider (1983). Chromopeptides from phytochrome and phycocyanin. NMR studies of the Pfr and Pr chromophore of phytochrome and E, Z isomeric chromophores of phycocyanin. *Z. Naturforsch., Teil C*, **38**, 359–368.

99. G. Wald (1968). The molecular basis of visual excitation. *Nature*, **219**, 800–807.

100. S.K. Hanks, T. Hunter (1995). The eukaryotic protein kinase superfamily: kinase (catalytic) domain structure and classification. *FASEB J.*, **9**, 576–596.

101. W.R. Briggs (1996). Signal transduction in phototropism. In: *UV/Blue Light: Perception and Responses in Plants and Microorganisms (Abstract)*, (pp. 49). University of Marburg, Germany.

102. Gene Bank accession number: U83281.

103. M. Salomon, E. Knieb, T. vonZeppelin, W. Rüdiger (2003). Mapping of low- and high-fluence autophosphorylation sites in photoprotein 1. *Biochemistry*, **42**, 4217–4225.

104. M. Salomon, W. Eisenreich, H. Dürr, E. Schleicher, E. Knieb, V. Massey, W. Rüdiger, F. Müller, A. Bacher, G. Richter (2001). An optomechanical transducer in the blue light receptor phototropin from Avena Sativa. *Proc. Natl. Acad. Sci. USA*, **98**, 12357–12361.

Chapter 10

Cryptochromes and their functions in plant development

May Santiago-Ong and Chentao Lin

Table of contents

Abstract

Cryptochromes are blue light receptors that are evolutionarily related to the light-dependent DNA-repair enzyme photolyase. The chromophore-binding characteristics of photolyases are conserved in cryptochromes but they do not possess DNA repair activity. Since the identification of cryptochromes in the model plant *Arabidopsis*, this type of blue/UV-A-light receptor has also been found in animals including human. Cryptochromes usually have a C-terminal extension not found in photolyases and this extension has been shown to be needed for the function of at least the *Arabidopsis* cryptochromes. It has been suggested that plant and animal cryptochromes may have arisen independently in evolution although plant and animal blue light receptors share some noticeable similarities. Cryptochromes play important roles in regulating light-dependent development (photomorphogenesis) in plants; they are known to regulate the circadian clock in both plants and animals, and they are nuclear proteins in different organisms. The signal transduction mechanism of the cryptochromes is largely unclear, but it has been suggested that they interact with signaling proteins or other photoreceptors and catalyze redox reactions. This chapter will focus on our current understanding of how cryptochromes mediate plant responses to light, such as changes of gene expression, inhibition of hypocotyl elongation, and regulation of floral initiation. Because these same processes are also regulated by the phytochrome family of photoreceptors, we will also discuss how these two major types of photoreceptor in plants may interact in mediating light signals.

10.1 Introduction

Exposure to various regions of the light spectrum elicits, through the action of photoreceptors, a wide range of responses in plants. The regions of the solar spectrum that are most critical for plant development are red/far-red (~600–750 nm), and UV-A/blue (~350–500 nm) light. From the time a seed germinates, itself a light-regulated event, the plant has constantly to interpret the diverse cues that can be gleaned from the environment. The intensity, direction and spectral quality of light can change considerably over the course of a day and the plant has to monitor and respond to these changes to optimize growth and to regulate its development. Changes in day-length also occur seasonally, giving plants cues for detecting the coming winter or summer before a drastic temperature change takes place. The ability to perceive and to measure these changes, whether they are random or periodic fluctuations, is conferred upon the plant by the combined action of its photoreceptors.

Three classes of sensory photoreceptors have been identified in higher plants: phytochromes, cryptochromes and phototropins [1–7]. Phytochromes are red/far-red-light receptors, which, in *Arabidopsis*, are encoded by a family of five genes, *PHYA* through *PHYE* (see Chapter 6). The absorption spectra of phytochromes show maxima not only in the red and far-red regions but also in the blue region, indicating that phytochromes can photochemically act as blue light receptors. Indeed, certain phytochrome species, such as *Arabidopsis* phyA, have been shown to be active in blue light, and it may be regarded as a blue light receptor. Phototropins are flavin-containing proteins that mediate phototropic responses, arguably the first blue light response documented in plants [3,8] (see also the Chapter 9). *Arabidopsis* has at least two phototropins: PHOT1 and PHOT2 (formerly NPH1 and NPL1), which function in a partially redundant manner [7,9], regulating not only curvature of hypocotyls but also chloroplast movement in response to blue light [7,9–12]. Cryptochromes are photolyase-like flavoproteins that mediate various growth and developmental responses to the blue/UV-A regions of the light spectrum. This review will focus on a discussion of cryptochromes, the function of cryptochromes in photomorphogenesis and the functional interaction of cryptochromes and phytochromes that often regulate similar photomorphogenic responses. Since most of the findings are from studies of the model plant *Arabidopsis*, our discussion will be largely limited to the *Arabidopsis* photoreceptors.

10.2 Cryptochromes are photolyase-like blue light receptors

Cryptochromes are photoreceptors that absorb predominantly in the blue and the ultra-violet-A (UV-A) regions of the spectrum. The founding member of the cryptochrome photoreceptor family is *Arabidopsis CRY1*, first identified in the laboratories of Koornneef and Cashmore, as the mutated locus responsible for the impairment of blue-light responses in the *hy4* mutant [13,14]. The apoprotein of cry1, designated as CRY1, has 681 amino acids (a.a.) with at least two discrete domains. The amino-terminal domain of approximately 500 a.a. of CRY1 exhibited amino acid sequence homology (~30% identical) to the microbial DNA photolyases. Most cryptochromes identified so far also contain a C-terminal extension of various lengths not found in photolyase. The C-terminal extensions of different cryptochromes share little sequence similarity to each other or to known protein motifs in the database. DNA photolyase is a class of light-dependent DNA-repairing enzymes, for which the structure and catalytic mechanism have been well studied [15]. There are two types of light-dependent DNA repairing enzymes: one (called photolyase or CPD photolyase) repairs cyclobutane pyrimidine dimers (CPD) of UV-damaged DNA and the other [called (6–4) photolyase] repairs pyrimidine-pyrimidone (6–4) photoproducts in UV-damaged DNA [15,16]. The CPD photolyase and (6–4) photolyase are apparently evolutionarily related as they share ~20–30% amino acid sequence identity [15,16]. Photolyases contain two chromophores, a pterin or deazaflavin and a flavin-adenine dinucleotide (FAD). The pterin/deazaflavin serves as the light-harvesting antenna chromophore whereas FAD serves as the catalytic cofactor in the DNA repair reaction [15]. Recombinant CRY1, whether expressed in *E. coli* or in insect cells, is found to be non-covalently associated with stoichiometric amounts of FAD [17,18], and pterin has also been identified in the recombinant CRY1 expressed in *E. coli* [17]. Study of recombinant CRY1 in *E. coli* clearly established the N-terminal photolyase-like domain as the chromophore-binding domain [17]. The finding that cry1 is a flavoprotein put to rest, at least for the moment, the long-standing debate of whether plant blue/UV-A-light receptors are flavoproteins. Cryptochromes are not photolyases as they show no DNA-repair activity in either in vitro assays or in the *E. coli* complementation tests [17,18]. Since the identification of the *hy4* mutant locus as a cryptochrome gene, a gene (*SA-PHH1*) previously identified as a photolyase in the white mustard *Sinapis alba* [19] has been re-designated as a cryptochrome gene because, like cry1, it has no photolyase activity [17,18,20]. Recently, five cryptochrome genes have been identified from the fern *Adiantum capillus-veneris*, and none of them can complement a photolyase mutant of *E. coli* [21] (see also Chapter 11).

A second *Arabidopsis* cryptochrome gene, *CRY2*, has been identified and studied [20,22]. The CRY2 protein is very similar (58% identity) to CRY1 in the N-terminal chromophore-binding region; it has thus been presumed that cry2 contains the same chromophores as cry1. Like CRY1, CRY2 has a C-terminal extension (~120 a.a.) not found in photolyases, but this extension is shorter than that in CRY1 (~185 a.a.). The white mustard *Sinapis alba* cryptochrome gene (*SA-PHH1*) product shares 89% amino acid sequence identity with *Arabidopsis* CRY2 [19], however, the mustard cryptochrome contains no C-terminal extension [19]. A 1900 nt *SA-PHH1* mRNA was detected in plants, excluding the possibility of it being a pseudogene [19]. It would be particularly interesting to find out whether *SA-PHH1* encodes a functional cryptochrome in plants because it will facilitate understanding of the functional role of the C-terminal domain of cryptochromes.

The *Arabidopsis CRY1* and *CRY2* genes have similar, if not identical, mRNA expression patterns. *CRY1* mRNA appeared to be constitutively expressed in 5-day-old seedlings, whether grown in the dark or in continuous light [14]. *CRY1* mRNA is also expressed in all organs of adult plants grown in continuous white light (cWL) or grown in cWL and then kept in the dark for two days [14]. Similarly, *CRY2* mRNA expression is not significantly affected by light [23] as was found in all the organs examined (Guo and Lin, unpublished results). Recently, a more detailed analysis of RNA accumulation using a DNA microarray technique revealed that the patterns of both *CRY1* and *CRY2* mRNA accumulation follow a circadian rhythm; that is, under constant light conditions, the abundance of the RNAs oscillate with approximately 24-hour periodicity [24]. *CRY1* and *CRY2* mRNAs accumulate to roughly the same levels and the mRNAs peak within hours of each other 4–8 h after lights on. Both mRNAs drop to similar levels that are about 50% of the peak levels. Thus, the abundance of *CRY1* and *CRY2* mRNA fluctuates rhythmically but is still readily detectable at those times when the levels are at their lowest.

At the protein level, *Arabidopsis CRY1* and *CRY2* are regulated differently. The cry1 protein, like *CRY1* mRNA, is found in all plant organs, and in either light or dark conditions [23,25,26]. In contrast to cry1 for which the protein level is not apparently regulated by light, the cellular level of cry2 protein is negatively regulated by blue light [23,26]. Seedlings grown in the dark or in red light accumulate cry2 protein; when these seedlings are exposed to blue light, a decrease in the level of cry2 protein is detectable within 10 min and the level of cry2 protein drops dramatically within 1 h to a new steady state-level [23]. Two lines of evidence suggest that a protein degradation mechanism is responsible for the blue light-induced down-regulation of cry2 protein [27,28]. It was found that cycloheximide treatment does not alter this response, which, together with the lack of light-induced change in *CRY2* mRNA, indicates an involvement of the post-translational mechanism [27]. Secondly, it was found that the cry2 protein derived from a *CRY2* transgene that does not contain the non-coding sequence of the native *CRY2* gene underwent similar blue light regulation [28]. Red light has no effect on the cellular cry2 protein level but UV-A and green light have a similar effect to blue light [23], demonstrating the wavelength specificity of the cry2 degradation response. The blue light-induced cry2 degradation in the *cry1* mutant is similar to that in the wild type, indicating that cry1 is not likely to be the photoreceptor for the degradation response, leaving cry2 as the most likely photoreceptor mediating the light regulation of its own stability.

Light regulation of the cry2 protein is likely to be complex as cry2 proteins can accumulate in light. A time-course of dark-grown seedlings given extended blue light treatment showed a drop in the level of cry2 at the beginning, followed by a gradual re-accumulation of the cry2 protein (Hongwei Guo and Chentao Lin, unpublished data). Seedlings grown in continuous white or blue light have clearly detectable levels of cry2 proteins (Hongyun Yang and Chentao Lin, unpublished data). These preliminary studies suggest that the maintenance of cry2 protein levels in the natural environment involves adaptation to light. Possible mechanisms could involve light-induced temporal or spatial separation of protein degradation components, or a feedback stimulation of CRY2 synthesis.

Two independent studies have started to delineate the domains of cry2 that are important for its degradation. In one study, the N- and C-terminal domains of cry1 and cry2 were swapped and the hybrid fusion proteins overexpressed in *hy4* mutants [27]. The stability of each fusion protein was determined by analyzing the levels of the fusion proteins in transgenic seedlings grown in blue light for five days or grown in blue light for four days then transferred to red light for one day before harvesting. A hybrid fusion protein which consisted of the cry2 N-terminal region (a.a. 1–366) and the cry1 C-terminal region, referred to as C2(366)C1, was undetectable in blue or red light, suggesting it was unstable or not expressed. When more cry2 N-terminal sequence of cry2 (a.a. 1–505) was included the hybrid protein, referred to as C2(505)C1, accumulated to higher levels under red than blue light, similar to the endogenous cry2 protein. One interpretation of these results is that the region between a.a. 366–505 of cry2 contains sequences needed for stability. However, fusion proteins containing the N-terminal region of cry1 (a.a. 1–505) and the C-terminal domain of CRY2 (a.a. 505–611), referred to as C1C2, also showed blue light dependent instability. Since the protein stability of both domain-swapped fusion proteins, C2(505)C1 and C1(505)C2, behaved like endogenous cry2 it appears that the N- and C-terminal domains of cry1 and cry2 are functionally interchangeable with respect to degradation. In a second study, full-length CRY2 coding region and the CRY2 C-terminal fragment (a.a. 480–612, CRY2C) were each fused to the coding sequence of GUS (β-*glucuronidase*, referred to as GUS-CRY2 and GUS-CRY2C, respectively) and transformed into *hy4* mutants [28]. To analyze the kinetics of degradation of the fusion proteins, transgenic plants were grown for ten days in continuous red light, and then exposed to blue light for varying lengths of time. Much like endogenous cry2, the abundance of GUS-CRY2 decreased to undetectable levels within half an hour of blue light exposure. Thus, the CRY2 protein was able to confer upon the GUS protein, by itself a very stable protein [29], the blue light-dependent instability. The abundance of the GUS-CRY2C fusion protein, on the other hand, remained relatively constant across the time-course of blue light treatment. These results are also consistent with the notion that the amino-terminal domain of CRY2 is required for degradation of CRY2. Taken together, the results of Ahmad et al. [27] and Guo et al. [28] indicate that the N-terminal photolyase-like domain, whether from cry1 or from cry2, confers instability in light, the C-terminal domain of cry1, but not cry2, can somehow suppress the light-dependent "destabilizing" effect of its N-terminal domain, rendering cry1, but not cry2, stable in blue light. Further studies are needed to investigate exactly how the cry1 and cry2 proteins are so differently regulated.

10.3   Modes of action of cryptochromes

Photoperception is the initial event in photoreceptor signal transduction. For a light signal to be interpreted by the plant, light energy transmitted to a photo-receptor needs to be transformed into chemical energy, a currency that plant proteins can transact. For phytochromes, light induces isomerization of the chromophore, which is accompanied by rearrangements of the protein back-bone and causes photoconversion between the P_r and P_{fr} forms [30]. For the blue light receptor phototropin, light induces the formation of a thiol linkage between the chromophore FMN and a cysteine residue [31] (see also the Chapter 9). The initial event of cryptochrome signaling, however, remains unclear at present. Given their evolutionary origins, it has been speculated that redox reactions may be involved in the initial event upon light absorption by a cryptochrome [4,17,32]. In photolyases, photon absorption of the pterin/deazaflavin leads to the transfer of excitation energy to the catalytic flavin, initiating electron-transfer events which culminate in the restoration of pyrimidines during DNA repair [15]. Since plant cryptochromes still possess FAD, the catalytic chromophore of photolyases, it is thus possible that analogous electron transfer events between cry FAD and an intramolecular or intermolecular partner could serve to initiate the signal transduction chain.

The C-terminal extension is found in many cryptochromes but there is little similarity between them. Cryptochromes of such diverse organisms as *Arabidopsis*, humans and *Drosophila* are more similar to each other (up to 60% sequence identity) than the cryptochrome C-terminal extensions of different cryptochromes within a species [33]. The *Arabidopsis CRY1* and *CRY2* genes share only 14% sequence identity across their C-terminal extensions. The two human *CRY* genes, which are 73% identical at the protein level, are also highly divergent in their C-terminal extensions [34]. These differences in the C-terminal domains may provide the specificity of function between cryptochromes of the same organism and make this region the one most likely to mediate interactions with their signal transducing partner proteins. Direct evidence that the extension binds effector molecules was provided first by Zhao and Sancar [35] when they showed that the C-terminal region of human cry2 (hcry2) interacts with and modulates the activity of the serine/threonine phosphatase PP5. There is also genetic evidence that the C-terminal extension of a cryptochrome is functionally relevant. Several mutations in this region of *Arabidopsis* cry1 cause abnormal photomorphogenic phenotypes [14].

The Cashmore group directly tested the role of the C-terminal extension of cryptochromes in *Arabidopsis* plants [36]. They fused GUS to the cry1 C-terminal domain (GUS-CCT1) or to the cry2 C-terminal domain (GUS-CCT2) and overexpressed these fusion proteins in *Arabidopsis*. Interestingly, transgenic plants overexpressing GUS-CCT1 or GUS-CCT2 exhibited a constitutive photomorphogenic phenotype; that is, transgenic seedlings displayed light-grown characteristics in the absence of light, similar to the _c_onstitutive _p_hotomorphogenic (*cop*) and _deet_iolated (*det*) mutants [2,5]. These phenotypes included short hypocotyls, open cotyledons, plastid development and enhancement of anthocyanin pigmentation. The specificity of these phenotypes to the C-terminal region of cryptochromes was demonstrated by using other GUS-fusion proteins. Mutant forms of CCT1, which in the context of the full-length CRY1 conferred loss-of-function phenotypes [14,37], were fused to GUS and studied. When overexpressed, these mutant CCT1 failed to confer the *cop/det* phenotype. Fusion proteins using the N-terminal domains of either cry1 or cry2, or the C-terminal domains of *Drosophila* or human cryptochromes, also failed to confer the *cop/det* phenotype when overexpressed. The dominant nature of the transgene effect suggests that the overexpressed C-terminal domains of *Arabidopsis* crys interact with, and thereby block, the normal function of the endogenous cry signaling partner protein(s). The Cashmore group interpreted these findings as being consistent with a role for crys in a light-mediated redox reaction and a light-induced de-repression of cry activity. Results consistent with this interpretation come from their experiments showing that cry1 interacts with a PAS-domain protein ADAGIO1 (ADO1) and that the region of cry1 important for the interaction falls in the C-terminal domain [38]. ADO1 is identical to ZTL1/LKP1, which was previously identified by its function in either the circadian clock [39] or the control of flowering time [40].

10.3.1 Cryptochromes are nuclear proteins

Cryptochromes translocate to the nucleus. Using fractionation and immunoblot analyses, CRY2 was found to be greatly enriched in the nuclear fraction [26,28]. Transgenic *Arabidopsis* plants expressing a fusion protein of CRY2 with the reporter protein GUS (GUS-CRY2) or green fluorescent protein (GFP) accumulated the fusion proteins exclusively in the nucleus [26,28]. Using a transient assay, a fusion protein of CRY1 and GFP was also found to be nuclear [4]. GUS-CCT1 was found to be nuclear localized in dark-grown seedlings and cytoplasmic in light-grown seedlings [36]. In the case of GUS-CRY2, light regulation of the localization was specifically tested and the authors found no evidence that localization differed under dark or light conditions [28]. These authors also found that, compared with CRY2, only a relatively small amount of cellular CRY1 could be found in the nuclear fraction in light-grown plants. Of the five *CRY* genes in the fern *Adiantum capillus-veneris*, two have been found to localize to the nucleus, one of which has a light-regulated nucleocytoplasmic distribution [21] (see Chapter 11). Thus, while the regulation by light and the kinetics of translocation to the nucleus of cry1 or cry2 remains to be studied further, the conclusion is that they can translocate to the nucleus. The finding that the cryptochrome family of photoreceptors localizes to the nucleus is particularly intriguing in light of the finding that another family of photoreceptors, the phytochromes, are capable of nuclear localization (see Chapter 6).

10.3.2 Cryptochromes may physically interact with other proteins

Cryptochromes and phytochromes have been reported to interact in vitro and in vivo. Ahmad et al. [41] first reported that cry1 interacts with phyA in vitro, and that this interaction leads to cry1 phosphorylation. By co-incubating the CRY1 apoprotein purified from insect cells with recombinant oat phyA purified from yeast, it was found that the appearance of a phosphorylated form of CRY1 correlated with the addition of phyA. This interaction was investigated further by giving plants overexpressing CRY1 (CRY1ox) red/far-red light treatments diagnostic for phytochrome action. The level of CRY1 protein was similar regardless of whether harvested CRY1ox plants are incubated with ^{32}P-γATP in the dark, left in the dark and then given red light, or left in dark and then given red followed by far-red light. However, only under red light treatment was a phosphorylated protein corresponding to the size of cryptochromes detected. The in vitro interaction was also shown in yeast cells heterologously expressing plant proteins. A phyA fragment (corresponding to a.a 624–1100) as the bait and a CRY1 C-terminal fragment (corresponding to a.a. 505–661) as the prey showed a positive interaction in a yeast two-hybrid experiment.

More recently, functional interaction between phyB and cry2 was demonstrated [42]. First, a direct, physical interaction in vitro was shown by co-immunoprecipitation studies. Then, co-localization of cry2 and phyB in the

nucleus was demonstrated using a fusion protein of cry2 with the red fluorescent protein (cry2-RFP) and phyB with the green fluorescent protein
(phyB-GFP) transfected into protoplasts of tobacco BY-2 cells. Intriguingly,
blue-light irradiation caused formation of cry2-RFP nuclear speckles. Nuclear
speckles are localized areas of high protein concentration within the nucleus.
In this study, a homogenous distribution for cry2-RFP was found in the dark,
arguing that speckling was light-specific rather than an artifact of overexpression. PhyB-GFP also showed nuclear speckling [42], a pattern described by
previous studies on the localization of phyB [43,44]. Superimposition of images
optimized for visualizing cry2-RFP and phyB-GFP revealed that both fusion
proteins co-localized in some of these speckles. A direct, in vivo interaction
was also shown using the fluorescence resonance energy transfer (FRET)
method. This technique is based on the fluorophores of GFP and RFP being in
close enough proximity to act as a donor–acceptor pair to allow the transfer of
energy. By this method, a direct molecular interaction of phyB and cry2 was
demonstrated.

The localization of cryptochromes to the nucleus raises the question of the
nature of their function in that organelle. Recalling the evolutionary ancestry
of cryptochromes, it appears that the nuclear localization, but not the DNA
repair function, of photolyases has been conserved in cryptochromes. No crystal structure for a cryptochrome has yet been solved. However, those for the
photolyases from E. coli [45] and Anacystis nidulans [46] are available. Given
the low sequence identity (30%) between the two photolyases and the different
second chromophores used by each photolyase, it was surprising that the
general structures of these photolyases are "virtually superimposable" [47]. It
is not unreasonable to predict that the amino terminal regions of cryptochromes will be of similar structure to photolyases given their degree of similarity.
Based on molecular modeling studies, Sancar [47] also noted that human
CRY1 (hCRY1) and CRY2 (hCRY2) have retained features of E. coli
photolyase, such as the cavity used by the enzyme for dinucleotide binding.
Imaizumi et al. [21] found that, much like Arabidopsis cryptochromes, the five
fern cryptochromes contain the amino acid residues that have been shown to
be important for DNA-binding in E. coli photolyase. Thus, even though the
DNA repair function of photolyases has been lost by cryptochromes, the DNA
binding function may have been retained [47].

A recent report showed that Arabidopsis cry2 is associated with DNA.
Cutler et al. [48] generated transgenic Arabidopsis lines containing random
cDNAs fused to the 3' end of GFP. They found one line, m253, which showed
GFP co-localizing with condensed chromosomes in cells undergoing mitosis.
The cDNA fused downstream of GFP turned out to encode the last 105
residues of cry2. This finding raises the possibility that cryptochromes may be
involved in chromatin remodeling. Intriguingly, light was recently shown to
induce chromatin modification in neurons of the hypothalamic suprachiasmatic nucleus, the site in mammals where the circadian clock is found [49]. In
humans and mice, cryptochromes are associated with the function of circadian
clocks and recent studies suggest that cryptochromes also act as circadian
photoreceptors in mammals [47].

Another obvious possible function for nuclear localized proteins is to regulate transcription. At present, evidence for cryptochrome involvement in transcription comes by its association with phytochrome, which can, in turn, bind transcription factors (see Chapter 6).

10.4 Cryptochrome regulation of gene expression

Light regulation of gene expression has been a well-studied area of research and the involvement of cryptochrome in transcriptional regulation is well known [50,51]. In this section, we will discuss first blue light-regulated *CHS* gene expression, and then the co-action of cryptochromes and phytochromes in light regulation of gene expression. The best studied gene for blue light regulation is arguably the chalcone synthase (*CHS*) gene. Chalcone synthase is the enzyme catalyzing the first committed step in anthocyanin biosynthesis. The subsequent steps are carried out by chalcone isomerase (*CHI*) and dihydroflavonol 4-reductase (*DFR*). The light regulation of the *CHI* and *DFR* genes is very similar to that of *CHS*. Anthocyanins belong to a class of compounds, the flavonoids, which are red or purple pigments, and which serve as repellants, phytoalexins and/or signal molecules in plant defense responses, signal molecules in plant–microbe interactions and UV protectants [52]. Not surprisingly, *CHS* and the anthocyanin pathway are not only regulated by light but also by endogenous factors and environmental stimuli such as stress or pathogen attack [53–55]. Given the complexity of its regulation and the diversity of its expression patterns in different plants, this review will focus on blue light regulation of *CHS* in *Arabidopsis* and will integrate findings from other plants such as maize (*Zea mays*), parsley (*Petroselinum crispum*), white mustard (*Sinapis alba*) and petunia (*Petunia hybrida*) only in so far as they might help to understand findings in *Arabidopsis*.

Unlike other plants, *Arabidopsis* has a single *CHS* gene which is differentially induced by blue light at the seedling and adult stages. Feinbaum and Ausubel [56] found that 3–4-week-old plants responded to exposure to high intensity white light with increased anthocyanins and increased chalcone synthase expression that was induced at the transcriptional level. Further studies by Feinbaum et al. [57] showed that 9-day-old seedlings grown in the dark and then transferred to 24 h of red, blue, white or UV light accumulated high levels of *CHS* mRNA under blue, white or UV light and not red. To test whether a blue light receptor independent of phytochrome was involved, seedlings were grown in low intensity red (4 μmol m^{-2} s^{-1}) then exposed to a 15 min pulse of blue light (50 μmol m^{-2} s^{-1}). It was presumed that the activity of a non-phytochrome blue light-photoreceptor would be revealed under conditions saturating for phytochrome activity. Under these conditions, a transient accumulation of *CHS* mRNA was observed, indicating the involvement of a blue light receptor.

The blue light receptor was identified as cryptochrome. Until the regulation of genes, or any other aspect of physiology for that matter, was analyzed in photoreceptor mutants, interpretation of studies on blue light phenomenology had been dogged by the fact that phytochromes themselves can absorb in the blue and may be involved, at least partially, in the observed response. It has therefore been very informative to study blue and red light regulation of the response, such as gene expression, in single or double photoreceptor mutants. *hy4* mutant seedlings grown for three days in red light then transferred to blue showed no increase in *CHS* mRNA as is normally seen in wild-type seedlings [37]. A similar result was obtained by Jackson and Jenkins [58] using 3–4-week-old *hy4* mutant plants grown in low intensity white light transferred to high intensity blue. The blue light induction of *CHI* and *DFR* mRNA was also lost in *hy4* mutants [58]. These reductions in mRNA accumulation of anthocyanin biosynthetic genes correlate with the reduced anthocyanin accumulation observed in different *hy4* mutants [37,58,59]. Accumulation of *CHS* mRNA generally correlates well with anthocyanin levels [54,58].

Of some controversy is the role that phytochrome plays in this blue light response in *Arabidopsis*. Direct phytochrome involvement in anthocyanin synthesis and induction of anthocyanin biosynthesis genes was established in white mustard, in which a family of four genes encode chalcone synthase [60]. The role of blue and red light in anthocyanin induction is complex. Mohr invoked the idea of "coaction" between phytochrome and UV/blue light signaling pathways to explain the extent to which one photoreceptor system modifies the effects of another [61]. Kubasek et al. [53,54] showed that *CHS*, *CHI*, and *DFR* mRNAs all accumulate within hours of blue light exposure in 3-day-old seedlings. They also found that the blue light-induced *CHS* gene expression is suppressed in older (7-day) seedlings, by probably a developmental signal. Using 4-day-old seedlings grown in continuous red or blue (40 and 36 μmol m^{-2} s^{-1}, respectively), Ahmad and Cashmore [62] showed that wild-type seedlings exhibited elevated levels of anthocyanin only under blue light, whereas the *hy4* mutant was deficient in this response. They found that *phyB* and *hy1* mutants behaved similarly to the *hy4* mutant. In contrast, Neff and Chory [63], using single, double and triple null combinations of *phyA*, *phyB*, and *cry1* mutants, found that the *phyB* mutant behaved like wild-type in blue light whereas the *phyA* mutant did not, reinforcing previous observations that phyA has biological activity in blue light [64–66]. This result was independently confirmed by Poppe et al. [67], who showed that under relatively low fluence blue light (16 μmol m^{-2} s^{-1}), phyA also contributes to anthocyanin accumulation. A role for phyA had previously been shown using far-red light treatment. Kaiser et al. [68] showed that exposure to 24 h of far-red (FR) light induces *CHS* in 6-day-old seedlings, a response that is absent in 8- and 10-day-old seedlings. This change in the responsiveness of *CHS* to FR has been observed in other plants [69,70] and likely results from developmental regulation of *CHS* [54]. Consistent with the effect of FR on *CHS* mRNA accumulation, Neff and Chory [63] found that 5-day-old seedlings exposed to continuous FR (cFR) accumulated high levels of anthocyanins. They also found that a *phyB/cry1* mutant had reduced anthocyanin accumulation in cFR, a response neither single mutant parent exhibited. Thus, a requirement for phyB or cry1 in FR-induced anthocyanin accumulation was unmasked.

The study of Wade et al. [71] using 3-week-old *Arabidopsis* plants provides further insight into the role of phytochromes and correlates with some of the observations made with seedlings. It should be noted, however, that, unlike the other experiments summarized here, their study used a combination of blue and UV-A light, which they found to be much more effective in inducing *CHS* expression than using blue light alone. They discovered that cry1-mediated induction (via exposure to UV-A/blue light) of *CHS* is enhanced by a red light pretreatment and this enhancement is reduced in *phyA* or *phyB* mutants. Thus, phyA or phyB may potentiate the light response mediated by cry1. Conversely, UV-A/blue light can induce *CHS* in the absence of red light pretreatment; this aspect of cry1-mediated induction is reduced in the *phyB* mutant, indicating a coaction between cry1 and phyB as well. Based on these experiments, Jenkins and his colleagues proposed that there are at least two pathways mediating blue light regulation of *CHS* expression: one shows coaction between phytochrome and cryptochrome, and the other shows phy modulation of the blue-light signaling pathway. These results may be summarized as follows. Although red light does not efficiently induce *CHS*, FR light can. As far as known, phyA mediates all tested responses to FR [72]. cFR-induction of anthocyanin accumulation requires cry1 or phyB for the full effect [63]. Phytochromes also contribute to anthocyanin accumulation in continuous blue at the seedling stage. PhyA and phyB possibly have overlapping roles in poten-tiating the cry1-signaling pathway. PhyB is clearly required to get the full *CHS* induction normally mediated by cry1. To take into account recent findings on photoreceptor localization and biochemistry, the binding of cryptochrome to a complex containing phytochrome is one plausible mechanism for coaction between cry1 and phyB in *CHS* induction.

Analyses of various *CHS* promoters have revealed potential light-responsive motifs. To identify the region of the *Arabidopsis CHS* promoter responsible for light-induced expression, different fragments of the *CHS* promoter fused to the reporter gene *GUS* were stably transformed into *Arabidopsis* [57]. The pattern of *GUS* mRNA accumulation from fusion constructs containing either 1975 bp or 523 bp of the promoter was similar to endogenous *CHS* mRNA under all light conditions, suggesting that the 523 bp promoter contains sufficient light-responsive elements. Fusion constructs containing 186 bp of the pro-moter sequence showed weak expression, indicating the loss of positive-acting DNA elements, but was, to a lesser extent, still inducible by blue light. Three-week-old plants showed strong induction in high white light. These results indicated that there are sequences between –523 and –186 that are required for optimal blue light induction of *CHS*. Because the –186 promoter fragment was still weakly responsive to blue, there must be other blue light-responsive elements within that segment of the promoter. At that time, Feinbaum et al. [57] pointed out that there are sequences, CACGTG at –442 and CACCTG at –265, which fit the E box consensus sequence CANNTG defined from studies of genes from mammals and *Drosophila*, and the G boxes identified from many light- and stress-regulated genes in plants.

In a different system, Hartmann et al. [73] used *CHS* promoter constructs transiently transformed into *Arabidopsis* protoplasts and blue light supplemented with UV as the inducing signal, a condition that in general gives greater *CHS* induction than blue alone [74]. In these experiments, the protoplasts were prepared from cell suspension cultures maintained either in darkness or in low white light. Constructs containing progressively longer 5′ deletions of the *CHS* promoter fused to the reporter gene *GUS* caused progressively reduced promoter activity in blue/UV-A light. Consistent with the findings of Feinbaum et al. [57], they, too, found a light-responsive unit (LRU) between −570 and −335; but surprisingly, mutating the G-box-like sequence at −442 in the context of a 668 bp promoter fragment had no effect on light regulation. They identified another LRU between −164 and −61, a region that contains sequences very similar to an LRU defined in the parsley *CHS* promoter [75]. This LRU contains two motifs, an ACGT-containing element (ACE) and a Myb-recognition element (MRE), which is conserved in *CHS* genes from parsley, *Arabidopsis* and white mustard. It should be noted that what these authors call ACE is also known as a G-box. Mutation of either or both of these elements in the context of a 164 bp fragment of the *Arabidopsis* *CHS* promoter abolished light-regulated activity. Furthermore, fusing a tetramer of positions −59 to −106 of the promoter, which spans the LRU, to the minimal CaMV 35S promoter conferred light responsiveness to a reporter gene, a result similar to studies of *CHS* genes from other plants [68,76,77]. Therefore, what is implicated in light-responsiveness of the *Arabidopsis CHS* promoter are sequences that contain a G-box-/E-box-like motif similar to the site to which PIF3 can bind to a basic helix-loop-helix (bHLH) transcription factor that binds to phyA and phyB [78,79].

One might speculate that the coaction of blue and red light in regulating *CHS* in *Arabidopsis* can arise from the binding of cryptochrome and/or phytochrome to a PIF3-like transcription factor. Although it is not clear whether PIF3 can bind to the *Arabidopsis CHS* promoter, it has been demonstrated that *CHS* induction and anthocyanin biosynthesis in plants is regulated by Myb-related proteins and bHLH-like transcription factors, exemplified by the B and R genes from maize, and their orthologues in snapdragon and petunia [80,81]. It has been recognized that anthocyanin biosynthesis is generally brought about by the complex interplay between different combinations of bHLH-type and Myb-type transcription factors [82]. The maize B and R genes can turn on anthocyanin genes when heterologously transformed into *Arabidopsis* or tobacco [83]. Interestingly, the bHLH region of PIF3 is closely related to the bHLH from B-Peru, an allele of the maize B protein [78]. Furthermore, in an evolutionary analysis of bHLH transcription factors from different organisms, Atchley and Fitch [84] found that R genes form a natural group with other bHLH families which contain a leucine zipper dimerization motif immediately upstream of the bHLH motif. Previous studies have shown that HY5, a bZIP transcription factor that acts in blue and red signaling pathways, regulates *CHS* [85,86]. It is possible that bZIP-type and bHLH-type transcription factors compete for the same site on the *CHS* promoter.

10.5   Cryptochromes mediate blue light inhibition of hypocotyl elongation

A breakthrough in the study of blue light-dependent inhibition of hypocotyl elongation was the identification of the *CRY1* gene as the mutated locus in the *hy4* mutant. In a seminal study using *Arabidopsis*, Koornneef et al. [13] identified a number of long hypocotyl (*hy*) mutants that grew taller than wild-type seedlings under white light. This genetic screen was based on seedling height, an easily quantifiable aspect of photomorphogenesis, the development of a seedling in light. A seedling that does not receive light signals grows in an etiolated form, extending its hypocotyl and keeping its cotyledons closed, a morphology optimized for a seedling to emerge from the soil without damaging its meristem. Upon exposure to light, seedlings undergo a developmental program that is strikingly distinct: hypocotyls are short and cotyledons are green and open. Chloroplast development commences as transcription and translation of nuclear- and chloroplast-encoded genes, many of which have functions related to photosynthesis, is initiated. The *hy4* mutant was specifically affected in blue light-dependent inhibition of hypocotyl elongation [13,14,58,87]. Other alleles of *hy4* since the cloning of *CRY1* are now denoted as *cry1* mutants. The primary role that cry1 plays in blue-light mediated extension inhibition was confirmed when transgenic *Arabidopsis* plants overexpressing *CRY1* (CRY1ox) displayed enhanced blue light sensitivity; that is, CRY1ox seedlings have shorter hypocotyls than wild-type not only under white but also blue, green and UV/A light, the wavelengths where cryptochromes are active [88,89].

The role of cry1 in mediating blue-light inhibition of hypocotyl elongation appears to be conserved in other plant species. Tomato CRY1 has 78% amino acid sequence identity to *Arabidopsis* CRY1 [90]. Similar to the *cry1/hy4* mutant phenotype, transgenic tomato plants expressing antisense tomato *CRY1* exhibited long hypocotyls when grown in blue light. This result was confirmed with the finding that tomato *cry1* mutants have long hypocotyls in blue light [91]. Furthermore, transgenic tobacco plants overexpressing *Arabidopsis CRY1* displayed hypersensitivity phenotypes similar to *Arabidopsis* plants overexpressing *CRY1* [88]. Other responses known to be regulated by blue light were also impaired in *hy4/cry1* mutants. These are stimulation of cotyledon expansion [63], stimulation of anthocyanin biosynthesis [37,58], entrainment of the circadian clock [92] and regulation of floral initiation [93,94].

The cry2 photoreceptor also plays a role in the de-etiolation process, albeit a more subtle one. Under blue light, transgenic *Arabidopsis* lines overexpressing *CRY2* displayed hypocotyl lengths shorter than wild-type [23]. This was interpreted to be a hypersensitivity response and became the basis for a genetic screen for loss-of-function mutants. As predicted, *cry2* deletion-mutant alleles exhibited long hypocotyls and small or unopened cotyledons under low-fluence blue light, similar to the phenotype of dark-grown seedlings. It was found that *cry2* showed more pronounced phenotype in low fluence blue light than in high fluence blue light, suggesting that, with respect to de-etiolation responses, cry1 and cry2 play dominant roles in high and low light, respectively, allowing plants to perceive a full range of light intensity.

The cellular mechanism underlying light inhibition of hypocotyl elongation has been proposed to involve membrane depolarization and has been reviewed recently [95]. Kinetic analysis of blue light-induced hypocotyl elongation revealed a rapid phase and a slow phase. The rapid response is transient and is initiated within seconds of a blue light pulse. The slow phase occurs later and lasts longer [96,97]. The rapid growth inhibition is preceded by a transient plasma membrane depolarization that involves activation of an anion channel [97,98]. The *cry1* mutant exhibited a reduced response to blue light-induced membrane depolarization and defective inhibition of hypocotyl growth in the slow phase, indicating that cry1 is one photoreceptor that mediates this response. The identity of the photoreceptor mediating the rapid phase of the response to blue light is still an open question. Preliminary results from the Spalding group indicate that the *cry2* and *nph1* mutants exhibit normal rapid phase responses as well [97]. As these authors suggested, it is possible that cry1 and cry2 may have overlapping roles that can be revealed by analyzing the rapid responses in a *cry1cry2* double mutant.

One protein that likely mediates blue light signals and the inhibition of hypocotyl elongation is HY5. The *hy5* mutant was first identified in the same mutant screen by Koornneef that identified the *hy3* and *hy4* mutants, since shown to be mutated at the *PHYB* and *CRY1* loci respectively. *hy5* mutants display long hypocotyls under continuous white, blue, red, or far-red light [13,99]. The *HY5* locus encodes a basic leucine zipper (bZIP) transcription factor, a member of a large family of DNA binding proteins in *Arabidopsis* which typically bind G-boxes [55]. HY5 is a positive regulator of photomorphogenesis: the HY5 protein undergoes COP1-dependent degradation in the dark, and the level of HY5 protein directly correlates with the degree of hypocotyl inhibition [100]. HY5 has been shown to bind in vitro to a 180 bp fragment of the white mustard *CHS* promoter and light-induced expression of *CHS::GUS* constructs in *hy5* mutants were reduced relative to wild type [85]. Elucidation of the complete set of HY5 targets would significantly enhance our understanding of the cellular mechanism underlying light inhibition of hypocotyl elongation.

Another protein that mediates blue and red light signaling is the SUB1 protein [101]. The *Arabidopsis sub1* was identified as a "*s*hort *u*nder *b*lue light" mutant. It also showed short hypocotyl under continuous far-red light, indicating some genetic interaction with phyA. Unlike the *det/cop* mutants, *sub1* behaves like wild-type in the dark. The enhanced response to light (greater inhibition of hypocotyl length than wild type) was also reflected at the level of gene expression. Accumulation of *CHS* and *CHI* mRNA was also increased in the *sub1* mutant grown in blue light. Genetic epistasis analysis between the *sub1*, *cry1*, *cry2*, *hy5* and *phyA* mutants indicates that while the activities of cry1 and cry2 are at least partially dependent on SUB1, the activities of HY5 and phyA are not. In contrast, the activity of SUB1 in far-red light is dependent on phyA. These results suggest that SUB1 is part of the crypto-chrome signal transduction, whereas it modulates phyA signal transduction, thus putting SUB1 as a molecule directly involved in the cryptochrome/phytochrome coaction. Consistent with the genetic results, SUB1 function is required for light-induced HY5 accumulation. Blue light-dependent accumulation of HY5 protein is accelerated in the *sub1* mutant. The *SUB1* locus encodes a calcium-binding protein whose localization is enriched at the nuclear periphery. The SUB1 protein can bind calcium in an in vitro assay, providing genetic and molecular evidence to pharmacological and physiological studies in the past which have pointed to the change in calcium levels as a possible second messenger in photoreceptor signal regulation [102,103].

10.6   The role of cryptochromes in photoperiodic flowering and regulation of the circadian clock

Attempts to identify the photoreceptors controlling floral initiation established very early on that the phytochrome family of photoreceptors are key regulators of flowering [104]. The ability of phytochromes to absorb in the blue region of the spectrum made it difficult to clearly ascertain whether the effects induced by blue light were due to the action of phytochrome or another family of blue light receptors. The identification of the *cry1* and *cry2* mutants of *Arabidopsis* and the characterization of flowering time in these mutants demonstrated that cryptochromes also regulated flowering time. A more detailed discussion of the roles of cryptochromes in flowering-time regulation can be found in a recent review [105] and in Chapter 12. The *cry2* mutant flowers late under continuous white light or continuous blue-plus-red light whereas it flowers at the same time as wild-type under either continuous red or blue light [25]. To explain why the function of a blue light receptor is dependent not only on blue light but also on red light, it was hypothesized that phyB mediates a red light-dependent inhibition of floral initiation, whereas cry2 mediate a blue light-dependent suppression of the phyB function [25]. To further test this hypothesis, Mockler et al. analyzed flowering time of *cry2*, *phyB*, and *cry2phyB* double mutants in continuous red light, blue light, red-plus-blue light, or white light, and found that the cry2 function in regulating flowering time was indeed dependent on phyB [94]. In another study, Mas et al. analyzed flowering time of the *cry2* mutant by supplementing the white light with far-red light (W + FR), a light condition which decreases the amount of active phytochromes [42]. Interestingly, in W + FR, the *cry2* mutant flowers at the same time as wild-type, indicating that the late flowering response of *cry2* mutants in white light can be abrogated by FR light. This characteristic is reminiscent of the reversal of phytochrome-mediated responses by FR light, suggesting again an involvement of phytochrome in the cry2 function with respect to the regulation of flowering time. These genetic and photophysiological interactions of cry2 and phyB were later shown to likely result from a molecular interaction between the two photoreceptors as discussed in the previous section [42].

In addition to the phyB-dependent pathway regulating flowering time, cry2 also functions in a phyB-independent pathway regulating flowering time, as suggested by the finding that the *cry1cry2* double mutant flowered later than wild type in continuous blue light [94]. Because neither mutant parent exhibits this phenotype, this result points to a functional redundancy of cry1 and cry2 in this phyB-independent pathway. Therefore, the functions of cry1 and cry2 in flowering time regulation also overlap, just as they do in light inhibition of hypocotyl growth. It should be pointed out that those experiments under continuous light conditions only showed that cryptochromes play roles in determining flowering time even in the absence of photoperiods.

Cryptochromes also regulate flowering under photoperiodic conditions. Photoperiods, or light/dark cycles, change seasonally and these changes can induce flowering in plants. The ability to measure the length of the dark and light periods in a cycle as well as to detect the changes in this photoperiod is conferred upon the plant by the combined action of its photoreceptors and its timekeeper–the circadian clock. Regulation of flowering by the circadian clock has been well known, partly based on the observation that the responsiveness of plants to floral inductive light conditions varied across a 24-hour cycle. For example, soybeans grown under 8 h light : 16 hour dark cycles (SD) and transferred to constant dark can be induced to flower by interrupting the dark period with a 4 h light treatment. The effectiveness of this "night break" to promote flowering shows a circadian rhythm [106]. This photoperiodic response, the response to the timing and duration of light and dark conditions, indicates that the circadian clock modulates the signaling pathway from photoreceptors to floral induction. In *Arabidopsis*, flowering responses to continuous irradiation conditions are similar to its responses to long day conditions. That is, the plant interprets a long day the same as it would in continuous light. Mutations or overexpression of cry2 lead to reduced sensitivity to photoperiods. The *cry2* mutants flower late in LD but not in SD. Overexpression of cry2 leads to early flowering only under SD [25]. Thus, aberrant expression of cry2 leads to abnormal flowering time regulation, resulting in an inability to respond correctly to time cues given by day length. The *cry2* mutant turned out to be allelic to *fha*, a photoperiodic insensitive mutant previously identified by Koornneef [24,80]. In addition to cry2, cry1 is also involved in the regulation of photoperiodic flowering, some alleles of *hy4/cry1* exhibit delayed flowering in SD [93,107]. However, based on the mutant phenotypes, it appears that cry2 plays a more prominent role in flowering time regulation than cry1.

Based on the external coincidence hypothesis, initially proposed in the 1930s [108,109], the function of a photoreceptor in photoperiodic flowering is twofold: a photoreceptor can provide light input to the circadian clock that regulates flowering time, and the photoreceptor can also directly regulate the floral initiation process that in turn is regulated by the circadian clock [6]. Cry2 appears to be involved in both pathways. The role of cryptochromes in mediating light input to the clock will be discussed briefly here and in greater detail elsewhere in this volume (see Chapter 12). Elucidation of plant circadian clock function has been greatly facilitated by the use of *Arabidopsis* photoreceptor mutants and two relatively convenient assays, rhythmic gene expression and rhythmic leaf movements [110,111]. These rhythms behave with a periodicity of exactly 24 h when a 24 h diurnal cycle is imposed on it or exactly 22 h when a 22 h diurnal cycle is imposed upon it. The length of this clock's period is revealed when plants are released into continuous light or dark, a condition under which the clock "free-runs", or runs in the absence of entrainment signals or light inputs. Under these conditions, the endogenous rhythm runs with a period of not exactly, but approximately 24 h. Under free-running conditions, the period length of the plant clock decreases with light intensity. For example, rhythms of gene expression run with the short period of approximately 18 h under continuous high intensity light and the long period of approximately 32 h under continuous dark [112]. A role for cry1 in the maintenance of rhythms was demonstrated by Somers et al. [92] using period length measurements of gene expression under increasing intensities of blue light. They showed that the *cry1* mutant exhibited longer periods at either the lower and higher fluence blue light. But across the same range of blue light intensity, the *cry2* single mutant behaves like wild type. Subsequent studies revealed that cry2 is also a circadian photoreceptor [113]. It was found that the *cry1cry2* double mutant exhibits longer periods (than the *cry1* monogenic mutant) across a whole range of blue light intensities [113]. Thus, the absence of cry1 unmasks a role for cry2 as a photoreceptor for a clock response such as rhythmic gene expression. This result, indicating functional redundancy between cry1 and cry2 for period length control under continuous blue light, parallels that found by Mockler et al. [94] for the redundant action of the two cryptochromes in flowering under continuous blue light. It was further found that the *cry2* mutant exhibits longer periods across a range of white light intensities (2–40 μmol m^{-2} s^{-1}) [42] whereas a similar change was not found in red light or blue light with the overlapping fluence rate (2–50 μmol m^{-2} s^{-1}) [113]. This finding also appears to be consistent with the results of Lin's lab that cry2 mutation affects flowering time in white light or in red-plus-blue light, but not in blue light or red light. More importantly, the discovery of the function of cry2 in regulating the circadian clock in white light suggests that cry2 can regulate photoperiodic flowering through its input to the circadian clock, via its interaction with phytochromes. It should be noted, however, that the effect of *cry2* mutation on the period length of the circadian clock diminished at higher fluence range of white light (>40 μmol m^{-2} s^{-1}) [42], whereas the late flowering phenotype of the *cry2* mutant could be clearly demonstrated at such or higher fluences of white light [25,94]. Therefore, a direct function of cry2 in the floral initiation process, as predicted by the external coincidence model, must also play a role in cry2-mediated photoperiodic flowering [6].

10.7   Closing remarks

The identification of cryptochromes as blue light photoreceptors in *Arabidopsis* has led to advances in photoreceptor study not only in blue light signaling but also in the mechanism of phytochrome signaling. Analyses of the biological responses of the *Arabidopsis* cryptochrome mutants *cry1* and *cry2* allowed the delineation of the independent and overlapping roles that each cryptochrome plays. An intriguing finding from recent years is that both cryptochromes and phytochromes can be localized to the nucleus, and most significantly, that cryptochromes and phytochromes physically interact to mediate physiological responses. It remains to be tested directly whether phytochrome can actually phosphorylate cryptochrome in vivo by examining whether cryptochromes indeed undergo blue light-dependent phosphorylation in plants and whether such a reaction is impaired in the phytochrome mutants. Taking together recent discoveries in phytochrome signaling, a physical interaction of phytochrome and cryptochrome implies that cryptochromes may also trigger a short signal transduction pathway leading to light-regulated gene expression. It is not clear how many proteins, in addition to phytochrome, may physically interact with cryptochromes. Identification and characterization of additional cryptochrome-interacting proteins should further strengthen our understanding of the molecular mechanism of cryptochrome signal transduction and regulation.

References

1. P.H. Quail, M.T. Boylan, B.M. Parks, T.W. Short, Y. Xu, et al. (1995). Phytochromes: Photosensory perception and signal transduction. *Science*, **268**, 675–680.
2. C. Fankhauser, J. Chory (1997). Light control of plant development. *Annu. Rev. Cell. Dev. Biol.*, **13**, 203–29.
3. W.R. Briggs, E. Huala (1999). Blue-light photoreceptors in higher plants. *Annu. Rev. Cell Dev. Biol.*, **15**, 33–62.
4. A.R. Cashmore, J.A. Jarillo, Y.J. Wu, D. Liu (1999). Cryptochromes: blue light receptors for plants and animals. *Science*, **284**, 760–765.
5. X.W. Deng, P.H. Quail (1999). Signalling in light-controlled development. *Semin. Cell. Dev. Biol*, **10**, 121–129.
6. C. Lin (2000). Plant blue-light receptors. *Trends Plant Sci.*, **5**, 337–342.
7. J.M. Christie, W.R. Briggs (2001). Blue light sensing in higher plants. *J. Biol. Chem.*, **276**, 11457–11460.
8. C. Darwin (1881). *The power of movement in plants*. D. Appleton and Company, New York.
9. W.R. Briggs, C.F. Beck, A.R. Cashmore, J.M. Christie, J. Hughes, et al. (2001). The phototropin family of photoreceptors. *Plant Cell*, **13**, 993–997.
10. J.A. Jarillo, H. Gabrys, J. Capel, J.M. Alonso, J.R. Ecker, et al. (2001). Phototropin-related NPL1 controls chloroplast relocation induced by blue light. *Nature*, **410**, 952–954.
11. T. Kagawa, T. Sakai, N. Suetsugu, K. Oikawa, S. Ishiguro, et al. (2001). *Arabidopsis* NPL1: a phototropin homolog controlling the chloroplast high-light avoidance response. *Science*, **291**, 2138–2141.
12. T. Sakai, T. Kagawa, M. Kasahara, T.E. Swartz, J.M. Christie, et al. (2001). *Arabidopsis* nph1 and npl1: blue light receptors that mediate both phototropism and chloroplast relocation. *Proc. Natl. Acad. Sci. U.S.A*, **98**, 6969–6974.

13. M. Koornneef, E. Rolff, C.J.P. Spruit (1980). Genetic control of light-inhibited hypocotyl elongation in *Arabidopsis thaliana* (L.) Heynh. *Z. Pflanzenphysiol.*, **100**, 147–160.

14. M. Ahmad, A.R. Cashmore (1993). HY4 gene of *A. thaliana* encodes a protein with characteristics of a blue-light photoreceptor. *Nature*, **366**, 162–166.

15. A. Sancar (1994). Structure and function of DNA photolyase. *Biochemistry*, **33**, 2–9.

16. T. Todo, H. Takemori, H. Ryo, M. Ihara, T. Matsunaga, et al. (1993). A new photoreactivating enzyme that specifically repairs ultraviolet light-induced (6-4) photoproducts. *Nature*, **361**, 371–374.

17. K. Malhotra, S.T. Kim, A. Batschauer, L. Dawut, A. Sancar (1995). Putative blue-light photoreceptors from *Arabidopsis thaliana* and *Sinapis alba* with a high degree of sequence homology to DNA photolyase contain the two photolyase cofactors but lack DNA repair activity. *Biochemistry*, **34**, 6892–6899.

18. C. Lin, D.E. Robertson, M. Ahmad, A.A. Raibekas, M.S. Jorns, et al. (1995). Association of flavin adenine dinucleotide with the *Arabidopsis* blue light receptor CRY1. *Science*, **269**, 968–70.

19. A. Batschauer (1993). A plant gene for photolyase: An enzyme catalyzing the repair of UV-light-induced DNA damage. *Plant J.*, **4**, 705–709.

20. P.D. Hoffman, A. Batschauer, J.B. Hays (1996). PHH1, a novel gene from *Arabidopsis thaliana* that encodes a protein similar to plant blue-light photoreceptors and microbial photolyases. *Mol. Gen. Genet.*, **253**, 259–265.

21. T. Imaizumi, T. Kanegae, M. Wada (2000). Cryptochrome nucleocytoplasmic distribution and gene expression are regulated by light quality in the fern *Adiantum capillus-veneris. Plant Cell*, **12**, 81–96.

22. C. Lin, M. Ahmad, J. Chan, A.R. Cashmore (1996). CRY2, a second member of the *Arabidopsis* cryptochrome gene family. *Plant Physiol.*, **110**, 1047.

23. C. Lin, H. Yang, H. Guo, T. Mockler, J. Chen, et al. (1998). Enhancement of blue-light sensitivity of *Arabidopsis* seedlings by a blue light receptor cryptochrome 2. *Proc. Natl. Acad. Sci. U.S.A.*, **95**, 2686–2690.

24. S.L. Harmer, J.B. Hogenesch, M. Straume, H.S. Chang, B. Han, et al. (2000). Orchestrated transcription of key pathways in *Arabidopsis* by the circadian clock. *Science*, **290**, 2110–2113.

25. H. Guo, H. Yang, T.C. Mockler, C. Lin (1998). Regulation of flowering time by *Arabidopsis* photoreceptors. *Science*, **279**, 1360–1363.

26. O. Kleiner, S. Kircher, K. Harter, A. Batschauer (1999). Nuclear localization of the *Arabidopsis* blue light receptor cryptochrome 2. *Plant J.*, **19**, 289–296.

27. M. Ahmad, J.A. Jarillo, A.R. Cashmore (1998). Chimeric proteins between cry1 and cry2 *Arabidopsis* blue light photoreceptors indicate overlapping functions and varying protein stability. *Plant Cell*, **10**, 197–208.

28. H. Guo, H. Duong, N. Ma, C. Lin (1999). The *Arabidopsis* blue light receptor cryptochrome 2 is a nuclear protein regulated by a blue light-dependent post-transcriptional mechanism. *Plant J.*, **19**, 279–287.

29. R.A. Jefferson, T.A. Kavanagh, M.W. Bevan (1987). GUS fusions: β-glucoronidase as a sensitive and versatile gene fusion marker in higher plants. *EMBO J.*, **6**, 3901–3907.

30. P.H. Quail (1994). Phytochrome genes and their expression. In: R.E. Kendrick, G.H.M. Kronenberg, (Eds), *Photomorphogenesis in Plants* (2nd Edn., pp. 71–104). Kluwer Academic Publishers, Dordrecht.

31. M. Salomon, J.M. Christie, E. Knieb, U. Lempert, W.R. Briggs (2000). Photochemical and mutational analysis of the FMN-binding domains of the plant blue light receptor, phototropin. *Biochemistry*, **39**, 9401–9410.

32. G.I. Jenkins, J.M. Christie, G. Fuglevand, J.C. Long, J.A. Jackson (1995). Plant responses to UV and blue light: biochemical and genetic approaches. *Plant Sci.*, **112**, 117–38.

33. T. Todo (1999). Functional diversity of the DNA photolyase/blue light receptor family. *Mutat. Res.*, **434**, 89–97.

34. D.S. Hsu, X. Zhao, S. Zhao, A. Kazantsev, R.P. Wang, et al. (1996). Putative human blue-light photoreceptors hCRY1 and hCRY2 are flavoproteins. *Biochemistry*, **35**, 13871–13877.

35. S. Zhao, A. Sancar (1997). Human blue-light photoreceptor hCRY2 specifically interacts with protein serine/threonine phosphatase 5 and modulates its activity. *Photochem. Photobiol.*, **66**, 727–731.

36. H.-Q. Yang, Y.-J. Wu, R.-H. Tang, D. Liu, Y. Liu, A.R. Cashmore (2000). The C termini of *Arabidopsis* cryptochromes mediate a constitutive light response. *Cell*, **103**, 815–827.

37. M. Ahmad, C. Lin, A.R. Cashmore (1995). Mutations throughout an *Arabidopsis* blue-light photoreceptor impair blue-light-responsive anthocyanin accumulation and inhibition of hypocotyl elongation. *Plant J.*, **8**, 653–658.

38. J.A. Jarillo, J. Capel, R.H. Tang, H.Q. Yang, J.M. Alonso, et al. (2001). An *Arabidopsis* circadian clock component interacts with both CRY1 and phyB. *Nature*, **410**, 487–490.

39. D.E. Somers, T.F. Schultz, M. Milnamow, S.A. Kay (2000). ZEITLUPE encodes a novel clock-associated PAS protein from *Arabidopsis. Cell,* **101**, 319–329.

40. T. Kiyosue, M. Wada (2000). LKP1 (LOV kelch protein 1): a factor involved in the regulation of flowering time in *arabidopsis. Plant J.*, **23**, 807–815.

41. M. Ahmad, J.A. Jarillo, O. Smirnova, A.R. Cashmore (1998). The CRY1 blue light photoreceptor of *Arabidopsis* interacts with phytochrome A in vitro. *Mol. Cell,* **1**, 939–948.

42. P. Mas, P.F. Devlin, S. Panda, S.A. Kay (2000). Functional interaction of phytochrome B and cryptochrome 2. *Nature*, **408**, 207–211.

43. S. Kircher, L. Kozma-Bognar, L. Kim, E. Adam, K. Harter, et al. (1999). Light quality-dependent nuclear import of the plant photoreceptors phytochrome A and B. *Plant Cell*, **11**, 1445–1456.

44. R. Yamaguchi, M. Nakamura, N. Mochizuki, S.A. Kay, A. Nagatani (1999). Light-dependent translocation of a phytochrome B-GFP fusion protein to the nucleus in transgenic *Arabidopsis. J. Cell Biol.*, **145**, 437–445.

45. H.W. Park, S.T. Kim, A. Sancar, J. Deisenhofer (1995). Crystal structure of DNA photolyase from *Escherichia coli. Science*, **268**, 1866–1872.

46. T. Tamada, K. Kitadokoro, Y. Higuchi, K. Inaka, A. Yasui, et al. (1997). Crystal structure of DNA photolyase from *Anacystis nidulans. Nat. Struct. Biol*, **4**, 887–891.

47. A. Sancar (2000). Cryptochrome: the second photoactive pigment in the eye and its role in circadian photoreception. *Annu. Rev. Biochem.*, **69**, 31–67.

48. S.R. Cutler, D.W. Ehrhardt, J.S. Griffitts, C.R. Somerville (2000). Random GFP::cDNA fusions enable visualization of subcellular structures in cells of *Arabidopsis* at a high frequency. *Proc. Natl. Acad. Sci. U.S.A.*, **97**, 3718–3723.

49. C. Crosio, N. Cermakian, C.D. Allis, P. Sassone-Corsi (2000). Light induces chromatin modification in cells of the mammalian circadian clock. *Nat. Neurosci.*, **3**, 1241–1247.

50. W.B. Terzaghi, A.R. Cashmore (1995). Light-regulated transcription. *Annu. Rev. Plant Physiol. Plant Mol. Biol.*, **46**, 419–444.
51. G.I. Jenkins (1997). UV-A and blue light signal transduction in *Arabidopsis. Plant Cell & Environ.*, **20**, 773–778.
52. D. Scheel, K. Hahlbrock (1989). Physiology and molecular biology of phenyl-propanoid metabolism. *Annu. Rev. Plant Physiol. Plant Mol. Biol.*, **40**, 347–369.
53. W.L. Kubasek, B.W. Shirley, A. McKillop, H.M. Goodman, W. Briggs, et al. (1992). Regulation of flavonoid biosynthetic genes in germinating *Arabidopsis* seedlings. *Plant Cell*, **4**, 1229–1236.
54. W.L. Kubasek, F.M. Ausubel, B.W. Shirley (1998). A light-independent develop-mental mechanism potentiates flavonoid gene expression in *Arabidopsis* seedlings. *Plant Mol. Biol.*, **37**, 217–223.
55. B.W. Shirley (1996). Flavonoid biosynthesis: "new" functions for an "old" pathway. *Trends Plant Sci.*, **1**, 377–382.
56. R.L. Feinbaum, F.M. Ausubel (1988). Transcriptional regulation of the *Arabidopsis thaliana* chalcone synthase gene. *Mol. Cell Biol.*, **8**, 1985–1992.
57. R.L. Feinbaum, G. Storz, F.M. Ausubel (1991). High intensity and blue light regulated expression of chimeric chalcone synthase genes in transgenic *Arabidopsis thaliana* plants. *Mol. Gen. Genet.*, **226**, 449–456.
58. J.A. Jackson, G.I. Jenkins (1995). Extension-growth responses and expression of flavonoid biosynthesis genes in the *Arabidopsis hy4* mutant. *Planta*, **197**, 233–239.
59. J. Chory (1992). A genetic model for light-regulated seedling development in *Arabidopsis. Development*, **115**, 337–354.
60. A. Batschauer, P.M. Gilmartin, F. Nagy, E. Schäfer (1994). The molecular biology of photoregulated genes. In: R.E. Kendrick, G.H.M. Kronenberg (Eds), *Photomorphogenesis in Plants* (2nd Edn., pp. 559–599). Kluwer Academic Publishers, Dordrecht.
61. H. Mohr (1994). Coaction between pigment systems. In: R.E. Kendrick, G.H.M. Kronenberg (Eds), *Photomorphogenesis in Plants* (2nd Edn., pp. 353–373). Academic Publisher, Dordrecht.
62. M. Ahmad, A.R. Cashmore (1997). The blue-light receptor cryptochrome 1 shows functional dependence on phytochrome A or phytochrome B in *Arabidopsis thaliana, Plant J.*, **11**, 421–427.
63. M.M. Neff, J. Chory (1998). Genetic interactions between phytochrome A, phytochrome B, and cryptochrome 1 during *Arabidopsis* development. *Plant Physiol.*, **118**, 27–35.
64. G.C. Whitelam, E. Johnson, J. Peng, P. Carol, M.L. Anderson, et al. (1993). Phytochrome A null mutants of *Arabidopsis* display a wild-type phenotype in white light. *Plant Cell*, **5**, 757–768.
65. T. Shinomura, A. Nagatani, H. Hanzawa, M. Kubota, M. Watanabe, et al. (1996). Action spectra for phytochrome A- and B-specific photoinduction of seed germination in *Arabidopsis* thaliana. *Proc. Natl. Acad. Sci. U.S.A.*, **93**, 8129–8133.
66. F. Hamazato, T. Shinomura, H. Hanzawa, J. Chory, M. Furuya (1997). Fluence and wavelength requirements for *Arabidopsis* CAB gene induction by different phytochromes. *Plant Physiol.*, **115**, 1533–1540.
67. C. Poppe, U. Sweere, H. Drumm-Herrel, E. Schafer (1998). The blue light receptor cryptochrome 1 can act independently of phytochrome A and B in *Arabidopsis thaliana. Plant J.*, **16**, 465–471.
68. T. Kaiser, K. Emmler, T. Kretsch, B. Weisshaar, E. Schafer, et al. (1995). Promoter elements of the mustard *CHS1* gene are sufficient for light regulation in transgenic plants. *Plant Mol. Biol.*, **28**, 219–229.

69. A. Batschauer, B. Ehmann, E. Schafer (1991). Cloning and characterization of a chalcone synthase gene from mustard and its light-dependent expression. *Plant Mol. Biol.*, **16**, 175–185.

70. H. Frohnmeyer, B. Ehmann, T. Kretsch, M. Rocholl, K. Harter, et al. (1992). Differential usage of photoreceptors for chalcone synthase gene expression during plant development. *Plant J.*, **2**, 899–906.

71. H.K. Wade, T.N. Bibikova, W.J. Valentine, G.I. Jenkins (2001). Interactions within a network of phytochrome, cryptochrome and UV-B phototransduction pathways regulate chalcone synthase gene expression in *Arabidopsis* leaf tissue. *Plant J.*, **25**, 675–685.

72. J.J. Casal (2000). Phytochromes, cryptochromes, phototropin: photoreceptor interactions in plants. *Photochem. Photobiol.*, **71**, 1–11.

73. U. Hartmann, W.J. Valentine, J.M. Christie, J. Hays, G.I. Jenkins, et al. (1998). Identification of UV/blue light-response elements in the *Arabidopsis thaliana* chalcone synthase promoter using a homologous protoplast transient expression system. *Plant Mol. Biol.*, **36**, 741–754.

74. G. Fuglevand, J.A. Jackson, G.I. Jenkins (1996). UV-B, UV-A, and blue light signal transduction pathways interact synergistically to regulate chalcone synthase gene expression in *Arabidopsis. Plant Cell*, **8**, 2347–2357.

75. P. Schulze-Lefert, M. Becker-Andre, W. Schulz, K. Hahlbrock, J.L. Dangl (1989). Functional architecture of the light-responsive chalcone synthase reporter from parsley. *Plant Cell*, **1**, 707–714.

76. B. Weisshaar, G.A. Armstrong, A. Block, O. de Costa e Silva, K. Hahlbrock (1991). Light-inducible and constitutively expressed DNA-binding proteins recognizing a plant promoter element with functional relevance in light responsiveness. *EMBO J.*, **10**, 1777–1786.

77. M. Rocholl, C. Talke-Messerer, T. Kaiser, A. Batschauer (1994). Unit 1 of the mustard chalcone synthase promoter is sufficient to mediate light responses from different photoreceptors. *Plant Sci.*, **97**, 189–198.

78. M. Ni, J.M. Tepperman, P.H. Quail (1998). PIF3, a phytochrome-interacting factor necessary for normal photoinduced signal transduction, is a novel basic helix-loop-helix protein. *Cell*, **95**, 657–667.

79. M. Ni, J.M. Tepperman, P.H. Quail (1999). Binding of phytochrome B to its nuclear signalling partner PIF3 is reversibly induced by light. *Nature*, **400**, 781–784.

80. M. Koornneef, C.J. Hanhart, J.H. van der Veen (1991). A genetic and physiological analysis of late flowering mutants in *Arabidopsis thaliana. Mol. Gen. Genet.*, **229**, 57–66.

81. F. Quattrocchio, J.F. Wing, K. van der Woude, J.N. Mol, R. Koes (1998). Analysis of bHLH and MYB domain proteins: species-specific regulatory differences are caused by divergent evolution of target anthocyanin genes. *Plant J.*, **13**, 475–488.

82. J. Mol, E. Grotewold, R. Koes (1998). How genes paint flowers and seeds. *Trends Plant Sci.*, **3**, 212–217.

83. A.M. Lloyd, V. Walbot, R.W. Davis (1992). *Arabidopsis* and Nicotiana anthocyanin production activated by maize regulators R and C1. *Science*, **258**, 1773–1775.

84. W.R. Atchley, and W.M. Fitch (1997). A natural classification of the basic helix-loop-helix class of transcription factors. *Proc. Natl. Acad. Sci. U.S.A.*, **94**, 5172–5176.

85. L.H. Ang, S. Chattopadhyay, N. Wei, T. Oyama, K. Okada, et al. (1998). Molecular interaction between COP1 and HY5 defines a regulatory switch for light control of *Arabidopsis* development. *Mol. Cell*, **1**, 213–222.

86. J.F. Martinez-Garcia, E. Huq, P.H. Quail (2000). Direct targeting of light signals to a promoter element-bound transcription factor. *Science,* **288**, 859–863.

87. J.C. Young, E. Liscum, R.P. Hangarter (1992). Spectral-dependence of light-inhibited hypocotyl elongation in photomorphogenic mutants of *Arabidopsis*: Evidence for a UV-A photosensor. *Planta*, **188**, 106–114.

88. C. Lin, M. Ahmad, D. Gordon, A.R. Cashmore (1995). Expression of an *Arabidopsis* cryptochrome gene in transgenic tobacco results in hypersensitivity to blue, UV-A, and green light. *Proc. Natl. Acad. Sci. U.S.A.*, **92**, 8423–8427.

89. C. Lin, M. Ahmad, A.R. Cashmore (1996). *Arabidopsis* cryptochrome 1 is a soluble protein mediating blue light-dependent regulation of plant growth and development. *Plant J.*, **10**, 893–902.

90. L. Ninu, M. Ahmad, C. Miarelli, A.R. Cashmore, G. Giuliano (1999). Cryptochrome 1 controls tomato development in response to blue light. *Plant J.*, **18**, 551–556.

91. J.L. Weller, G. Perrotta, M.E. Schreuder, A. Van Tuinen, M. Koornneef, et al. (2001). Genetic dissection of blue-light sensing in tomato using mutants deficient in cryptochrome 1 and phytochromes A, B1 and B2. *Plant J.*, **25**, 427–440.

92. D.E. Somers, P.F. Devlin, S.A. Kay (1998). Phytochromes and cryptochromes in the entrainment of the *Arabidopsis* circadian clock. *Science*, **282**, 1488–1490.

93. D.J. Bagnall, R.W. King, R.P. Hangarter (1996). Blue-light promotion of flowering is absent in *hy4* mutants of *Arabidopsis*. *Planta*, **200**, 278–280.

94. T.C. Mockler, H. Guo, H. Yang, H. Duong, C. Lin (1999). Antagonistic actions of *Arabidopsis* cryptochromes and phytochrome B in the regulation of floral induction. *Development,* **126**, 2073–2082.

95. E.P. Spalding (2000). Ion channels and the transduction of light signals. *Plant Cell Environ.*, **23**, 665–674.

96. M. Laskowski, W.R. Briggs (1989). Regulation of pea epicotyl elongation by blue light. *Plant Physiol.*, **89**, 293–298.

97. B.M. Parks, M.H. Cho, E.P. Spalding (1998). Two genetically separable phases of growth inhibition induced by blue light in *Arabidopsis* seedlings. *Plant Physiol.*, **118**, 609–615.

98. M.H. Cho, E.P. Spalding (1996). An anion channel on *Arabidopsis* hypocotyls activated by blue light. *Proc. Natl. Acad. Sci. U.S.A.*, **93**, 8134–8138.

99. T. Oyama, Y. Shimura, K. Okada (1997). The *Arabidopsis* HY5 gene encodes a bZIP protein that regulates stimulus-induced development of root and hypocotyl. *Genes Dev.*, **11**, 2983–2995.

100. M.T. Osterlund, C.S. Hardtke, N. Wei, X.W. Deng (2000). Targeted destabilization of HY5 during light-regulated development of *Arabidopsis*. *Nature*, **405**, 462–466.

101. H. Guo, T.C. Mockler, H. Duong, C. Lin (2001). SUB1, an *Arabidopsis* Ca^{2+}-binding protein involved in cryptochrome and phytochrome coaction. *Science*, **291**, 487-490.

102. C. Bowler, G. Neuhaus, H. Yamagata, N.-H. Chua (1994). Cyclic GMP and calcium mediate phytochrome phototransduction. *Cell*, **77**, 73–81.

103. J.M. Christie, G.I. Jenkins (1996). Distinct UV-B and UV-A/blue light signal transduction pathways induce chalcone synthase gene expression in *Arabidopsis* cells. *Plant Cell*, **8**, 1555–1567.

104. P.J. Lumsden (1991). Circadian rhythms and phytochrome. *Annu. Rev. Plant Physiol. Plant Mol. Biol.*, **42**, 351–371.

105. C. Lin (2000). Photoreceptors and regulation of flowering time. *Plant Physiol.*, **123**, 39–50.

106. M.W. Coulter, K.C. Hamner (1964). Photoperiodic flowering response of Biloxi soybean in 72 hour cycles. *Plant Physiol.*, **39**, 848–856.

107. D. Mozley, B. Thomas (1995). Developmental and photobiological factors affecting photoperiodic induction in *Arabidopsis thaliana* Heynh. Landsberg erecta. *J. Exp. Botany*, **46**, 173–179.

108. I.A. Carre (1998). Genetic dissection of the photoperiod-sensing mechanism in the long-day plant *Arabidopsis thaliana.* In: P.J. Lumsden, A.J. Millar (Eds), *Biological Rhythms and Photoperiodism in Plants* (pp. 257–269). BIOS Science Publishers Ltd.,Oxford.

109. B. Thomas, D. Vince-Prue (1997). *Photoperiodism in Plants.* Academic Press, New York.

110. D.E. Somers (1999). The physiology and molecular bases of the plant circadian clock. *Plant Physiol.*, **121**, 9–20.

111. C.R. McClung (2001). Circadian rhythms in plants. *Annu. Rev. Plant. Physiol. Plant. Mol. Biol.,* **52**, 139–162.

112. A.J. Millar, M. Straume, J. Chory, N.H. Chua, S.A. Kay (1995). The regulation of circadian period by phototransduction pathways in *Arabidopsis. Science*, **267**, 1163–1166.

113. P.F. Devlin, S.A. Kay (2000). Cryptochromes are required for phytochrome signaling to the circadian clock but not for rhythmicity. *Plant Cell*, **12**, 2499–2510.

Chapter 11

Blue light receptors in fern and moss

Masamitsu Wada

Table of contents

Abstract

Blue light responses in ferns and mosses were studied for many years mainly in Europe [1]. The transition from one- to two-dimensional growth for example, was documented as the ratio of length and width of gametophyte cells by Mohr's group [2]. Protonemata of the fern *Dryopteris* grow filamentously under red light, but under blue light they spread in a two-dimensional shape. A blue light absorbing pigment other than phytochrome was proposed to be involved in this process [3]. Sugai and Furuya [4] found that red light-mediated spore germination of *Pteris vittata* was strongly inhibited by a short pulse of blue light irradiation even if the blue light was given either before or after the red light irradiation. They also gave far-red light after blue light to test whether the blue-light effect could be cancelled by subsequent irradiation with far-red light. If the blue-light effect is cancelled by far-red light, the response must be mediated by phytochrome, but their results showed the opposite, demonstrating that there must exist blue-light receptor(s) other than phytochrome in ferns. Thereafter, a number of physiological responses induced by blue light absorbing pigment(s), such as phototropism [5], cell division [6], and chloroplast movement [7] were tested to see whether they are regulated by real blue light pigment(s) or by phytochrome.

11.1   Photoreceptive sites

The intracellular localization of blue light receptors regulating cell division has been studied by partial cell irradiation with a blue microbeam in ferns [8,9]. Red light-grown, single-celled protonemata of *Adiantum* have a nucleus placed 60 μm from the tip of the protonemata. When the nuclear region was irradiated with a blue microbeam, cell division could be induced much faster than in protonemata irradiated at non-nuclear regions [8]. Furthermore, similar experiments with microbeam irradiation were performed using centrifuged protonemata. If a protonema is centrifuged basipetally, the nucleus in the cell moves downward. It was revealed in this experiment that the blue light receptive site also moved downward, suggesting that blue light receptors may be localized in, or very close to, the nucleus [9]. Inhibition of red light-induced spore germination could be induced by microbeam irradiation of the nuclear region with blue light [10]. It is therefore reasonable to assume that the blue-light receptors are localized in the nucleus and there control the expression of genes.

   Blue light effects are not restricted to the nuclear region of fern protonemata. When a filamentous protonema growing under red light is irradiated with blue or white light, the tip of the cell swells, and, later, cell division occurs. The apical cell swelling of protonema was induced by local irradiation at the tip where the nucleus is not positioned [11]. Irradiation of the nucleus had no effect. Moreover, blue light irradiation shows dichroic effects. Irradiation of the apical part of the cell with a microbeam of polarized blue light vibrating parallel to the cell membrane was shown to be more effective than light vibrating perpendicularly. These results indicate that the blue light photoreceptors may be located on, or close to, the plasma membrane so that the transition moment of the photoreceptors is somewhat parallel to the membrane [11]. Chloroplasts in protonemal cells move to the area of highest absorption when irradiated with a microbeam of blue light of moderate intensity (Figure 1) [7,12,13]. Also, chloroplast movement shows a dichroic effect, suggesting again that the photoreceptors are localized close to (or on) the plasma membrane.

Taken, together, there are at least two different distributions of blue light receptors in fern protonemata on the experimental bases of microbeam or polarized light irradiation – one is in or close to the nucleus and the other is on or close to the plasma membrane (Table 1). In *Arabidopsis*, cryptochromes and phototropins were shown to be localized in the nucleus or attached to the plasma membrane, respectively [14–17]. Thus, the localization of blue light receptors in *Arabidopsis* fits the experimental results obtained from fern gametophytes.

In the mosses *Funaria hygrometrica*, *Ceratodon purpureus* and *Physcomitrella patens* chloroplast movement is induced by polarized blue light [18,19] (Y. Sato, personal communication). Recently, the branching of protonemal cells of *Physcomitrella* was found to be controlled by cryptochrome localized in the nucleus [20]. Although blue light effects in mosses are not well understood, these findings indicate several similarities to ferns.

11.2   Cryptochrome in the fern *Adiantum*

11.2.1   *Photolyase-related genes*

The first blue-light photoreceptor genes cloned from ferns encode cryptochromes from *Adiantum capillus-veneris* [21]. In *Adiantum*, five cryptochrome genes (*AcCRY1-AcCRY5*) have been isolated (Table 2). Their N-terminal regions are photolyase-related and they are all very similar in their amino acid sequences but have different C-terminal extensions [21,22]. A phylogenic tree made of the N-terminal regions showed that *AcCRY1* and *AcCRY2* are in one cluster, while *AcCRY3* and *AcCRY4* are in another, indicating that gene duplications might have occurred during evolution. *AcCRY5* is in a different position in the phylogenic tree, suggesting that a sixth *CRY* gene similar to *AcCRY5* may exist, or did, in the progenitor of recent ferns, even though this gene might have become a pseudogene. No *Adiantum* CRYs showed photolyase activity when tested in photolyase-deficient *E. coli* strains [22]. These results are consistent with the fact that tryptophane −277, which is essential for *E. coli* photolyase activity, is not conserved in any *Adiantum* CRYs.

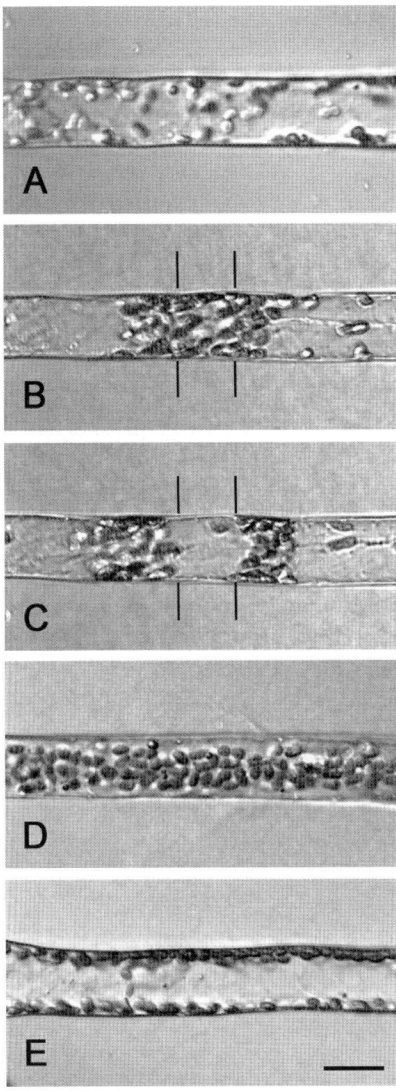

Figure 1. Blue light-induced chloroplast relocation movement in *Adiantum* protonemata. (A) Part of an *Adiantum* protonemal cell cultured under continuous red light for
six days, then irradiated with white light for 5 h to induce cell division, and, finally,
kept in the dark overnight for dark adaptation. (B) Part of a protonemal cell cultivated
as in (A), and then irradiated with a blue microbeam of 1 W m^{-2} and 20 µm width
(shown by bars) for 4 h. Chloroplasts accumulated in the irradiated area. (C) The same
part of the cell shown in (B) was irradiated with a blue microbeam of 10 W m^{-2} for 3 h.
Chloroplasts moved out of the irradiated area. (D, E) Protonemal cells cultured under
white light for several days were irradiated horizontally with polarized blue light vibrating either horizontally (D) or vertically (E). Chloroplasts accumulated at the top and
bottom of the cell in (D) and on both sides of the cell in (E). Bar in (E) represents
20 µm.

Table 1. Intracellular localization of blue-light sensitive regions in gametophytes of the fern *Adiantum* analyzed by microbeam and/or polarized light irradiation

Blue light effect	Localization	Dichroic effect	Ref.
Induction of cell division	Nucleus[a]	n.d.[b]	[8]
Induction of cell swelling	Cell tip plasma membrane[c]	+	[11]
Light attraction of chloroplasts	Whole cell plasma membrane[c]	+	[7]
Light repulsion of chloroplasts	Whole cell plasma membrane[c]	+	[7,42]
Inhibition of spore germination	Nucleus[a]	n.d.[b]	[10]
Induction of phototropism	Cell tip[d] Plasma membrane[c]	+	[5,43]

[a] Nucleus: It is not clear from these studies whether the photoreceptor is inside the nucleus or attached to the nuclear envelope. [b] n.d.: Not determined. [c] Plasma membrane: Dichroic effects suggest binding of the photoreceptor to the plasma membrane but this has not been proven by independent experiments. [d] Cell tip: The apical dome of the protonemal cells.

Table 2. Blue light receptors cloned in mosses and ferns

Photoreceptors	Localization	Function	Ref.
[*Adiantum*]			
Accry1	Cytoplasm	n.d.	[21,22]
Accry2	Cytoplasm	n.d.	[21,22]
Accry3	Nucleus (in darkness and red light)	n.d.	[21,22]
Accry4	Nucleus	n.d.	[22]
Accry5	Cytoplasm	n.d.	[22]
Acphot1	Plasma membrane[c], whole cell	Blue light-induced chloroplast avoidance movement	T. Kagawa et al., unpublished
Acphot2	n.d	n.d.	T. Kagawa et al., unpublished
Acphy3	Plasma membrane[c], whole cell for chloroplast movement, cell tip for phototropism	Red light induced phototropism and chloroplast movement	[7,39,44,45]
[*Physcomitrella*]			
Ppcry1a	Nucleus	Regulation of IAA induced genes; branching of protonemata etc.	[22]
Ppcry1b	Nucleus	As for Ppcry1a	[22]

n.d.: Not determined. [c] Plasma membrane; same as in Table 1.

11.2.2 Expression pattern

The expression patterns of all *Adiantum* *CRY*s were studied by RNA gel blot analysis using various tissues and developmental stages, such as spores imbibed for 1 day in the dark, protonemata grown under red light for 3 days, prothalli cultured under white light for 1 month, and young leaves grown under white light or in the dark [22]. *AcCRY1* and *AcCRY2* were expressed in a similar manner in all developmental stages tested. The amounts of these mRNAs rose slightly after spore germination and then remained constant in the gametophyte and in the sporophyte stages. The amount of *AcCRY3* mRNA was shown to be relatively higher in protonemata and sporophytes. *AcCRY4* was expressed in spores and dark-grown leaves, and only very small amounts were observed in other tissues and developmental stages. In contrast, *AcCRY5* was highly expressed in sporophytes irrespective of the light conditions but showed low expression in gametophytes. Although *AcCRY3* and *AcCRY4* are in the same cluster in the phylogenic tree, they showed different expression patterns. It will be interesting to see whether *Adiantum* CRY3 and CRY4 are functionally redundant.

The expression patterns of the five *Adiantum* CRYs during spore germination were studied more precisely by competitive RT-PCR [22]. *Adiantum* spores were imbibed 4 days in the dark and then transferred to continuous red or blue light. Two days after imbibition the germination rate was very high in red light but low in blue. It is known that blue light inhibits spore germination which is induced by red light in *Pteris vittata* [4] and in *Adiantum* [4,10,23]. The mRNA amounts of *AcCRY1*, *AcCRY2* and *AcCRY3* increased under red light two- to three-fold within 12 h and stayed at this level thereafter. Conversely, the *AcCRY4* mRNA level decreased about 50-fold in red light and five-fold in blue light within the first 24 h after onset of irradiation. *AcCRY5* mRNA levels increased 300–400-fold under red and blue light during the first 12 h and then decreased to a level 20–40-fold higher than in darkness. The increase of *AcCRY4* and *AcCRY5* mRNA levels under red light is regulated by phytochrome. It is presently unknown whether the blue light effect depends on phytochrome, blue light receptor(s) or both.

11.2.3 Intracellular distribution

The intracellular distribution of cryptrochromes in *Adiantum* is more complex compared with *Arabidopsis* where they seem to be localized only in the nucleus. *GUS* (β-glucuronidase)-*AcCRY* fusion genes bombarded into young gametophytes which were then cultured under red or blue light or in darkness for 16 h showed nuclear localization of GUS-Accry3 and GUS-Accry4 proteins but not GUS-Accry1, GUS-Accry2 and GUS-Accry5 [22]. The GUS-Accry 3 protein tended to accumulate in the nucleus in the dark and in red light but not in blue light. The difference in the nuclear localization patterns among Accrys probably depends on their C-termini, because the N-terminal regions are very similar among all Accrys. To analyze whether the C-terminal regions of *Adiantum* CRYs carry nuclear localization signals (NLS), these regions from *AcCRY3* and *AcCRY4* were fused with the *GUS* gene and the constructs introduced into gametophytes by particle bombardment [22]. The GUS-fusion proteins were found in the nucleus, demonstrating that the C-termini of Accry3 and Accry4 contain functional NLS.

Although the intracellular distribution of *Adiantum* CRYs and their expression patterns are known, their function as photoreceptors remains obscure. Although mutants could help to clarify their functions, mutant screening in ferns is difficult. Also, even if a mutant is identified the identification of the mutated gene is almost impossible. A way out of this dilemma is either homologous recombination which allows us to knockout genes or gene silencing by introducing antisense or RNAi constructs. Unfortunately, both techniques are not yet established in ferns.

11.3   Cryptochromes in the moss *Physcomitrella patens*

11.3.1   Genes and their structure

Physcomitrella is a species in which knockout of a target gene is available by homologous recombination. To elucidate the function of moss cryptochromes, Imaizumi et al. [20] screened for *Physcomitrella CRY* genes, knocked out the genes and clarified their functions. *Physcomitrella* has at least two *CRY* genes. The two identified genes, *PpCRY1a* and *PpCRY1b*, encode both proteins of 727 amino acids with only one base difference, that is the change in threonine at position 80 of Ppcry1a to methionine in Ppcry1b. The amino acid sequence indicates that FAD (flavin adenine dinucleotide) can bind to the apoprotein and could thus be the chromophore in moss cryptochrome as in other cryptochromes. In the C-terminal extension, a putative monopartite nuclear localization signal was found in a similar position as in the fern CRY3 and CRY4 proteins for which nuclear localization was demonstrated [22]. The intracellular localization of the moss CRYs was tested as GUS-fusion proteins. The genes encoding the GUS-fusions were introduced into protoplasts, and GUS staining showed that both cryptochromes are localized in the nucleus independent of the light conditions (red, blue, white light and darkness).

11.3.2   Gene knockout lines

Physcomitrella CRY disruptants were made by homologous recombination [20]. Besides single disruptants of *PpCRY1a* and *PpCRY1b*, double knockouts were also made since their nearly identical amino acid sequence indicates functional redundancy. The protonemal cells of the single and double knockout lines were cultured under different light conditions (red, blue, white light and darkness). In darkness, the protonemal cells did not grow, while under white and red light the size and shape of the colonies were the same for all mutants and the wild type. In contrast, under blue light the colonies of *Ppcry1a*, *Ppcry1b* and *Ppcry1aPpcry1b* knockouts were larger than that of the wild type and the density of protonemata in those colonies decreased in the mutants, particularly in the double mutant. Microscopic observation of the colonies revealed that side branch formation in protonemal cells of the mutants is very rare compared with the wild type. Overexpression of *PpCRY1b* in the *Ppcry1aPpcry1b* background recovered the branching frequency, confirming that the blue light signal is mediated by cryptochrome for this response.

Cryptochromes also control gametophore development in *Physcomitrella* [20]. The number of gametophores emerging from colonies is different between wild type and *PpCRY* disruptants. *CRY*-disruptants form more gametophores than the wild type when cultured under blue light but have the same number as wild type under other light conditions. Moreover, leaf and stem growth is also different between wild type and the disruptants under blue light, with shorter stems and narrower leaves in the disruptants. These results indicate that blue light controls not only gametophore differentiation but also gametophore growth.

11.3.3   Auxin and cryptochrome interaction

The development of moss gametophytes is known to be controlled by plant hormones, such as auxin and cytokinins [24]. Since the relationship between light effects and the effects of plant hormones is not well understood, Imaizumi et al. [20] analyzed this interplay in gametophyte development by applying the synthetic auxin, 1-naphthalene acetic acid (NAA). Adding NAA in agar medium on which the moss gametophytes were cultivated under blue light showed that the diameter of the colonies increased in relation to the NAA concentration. The growth rate was highest for the *Ppcry1aPpcry1b* double mutant, intermediate for the single mutants, and lowest for the wild type. This result indicates that cryptochromes inhibit gametophyte growth. Under red light where cryptochrome should not operate, the growth rate of wild type cultured on NAA is very high and almost the same as the growth rate of the *Ppcry1aPpcry1b* double mutant under white light. In summary it is suggested that blue light inhibits the auxin response through cryptochrome.

Moss protonemata have two cell types, chloronemata and caulonemata. When spores germinate, they form initially chloronemata. Later in development, chloronemata differentiate to caulonemata on which buds are formed which then differentiate to gametophores. Chloronemal cells are thinner and shorter than caulonemal cells and the septa between chloronemal cells are perpendicular to the cell axis, whereas they are oblique between caulonema cells. The differentiation of chloronemata to caulonemata occurs faster under red light than under white light, and is also induced by auxin added to the medium when chloronemata are cultivated under white or blue light. Under red light the auxin effect is not observed in any of the knockout lines, indicating that the auxin effect is mediated by cryptochrome.

11.3.4   Control of gene expression

Since moss cryptochromes have nuclear localization signals it is possible that the cryptochromes move into the nucleus and control gene expression. Imaizumi et al. [20] tested whether auxin-inducible gene expression is affected by blue-light irradiation through cryptochromes by using a soybean *GH3* promoter. Protoplasts prepared from cryptochrome single and double knock-out lines were transformed with *GH3::GUS* genes and the *GUS* expression was analyzed under different light conditions in the presence or absence of NAA (10 μM). The GUS activity was highest in the *Ppcry1aPpcry1b* double mutant and lowest in the wild type. Single mutants had intermediate GUS levels. These results suggest that blue light inhibits auxin signals through cryptochromes at the transcriptional level.

Since these results were obtained with a foreign promoter, auxin-inducible moss genes were cloned to reinvestigate the results obtained with the soybean promoter [20]. Genes related to soybean *GH3*, namely the *P. patens* GH3-like protein 1 (*PpGH3L1*), and indole-3-acetic acid (*IAA*) gene homologs (*PpIAA1a* and *PpIAA1b*) were cloned from *Physcomitrella* and sequenced. The accumulation of *PpGH3L1* and *PpIAA1a/b* transcripts was induced by exogenous auxin within 24 h. The highest accumulations were seen under 10 μM NAA, in particular in the *Ppcry1aPpcry1b* knockout strain, demonstrating again that NAA-induced gene expression in moss is under the negative control of cryptochromes.

11.4   Phototropin family in fern and moss

11.4.1   Phototropin in fern

Phototropin (formerly nph1, now phot1 [25]) was identified as a blue light receptor that mediates the phototropic response in *Arabidopsis* [26]. Soon thereafter, a homologue of the phototropin gene (*NPL1*, now *PHOT2*) was cloned from *Arabidopsis* [27], although its function was not known at that time. Recently, phot2 was shown to be a photoreceptor for the chloroplast avoidance response [28,29].

Phototropin genes were also found in *Adiantum* [30] (T. Kagawa, M. Kasahara unpublished data; Table 2) and in *Physcomitrella* (T. Kiyosue, unpublished data). In *Adiantum*, cDNAs for *PHOT1* and *PHOT2* and their corresponding genes were cloned and sequenced. It was found that Acphot2 mediates the chloroplast avoidance response (T. Kagawa, unpublished data). The *Adiantum PHOT1* cDNA clone (*Acphot1*) is 3492 bp in length and codes for 1092 amino acids. The amino acid sequence of *Acphot1* is 45–49% identical with other higher plants' phot1s. Thin-layer chromatography of the chromophore released from the LOV2 domain (see the Chapter 9 for further details about LOV domains) of Acphot1 expressed in *E. coli* showed that flavin mononucleotide (FMN) is attached to this domain as in other phototropins [30]. The sequence of the *AcPHOT2* gene (T. Kagawa, unpublished data) showed nearly the same pattern of exon–intron structure as the seed plant *PHOT*s with 22 introns, the same number as in *Arabidopsis PHOT1*. Given that chloroplast relocation movement and phototropic response can be induced by blue light in *Adiantum*, it is very likely that phots mediate both responses in *Adiantum* as in *Arabidopsis* [28,29,31–34].

11.4.2 Phototropin in moss

Physcomitrella patens has at least two-to-four or more *PHOT* genes (T. Kiyosue, unpublished data). Although their function is not revealed yet, it is possible that the Ppphots mediate blue light-induced chloroplast relocation movement because *Physcomitrella cry1a/cry1b* double mutants still showed chloroplast movement under blue light (Y. Sato, unpublished data). Moreover, the shape of the action spectrum for chloroplast relocation movement in the moss *Funaria* is very similar to the absorption spectrum of phototropins, as seen in a comparison of published data [35,36]. The chloroplast photo-relocation movement in *Physcomitrella* has very characteristic features (Figure 2) [37]. The chloroplast movement can be induced by red and blue light, as is the case for the fern *Adiantum* [38]. It is interesting that under blue light both actin filaments and microtubules are involved in the chloroplast movement of *Physcomitrella*. As far as is known, seed plants and ferns use only actin fila-ments for chloroplast relocation movement. Moreover, chloroplast movement in *Physcomitrella* is mediated only by microtubules under red light [37]. So far it is not known whether in *Physcomitrella* the same photoreceptor(s) are used

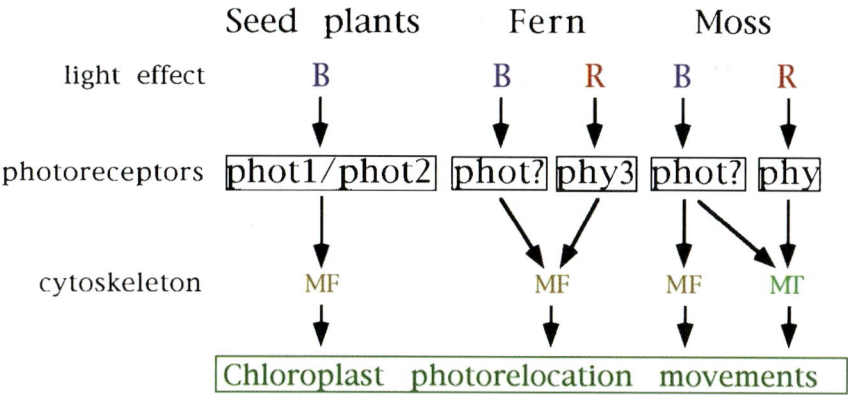

B: Blue light, R: Red light, MF: microfilament, MT, microtubule

Figure 2. Scheme of chloroplast photorelocation movement in plants. Seed plants use blue light through phot1 and phot2 for the movement by microfilaments. The fern *Adiantum* and the moss *Physcomitrella* use both red and blue light which are sensed by phytochrome and probably by phototropin(s), respectively. Ferns use microfilaments for the movement whereas mosses use both microfilaments and microtubules.

for the movement of chloroplasts towards and away from the light, and how the light signal is separated into actin- and microtubule-mediated movement.

11.5   Phytochrome 3 as a phototropin-related photoreceptor

Phytochrome 3 is a very special chimeric protein found in the fern *Adiantum capillus-veneris*. It consists of 1465 amino acids, and the N-terminal half carries a typical phytochrome chromophore binding domain. The C-terminal half, however, is almost a complete phototropin with two FMN-binding LOV domains [39]. The phytochrome domain consists of 564 amino acids and has 52% identity to the *Arabidopsis* phyA chromophore binding region. The C-terminal phot region has 57% identity to *Arabidopsis* phot1. The *AcPHY3* gene has no intron throughout the whole sequence, although *Adiantum* and other *PHOT* genes have many introns (see above). The lack of any intron might give some clues about how this gene arose during evolution.

The apoprotein of Acphy3 was expressed in yeast-bound phycocyanobilin, given as an analog of phytochromobilin, and showed the typical absorption-difference spectrum of phytochromes [39]. This result indicates that Acphy3 can function as a red/far-red light reversible photoreceptor. In addition, it was shown that the LOV domains of Acphy3 bind FMN as *Arabidopsis* phot1 and phot2 (see also Chapter 9) [36]. Taken together, Acphy3 has in principle the possibility to act both as a red/far-red light and as a blue light photo-receptor. Among the *Adiantum* photoresponses, some responses such as spore germination and cell division are antagonistically controlled by red and blue light [6,8,10]. Other responses such as chloroplast movement are controlled synergistically by blue and red light [7,31], suggesting the possibility that chloroplast movement is controlled by Acphy3. Young leaves of *Adiantum* show phototropic responses under red light as well as under blue light [40].

Red light aphototropic (*rap*) *Adiantum* mutants were identified in which the red light-induced chloroplast movement and the red-light-induced phototropic response are deficient [41]. However, blue-light-induced chloroplast movement and blue light-induced phototropism are normal in these mutants. Recently, *PHY3* genes of five *rap* mutants, which were created by ethylmethane sulfonate (EMS) mutagenesis, were sequenced and it was found that all of the *rap* mutants had some deletions, or duplications, or point mutations in the *PHY3* gene [44]. These data suggested strongly that Acphy3 is the photoreceptor for red light-induced chloroplast movement and phototropic response. Moreover, transient expression of the wild-type *PHY3* gene in *rap* mutants rescued the red light-induced chloroplast movement [44]. The question of why Acphy3 mediates only red light but not blue light responses, although it carries FMN, can not be answered yet. Since *Adiantum* has two blue light photoreceptors, Acphot1 and Acphot2, it is very likely that both mediate blue light-induced chloroplast movement. However, for a final answer about a possible blue light receptor function of Acphy3, a triple mutant (*Acphy3Acphot1Acphot2*) and double mutants in the various combinations have to be created and analyzed with respect to blue light controlled chloroplast movement.

11.6   Concluding remarks

Fern and moss gametophytes are excellent experimental systems to study photobiological responses due to their simple structure of one-dimensionally arranged cell chains or two-dimensionally-arranged single layered sheets of cells. The cells can easily be analyzed under the microscope and/or irradiated, because no other tissues cover the gametophyte cells. Methods for staining with chemicals and antibodies and for transformation (transient or stable) with reporter genes such as GUS are established for these species so that the elucidation of signal transduction pathways, including the analysis of the intracellular distribution of signaling components, is extremely facile.

The haploid stage of the gametophyte is also a big advantage for gene knockout studies in *Physcomitrella* and, hopefully, for gene silencing in ferns. In contrast, the molecular biology, in particular the cloning of target genes from mutant plants, is not easy in moss and fern plants. Since 66% of the ESTs from *Physcomitrella* are related to genes in *Arabidopsis* [46] it is tempting to assume that the study of gene function in fern and moss systems will also help us to understand the function of related genes in seed plants.

Acknowledgements

I thank Dr A. Batschauer for critical reading of the manuscript and H. Kawai for preparing the figures. I acknowledge Mr E. Sugiyama and Ms T. Yasuki for their technical assistance in fern culture. This work was supported in part by a Grant-in-Aid for Scientific Research (A) (13304061) and Scientific Research on Priority Areas (835), and by PROBRAIN to M.W.

References

1. M. Wada, A. Kadota (1989). Photomorphogenesis in lower green plants. *Annu. Rev. Plant Physiol. Plant Mol. Biol.*, **40**, 169–191.
2. H. Mohr (1980). Interaction between blue light and phytochrome in photo-morphogenesis. In: H. Senger (Ed), *The Blue Light Syndrome* (pp. 97–109). Springer-Verlag, Berlin.
3. H. Mohr (1956). Die abhängigkeit des protonemawachstums und der protonemapolarität bei farnen von licht. *Planta*, **47**, 1276–1580.
4. M. Sugai, M. Furuya (1967). Photomorphogenesis in *Pteris vittata* I. Phytochrome-mediated spore germination and blue light interaction. *Plant Cell Physiol.*, **8**, 737–748.
5. J. Hayami, A. Kadota, M. Wada (1986). Blue light-induced phototropic response and the intracellular photoreceptive site in *Adiantum* protonamta. *Plant Cell Physiol.*, **27**, 1571–1577.
6. M. Wada, M. Furuya (1972). Phytochrome action on the timing of cell division in *Adiantum* gametophytes. *Plant Physiol.*, **49**, 110–113.

7. H. Yatsuhashi, A. Kadota, M. Wada (1985). Blue- and red-light action in photoorientation of chloroplasts in *Adiantum* protonemata. *Planta*, **165**, 43–50.

8. M. Wada, M. Furuya (1978). Effects of narrow-beam irradiations with blue and far-red light on the timing of cell division in *Adiantum* gametophytes. *Planta*, **138**, 85–90.

9. A. Kadota, Y. Fushimi, M. Wada (186). Intracellular photoreceptive sites for blue light-induced cell division in protonemata of the fern *Adiantum* – Further analyses by polarized light irradiation and cell centrifugation. *Plant Cell Physiol.*, **27**, 989–995.

10. M. Furuya, M. Kanno, H. Okamoto, S. Fukuda M. Wada (1997). Control of mitosis by phytochrome and a blue-light receptor in *Adiantum* spores. *Plant Physiol.*, **113**, 677–683.

11. M. Wada, A. Kadota, M. Furuya (1978). Apical growth of protonemata in *Adiantum capillus-veneris*. II. Action spectra for the induction of apical swelling and the intracellular photoreceptive site. *Bot. Mag. Tokyo*, **91**, 113–120.

12. M. Wada, F. Grolig, W. Haupt (1993). Light-oriented chloroplast positioning. Contribution to progress in photobiology. *J. Photochem. Photobiol. B*, **17**, 3–25.

13. M. Wada, T. Kagawa (2001). Light induced chloroplast relocation. In: M. Lebert, D-P. Haeder (Eds), *ESP Comprehensive Series in Photoscience. Vol. 1 Photomovement* (pp. 895–922). Elsevier Science Publishers, Dordrecht.

14. A.R. Cashmore, J.A. Jarillo, Y.J. Wu, D. Liu (1999). Cryptochromes: blue light receptors for plants and animals. *Science*, **284**, 760–765.

15. H. Guo, H. Duong, N. Ma, C. Lin (1999). The *Arabidopsis* blue light receptor cryptochrome 2 is a nuclear protein regulated by a blue light-dependent post-transcriptional mechanism. *Plant J.*, **19**, 279–287.

16. O. Kleiner, S. Kircher, K. Harter, A. Batschauer (1999). Nuclear localization of the *Arabidopsis* blue light receptor cryptochrome 2. *Plant J.*, **19**, 289–296.

17. W.R. Briggs, E. Huala (1999). Blue-light photoreceptors in higher plants. *Annu. Rev. Cell Dev. Biol.*, **15**, 33–62.

18. J. Zurzycki (1967). Properties and localization of the photoreceptor active in displacements of chloroplasts in *Funaria hygrometrica*. II. Studies with polarized light. *Acta Soc. Bot. Pol.*, **36**, 133–142.

19. T. Kagawa, T. Lamparter, E. Hartmann, M. Wada (1997). Phytochrome-mediated branch formation in protonemata of the moss *Ceratodon purpureus*. *J. Plant Res.*, **110**, 363–370.

20. T. Imaizumi, A. Kadota, M. Hasebe, M. Wada (2002). Cryptochrome light signals control development to suppress auxin sensitivity in the moss *Physcomitrella patens*. *Plant Cell*, **14**, 373–386.

21. T. Kanegae, M. Wada (1998). Isolation and characterization of homologues of plant blue-light photoreceptor (cryptochrome) genes from the fern *Adiantum capillus-veneris*. *Mol. Gen. Genet.*, **259**, 345–353.

22. T. Imaizumi, T. Kanegae, M. Wada (2000). Cryptochrome nucleocytoplasmic distribution and gene expression are regulated by light quality in the fern *Adiantum capillus-veneris*. *Plant Cell*, **12**, 81–96.

23. M. Sugai, M. Furuya (1985). Action spectrum in ultraviolet and blue light regions for the inhibition of red-light-induced spore germination in *Adiantum capillus-veneris* L. *Plant Cell Physiol.*, **26**, 953–956.

24. B. Lehnert, M. Bopp (1983). The hormonal regulation of protonema development in mosses. I. Auxin-cytokinin interaction. *Z. Pflanzenphysiol.*, **110**, 379–391.

25. W.R. Briggs, C. Beck, A.R. Cashmore, J.M. Christie, J. Hughes, J.A. Jarillo, T. Kagawa, H. Kanegae, E. Liscum, A. Nagatani, K. Okada, M. Salomon, W. Rüdiger, T. Sakai, M. Takano, M. Wada, J.C. Watson (2001). The phototropin family of photoreceptors. *Plant Cell*, **13**, 993–997.

26. E. Huala, P.W. Oeller, E. Liscum, I.-S. Han, E. Larsen, W.R. Briggs (1997). *Arabidopsis* NPH1: A protein kinase with a putative redox-sensing domain. *Science*, **278**, 2120–2123.

27. J.A. Jarillo, M. Ahmad, A.R. Cashmore (1998). NPL1 (accession No. AF053941): a second member of the NPH serine/threonine kinase family of Arabidopsis. *Plant Physiol.*, **117**, 719.

28. T. Kagawa, T. Sakai, N. Suetsugu, K. Oikawa, S. Ishiguro, T. Kato, S. Tabata, K. Okada, M. Wada (2001). *Arabidopsis* NPL1: A phototropin homologue controlling the chloroplast high-light avoidance response. *Science*, **291**, 2138–2141.

29. J.A. Jarillo, H. Gabrys, J. Capel, J.M. Alonso, J.R. Ecker, A.R. Cashmore (2001). Phototropin-related NPL1 controls chloroplast relocation induced by blue light. *Nature*, **410**, 952–954.

30. K. Nozue, J. Christie, T. Kiyosue, W.R. Briggs, M. Wada (2000). Isolation and characterization of fern phototropin (Accession No. AB037188). A putative blue-light photoreceptor for phototropism (PGR00-039). *Plant Physiol.*, **122**, 1457.

31. T. Kagawa, M. Wada (1994). Brief irradiation with red or blue light induces orientational movement of chloroplasts in dark-adapted prothallial cells of the fern *Adiantum*. *J. Plant Res.*, **107**, 389–398.

32. T. Kagawa, M. Wada (1996). Phytochrome- and blue light-absorbing pigment-mediated directional movement of chloroplasts in dark-adapted prothallial cells of fern *Adiantum* as analyzed by microbeam irradiation. *Planta*, **198**, 488–493.

33. T. Kagawa, M. Wada (2000). Blue light-induced chloroplast relocation in *Arabidopsis thaliana* as analyzed by microbeam irradiation. *Plant Cell Physiol.*, **41**, 84–93.

34. T. Sakai, T. Kagawa, M. Kasahara, T.E. Swartz, J.M. Christie, W.R. Briggs, M. Wada, K. Okada (2001). *Arabidopsis* Nph1 and npl1: Blue-light receptors that mediate both phototropism and chloroplast relocation. *Proc. Natl. Acad. Sci. U.S.A.*, **98**, 6969–6974.

35. J. Zurzycki (1967). Properties and localization of the photoreceptor active in displacements of chloroplasts in *Funaria hygrometrica*. I. Action spectrum. *Acta Soc. Bot. Pol.*, **36**, 133–142.

36. J.M. Christie, M. Salomon, K. Nozue, M. Wada, W.R. Briggs (1999). LOV (light, oxygen, or voltage) domains of the blue-light photoreceptor phototropin (nph1): Binding sites for the chromophore flavin mononucleotide. *Proc. Natl. Acad. Sci. U.S.A.*, **96**, 8779–8783.

37. Y. Sato, M. Wada, A. Kadota (2001). Choice of tracks, microtubules and/or actin filaments for chloroplast photo-movement is differentially controlled by phytochrome and a blue light receptor. *J. Cell Sci.*, **114**, 269–279.

38. A. Kadota, Y. Sato, M. Wada (2000). Intracellular chloroplast photorelocation in the moss *Physcomitrella patens* is mediated by phytochrome as well as by a blue-light receptor. *Planta*, **210**, 932–937.

39. K. Nozue, T. Kanegae, T. Imaizumi, S. Fukuda, H. Okamoto, K.-C. Yeh, J.C. Lagarias, M. Wada (1998). A phytochrome from the fern *Adiantum* with

features of the putative photoreceptor NPH1. *Proc. Natl. Acad. Sci. U.S.A.*, **95**, 15826–15830.

40. M. Wada, H. Sei (1994). Phytochrome-mediated phototropism in *Adiantum cuneatum* young leaves. *J. Plant Res.*, **107**, 181–186.
41. A. Kadota, M. Wada (1999). Red light-aphototropic (*rap*) mutants lack red light-induced chloroplast relocation movement in the fern *Adiantum capillus-veneris*. *Plant Cell Physiol.*, **40**, 238–247.
42. H. Yatsuhashi, M. Wada (1990). High-fluence rate responses in the light-oriented chloroplast movement in *Adiantum* protonemata. *Plant Sci.*, **68**, 87–94.
43. J. Hayami, A. Kadota, M. Wada (1992). Intracellular dichroic orientation of the blue light-absorbing pigment and the blue-absorption band of red-absorbing form of phytochrome responsible for phototropism of the fern *Adiantum* protonemata. *Photochem. Photobiol.*, **56**, 661–666.
44. H. Kawai, T. Kanegae, S. Christensen, T. Kiyosue, Y. Sato, T. Imaizumi, A. Kadota, M. Wada (2003). Responses of ferns to red light are mediated by an unconventional photoreceptor. *Nature*, **421**, 287–290.
45. M. Wada, A. Kadota, M. Furnya (1981). Intracellular photoreceptive sites for polarotropsin in protonemata of the fern Adiantum capillus-venerisl. *Plant Cell Physiol.*, **22**, 1481–1488.
46. T. Nishiyama, T. Fujita, T. Shin-I, M. Seki, H. Nishide, I. Uchiyama, A. Kamiya, P. Carninci, Y. Hayashizaki, K. Shinozaki, Y. Kohara, M. Hasebe (2003). Comparative genomics of *Physcomitrella patens* gametophytic transcriptome and *Aribidopsis thaliana*: Implication for land plant evolution. *Proc. Natl. Acad. Sci. USA*, **100**, 8007–8012.

Chapter 12

Photoreceptors resetting the circadian clock

Paul F. Devlin

Table of contents

Abstract

Throughout the history of the earth, the cycle of day and night has continued. The relatively rapid transition from light to darkness and back again arguably forms the greatest environmental challenge to life. Plants can no longer photosynthesise at night and animals can no longer see. The ambient temperature often drops dramatically at night and rises dramatically again during the day. Adaptation to the day/night transition is something which has been developed in nearly all organisms exposed to this challenge and, in each case, this adaptation makes use of an internal timekeeper, the circadian clock. The circadian clock allows an organism to anticipate dawn and dusk and yet, to be useful, the clock must first be set to the correct time by these environmental cues. The most prominent time-of-day signal is light. Organisms make use of signals from an array of different photoreceptors in setting the clock. The mechanism of the clock itself varies greatly between plants, fungi, insects and mammals and, as would be expected, the photoreceptors used also vary greatly between these different groups. However, despite these differences, the principals of the clock mechanism and of the processes involved in light resetting of the clock have proved to be applicable throughout biology.

12.1   Introduction

The concept of preparing for expected changes in the environment is a familiar one. Within Northern and Southern latitudes ambient temperature changes dramatically with the seasons. Each spring many plants unfurl new leaves and animals emerge from hibernation or shed thick winter coats in preparation for the more moderate summer climate. Each autumn these plants lose their leaves again ready to over-winter in a dormant state, whilst animals grow think coats or hibernate to protect themselves from the harsher climate of winter. The preparation for dawn and dusk requires changes over a much more rapid timescale, yet equally dramatic changes in physiology and metabolism do occur on a daily basis. In plants, the photosynthetic machinery gears up prior to dawn ready to gain maximum benefit as light becomes available [1]. The leaves of many plants move to become more vertical at night and more horizontal during the day to protect them from chilling damage [2,3]. In insects, a rhythm in the timing of eclosion is observed whereby adult *Drosophila* will eclose from their pupae only at dawn [4]. In mammals the cycles of sleep and waking, of activity, and of body temperature all anticipate dawn and dusk [5]. All of these rhythms respond not to the appearance of light at dawn or its disappearance at dusk but to an internal clock. It is this "circadian" clock which allows anticipation of dawn and dusk and negates any need for a period of acclimatising.

If the circadian clock is to predict dawn and dusk accurately it must be precisely set to the correct time. Paradoxically, the very day/night cycles which the clock has evolved to anticipate provide the cues which set the clock. The dominant environmental signal involved in resetting the clock is light. Changes from light to dark or dark to light, thus, form "Zeitgebers" or time-givers [6]. The recent rapid progress in the field of photoreceptor biology has allowed us to begin to uncover just how this light resetting of the clock takes place: What are the photoreceptors perceiving the light signals? How are these signals transduced to the clock mechanism? How is this resetting regulated?

Interactions between photoreceptor signals and clock signals are of great interest in another phenomenon, that of photoperiodism, the measurement of day-length. In anticipating the changes associated with the change of the seasons, organisms use daylength as an indicator of time of year. A shortening of day-length occurs as we move towards winter and a lengthening occurs as we move toward spring. The circadian clock forms a timing mechanism for day-length. In conjunction with light input from photoreceptors, an organism can use the circadian clock to measure the length of the light period and so determine the time of year [7].

12.2   How is the circadian clock reset?

The circadian clock in all organisms will "free run" in constant environmental conditions. Often the "period length" under such conditions will deviate slightly from 24 h, illustrating the necessity for daily resetting. This in itself also provides a mechanism for maintaining accuracy. In addition, a capacity for clock resetting is essential as the timing of dawn and dusk advances and recedes throughout the year. This plasticity allows an organism to coordinate the regulation of physiology and metabolism in line with these changes in day-length.

Just as with a wristwatch, the resetting of the circadian clock requires that the mechanism of the clock is advanced or delayed so that the hands display the correct time. Once this is done, the clock must continue to run as before. The hands of the clock within circadian biology are the overt rhythms in physiology and metabolism that we observe from day to day, whilst the mechanism of the circadian clock resides within a core oscillator that maintains the self-sustaining rhythm.

The basic components of the clock mechanism have been established in several systems [8]. Central to each is a transcriptional feedback loop formed of both positive and negative acting elements (Figure 1). This results in an oscillation in the level of at least one of the components with a period of about 24 h. The level (and direction of change) of this component determines the time or "phase" of the clock. If the level of this component is caused to change the clock will be reset to a new time or "phase shifted" (Figure 1). Input and output pathways allow signals to be transmitted to and from the oscillator. Signals must be transduced to the oscillator from the environment, in particular from the photoreceptor signalling pathways, to cause such phase shifts and to set the clock to the right time. Signals must also be transduced from the

(a)

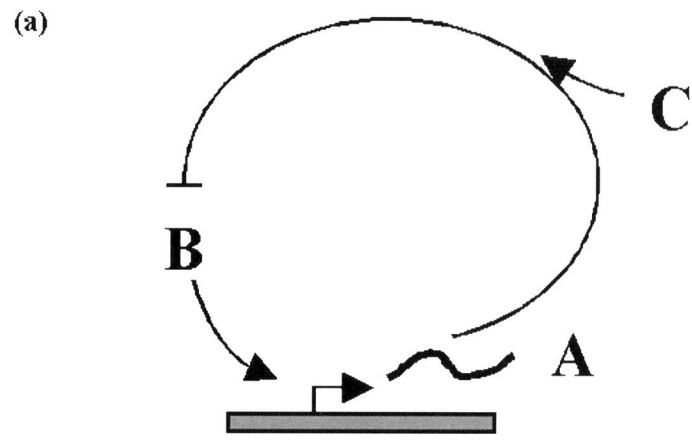

(b)

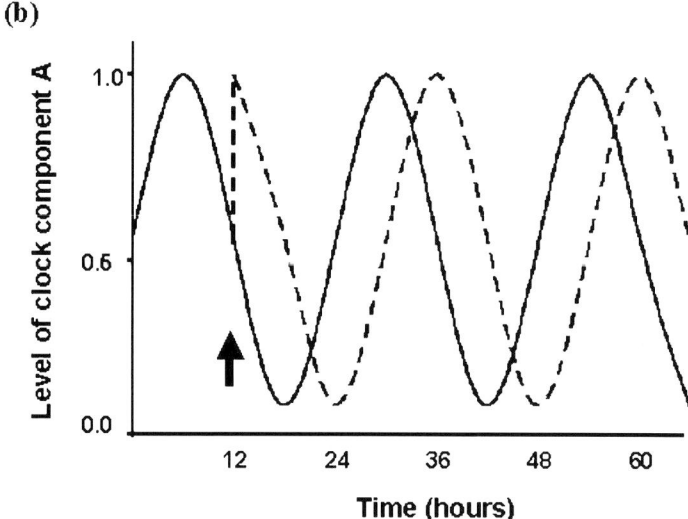

Figure 1. (a) Stylised representation of a transcriptional feedback loop capable of generating a circadian oscillation. Transcription of component A is promoted by component B. As component A accumulates it acts to suppress the activity of component B thereby negatively feeding back on its own transcription. A third factor, component C provides a delay, being necessary to modify component A before it can affect component B. This prevents the system from reaching a stable equilibrium, ensuring that oscillation is maintained with a period length of 24 h. (b) The solid line represents the oscillation in the level of component A with time for the circadian clock loop shown in (a). The dotted line represents the same trace in an individual which received a clock-resetting stimulus 12 h after recording commenced (indicated by an arrow). Clock resetting involves a rapid change in the level of a clock component, in this case component A. The clock is therefore reset or shifted to a new "phase" and continues to oscillate from that point.

oscillator to control the hands of the clock, the overt rhythms within the org-anism. Within animals, where the clock mechanism is well characterised, the control of overt rhythms via this "output" pathway from the clock is a rare example where behaviour has been dissected down to the level of the inter-actions of molecules within the cell [9]. The central mechanism of the clock and the control of clock outputs have been extensively reviewed elsewhere [10–13]. The study of light "input" pathways to the clock is the subject of this chapter.

12.3   Light input to the clock

In the natural environment, clock resetting occurs when an organism sees light at a time when it would normally expect to be seeing darkness, for example as dawn becomes earlier with the approach of summer. Such pulses of light prior to the expected dawn will advance the clock, causing a positive phase shift. Pulses of light perceived after expected dusk will delay the clock, causing a negative phase shift. The result is a characteristic phase response curve (PRC) [14] (Figure 2). During the subjective day when the organism would be expecting to perceive light there is a reduced response to pulses of light for clock resetting. Many organisms display a dead-zone where responses to light are completely absent during the day. Such modulation of responsiveness to light is termed "gating". The output from the clock rhythmically regulates the response to environmental input, effectively controlling a gate. When the "gate" is closed (during the middle of the subjective day) light input to the clock is greatly suppressed. When the "gate" is open (during the subjective night and around dawn and dusk) light signals are capable of clock resetting (Figure 2). Thus, the clock is not being continually reset throughout each light period but can maintain a meaningful rhythm regulating the various physio-logical and biochemical outputs. Such a close relationship between input, oscillator and output blurs the distinction between these divisions of the cir-cadian system, making it difficult to distinguish where one begins and another ends [15].

The mechanism of the circadian clock has been well established in animals. Much of the pioneering work on the animal clock came from the model orga-nisms, the fruit fly, *Drosophila melanogaster*, and the mouse, *Mus musculus*. The molecules interacting to form the feedback loop which makes up the central oscillator are well characterised [10,11]. Insect and mammalian clocks share considerable homology in that many component molecules are common to both systems. In plants, however, the mechanism of the clock is less well characterised and the components of the plant clock are just now beginning to be discovered in the model plant, *Arabidopsis thaliana* [16,17,17a]. The components of the established animal clock are absent in plants, suggesting that clocks have arisen more than once throughout the history of life on earth and that this has happened independently in plants and animals. This theory is supported by extensive research into fungal clocks where the mechanism has

(a)

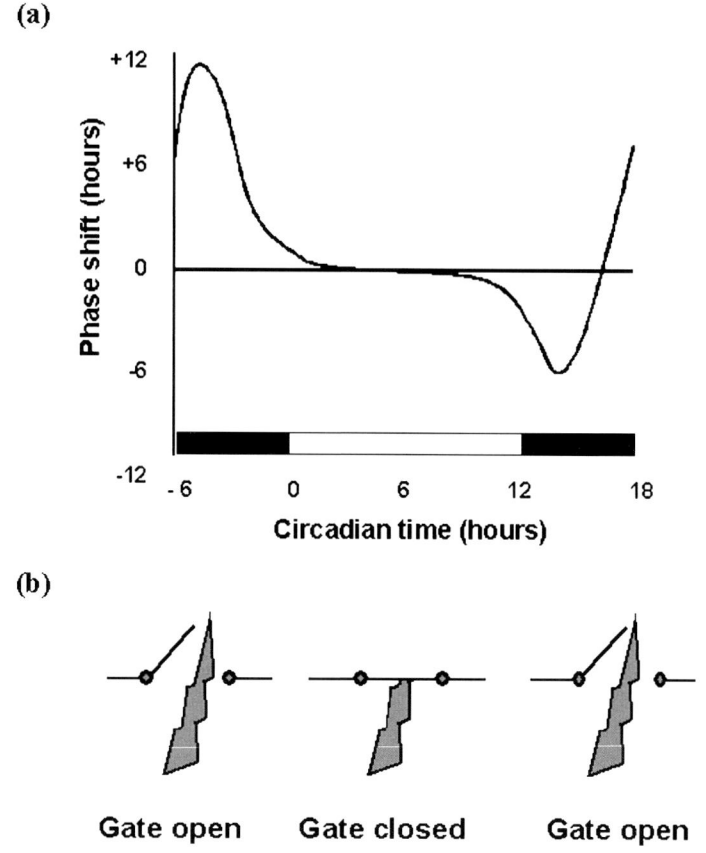

(b)

Figure 2. (a) Stylised phase response curve for clock resetting by light. Light received before subjective dawn results in phase advances (positive phase shifts), whilst light received after subjective dusk results in phase delays (negative phase shifts). Light received during the subjective day has little effect on the phase of the clock. (b) The gating model of clock resetting. The model relates to the phase response curve (a). At times when light causes phase shifts (dawn and dusk) the "gate" can be considered to be open, allowing light signals to act upon the clock mechanism. At times when the clock is insensitive to light, the "gate" can be considered to be closed to light signals.

also been demonstrated [13]. Although the central oscillator in fungi also consists of a feedback loop, the components of this loop are quite different from those in animal clocks. Some similarity between plant and animal clocks is found amongst the photoreceptors mediating light input to the clock but even this appears to be the result of an independent origin.

The earliest demonstration of the photoreceptors mediating light input to the clock came from studies of photoreceptor mutants of *Arabidopsis* [18]. Subsequently photoreceptors mediating light input to the clock in insects were characterised [19]. However, until recently, the mammalian circadian photoreceptor remains the subject of much conjecture.

12.4   Photoreceptors mediating light input to the clock in plants

Plants have a range of sensory mechanisms to detect environmental signals and are able to minutely modify their growth and developmental pattern to adapt to take maximum advantage of their habitat. Plants, as sessile organisms, need to be especially plastic in their development. The most important environmental factor in the life of a plant is light. As photoautotrophs, plants rely on energy from the sun for photosynthesis and consequently have developed a range of signal-transducing photoreceptors which give information about the light environment. Plants monitor the intensity, quality, direction and duration of light and the photoreceptors responsible for gathering this information have been well characterised [20]. Plant photoreceptors fall into three distinct classes: the red/far-red-absorbing phytochromes [21], the blue/UV-A-absorbing cryptochromes [22] and the blue/UV-A -absorbing phototropins [23]. More in-depth description of these classes of photoreceptor can be found in the preceding chapters of this volume. Much of our knowledge of plant photoreceptors has come from the study of photoperception in *Arabidopsis*. Five phytochrome genes are present in *Arabidopsis*, *PHYA–PHYE*, encoding the photoreceptors phyA–phyE [24,25]. The phytochromes consist of a protein moiety of about 124 kD with a covalently-attached, linear tetrapyrrole chromophore. Two such monomers dimerise within the cell. Phytochrome exists in two photo-interconvertible forms, Pr, absorbing maximally in the red region of the spectrum and Pfr, absorbing maximally in the far-red region of the spectrum. Absorption of a photon of light causes a reversible conformational change from one form to the other [26]. The phytochromes differ in both their function and their spectral sensitivity. The Pfr form of phyA is rapidly degraded, consequently, whilst levels of phyA are high in dark-grown seedlings, soon after emergence into bright sunlight, which causes the conversion of much of the phytochrome pool into the Pfr form, levels of phyA decline. PhyB–phyE are relatively light stable [27].

Two cryptochromes are present in *Arabidopsis*, cry1 and cry2. The N-terminus of the cryptochromes shares strong homology with the type II photolyases responsible for UV-mediated repair of pyrimidine dimers. Each of the cryptochromes possesses a unique C-terminal extension. Like the photolyases, the cryptochromes bind two chromophores, a light absorbing pterin and a catalytic flavin. The cryptochromes also differ in both their function and their spectral sensitivity: cry1 is light-stable whilst cry2 is light labile, being rapidly degraded in high intensity blue light [28,29].

Finally, the phototropins control very specific blue light responses within the plant. *Arabidopsis* possesses two phototropins, nph1 (non-phototropic hypocotyl, named after the mutant) and npl1 (nph1-like) and, whilst the phytochromes and the cryptochromes mediate a large range of growth and developmental responses to red and blue light respectively, the phototropins appear to have very limited roles. Nph1 is involved in the phototropic response whereby plants bend towards the brightest light source [30], whilst npl1 is involved in the movement of chloroplasts within the leaf mesophyll cells in response to the amount of light available for photosynthesis [31,32].

Our knowledge of the functions of the various photoreceptors has come from the study of mutants deficient in one or more photoreceptor species, particularly in *Arabidopsis*. For example, in seedling establishment, *phyA* mutant seedlings specifically show a deficiency in response to very low fluence rates of red light. *PhyB* mutants show wild-type responses to very low fluence rates of red light but show a deficiency in responses to higher fluence rates of red light. This is consistent with a role for phyA as an antenna pigment detecting small amounts of light as a seedling begins to emerge into daylight. Once in the light such a sensitive pigment would be rapidly saturated and would no longer be useful, hence its destruction in light. At this point the less-sensitive, light-stable phyB becomes the major red light photoreceptor [33]. The identification of the photoreceptors involved in light input to the circadian clock in *Arabidopsis* was similarly achieved by studying the effect of light on the clock in the various photoreceptor mutants.

Some of the earliest work on the circadian clock was performed in plants. Androsthenes, historian to Alexander the Great, noticed that the leaves of various species of tree moved from a more horizontal position during the day to a more vertical position during the night. The astronomer, De Marian [34] first identified the action of an endogenous oscillator. He discovered that such leaf movements continued even when the plants were placed in deep shade, away from any external cues as to time of day. In 1928, Erwin Bünning discovered that pulses of red light used during watering of Phaseolus plants otherwise maintained in darkness were capable of synchronising the clock, first demonstrating light-mediated clock resetting [see 35].

More recently Millar et al. [36] devised a reporter system to analyse circadian regulated gene expression in *Arabidopsis*. They attached the promoter of the gene encoding light-harvesting chlorophyll a/b protein (better known as chlorophyll a/b binding protein, CAB) to the firefly luciferase coding sequence (*LUC*). CAB forms part of the photosynthetic machinery of the cell and shows a circadian rhythm of expression with a peak in the early part of the day. By using a highly sensitive photon-counting camera to measure the bioluminescence due to the luciferase produced, they were able to follow the rhythm of *CAB* expression in vivo. This provided a very amenable system for the study of environmental input to the circadian clock. It was demonstrated that both red and blue light were capable of mediating light input to the clock and that mutants deficient in ability to manufacture the phytochrome chromophore, phytochromobilin, were unable to fully respond to light [37]. Somers et al. [18] used this system to specifically identify the phytochrome species involved and, furthermore, demonstrated the involvement of cryptochrome in blue light input to the clock in *Arabidopsis*.

The assay of Somers et al. made use of a phenomenon known as Aschoff's rule, whereby the period length of the circadian rhythm in constant light is dependent on the fluence rate (intensity) of the light [38]. Over the course of a day in constant light, the clock will be subject to phase advances and phase delays as determined by the phase response curve. In diurnal organisms, phase advances occurring during the early part of the subjective day are greater than phase delays occurring during the late part of the subjective day and, consequently, period length decreases in constant light. With increasing fluence rate this period-shortening effect becomes progressively greater. In nocturnal organisms, phase delays are greater than phase advances and, consequently, period length increases with increasing fluence rate. Somers et al. [18] demonstrated that *Arabidopsis* obeys Aschoff's rule, behaving as a typical diurnal organism, in that period length decreases with increasing fluence rate.

Previously, Miller et al. had demonstrated that both red and blue light were capable of causing a shortening of period length of the *CAB::LUC* rhythm relative to darkness in *Arabidopsis,* suggesting that both red and blue light photoreceptors were involved [37]. Somers et al. [18] analysed the response of mutants deficient in one or more photoreceptor species to increasing fluence rate of either red or blue light. It was found that *phyA* mutants specifically showed a deficiency in the perception of low fluence rate red light whilst *phyB* mutants showed a specific deficiency in the perception of higher fluence rates of red light. As in seedling establishment, this demonstrates a plasticity in recruitment of photoreceptors depending on the light environment. An additivity between the *phyA* and *phyB* mutant phenotypes was very recently demonstrated using the *phyA phyB* double mutant. *PhyA phyB* shows a deficiency in the perception of both low and high fluence rates of red light [39].

Mutants deficient in phyD and phyE have also been isolated. Both phyD and phyE show a strong conditional redundancy with phyB in the regulation of growth and development. Consistent with this, monogenic mutants of phyD and phyE show no effect on the period length of the clock in constant red light. However, the *phyA phyB phyD* triple mutant showed a deficiency in the perception of high fluence rate red light relative to the *phyA phyB* double mutant. The *phyA phyB phyE* triple mutant also showed a deficiency in the perception of high fluence rate red light relative to the *phyA phyB* double mutant. Hence phyD and phyE play a role in the perception of high fluence rate red light along with phyB [39]. Significantly, a response to fluence rate was still observed in the *phyA phyB phyD* and *phyA phyB phyE* triple mutants, suggesting the action of other phytochromes. Whether this represents the action of phyC awaits the creation of the *phyA phyB phyD phyE* quadruple mutant.

The *phyA* mutant also showed a deficiency in the perception of low fluence rate blue light. The absorption spectrum for phytochrome shows a peak in the blue region of the spectrum and *phyA* mutants have previously been shown to display a deficiency in response to blue light in seedling establishment. Although only a small amount of Pfr would be formed under blue light, the sensitivity to small amounts of phyA Pfr is such that a response is triggered. No effect of phyB deficiency was observed in blue light [39].

Both cry1 and cry2 were also demonstrated to play a role in blue light input to the clock. The *cry1* mutant shows a deficiency in perception of both low and high fluence rates of blue light but displays a wild-type phenotype at intermediate fluence rates. The *cry2* monogenic mutant shows a wild-type response to blue light. However, when the *cry1 cry2* double mutant was examined, a redundancy between cry1 and cry2 is revealed at intermediate fluence rates. Both cry1 and cry2 mediate blue light input to the clock over this fluence rate range and each can compensate for the loss of the other. At higher fluence rates of blue light cry2 would be degraded, leaving cry1 as the main blue light photoreceptor, explaining the phenotype observed in the *cry1* mutant at high fluence rates of blue [39]. A similar situation is observed in the action of the cryptochromes in seedling establishment. Both cry1 and cry2 act as photo-receptors at lower fluence rates of blue light whilst only cry1 is important at higher fluence rates of blue, where cry2 would be degraded [40]. Again, this demonstrates a plasticity in recruitment of photoreceptors as an adaptation to the light environment. In dim light greater sensitivity is achieved by the combined action of the two cryptochromes.

One further interesting point emerged from this study in that the cry1 mutant shows a deficiency in the perception of low fluence rate red light. The phyto-chrome and cryptochrome mutants show no phenotype in darkness, indicating that this is a light-dependent phenotype. However, cryptochrome shows no peak of absorption in the red region of the spectrum, suggesting cry1 is not acting as a photoreceptor in this response. The deficiency in response to red light shown by the cry1 mutant resembles that seen in the phyA mutant, suggesting that cry1 may be acting as a signal transduction component downstream of phyA. In white light, where both cryptochrome and phytochrome can act as photo-receptors, no additivity is observed between the *phyA* and *cry1* mutations, indi-cating that cry1 does not act as a photoreceptor in its own right at low fluence rates of either red or blue and confirming the proposal that cry1 acts as a signal transduction component downstream of phyA [39].

In summary, phyA is the low fluence rate photoreceptor for light input to the clock whilst cry1 is necessary for this phyA signalling to the clock. Phyto-chromes B, D and E act as photoreceptors for higher fluence rate red light-input, whilst cry1 and cry2 act as photoreceptors for higher fluence rate blue light input. The *nph1* mutant of *Arabidopsis* showed a wild-type response for period length under constant blue light, suggesting that nph1 is not involved in light input to the clock. However, it remains possible that nph1 may act redundantly with other blue-light photoreceptors (Figure 3).

12.5   Circadian photoreceptors as part of the clock output pathway in plants

Recent work has demonstrated that the circadian photoreceptors in plants are part of the output pathway from the clock as well as part of the input pathway. Bognar et al. demonstrated a circadian rhythm in expression of the *PHYB* gene [41], whilst Harmer et al. demonstrated that both phyB and the two cryptochromes, cry1 and cry2, show a cycling of transcript level [42]. Hence the levels of all of the major higher fluence rate photoreceptors, those still present following initial seedling emergence, are regulated by the clock. It is possible that the cycling of the photoreceptors involved in light input contri-butes to the gating or modulation of light input to the clock, the phenomenon which results in the dead-zone for the effect of light on clock resetting observed during subjective day in many diurnal organisms.

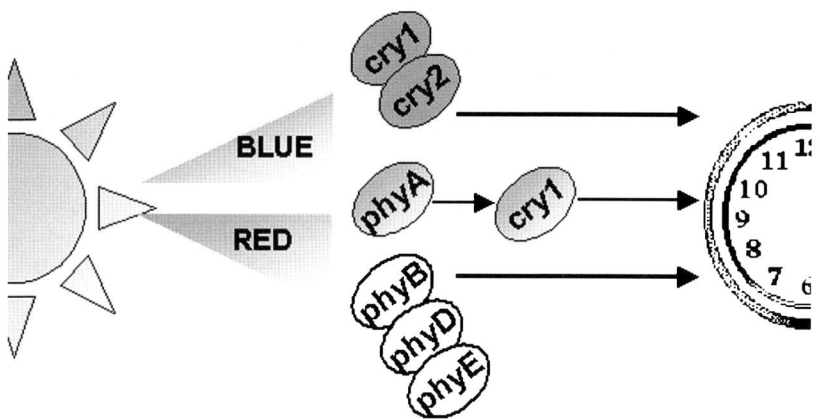

Figure 3. Photoreceptors mediating light input to the circadian clock in *Arabidopsis thaliana*. Phytochrome A mediates low fluence rate red and blue light input to the clock. Cryptochrome 1 and cryptochrome 2 mediate higher fluence rate blue light input, whilst phytochromes B, D and E mediate higher fluence rate red light input. Cryptochrome 1 is necessary for phytochrome A signalling to the clock.

Gating is also seen for other light regulated phenomena. *CAB* gene expression is regulated by light as well as by the clock. In wild-type seedlings, an acute induction of *CAB* expression in response to light is gated by the clock. When seedlings are transferred to constant darkness after entrainment in light/dark cycles, they show a rhythm in the degree of acute *CAB* induction by a light pulse. This rhythm shows a peak of induction early in the subjective day, coinciding with the normal circadian peak of *CAB* expression [43]. This finding has led to the proposal that the circadian regulation of *CAB* may, at least in part, be mediated by gating of the light signal (Figure 4). Recent analysis of an arrhythmic mutant of *Arabidopsis* called *elf3* (*early flowering3*) has revealed a component of this gating mechanism. The *elf3* mutant was demonstrated to be disrupted in the pathway by which light input is gated, and it is the absence of gating in this mutant which results in its arrhythmic phenotype. In *elf3*, acute, light-mediated induction of *CAB* expression continually occurs in the light, with the result that levels of *CAB* transcription are high throughout the day [44]. Clock resetting also continually occurs throughout the day in *elf3* so that the clock effectively stops whilst the seedlings are in the light. Following the light-to-dark transition the clock starts running normally again. Thus, in *elf3*, the clock is always set to the same circadian time by the light/dark transition regardless of the actual timing of dusk [44]. In a wild-type seedling, the normal circadian cycle continues through the light/dark transition. The normal ELF3 protein clearly plays an important role in photoreceptor input to the clock.

Very recently the mechanism by which these photoreceptors act to reset the plant clock has begun to be uncovered. A transcriptional feedback loop

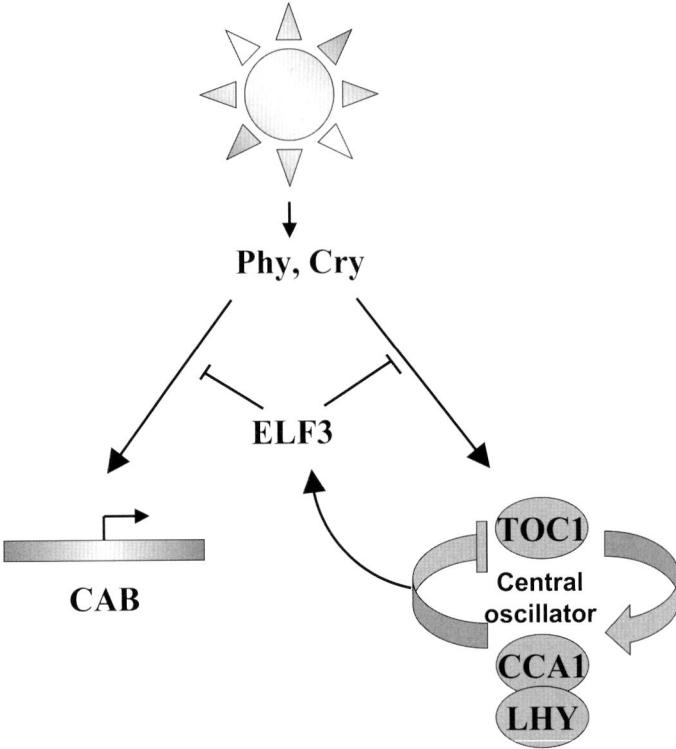

Figure 4. ELF3 is involved in the gating of light signalling in *Arabidopsis thaliana*. Light signals detected by the phytochrome and cryptochrome photoreceptors (Phy, Cry) act to induce expression of light-regulated genes such as chlorophyll a/b binding protein (CAB), as well as acting to reset the clock (Central oscillator). The ELF3 protein acts in the output pathway from the clock to modulate light signalling such that light-regulated gene expression or light-mediated clock resetting are only responsive to light at certain times during the day.

consisting of the proteins TIMING OF CAB 1 (TOC1), LATE ELON-GATED HYPOCOTYL (LHY) and CIRCADIAN CLOCK ASSOCIATED 1 (CCA1), has been demonstrated to be critical for clock function in *A. thaliana*. CCA1 and LHY oscillate with a peak at dawn, whilst TOC1 oscillates with a peak at dusk. It was recently demonstrated that TOC1 is responsible for the positive regulation of *CCA1* and *LHY* expression whilst both LHY and CCA1 bind to the *TOC1* promoter to negatively regulate *TOC1* expression, thus forming a self perpetuating oscillator [17a] (Figure 4).

Both *CCA1* and *LHY* show a pronounced increase in expression in response to light, regulated by phytochrome [44a]. Such a system would allow a pulse of light to reset the clock by triggering a change in *CCA1* and *LHY* message levels.

12.6   Photoreceptors mediating light input to the clock in fungi

The fungus, *Neurospora crassa*, was one of the first model systems in which a genetic analysis of the circadian clock was carried out. Although no photoreceptors have been clearly defined, it is known that all responses to light in *Neurospora* involve the action of two proteins, white collar 1 and white collar 2 (WC1 and WC2) [45–49]. WC1 and WC2 are members of the GATA family of transcription factors [50,51]. They contain a PAS/LOV domain similar to the domains which bind flavin-based chromophores in the phototropin photoreceptors in plants. It has been proposed that WC1 and WC2 may, similarly, bind a flavin chromophore and may, themselves, be photoreceptors. The *wc1* and *wc2* mutants of *Neurospora* are arrhythmic in constant darkness, indicating their close involvement with the circadian clock [48]. *Neurospora* shows a circadian rhythm of conidiation (asexual spore production). When inoculated onto one end of a tube, known as a race tube, the fungus will proceed to produce mycelia, growing at a constant rate along the tube. Approximately every 24 h this growth is punctuated by production of bands of conidia and measurement of this banding pattern allows the circadian rhythm to be monitored [52].

A screen for aberrant circadian rhythm in *Neurospora* yielded several mutations which mapped to the frequency (*frq*) locus [53,54]. *Frq* null mutants are arrhythmic in constant light, although both long period and short period alleles of *frq* were also identified. *Frq* forms part of a central oscillator in *Neurospora*: a transcriptional feedback loop generates a self-sustaining oscillation in FRQ protein levels that is essential for the observed rhythmic conidiation in *Neurospora* [55]. The WC1 and WC2 proteins were found to form a transcriptional activation complex, the white collar complex (WCC), that activates transcription of the *frq* gene [4,56,57]. As levels of FRQ protein rise, FRQ binds to the WCC and inhibits its action, thus suppressing *frq* transcription [57]. Subsequently, levels of FRQ fall again and, eventually, the FRQ-mediated inhibition of the WCC can no longer occur, allowing *frq* transcription to begin again (Figure 5).

Curiously, in the *frq* null mutant, a rhythm of conidiation can be entrained by temperature [58]. This possibly suggests the presence of a second, *frq*-less oscillator also operating within *Neurospora*, although light entrainment of this second oscillator is not possible.

Clock resetting by light in *Neurospora* involves the light-mediated induction of *frq* transcript. Induction of *frq* transcript phase shifts the clock to a point at which *frq* transcript is normally high and the cycle then continues from this point [59]. WC1 is essential for the light-induced transcription of *frq* which mediates clock resetting though it remains uncertain whether WC1 is itself the photoreceptor [60].

A recent study has examined the phenomenon of gating in *Neurospora*. At sub-saturating light levels, a modulation or gating of the light-induction of *frq* transcript can be observed [61]. A protein called VIVID (VVD) was demonstrated to be involved in this phenomenon. The *vvd* mutant of *Neurospora* was isolated as a high pigment mutant [62]. It shows a hypersensitivity to light for

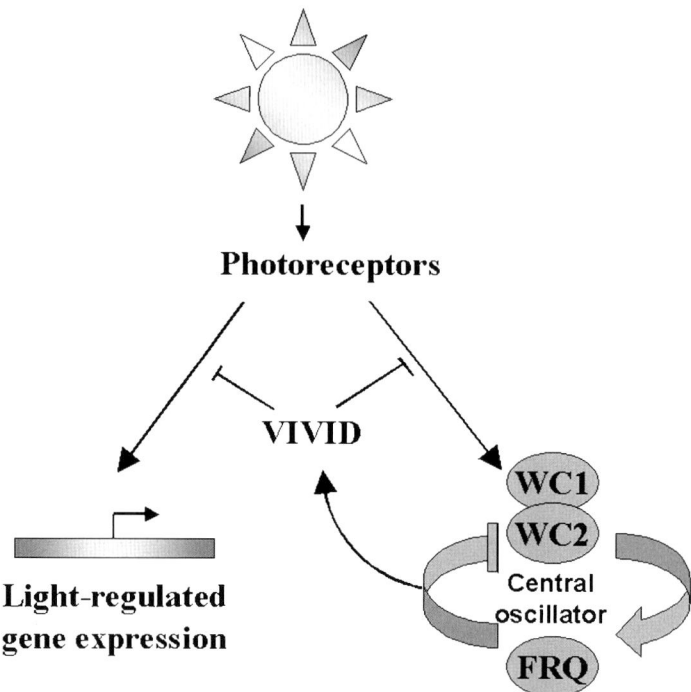

Figure 5. The Central oscillator in the fungus, *Neurospora crassa*, consists of a feedback loop involving the proteins FREQUENCY (FRQ), WHITE COLLAR 1 (WC1) and WHITE COLLAR 2 (WC2). WC1 and WC2 form a transcription-activating white collar complex (WCC) which promotes transcription of *frq*. As FRQ protein accumulates, FRQ interacts with WCC, negating the action of WCC and, thus, inhibiting *frq* transcription. Light signals reset the clock by directly inducing *frq* gene expression. VIVID is involved in the gating of this light signal. The VIVID protein acts in the output pathway from the clock to modulate the effectiveness of the light signal such that *frq* expression and clock resetting are only responsive to light at certain times during the day.

induction of carotenoid biosynthesis, suggesting that VVD acts as a suppressor of light signalling in *Neurospora*. Transcription of *vvd* is regulated both by light and by the clock [61]. On transfer to continuous light many light-regulated genes in *Neurospora* show an acute induction of expression followed by a suppression of the response. Transcription of *vvd*, itself, is induced by light. The VVD protein then acts to suppress the light signalling pathway, making VVD a strong candidate for a component in the pathway mediating this desen-sitisation to light. Transcription of *vvd* is also clock regulated, thus modulation of light signalling by VVD will vary depending on the time of day. Such a time-of-day-dependent modulation of light signalling fulfils the definition of gating and, indeed, the involvement of VVD in gating in *Neurospora* was clearly demonstrated by Heintzen et al. [61]. The gating of the light-induction of *frq* transcript is severely reduced in the *vvd* mutant, suggesting that VVD is a key component in the gating response. VVD is, thus, part of a loop whereby the output from the clock regulates light input to the clock (Figure 5). This VVD loop exemplifies the close link between the clock and the pathways of light signalling apparent in all systems: plants, fungi, insects and mammals.

12.7   Photoreceptors mediating light input to the clock in insects

The insect *Drosophila melanogaster* formed another early model system for the genetic dissection of the circadian clock and the clock mechanism is, consequently, well understood in insects. Analysis of *Drosophila* mutants which show an altered period length for the circadian rhythm of locomotor activity [63,64] revealed two of the key components of a transcriptional feedback loop making up the insect clock. These are PERIOD (PER) and TIMELESS (TIM). Levels of PER and TIM increase during the late afternoon and early evening in the cell cytoplasm where they begin to form PER-TIM dimers. These dimers are capable of entering the nucleus where they act to repress their own transcription by inhibiting the action of a transcriptional activation complex made up of the proteins CLOCK (CLK) and CYCLE (CYC) [65–67]. Levels of PER and TIM consequently fall during the late night and early morning to levels at which they no longer inhibit their own transcription and the cycle begins again. A second, interlocked negative feedback loop causes CLK to cycle in antiphase with PER and TIM. In this, CLK feeds back as a repressor of its own transcription whilst PER and TIM act as de-repressors [68] (Figure 6).

At the time of the completion of this clock loop, despite extensive knowledge of the mechanism of the clock, little was known about the photoreceptors involved in resetting of the circadian clock in *Drosophila*. The eyes of animals have a well-characterised array of visual photoreceptors. However, disruption of vision in *Drosophila*, via a mutation which knocks out part of the pathway of signal transduction involved in visual photoreception (*NorpA^{p41}*), did not prevent circadian photoperception [69,70]. Furthermore, when isolated body parts of *Drosophila* were maintained in culture, not only was each capable of maintaining a rhythm but this rhythm could be reset by light, indicating that the circadian photoreceptor was present throughout the body of the fly [71]. The discovery of a cryptochrome-like molecule in *Drosophila* provided a new candidate for the circadian photoreceptor.

Following the discovery of cryptochrome in plants, similar molecules were discovered in animals, in insects and mammals [72–74], bearing strong homology to the photolyase family of photoreceptors but showing no photolyase activity [75]. The animal cryptochromes, like the plant cryptochromes, possess an N-terminal domain very closely related to the photolyase molecule, which binds both the pterin and flavin chromophores, and a unique C-terminal extension [76,28]. However, whilst the plant cryptochromes are most closely related to the type I photolyases, involved in repair of pyrimidine dimers formed as a result of UV light damage to DNA, the animal cryptochromes more closely

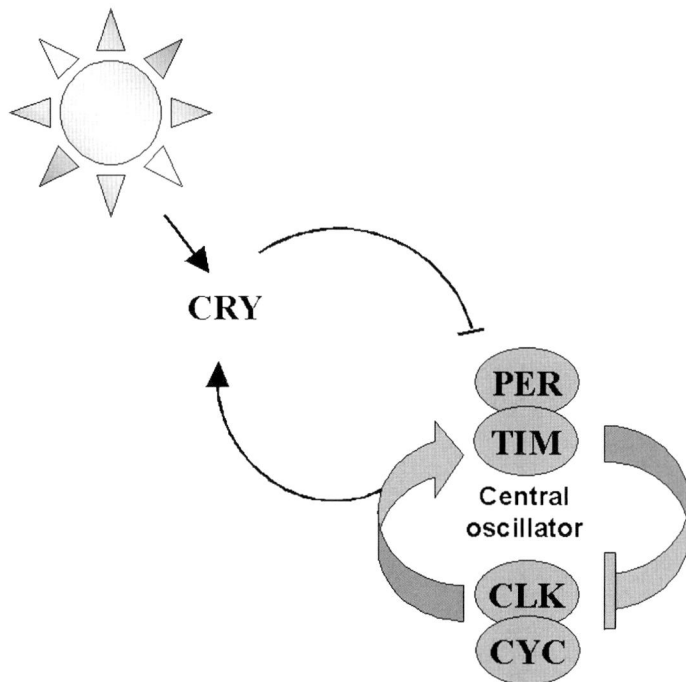

Figure 6. The central oscillator in the fruit fly, *Drosophila melanogaster*, consists of a feedback loop involving the proteins PERIOD (PER), TIMELESS (TIM), CLOCK (CLK) and CYCLE (CYC). CYC and CLK form a transcription-activating complex which promotes transcription of *per* and *tim*. As PER and TIM proteins accumulate in the cytoplasm they dimerise and re-enter the nucleus where they inhibit the action of the CLK-CYC complex, thus inhibiting their own transcription. In response to light, the photoreceptor, cryptochrome (CRY), interacts with TIM, inhibiting the action of the PER-TIM dimers. PER and TIM then no longer inhibit their own transcription and the clock is reset to a point at which *per* and *tim* transcription is high. Expression of the *cry* gene is, itself, regulated by the clock. It is possible that the clock modulates its own response to light by regulating the level of the photoreceptor, CRY.

resembled the 6-4 photolyases, involved in repair of 6-4 photoproducts [72]. A study of the phylogeny of the plant and animal cryptochromes revealed that the plant cryptochromes had probably diverged from the type I photolyases prior to the divergence of plants and animals. Animal cryptochrome probably independently diverged from 6-4 photolyase after the divergence of plants and animals, implying that cryptochromes related to type I photolyase have subsequently been lost in animals [29]. Although both plant and animal crypto-chromes act as circadian photoreceptors, this independent origin of crypto-chrome in plants and animals is consistent with an independent origin of the clock mechanism in plants and animals. None of the molecules involved in the *Drosophila* clock are present in *Arabidopsis*. The involvement of molecules derived from photolyases as circadian photoreceptors in both plants and animals appears to be a case of convergence.

The speculation as to cryptochrome being a good candidate for the circadian photoreceptor in *Drosophila* was supported by the finding that cryptochrome is expressed throughout the fly's body [74]. Proof of its role in light input to the clock in *Drosophila* came from studies of cryptochrome null mutants and cryptochrome overexpressors. Stanewsky et al. [19] designed a highly automated screen for new circadian rhythm mutants in *Drosophila* using a *per-luc* reporter construct. Wild-type flies show a peak of expression of the *per-luc* construct during the night consistent with the rhythm of expression of the endogenous *per* gene generated by the clock loop. A mutation called *cry^{baby}* (*cry^b*) was identified that resulted in an arrhythmic expression of *per-luc* in light/dark cycles. The mutation mapped to the position of the *Drosophila* cryptochrome (*dcry*) gene and sequencing identified a missense mutation in the C-terminal region of the protein. In light/dark cycles, the *cry^b* mutant fails to synchronise and is arrhythmic for *per* and tim expression. *Cry^b* flies also fail to show any response to pulses of light for phase shifting. However, in constant darkness, *per* and *tim* expression can be synchronised to temperature cycles, suggesting that the clock itself was not disrupted. Curiously, behavioural rhythms can be observed in *cry^b* flies entrained to light/dark cycles and this phenomenon was correlated with the persistence of a rhythm of *per* and *tim* expression in the lateral neurons. When the *cry^b* and *NorpA^{p41}* mutations were combined both the behavioural rhythm and the rhythm of *per* and *tim* expression in the lateral neurons was abolished, suggesting that some combination of visual input and light input via cryptochrome mediates entrainment in *Drosophila*. However, experiments similar to those used to identify the circadian photoreceptors in plants have subsequently demonstrated that cryptochrome is the key circadian photoreceptor in *Drosophila*. Wild-type *Drosophila* obey Aschoff's rule: in constant light, the circadian period length is dependent on the light intensity. Like other arthropods, *Drosophila* display a lengthening of circadian period with increasing light intensity to the extent that they become arrhythmic in bright light. The *cry^b* mutant fails to become arrhythmic in constant bright light and maintains a normal, wild-type rhythm, suggesting that there are no light signals reaching the clock in *cry^b*. Expression of *dcry* in lateral neuron cells of *cry^b* flies restored a wild-type response [77]. It is proposed that the behavioural rhythm observed in *cry^b* flies is initiated by the flies merely responding to the light/dark transitions and that this response eventually feeds back to entrain the clock.

Overexpressors of DCRY were also generated and were found to be disrupted in light signalling to the clock, although the responses of these overexpressors are somewhat contradictory. Emery et al. [74] observed an enhanced response to light pulses for phase shifting in overexpressors of *dcry*. Ishikawa et al. [78] observed a decreased response.

The action of cryptochrome in mediating light input to the *Drosophila* clock involves a direct interaction of *dcry* with the components of the clock mechanism itself. DCRY will bind to TIM in a light-dependent manner and prevent the action of the PER-TIM dimer in suppressing the *per* and *tim* transcription [79]. Thus, in response to light, the clock is re-set to the point at which *per* and *tim* transcription is de-repressed and *per* and *tim* transcripts begin to accumulate.

Like the plant cryptochromes, *dcry* shows a circadian rhythm of transcription with a peak of expression in the late part of the day. As in plants, an oscillation of levels of CRY may contribute to a modulation of light signalling over the course of a day [78].

12.8   Photoreceptors mediating light input to the clock in mammals

Despite the vast amount of research into mammalian visual photoperception [80], circadian photoperception in mammals remained something of a mystery until very recently. Enucleation, removal of the eyes, results in a loss of the ability to synchronise to light/dark cycles [81]. However, when both rods and cones are ablated by mutation, normal entrainment can still occur, indicating that the visual opsins are not required for circadian photoperception [82].

In mammals the clock appears to be centrally controlled by signals from the suprachiasmatic nucleus (SCN), a region of the hypothalamus [83]. A direct connection from the eyes to the SCN exists via the retinohypothalamic tract (RHT). Viral tract tracing experiments have demonstrated that the axons of the RHT have extensive dendritic arbours, diffusely branched over a very large area of the retinal surface [84]. Mammalian cryptochrome is highly expressed throughout the retinal ganglion cells spread evenly across the inner nuclear layer of the retina, and thus cryptochrome formed a strong initial candidate for the photoreceptor [85]. However, the peak of the action spectrum for clock resetting in mammals more closely resembles the spectrum of an opsin (~500 nm) [86,87] rather than that of a cryptochrome (370–440 nm) [88].

Attempts to find a role for cryptochrome in circadian photoperception in mammals revealed that cryptochrome in fact plays a role as a component of the central oscillator in mammals. The mammalian clock consists of a transcriptional feedback loop related to that found in flies. However, it is apparent that CRY replaces TIM within the clock loop in mammals [89]. Mice, like other mammals, possess two cryptochromes, mCRY1 and mCRY2. A CLK-CYC transcriptional activation complex promotes transcription of the two mammalian *Cry* genes and three mammalian *Per* genes [10] (Figure 7). CRY-PER dimers and CRY-CRY dimers form in the cytoplasm and then enter the nucleus where they inhibit the action of the CLK-CYC complex, inhibiting expression of the *Cry* and *Per* genes. *mCry1*[-/-] *mCry2*[-/-] double mutant mice, consequently, show an absence of any circadian rhythm [90].

Experiments using *mCry1*[-/-] *mCry2*[-/-] double mutant mice to investigate whether cryptochrome also plays a role in light input to the clock are not possible given the absence of a functioning clock to test in these mutants. However, a few pieces of evidence do point to some role for cryptochrome in clock resetting by light. Firstly, monogenic mutants deficient in mCRY2 show

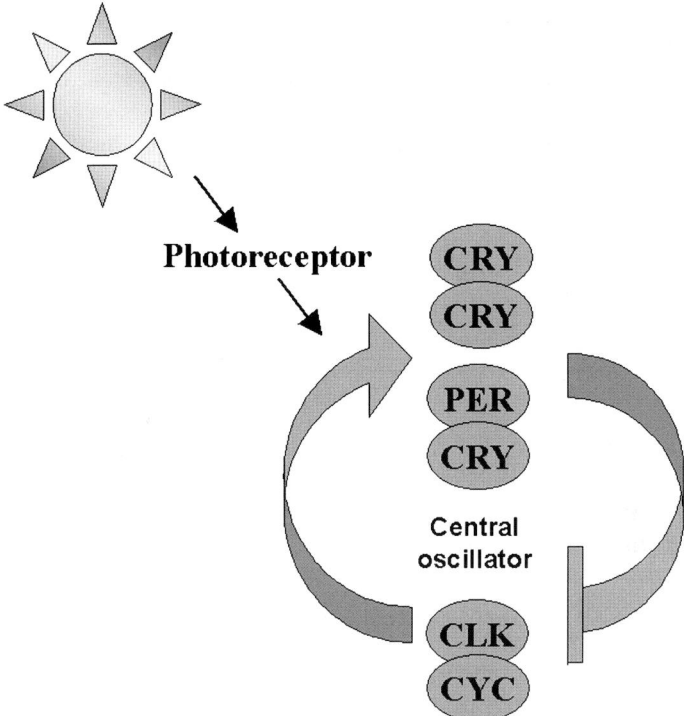

Figure 7. The central oscillator in the mouse, *Mus musculus*. The clock consists of a feedback loop involving the proteins CRYPTOCHROME (CRY), PERIOD (PER), CLOCK (CLK) and CYCLE (CYC). CYC and CLK form a transcription-activating complex which promotes transcription of *per* and *cry*. As PER and CRY proteins accumulate in the cytoplasm they dimerise and re-enter the nucleus where they inhibit the action of the CLK-CYC complex, thus inhibiting their own transcription. Light-mediated clock resetting involves the induction of one of three *per* genes, *mPer1*. Whilst the photoreceptor involved remains unknown, evidence suggests some involvement of the cryptochromes, mCRY1 and mCRY2, as photoreceptors in this response.

an enhanced response to a phase-shifting light pulse during the subjective night [91]. Secondly, in *mCry1*[-/-] *mCry2*[-/-] double mutant mice maintained in light/dark cycles, light pulses given during the dark period fail to induce *mper1* transcript, a key light-regulated step in clock resetting in wild-type mice [92]. It was noted, however, that if *mCry1*[-/-] *mCry2*[-/-] double mutant mice are maintained in darkness for 52 h prior to a light pulse, light-mediated induction of *mPer1* is still observed [93], perhaps suggesting the involvement of a photoreceptor that accumulates slowly in darkness.

Furthermore, a role for cryptochrome as a non-circadian photoreceptor in mice has been demonstrated in the perception of light signals that trigger behavioural responses. As well as displaying a circadian regulation of activity, mice will respond to light and darkness, becoming inactive at times when the light is on. A study of rd[-/-] *mCry1*[-/-] *mCry2*[-/-] triple mutant mice, lacking rods and most cones in the retina, as well as the two cryptochromes, shows that these triple mutants no longer display this modification of behaviour in response to light. rd$^{-/-}$ monogenic mutants or $mCry1^{-/-}$ $mCry2^{-/-}$ double mutants still show a normal response, suggesting a redundancy between the visual photoreceptors and cryptochrome in the perception of light mediating this response [94].

However, the conjecture as to nature of the mammalian circadian photoreceptor was finally solved following the discovery of a novel opsin, melanopsin, in *Xenopus laevis* and in Salmon [95,96]. Melanopsin was subsequently discovered in mammals in retinal ganglion cells, the very cells shown to link to the SCN via the RHT. It was demonstrated that Melanopsin was required for normal light-induced circadian phase shifting in mice [96a] and that mutant mice lacking rods, cones and melanopsin fail to entrain at all to light/dark cycles [96b] indicating that together melanopsin and the visual photoreceptors can entirely account for circadian photoperception in mice.

12.9   Discussion

The identification of photoreceptors resetting the circadian clock and the elucidation of their mechanism of action continues to be an exciting area of research. The earliest genetic dissections of the clock were carried out in *Drosophila* and in *Neurospora* following the isolation of circadian clock mutants over twenty years ago. Since that time, *Drosophila* and *Neurospora* have continued to serve as excellent model systems for circadian research. Both the *Drosophila* and *Neurospora* clocks have simple mechanisms and direct links between the photoreceptors mediating clock resetting and the clock itself have been established for each. Mammalian and plant systems appear more complicated, although many parallels with *Drosophila* and *Neurospora* are apparent.

In mammals, the mechanism of the clock is well established. However, the photoreceptors mediating light input to the clock are less well understood. In humans, with increasing international travel, jetlag is becoming more of a problem. Dangers associated with shift work are becoming more apparent, particularly the increased accident risk due to tiredness [97]. Sleep disorders such as advanced sleep phase syndrome or delayed sleep phase syndrome can lead to extreme tiredness or psychological problems due to being "out of synch" with the rest of the world [98–100]. Similarly, in a number of blind people, the absence of light cues means that their circadian clock "free runs" rather than being reset slightly to the correct time each day and they soon become desynchronised with the cycle of day and night [101,102].

The hormone melatonin has been proposed to be an early target for light-mediated clock resetting in mammals. Melatonin has been used to successfully synchronise the circadian rhythm in blind people who are otherwise "free running" and to restore a normal sleep pattern. However, the mechanism of clock resetting remains unclear and the effects of melatonin application have only been observed in a small number of subjects [103]. A full understanding of the synchronisation of the clock would greatly improve the quality of life for many people.

Within plants, a great deal is known about the photoreceptors involved in clock resetting and, in fact, plants formed the first system in which specific photoreceptors involved in light input to the clock were identified. However, with only an initial picture of the central clock mechanism, the mode of action of these photoreceptors in resetting the clock is less well understood. Germination, flowering and bud dormancy are all determined by the photoperiod and this, in turn, determines the growing season and the latitude at which crops can be grown [7]. The measurement of photoperiod involves not only the correct synchronisation of the clock with the day/night cycle but also an interaction between the clock and the light signalling pathways to determine whether light is present at a given time of day. Here too research is proceeding apace. A recent paper by Suarez-Lopez identifies the expression of the CONSTANS gene as a key target of the mechanism of measuring daylength, downstream of the integration of signals from the photoreceptors and from the circadian clock [104].

Clearly, an understanding of the mechanism of clock resetting by light in mammals and plants could have great significance for us all and a great deal of further research is called for.

References

1. A.J. Millar, S.A. Kay (1991). Circadian control of *cab* gene transcription and mRNA accumulation in *Arabidopsis. Plant Cell*, **3**, 541–550.
2. C. Darwin. *The Power of Movement in Plants* [*1895*], (1981), D. Appleton and Co., New York.
3. J.T. Enright (1982). Sleep movements of leaves: in defense of Darwin's interpretation. *Oecologia*, **54**, 253–259.
4. C.S. Pittendrigh (1954). On temperature independence in the clock system controlling emergence time in *Drosophila. Proc. Natl. Acad. Sci. U.S.A.*, **40**, 1018–1029.
5. C.M. Moore-Ede, F.M. Sulzman, C.A. Fuller (1982). *The Clocks That Time Us.* Harvard University Press, Cambridge.
6. T. Roenneberg, R.G. Foster (1997). Twilight times: Light and the circadian system. *Photochem. Photobiol.*, **66**, 549–561.
7. B. Thomas, D. Vince-Prue (1997). *Photoperiodism in Plants* (2nd Edn). Academic Press, London.
8. P.F. Devlin (2002). Signs of the time – Environmental input to the circadian clock. *J. Exp. Bot.*, **53**, 1535–1550.
9. K. Wager-Smith, S.A. Kay (2000). Circadian Rhythm Genetics: from files to mice to humans. *Nature Gen.*, **26**, 23–27.
10. S.M. Reppert, D.R. Weaver (2001). Molecular analysis of mammalian circadian rhythms. *Annu. Rev. Physiol.*, **63**, 647–676.
11. J.A. Williams, A. Sehgal (2001). Molecular components of the circadian system in Drosophila. *Annu. Rev. Physiol*, **63**, 729–755.
12. C.H. Johnson (2001). Endogenous timekeepers in photosynthetic organisms. *Annu. Rev. Physiol.*, **63**, 695–728.

13. J.J. Loros, J.C. Dunlap (2001). Genetic and molecular analysis of circadian rhythms in *Neurospora. Annu. Rev. Physiol.*, **63**, 757–794.

14. C.H. Johnson (1990). PRC Atlas. http://johnsonlab.biology.vanderbilt.edu/prcatlas/prcatlas.html

15. P.F. Devlin, S.A. Kay (2001). Circadian photoperception. *Annu. Rev. Physiol.*, **63**, 677–694.

16. C. Strayer, T. Oyama, T.F. Schultz, R. Raman, D.E. Somers, P. Mas, S. Panda, J.A. Kreps, S.A. Kay (2000). Cloning of the Arabidopsis clock gene TOC1, an autoregulatory response regulator homolog. *Science*, **289**, 768–771.

17. D.E. Somers, T.F. Schultz, M. Milnamow, S.A. Kay (2000). *ZEITLUPE*, a novel clock associated PAS protein from *Arabidopsis. Cell*, **101**, 319–329.

17a. D. Alabadi, T. Oyama, M.J. Yanovsky, F.G. Harmon, P. Mas, S.A. Kay (2001). Reciprocal regulation between TOC1 and LHY/CCA1 within the Arabidopsis circadian clock. *Science.*, **293**, 880–883.

18. D.E. Somers, P.F. Devlin, S.A. Kay (1998). Phytochromes and cryptochromes in the entrainment of the Arabidopsis circadian clock. *Science*, **282**, 1488–1490.

19. R. Stanewsky, M. Kaneko, P. Emery, B. Beretta, K. Wager-Smith, S.A. Kay, M. Rosbash, J.C. Hall (1998). The *cry^b* mutation identifies cryptochrome as a circadian photoreceptor in Drosophila. *Cell*, **95**, 681–692.

20. G.C. Whitelam, P.F. Devlin (1998). Light signalling in *Arabidopsis. Plant Physiol. Biochem.*, **36**, 125–133.

21. Chapters 5 and 6, this volume

22. Chapter 10, this volume

23. Chapter 9, this volume

24. R.A. Sharrock, P.H. Quail (1989). Novel phytochrome sequences in *Arabidopsis thaliana*: Structure, evolution, and differential expression of a plant regulatory photoreceptor family. *Genes Dev.*, **3**, 1745–1757.

25. T. Clack, S. Mathews, R.A. Sharrock (1994). The phytochrome apoprotein family in *Arabidopsis* is encoded by five genes: The sequences and expression of *PHYD* and *PHYE. Plant Mol. Biol.*, **25**, 413–427.

26. P.H. Quail (1997). An emerging molecular map of the phytochromes. *Plant Cell Environ.*, **20**, 657–665.

27. M. Hirschfeld, J.M. Tepperman, T. Clack, P.H. Quail, R.A. Sharrock (1998). Coordination of phytochrome levels in *phyB* mutants of Arabidopsis as revealed by apoprotein-specific monoclonal antibodies. *Genetics*, **149**, 523–535.

28. P.F. Devlin, S.A. Kay (1999). Cryptochromes–bringing the blues to circadian rhythms. *Trends. Cell Biol.*, **9**, 295–298.

29. A.R. Cashmore, J.A. Jarillo, Y.J. Wu, D. Liu (1999). Cryptochromes: Blue light receptors for plants and animals. *Science*, **284**, 760–765.

30. J.M. Christie, P. Reymond, G.K. Powell, P. Bernasconi, A.A. Raibekas, E. Liscum, W.R. Briggs (1998). *Arabidopsis* NPH1: A flavoprotein with the properties of a photoreceptor for phototropism. *Science*, **282**, 1698–1701.

31. T. Kagawa, T. Sakai, N. Suetsugu, K. Oikawa, S. Ishiguro, T. Kato, S. Tabata, K. Okada, M. Wada (2001). Arabidopsis NPL1: a phototropin homolog controlling the chloroplast high-light avoidance response. *Science*, **291**, 2138–2141.

32. J.A. Jarillo, H. Gabrys, J. Capel, J.M. Alonso, J.R. Ecker, A.R. Cashmore. (2001) Phototropin-related NPL1 controls chloroplast relocation induced by blue light. *Nature*, **410**, 952–954.

33. G.C. Whitelam, S. Patel, P.F. Devlin (1998). Phytochromes and photomorphogenesis in *Arabidopsis. Philos. Trans. R. Soc. Lond. [Biol.]*, **353**, 1445–1453.

34. J. de Marian (1729). Observation botanique. *Histoire de l'Academie Royale des Sciences.* 35–36.

35. E. Bünning (1970). Potato cellars, trains, and dreams: discovering the biological clock. In: F.B. Salisbury, C.W. Ross (Eds), *Plant Physiology*, (3rd Edn, pp. 396–397). Wadsworth, Belmont, CA.

36. A.J. Millar, S.R. Short, N.-H. Chua, S.A. Kay (1992). A novel circadian phenotype based on firefly luciferase expression in transgenic plants. *Plant Cell*, **4**, 1075–1087.

37. A.J. Millar, M. Straume, J. Chory, N.-H. Chua, S.A. Kay (1995). The regulation of circadian period by phototransduction pathways in Arabidopsis. *Science*, **267**, 1163–1166.

38. J. Aschoff (1979). Circadian rhythms: influences of internal and external factors on the period measured in constant conditions. *Z. Tierpsychol.*, **49**, 225–249.

39. P.F. Devlin, S.A. Kay (2000). Cryptochromes are required for phytochrome signaling to the circadian clock but not for rhythmicity. *Plant Cell*, **12**, 2499–2510.

40. C. Lin, H.Y. Yang, H.W. Guo, T. Mockler, J. Chen, A.R. Cashmore (1998). Enhancement of blue-light sensitivity of Arabidopsis seedlings by a blue light receptor cryptochrome 2. *Proc. Natl. Acad. Sci. U.S.A.*, **95**, 2686–2690

41. L.K. Bognar, A. Hall, E. Adam, S.C. Thain, F. Nagy, A.J. Millar (1999). The circadian clock controls the expression pattern of the circadian input photoreceptor, phytochrome B. *Proc. Natl. Acad. Sci. U.S.A.*, **96**, 14652–14657.

42. S.L. Harmer, J.B. Hogenesch, M. Straume, H.S. Chang, B. Han, T. Zhu, X. Wang, J.A. Kreps, S.A. Kay (2000). Orchestrated transcription of key pathways in Arabidopsis by the circadian clock. *Science*, **290**, 2110–2113.

43. A.J. Millar, S.A. Kay (1996). Integration of circadian and phototransduction pathways in the network controlling *CAB* gene transcription in *Arabidopsis. Proc. Natl. Acad. Sci. U.S.A.*, **93**, 15491–15496.

44. H.G. McWatters, R.M. Bastow, A. Hall, A.J. Millar (2000). The ELF3 Zeitnehmer regulates light signalling to the circadian clock. *Nature*, **408**, 716–720.

44a. J.F. Martinez-Garcia, E. Huq, P.H. Quail (2000). Direct targeting of light signals to a promoter element-bound transcription factor. *Science.*, **288**, 859–863.

45. G. Arpaia, J.J. Loros, J.C. Dunlap, G. Morelli, G. Macino (1993). The interplay of light and the circadian clock. *Plant Physiol.*, **102**, 1299–1305.

46. F.R. Lauter, V.E Russo (1991). Blue light induction of conidiation-specific genes in *Neurospora crassa. Nucleic Acids Res.*, **19**, 6883–6886.

47. T. Sommer, J.A. Chambers, J. Eberle, F.R. Lauter, V.E. Russo (1989). Fast light-regulated genes of *Neurospora crassa. Nucl Acids Res.*, **17**, 5713–5723.

48. V.E. Russo (1988). Blue light induces circadian rhythms in the *bd* mutant of Neurospora: double mutants *bd,wc-1* and *bd,wc-2* are blind. *J. Photochem. Photobiol.*, **2**, 59–65.

49. R.W. Harding, S. Melles (1983). Genetic analysis of phototropism of *Neurospora crassa* perithecial beaks using white collar and albino mutants. *Plant Physiol.*, **72**, 996–1000.

50. H. Linden, G. Macino (1997). White collar 2, a partner in blue-light signal transduction, controlling expression of light-regulated genes in *Neurospora crassa. EMBO J.*, **16**, 98–109.

51. P. Ballario, P. Vittorioso, A. Magrelli, C. Talora, A. Cabibbo, G. Macino (1996). White collar-1, a central regulator of blue light responses in *Neurospora*, is a zinc finger protein. *EMBO J.*, **15**, 1650–1657.

52. J.F. Feldman (1983). Genetics of circadian clocks. *Bioscience*, **33**, 426–431.

53. J.F. Feldman, M.N. Hoyle (1973). Isolation of circadian clock mutants of *Neurospora crassa. Genetics*, **75**, 605–613.

54. J.F. Feldman, G.F. Gardner, R.A. Denison (1979). Genetic analysis of the circadian clock of Neurospora. In: M. Suda, I.O. Hayaishi, H. Nakagawa (Eds), *Biological Rhythms and their Central Mechanism* (pp. 56–66). Elsevier/North Holland Biomedical Press, Amsterdam,

55. B.D. Aronson, K.A Johnson, J.J. Loros, J.C. Dunlap (1994). Negative feedback defining a circadian clock: autoregulation of the clock gene frequency. *Science*, **263**, 1578–1584.

56. C. Talora, L. Franchi, H. Linden, P. Ballario, G. Macino (1999). Role of a white collar-1-white collar-2 complex in blue-light signal transduction. *EMBO J.*, **18**, 4961–4968.

57. D.L. Denault, J.J. Loros, J.C. Dunlap (2001). WC-2 mediates WC-1-FRQ interaction within the PAS protein-linked circadian feedback loop of Neurospora. *EMBO J.*, **20**, 109–117.

58. M. Merrow, M. Brunner, T. Roenneberg (1999). Assignment of circadian function for the Neurospora clock gene frequency. *Nature*, **399**, 584–586.

59. S.K. Crosthwaite, J.J. Loros, J.C. Dunlap (1995). Light-induced resetting of a circadian clock is mediated by a rapid increase in frequency transcript. *Cell*, **81**, 1001–1012.

60. S.K. Crosthwaite, J.C. Dunlap, J.J. Loros (1997). *Neurospora wc-1* and *wc-2*: Transcription, photoresponses, and the origins of circadian rhythmicity. *Science*, **276**, 763–769.

61. C. Heintzen, J.J. Loros, J.C Dunlap (2001). The PAS protein VIVID defines a clock-associated feedback loop that represses light input, modulates gating, and regulates clock resetting. *Cell*, **104**, 453–464.

62. M.D. Hall, S.N. Bennett, W.A. Krissinger (2001). Characterization of a newly isolated pigmentation mutant of *Neurospora crassa. Georgia J. Sci.*, **51**, 27–27.

63. R.J. Konopka, S. Benzer (1971). Clock mutants of *Drosophila melanogaster. Proc. Natl. Acad. Sci. U.S.A.*, **68**, 2112–2116.

64. A. Sehgal, J.L. Price, B. Man, M.W. Young (1994). Loss of circadian behavioral rhythms and *per* RNA oscillations in the *Drosophila* mutant *timeless. Science*, **263**, 1603–1605.

65. T.K. Darlington, K. Wager-Smith, M.F. Ceriani, D. Staknis, N. Gekakis, T.D.L. Steeves, C.J. Weitz, J.S. Takahashi, S.A. Kay (1998). Closing the circadian loop: CLOCK-induced transcription of its own inhibitors *per* and *tim. Science*, **280**, 1599–1603.

66. R. Allada, N.E. White, W.V. So, J.C. Hall, M. Rosbash (1998). A mutant *Drosophila* homolog of mammalian *Clock* disrupts circadian rhythms and transcription of *period* and *timeless. Cell*, **93**, 791–804.

67. J.E. Rutila, V. Suri, M. Le, W.V. So, M. Rosbash, J.C. Hall (1998). CYCLE is a second bHLH-PAS clock protein essential for circadian rhythmicity and transcription of *Drosophila period* and *timeless. Cell*, **93**, 805–814.

68. N.R. Glossop, L.C Lyons, P.E. Hardin (1999). Interlocked feedback loops within the Drosophila circadian oscillator. *Science*, **286**, 766–768.

69. D.A. Wheeler, M.J. Hamblen-Coyle, M.S. Dushay, J.C. Hall (1993). Behavior in light-dark cycles of Drosophila mutants that are arrhythmic, blind, or both, *J. Biol. Rhythms*, **8**, 67–94.

70. Z. Yang, M. Emerson, H.S. Su, A. Sehgal (1998). Response of the timeless protein to light correlates with behavioral entrainment and suggests a nonvisual pathway for circadian photoreception. *Neuron*, **21**, 215–223.

71. J.D. Plautz, M. Kaneko, J.C. Hall, S.A. Kay (1997). Independent photoreceptive circadian clocks throughout Drosophila. *Science*, **278**, 1632–1635.

72. T. Todo, H. Ryo, K. Yamamoto, H. Toh, T. Inui, H. Ayaki, T. Nomura, M. Ikenaga (1996). Similarity among the Drosophila (6–4) photolyase, a human photolase homolog, and the DNA photolyase-blue-light receptor family. *Science*, **272**, 109–112.

73. P.J. Van Der Spek, K. Kobayashi, D. Bootsma, M. Takao, A.P.M. Eker, A. Yasui (1996). Cloning, tissue expression and mapping of a human photolyase homolog with similarity to plant blue light receptors. *Genomics*, **37**, 177–182.

74. P. Emery, W.V. So, M. Kaneko, J.C. Hall, M. Rosbash (1998). CRY, a *Drosophila* clock and light-regulated cryptochrome, is a major contributor to circadian rhythm resetting and photosensitivity. *Cell*, **95**, 669–679.

75. S. Okano, S. Kanno, M. Takao, A.P. Eker, K. Isono, Y. Tsukahara, A. Yasui (1999). A putative blue-light receptor from *Drosophila melanogaster*. *Photochem. Photobiol.*, **69**, 108–113.

76. D.S. Hsu, X.D. Zhao, S.Y. Zhao, A. Kazantsev, R.P. Wang, T. Todo, Y.F. Wei, A. Sancar (1996). Putative human blue light photoreceptors hCRY1 and hCRY2 are flavoproteins. *Biochemistry*, **35**, 13871–13877.

77. P. Emery, R. Stanewsky, J.C. Hall, M. Rosbash (2000). A unique circadian-rhythm photoreceptor. *Nature*, **404**, 456–457.

78. T. Ishikawa, A. Matsumoto, T. Kato, Jr., S. Togashi, H. Ryo, M. Ikenaga, T. Todo, R. Ueda, T. Tanimura (1999). DCRY is a Drosophila photoreceptor protein implicated in light entrainment of circadian rhythm. *Genes Cells*, **4**, 57–65.

79. M.F. Ceriani, T.K. Darlington, D. Staknis, P. Mas, A.A. Petti, C.J. Weitz, S.A. Kay (1999). Light-dependent sequestration of Timeless by cryptochrome. *Science*, **285**, 553–556.

80. Chapter 3, this volume.

81. R.G. Foster (1998). Shedding light on the biological clock. *Neuron*, **20**, 829–832.

82. M.S. Freedman, R.J. Lucas, B. Soni, M. von Schantz, M. Munoz, Z. David-Gray, R. Foster (1999). Regulation of mammalian circadian behaviour by non-rod, non-cone ocular photoreceptors. *Science*, **284**, 502–504.

83. S. Yamazaki, R. Numano, M. Abe, A. Hida, R. Takahashi, M. Ueda, G.D. Block, Y. Sakaki, M. Menaker, H. Tei (2000). Resetting central and peripheral circadian oscillators in transgenic rats. *Science*, **288**, 682–685.

84. I. Provencio, H.M. Cooper, R.G. Foster (1998). Retinal projections in mice with inherited retinal degeneration: implications for circadian photoentrainment. *J. Comp. Neurol.*, **395**, 417–439.

85. Y. Miyamoto, A. Sancar (1998). Vitamin B2-based blue-light photoreceptors in the retinohypothalamic tract as the photoactive pigments for setting the circadian clock in mammals. *Proc. Natl. Acad. Sci. U.S.A.*, **95**, 6097–6102.

86. I. Provencio, R.G. Foster (1995). Circadian rhythms in mice can be regulated by photoreceptors with cone-like characteristics. *Brain Res.*, **694**, 183–190.

87. T. Yoshimura, S. Ebihara (1996). Spectral sensitivity of photoreceptors mediating phase-shifts of circadian rhythms in retinally degenerate CBA/J (rd/rd) and normal CBA/N (+/+) mice. *J. Comp. Physiol. A-Sensory Neural Behav. Physiol.*, **178**, 797–802.

88. A. Sancar (2000). Cryptochrome: The second photoactive pigment in the eye and its role in circadian photoreception. *Annu. Rev. Biochem.*, **69**, 31–67.

89. K. Kume, M.J. Zylka, S. Sriram, L.P. Shearman, D.R. Weaver, X. Jin, E.S. Maywood, M.H. Hastings, S.M. Reppert (1999). mCRY1 and mCRY2 are essential components of the negative limb of the circadian clock feedback loop. *Cell*, **98**, 193–205.

90. G.T. van der Horst, M. Muijtjens, K. Kobayashi, R. Takano, S. Kanno, M. Takao, J. de Wit, A. Verkerk, A.P. Eker, D. van Leenen, R. Buijs, D. Bootsma,

J.H. Hoeijmakers, A. Yasui (1999). Mammalian Cry1 and Cry2 are essential for maintenance of circadian rhythms. *Nature*, **398**, 627–630.

91. R.J. Thresher, M.H. Vitaterna, Y. Miyamoto, A. Kazantsev, D.S. Hsu, C. Petit, C.P. Selby, L. Dawut, O. Smithies, J.S. Takahashi, A. Sancar (1998). Role of mouse cryptochrome blue-light photoreceptor in circadian photoresponses. *Science*, **282**, 1490–1494.

92. M.H. Vitaterna, C.P. Selby, T. Todo, H. Niwa, C. Thompson, E.M. Fruechte, K. Hitomi, R.J. Thresher, T. Ishikawa, J. Miyazaki, J.S. Takahashi, A. Sancar (1999). Differential regulation of mammalian period genes and circadian rhythmicity by cryptochromes 1 and 2. *Proc. Natl. Acad. Sci. U.S.A.*, **96**, 12114–12119.

93. H. Okamura, S. Miyake, Y. Sumi, S. Yamaguchi, A. Yasui, M. Muijtjens, J.H. Hoeijmakers, G.T. van der Horst (1999). Photic induction of mPer1 and mPer2 in cry-deficient mice lacking a biological clock. *Science*, **286**, 2531–2534.

94. C.P. Selby, C. Thompson, T.M. Schmitz, R.N. Van Gelder (2000). A. Sancar, Functional redundancy of cryptochromes and classical photoreceptors for nonvisual ocular photoreception in mice. *Proc. Natl. Acad. Sci. U.S.A.*, **97**, 14697–14702.

95. I. Provencio, G. Jiang, W.J. De Grip, W.P. Hayes, M.D. Rollag (1998). Melanopsin: An opsin in melanophores, brain, and eye. *Proc. Natl. Acad. Sci. U.S.A.*, **95**, 340–345.

96. B.G. Soni, A.R. Philp, R.G. Foster, B.E. Knox (1998). Novel retinal photoreceptors. *Nature*, **394**, 27–28.

96a. S. Panda, T.K. Sato, A.M. Castrucci, M.D. Rollag, W.J. DeGrip, J.B. Hogenesch, I. Provencio, S.A. Kay (2002). Melanopsin (Opn4) requirement for normal light-induced circadian phase shifting. *Science.*, **298**, 2213–2216.

96b. S. Hattar, R.J. Lucas, N. Mrosovsky, S. Thompson, R.H. Douglas, M.W. Hankins, J. Lem, M. Biel, F. Hofmann, R.G. Foster, K.W. Yau (2003). Melanopsin and rod-cone photoreceptive systems account for all major accessory visual functions in mice. *Nature.*, **424**, 75–81.

97. L. Smith, S. Folkard, C.J. Poole (1994). Increased injuries on night shift. *Lancet*, **344**, 1137–1139.

98. K.L. Toh, C.R. Jones, Y. He, E.J. Eide, W.A. Hinz, D.M. Virshup, L.J. Ptacek, Y.H. Fu (2001). An hPer2 phosphorylation site mutation in familial advanced sleep phase syndrome. *Science*, **291**, 1040–1043.

99. W.E. Bunney, B.G. Bunney (2000). Molecular clock genes in man and lower animals, possible implications for circadian abnormalities in depression. *Neuropsychopharmacology*, **22**, 335–345.

100. M.H. Kryger, T. Roth, W.C. Dement (Eds) (2000). *Principles and Practice of Sleep Medicine. Section 8. Disorders of Chronobiology,* (3rd Edn., pp. 589–614). W.B. Saunders, Philadelphia.

101. S.W. Lockley, D.J. Skene, K. James, K. Thapan, J. Wright, J. Arendt (2000). Melatonin administration can entrain the free-running circadian system of blind subjects. *J. Endocrinol.*, **164**, R1-R6.

102. R.L. Sack, R.W. Brandes, A.R. Kendall, A.J. Lewy (2000). Entrainment of free-running circadian rhythms by melatonin in blind people. *N. Engl. J. Med.*, **343**, 1070–1077.

103. J. Arendt (2000). Melatonin, circadian rhythms, and sleep. *New. Engl. J. Med.*, **343**, 1114–1116.

104. P. Suarez-Lopez, K. Wheatley, F. Robson, H. Onouchi, F. Valverde, G. Coupland (2001). CONSTANS mediates between the circadian clock and the control of flowering in Arabidopsis. *Nature*, **410**, 1116–1120.

Subject Index